Offense, Defense, and War

International Security Readers

Strategy and Nuclear Deterrence (1984)
Military Strategy and the Origins of the First World War (1985)
Conventional Forces and American Defense Policy (1986)
The Star Wars Controversy (1986)
Naval Strategy and National Security (1988)
Military Strategy and the Origins of the First World War, revised and expanded edition (1991)
—published by Princeton University Press

Soviet Military Policy (1989)
Conventional Forces and American Defense Policy, revised edition (1989)
Nuclear Diplomacy and Crisis Management (1990)
The Cold War and After: Prospects for Peace (1991)
America's Strategy in a Changing World (1992)
The Cold War and After: Prospects for Peace, expanded edition (1993)
Global Dangers: Changing Dimensions of International Security (1995)
The Perils of Anarchy: Contemporary Realism and International Security (1995)
Debating the Democratic Peace (1996)
East Asian Security (1996)
Nationalism and Ethnic Conflict (1997)
America's Strategic Choices (1997)
Theories of War and Peace (1998)
America's Strategic Choices, revised edition (2000)
Rational Choice and Security Studies: Stephen Walt and His Critics (2000)
The Rise of China (2000)
Nationalism and Ethnic Conflict, revised edition (2001)
Offense, Defense, and War (2004)
New Global Dangers: Changing Dimensions of International Security (2004)
—published by The MIT Press

Offense, Defense, and War

AN *International Security* READER

EDITED BY
Michael E. Brown
Owen R. Coté Jr.
Sean M. Lynn-Jones
and Steven E. Miller

THE MIT PRESS
CAMBRIDGE, MASSACHUSETTS
LONDON, ENGLAND

The contents of this book were first published in *International Security* (ISSN 0162-2889), a publication of The MIT Press under the sponsorship of the Belfer Center for Science and International Affairs at Harvard University.

Stephen Van Evera, "The Cult of the Offensive and the Origins of the First World War," 9:1 (Summer 1984); Jack Snyder, "Civil-Military Relations and the Cult of the Offensive, 1914 and 1984," 9:1 (Summer 1984); Scott D. Sagan, "1914 Revisited: Allies, Offense, and Instability," 11:2 (Fall 1986); Jack Snyder and Scott D. Sagan, "Correspondence: The Origins of Offense, and the Consequences of Counterforce," 11:3 (Winter 1986/87); Jonathan Shimshoni, "Technology, Military Advantage, and World War I: A Case for Military Entrepreneurship" 15:3 (Winter 1990/91); Stephen Van Evera, "Offense, Defense, and the Causes of War," 22:4 (Spring 1998); Chaim Kaufmann and Charles L. Glaser, "What is the Offense-Defense Balance and Can We Measure It?" 22:4 (Spring 1998); James W. Davis Jr., Bernard I. Finel, Stacie E. Goddard, Stephen Van Evera, Charles L. Glaser, and Chaim Kaufmann, "Correspondence: Taking Offense at Offense-Defense Theory," 23:3 (Winter 1998/99); Richard K. Betts, "Must War Find a Way? A Review Essay," 24:2 (Fall 1999); Keir A. Lieber, "Grasping the Technological Peace: The Offense-Defense Balance and International Security," 25:1 (Summer 2000); Karen Ruth Adams, "Attack and Conquer? International Anarchy and the Offense-Defense-Deterrence Balance," 28:3 (Winter 2003/04).

Library of Congress Cataloging-in-Publication Data

Offense, defense, and war / edited by Michael E. Brown . . . [et al.]
p. cm.—(An international security reader)
Includes bibliographical references.
ISBN 0-262-52316-7 (pbk. : alk. paper)
1. Offensive (Military science) 2. Defensive (Military science) 3. World War, 1914–1918—Causes. I. Brown, Michael E. (Michael Edward), 1954– II. International security readers.

U162.O36 2004
355′.03—dc22

2004050499

Contents

The Contributors

MICHAEL E. BROWN is Editor of *International Security* and Director of the Security Studies Program and the Center for Peace and Security Studies at the Edmund A. Walsh School of Foreign Service, Georgetown University.

OWEN R. COTÉ JR. is Editor of *International Security* and Associate Director of the Security Studies Program at the Massachusetts Institute of Technology.

SEAN M. LYNN-JONES is Editor of *International Security* and a Research Associate at the Belfer Center for Science and International Affairs (BCSIA), John F. Kennedy School of Government, Harvard University.

STEVEN E. MILLER is Editor-in-Chief of *International Security* and Director of the International Security Program at BCSIA.

KAREN RUTH ADAMS is Assistant Professor of Political Science at the University of Montana.

RICHARD K. BETTS is Arnold A. Saltzman Professor of War and Peace Studies in the Political Science Department, Director of the Saltzman Institute of War and Peace Studies, and Director of the International Security Policy program in the School of International and Public Affairs, Columbia University.

JAMES W. DAVIS JR. is Assistant Professor of International Politics at the Ludwig-Maximilians-Universität, Germany and Associate Editor of the *European Journal of International Relations.*

BERNARD I. FINEL is Executive Director of the M.A. in Security Studies Program and the Center for Peace and Security Studies, Georgetown University.

CHARLES L. GLASER is Deputy Dean and Professor at the Irving B. Harris Graduate School of Public Policy Studies and co-director of the Program on International Security Policy at the University of Chicago.

STACIE E. GODDARD is Lecturer in Political Science at Princeton University.

ROBERT JERVIS is Adlai E. Stevenson Professor of International Politics at Columbia University.

CHAIM KAUFMANN is Associate Professor of International Relations at Lehigh University.

KEIR A. LIEBER is Assistant Professor of Political Science in the Joan B. Kroc Center for International Peace Studies at the University of Notre Dame.

GEORGE H. QUESTER is Professor of Government and Politics at the University of Maryland.

SCOTT D. SAGAN is Professor of Political Science and Co-Director of the Center for International Security and Cooperation at Stanford University.

JONATHAN SHIMSHONI is a partner with PwC Consulting in Israel, responsible for business strategy and public policy advisory services.

JACK SNYDER is Robert and Renée Belfer Professor of International Relations in the Political Science Department at Columbia University.

STEPHEN VAN EVERA is Professor of Political Science at the Massachusetts Institute of Technology and a member of MIT's Security Studies Program.

Acknowledgments

The editors gratefully acknowledge the assistance that has made this book possible. A deep debt is owed to all those at the Belfer Center for Science and International Affairs (BCSIA), Harvard University, who have played an editorial role at *International Security*. We are grateful for support from the Carnegie Corporation of New York. Special thanks go to Diane McCree, Sarah Buckley, Jennifer Cook, Michelle Von Euw, and Meara Keegan Zaheer at BCSIA for their invaluable help in preparing this volume for publication. We thank Corentin Brustlein and Stephen Van Evera for their helpful suggestions. We are grateful to the Johns Hopkins University Press and to George Quester for granting permission to reprint material here.

Preface

Sean M. Lynn-Jones

Offense-defense theory argues that the character of international politics is influenced by whether offensive military operations are easy or difficult. When offensive operations have the advantage over defensive operations, the theory suggests that war and conflict will become more likely; when defensive operations have the advantage, peace and cooperation are more likely. Most offense-defense theorists describe the relative efficacy of offense and defense as the offense-defense balance. This balance may change as new technologies emerge and are applied to military operations. Some offense-defense theorists argue that a broader set of political, economic, and social factors also influence whether the offense-defense balance favors offense or defense.

Although earlier writers developed rudimentary elements of offense-defense theory, the theory in its present form is largely the result of scholarship published since the late 1970s. During the 1980s and 1990s, several major books and articles elaborated, refined, and applied offense-defense theory. At the same time, critics published objections to offense-defense theory, and a vigorous debate emerged.[1]

Offense-defense theory has been used to explain many aspects of international politics and foreign policy. The most general prediction of the theory is that international conflict and war are more likely when offense has the advantage, while peace and cooperation are more probable when defense has the advantage. The theory thus has been used to explain the propensity for war (or peace) in various international systems, ranging from ancient China to Europe in the 19th century. Variants of offense-defense theory also have been used to explain important issues in security studies, including alliance formation, grand strategy and military doctrine, arms racing, the international consequences of revolutions, deterrence and nuclear strategy, and escalation. The theory also has been used to explain specific events, such as the outbreak of World War I and the conflicts that erupted in the former Yugoslavia in the 1990s.

The basic elements of offense-defense theory have been invoked or endorsed by many scholars who have not participated in its development and initial applications. For example, Kenneth Waltz writes that "Weapons and strategies

1. For an overview of this literature, see the suggestions for further reading at the end of this volume.

that make defense and deterrence easier, and offensive strikes harder to mount, decrease the likelihood of war."[2] Robert Keohane and Lisa Martin argue that the importance of relative gains "is *conditional* on factors such as . . . whether military advantage favors offense or defense."[3]

The theory also has made important contributions to more general theories of international politics. The different variants of offense-defense theory should be classified as structural-realist (or neorealist) theories. Offense-defense theory resolves many problems in the standard neorealist theory of Kenneth Waltz and enhances its explanatory range and power.

Offense-defense theories share the basic assumptions and approach of structural realism, as Charles Glaser and Chaim Kaufmann point out in their contribution to this volume. Like other structural theories, they focus on the international incentives and constraints that states face as they pursue their goals—which often require the threat or use of military capabilities. They share the main assumptions of other structural realist theories: states seek to maximize their security by employing self-help strategies in an anarchic world.[4]

Offense-defense theory offers several important theoretical contributions to structural realism. It enables realism to explain a wider range of behavior than can be explained by changes in the distribution of power alone. Unlike structural-realist theories that argue that the international distribution of power is the only important element of international structure, offense-defense theories contend that other factors, particularly the offense-defense balance, are important determinants of state behavior. The most parsimonious versions of structural-realist theory offer very general predictions on the basis of changes in the aggregate distribution of capabilities. Waltz's argument that war is more likely

---

2. Kenneth N. Waltz, "The Origins of War in Neorealist Theory," in Robert I. Rotberg and Theodore K. Rabb, eds., *The Origin and Prevention of Major Wars* (Cambridge, U.K.: Cambridge University Press, 1989), p. 50. See also Waltz, "Toward Nuclear Peace," in Robert J. Art and Kenneth N. Waltz, eds., *The Use of Force: Military Power and International Politics*, 4th ed. (Lanham, Md.: University Press of America, 1993), pp. 528–529, 555.
3. Robert O. Keohane and Lisa L. Martin, "The Promise of Institutionalist Theory," *International Security*, Vol. 20, No. 1 (Summer 1995), p. 44. Emphasis in original. See also Charles Lipson, "International Cooperation in Economic and Security Affairs," *World Politics*, Vol. 37, No. 1 (October 1984), pp. 15–16.
4. Lists of the assumptions of structural realism vary slightly, but these assumptions are prominent and widespread. See Kenneth N. Waltz, *Theory of International Politics* (Reading, Mass.: Addison-Wesley, 1979); Joseph M. Grieco, *Cooperation Among Nations: Europe, America, and Non-Tariff Barriers to Trade* (Ithaca, N.Y.: Cornell University Press, 1990); Robert O. Keohane, "Realism, Neorealism and the Study of World Politics," in Keohane, ed., *Neorealism and its Critics* (New York: Columbia University Press, 1986), pp. 1–26; and John J. Mearsheimer, *The Tragedy of Great Power Politics* (New York: W.W. Norton, 2001), pp. 30–32.

in multipolar systems and less likely under bipolarity is the most prominent prediction. Integrating the offense-defense balance into structural realism makes it possible to explain and predict particular wars.[5]

Adding the offense-defense balance to structural-realist theories also enables structural realism to identify the conditions under which peace and cooperation become more likely, thereby countering the pessimism of many realist theories and removing the need to use nonrealist theories to explain such outcomes. Traditionally, realists have been regarded as pessimists who depict international politics as an unending cycle of conflict, hostility, and war. This image of realism now applies only to offensive realists, who generally argue that the international system fosters conflict and aggression.[6] Security is scarce, making international competition and war likely. Rational states often are compelled to adopt offensive strategies in their search for security.[7] On the other hand, defensive realists—many of whom embrace offense-defense theory—argue that the international system does not necessarily generate intense conflict and war, and that defensive strategies are often the best route to security.[8]

Proponents of offense-defense theory also suggest that the theory has much relevance to security policy. First, some advocates of the theory claim that it

5. Stephen Van Evera calls the offense-defense balance and other related variables the "fine-grained structure of power" and argues that "Realism becomes far stronger when it includes these fine-grained structures and perceptions of them." Stephen Van Evera, *Causes of War: Power and the Roots of Conflict* (Ithaca, N.Y.: Cornell University Press, 1999), p. 256.
6. For explications of the differences between offensive realists, defensive realists, and neoliberals, see Robert Jervis, "Realism, Neoliberalism, and Cooperation: Understanding the Debate," *International Security*, Vol. 24. No. 1 (Summer 1999), pp. 42–63; Jack Snyder, *Myths of Empire: Domestic Politics and International Ambition* (Ithaca, N.Y.: Cornell University Press, 1991), pp. 11–12; Sean M. Lynn-Jones and Steven E. Miller, "Preface," in Michael E. Brown, Sean M. Lynn-Jones, and Steven E. Miller, eds., *The Perils of Anarchy: Contemporary Realism and International Security* (Cambridge, Mass.: MIT Press, 1995), pp. ix–xiii; Benjamin Frankel, "Restating the Realist Case: An Introduction," *Security Studies*, Vol. 5, No. 3 (Spring 1996), pp. xiv–xx; Fareed Zakaria, *From Wealth to Power: The Unusual Origins of America's World Role* (Princeton, N.J.: Princeton University Press, 1998); Sean M. Lynn-Jones, "Realism and America's Rise: A Review Essay," *International Security*, Vol. 23, No. 2 (Fall 1998), pp. 157–182; and Gideon Rose, "Neoclassical Realism and Theories of Foreign Policy," *World Politics*, Vol. 51, No. 1 (October 1998), pp. 144–172.
7. Mearsheimer, *The Tragedy of Great Power Politics* is the most definitive statement of offensive realism.
8. Van Evera, *Causes of War* is often regarded as the clearest statement of defensive realism. For an explicit discussion of the connection between offense-defense theory and a more optimistic version of realist theory, see Charles L. Glaser, "Realists as Optimists: Cooperation as Self-Help," *International Security*, Vol. 19, No. 3 (Winter 1994/95), pp. 50–90. For a defense of defensive realism, see Jeffrey W. Taliaferro, "Security Seeking under Anarchy: Defensive Realism Revisited," *International Security*, Vol. 25, No. 3 (Winter 2000/01), pp. 128–161.

provides a basis for unilateral or multilateral efforts to control weapons that make offense easier.[9] Limiting or banning such weapons might reduce international tensions and the risk of war. This aspiration was central to some of the earliest writings on offense and defense in international politics. The 1932 World Disarmament Conference unsuccessfully attempted to limit or prohibit "offensive" weapons. During the 1980s, West European proponents of nonoffensive defense who argued that NATO and the Warsaw Pact should limit themselves to defensive weapons and doctrines embraced this idea.[10] To some extent, both superpowers adopted aspects of this approach. The Soviet Union under Gorbachev attempted to reduce tensions with the West by proclaiming that it had adopted a defensive military doctrine. In the negotiations on Conventional Forces in Europe (CFE), NATO sought to limit weapons with offensive capabilities. Even if it was hard to define which weapons were offensive, some (e.g., bridging equipment) would only be useful for offensive action.[11] Limitations on offensive potential can be pursued multilaterally or unilaterally. States might attempt to limit offensive weapons through negotiated agreements with other states, or they might unilaterally adopt defensive postures to signal their benign intent and reduce tensions.

Second, even if offense-defense theory does not offer a basis for multilateral or unilateral arms control, the theory might be used to reduce the likelihood of war. If defense has an advantage—particularly if that advantage is large—states that understand this fact are likely to conclude that war is unprofitable. In practice, however, states often exaggerate the strength of offense. Offense-defense theory might therefore reduce the likelihood of war by offering accurate assessments of the offense-defense balance and correcting these misperceptions.

Third, offense-defense theory could be used to guide military policies. If it is

9. It is often difficult to classify weapons as offensive or defensive and offense-defense theory does not depend on this distinction. Thus the idea of controlling offensive weapons may not be completely consistent with offense-defense theory. In fact, not all proponents of offense-defense theory are strong advocates of arms control. Nevertheless, the basic notion of reducing the risk of war by reducing the prospects for offensive action is compatible with the main thrust of offense-defense theory.

10. For overviews of nonoffensive defense, see Bjørn Møller, *Common Security and Nonoffensive Defense* (Boulder, Colo.: Lynne Rienner, 1992); and Stephen J. Flanagan, "Nonprovocative and Civilian-Based Defenses," in Joseph S. Nye Jr., Graham T. Allison, and Albert Carnesale, eds., *Fateful Visions: Avoiding Nuclear Catastrophe* (Cambridge, Mass.: Ballinger, 1988), pp. 93–109.

11. I am indebted to George Quester for reminding me of the virtually unambiguous offensive character of bridging equipment, which is mainly useful for enabling attacking forces to cross rivers inside enemy territory.

possible to assess the offense-defense balance, the results of such assessments could be used to help states adopt optimal military postures. For example, the existence of a large defensive advantage might imply that a given state should avoid offensive action unless it has a very large advantage in capabilities. At the very least, awareness of the offense-defense balance would enable states to avoid gross blunders, such as being overly confident of fighting a successful defensive war when offense is relatively strong. Recently, many analysts have claimed that a Revolution in Military Affairs (RMA) has tilted the offense-defense balance back toward offense. Although other analysts question this conclusion, it is a prime contemporary example of how an element of offense-defense theory can influence defense and military policy.[12]

Critics and skeptics argue that offense-defense theory is far less promising than its proponents claim. The standard litany of criticisms includes the arguments that weapons cannot be classified as offensive or defensive, that states fail to perceive the offense-defense balance correctly, that other variables may be more important than the offense-defense balance, that offense-defense theory explains little because the offense-defense balance always favors the defense, and that states manipulate the offense-defense balance to create offensive and defensive advantages to serve their strategic goals.[13]

More recent criticisms argue that some of the most prominent attempts to define the offense-defense balance are seriously flawed. For example, several critics have suggested that Stephen Van Evera's broad definition of the offense-defense balance, which includes technological, doctrinal, geographical, domestic, and diplomatic factors, is an inadequate basis for further research.[14] They have suggested that this definition includes too many variables, thereby conflating the offense-defense balance with other factors and making it impossible to measure, and that its primary prediction—that war is more likely when conquest is easy—becomes tautological.[15] The alternative definition of the offense-defense balance offered by Kaufmann and Glaser has been criticized on

12. For a skeptical view of the RMA, see Stephen Biddle, "Victory Misunderstood: What the Gulf War Tells Us about the Future of Conflict," *International Security*, Vol. 21, No. 2 (Fall 1996), pp. 139–179.
13. In addition to the relevant essays in this volume, many seminal works that criticize offense-defense theory are listed in the suggestions for further reading at the end of the book.
14. Van Evera suggests that the offense-defense balance is synonymous with "the feasibility of conquest." For Van Evera's definition and list of the factors that determine the offense-defense balance, see his "Offense, Defense, and the Causes of War" in this volume and Van Evera, *Causes of War*, pp. 160–166.
15. See the contributions of Betts, Lieber, Davis, Goddard, and Finel in this volume.

the grounds that it ignores interaction effects in warfare and conflates the offense-defense balance with other variables, such as power and skill.[16]

Other critics argue that offense-defense theory lacks empirical support. They have attempted to test whether various definitions of the offense-defense balance can explain the outcomes of battles and the frequency of wars—the most important predictions of the theory—and have concluded that offense-defense theory fails these tests.[17]

The debate over offense-defense theory has raged for over two decades, but proponents and critics often have talked past one another. Reading the literature, one gets the sense that proponents of offense-defense theory regard it as an established theory that should take its place alongside deterrence theory, balance-of-power theory, and other major theories in international security studies. The writings of the critics, on the other hand, give the impression that offense-defense theory is dead or dying, and convey a sense of surprise and bewilderment that the theory has any advocates at all. This volume collects important essays that have developed and challenged offense-defense theory so that readers will have an opportunity to understand and assess the debate.[18]

The first two essays in this volume present the foundations of modern offense-defense theory. In "Cooperation under the Security Dilemma," Robert Jervis develops many of the basic concepts and insights that have inspired subsequent scholarship on offense and defense in international relations.[19] Jervis argues that the security dilemma makes it difficult for states to cooperate in international politics. In the absence of a common sovereign and any assurances of the benign intentions of other states, states frequently attempt to expand their control and influence over resources or land outside their own territory.

---

16. See the contributions from Finel and Goddard, and the reply from Kaufmann and Glaser, in this volume.
17. In addition to Keir Lieber's essay in this volume, see James D. Fearon, "The Offense-Defense Balance and War Since 1648," paper prepared for the 1995 annual convention of the International Studies Association, Chicago, Illinois, February 21–25, 1995; and Stephen D. Biddle, "Rebuilding the Foundations of Offense-Defense Theory," *Journal of Politics*, Vol. 63, No. 3 (August 2001), pp. 741–774.
18. Most previous *International Security* Readers have contained only articles that were originally published in *International Security*. This volume, however, includes works by Robert Jervis and George Quester that first appeared elsewhere. These expositions of offense-defense theory are so seminal that they merit inclusion in any volume that attempts to offer an overview of offense-defense theory.
19. For a retrospective analysis of Jervis's article and its influence, see Charles L. Glaser, "The Security Dilemma Revisited," *World Politics*, Vol. 50, No. 1 (October 1997) pp. 171–201.

These attempts to increase a state's security often reduce the security of other states.

According to Jervis, the security dilemma can vary in severity. When states are less vulnerable, they are more likely to be able to trust others and to cooperate. Under such conditions, states do not need to launch preemptive attacks or to match every real or potential military capability of other states. States can attain security more easily in such circumstances and they can pursue more passive foreign policies.

Jervis argues that "this situation is approximated when it is easier for states to defend themselves than to attack others, or when mutual deterrence obtains because neither side can protect itself." Jervis enumerates many other factors that affect the severity of the security dilemma and the probability of international cooperation, including subjective security requirements, threat perceptions, the existence of collective security systems, the magnitude of the advantages of cooperation, and whether leaders understand the operation of the security dilemma and recognize that their adversaries may have defensive motivations. Ultimately, however, he focuses primarily on two variables that influence whether an increase in one state's security reduces the security of others: (1) "whether the defense or the offense has the advantage," and (2) "whether defensive weapons and policies can be distinguished from offensive ones." When defense has an advantage over offense, a large increase in one state's security only slightly decreases the security of other states, enabling all states to enjoy a high level of security.

Jervis defines offensive advantage as a condition in which "it is easier to destroy the other's army and take its territory than it is to defend one's own." Conversely, "when defense has the advantage it is easier to protect and to hold than it is to move forward, destroy, and take." Offensive advantages make the security dilemma severe, because the best route to security is aggression and expansion. When defense has the advantage, aggression becomes less feasible and states can make themselves secure without threatening others.

Jervis enumerates what is likely to happen when it is easier to attack than to defend. Cooperation will be less likely because wars will be short and profitable for the winner. States will maintain high levels of arms and recruit allies in advance. Decisionmakers will perceive ambiguous information as threatening because they cannot afford to be the victim of an attack. The opposite results will emerge when defense has the advantage.

Geography and technology are the two main variables that influence the offense-defense balance. Geographical factors such as buffer zones or natural barriers like mountain ranges, large rivers, or oceans make successful offensives less feasible. Under these conditions, both sides can have sufficient defensive capabilities without threatening the other. On the other hand, when it is easy to attack across borders, frequent wars are much more likely.

Technology determines the characteristics of weapons and thereby affects the offense-defense balance. When weapons are highly vulnerable, they must be used first, making offense more effective and attacks more likely. When fortifications are impregnable, a small number of defenders can hold off many attackers. Highly mobile forces may give offense the advantage, but the implications of particular weapons are often unclear. Tanks and tactical airpower did not appear to enhance the offense in the 1973 Arab-Israeli War. Jervis argues that second-strike nuclear deterrence creates a "result [that] is the equivalent of the primacy of defense." Mutual nuclear deterrence renders security relatively cheap, makes each side secure, can be maintained even with highly unequal force levels, and reduces incentives to strike in a crisis.

When offensive and defensive weapons can be distinguished, a state can increase its security without reducing the security of others. If defense is at least as potent as the offense, a clear differentiation between the two enables status-quo powers to signal their benign intentions and to identify one another. It will be easy to identify aggressor states before they attack, and states will find it easier to negotiate arms control agreements.

Jervis recognizes that many observers argue that it is impossible to distinguish between offensive and defensive weapons. Whether a weapon is offensive or defensive often depends on the situation in which it is used. The offensiveness or defensiveness of many weapons is ambiguous, but forces that are not mobile tend to be defensive, whereas those that can reduce fortifications or depend on surprise tend to be offensive.

Even if weapons are distinguishable, states may choose offensive weapons if the offense has a large advantage over the defense, if offensive forces are necessary to recapture territory, or if only offensive action will compel an adversary to seek peace. In addition, aggressor states will often include some defensive weapons in their arsenals.

Jervis determines that the two key variables—the offense-defense balance and whether offense and defense can be distinguished—combine to form four potential worlds: (1) When offense has the advantage and weapons cannot be differentiated, arms races and wars are likely. (2) When defense has the advantage but offensive and defensive weapons cannot be distinguished, the security dilemma operates but is not insurmountable. Status-quo states may be able to adopt compatible security policies under these conditions. (3) When offense has the advantage but weapons can be differentiated, states should be able to remove the security dilemma by procuring nonthreatening defensive weapons. The offensive advantage, however, may tempt states into launching aggressive wars. (4) The safest world of all is one in which defense has the advantage and offensive and defensive weapons can be distinguished.

In "Offense and Defense in the International System," which consists of the introductory and concluding chapters of his book, *Offense and Defense in the International System,* George Quester summarizes some of his hypotheses relevant to offense-defense theory.[20] Noting that the terms "offense" and "defense" have created semantic confusion because most states prefer to claim that their actions are defensive, Quester argues that objective factors, such as military technology, often determine whether it is easier for states to defend than to attack. The central questions of his work are: "Can we identify the technical, political, and social factors that make it advantageous to strike out offensively at the enemy, rather than to sit in prepared defensive positions waiting for him to strike? And what then are the military and political consequences of such an offensive or defensive preponderance?"

Quester suggests that offensive actions become more likely when attacking forces can inflict more casualties on the enemy than they suffer themselves. In general, when weapons and forces are more mobile, offense becomes more advantageous. Weapons that are only potent for temporary durations also create incentives for offensive action. Thus a mobilization of forces may necessitate an attack because the forces cannot be kept mobilized indefinitely.

In Quester's view, other technological and geographical conditions make defensive actions more likely. If, for example, light weapons are mainly useful in ambushes, they will offer a defensive advantage to forces familiar with the territory they are defending.

Quester argues that the extreme "countervalue" capabilities associated with nuclear weapons—the ability to kill many enemy civilians without first defeating the enemy's armed forces—appear to be offensive, but this ability to retaliate actually tends to support peace and prevent war. He thus uses "offensive"

20. In his book, *Offense and Defense in the International System* (New York: Wiley, 1977), Quester develops these ideas in more detail and illustrates them with many historical examples. Unfortunately, space constraints make it impossible to reproduce all of this important material in this volume.

to "refer to technology and techniques that reward counter*force* initiatives" (emphasis in original).

Defensive advantages tend to make peace more likely. States will not attack other states if offensive military action does not produce victory. In crises, a defensive advantage will encourage states to avoid initiating war. Quester notes that this conclusion is not new. Karl von Clausewitz and Sun Tzu recognized the advantages of remaining on the defensive and the tendency of defense to cause peace.

A defensive advantage, according to Quester, tends to render states unconquerable and thereby enables small political units to remain independent. An offensive advantage, on the other hand, makes a universal empire more possible. Paradoxically, either situation may be peaceful. States may avoid wars of conquest when there is a large defensive advantage. At the other end of the spectrum, peace would exist in a universal empire because conquest on such a scale would abolish war and international politics.

Quester recognizes that the human race often values the autonomy of its political units more than it values peace. He also acknowledges that offensive and defensive advantages are not the only causes of war. Political disagreements, exaggerated estimates of the chances of victory, and mistaken assumptions about the opponent's resolve or hostility all can cause war. Quester suggests, however, that the "state of military technique that rewards taking the offense . . . is now the most worthy of attention, because it may indeed be more manageable than the others, because it in recent times has received less analysis than the others, and because it may increase or decrease the problems caused by the others." Quester also points out that "What statesmen believe about war and weapons is crucial" and that illusions of an offensive advantage helped to cause war in 1914.

Although Quester's analysis suggests that a defensive advantage is preferable, he qualifies his general endorsement of defense by pointing out that the high level of political fragmentation that might accompany a defensive advantage is not always desirable. Effective defenses may also prolong wars. Although peace is better than a short war, a short war is better than a long one. Even if a defensive advantage might be conducive to peace, it might be better to have an offensive advantage during war so that the conflict could be terminated quickly.

Quester also qualifies his arguments by pointing out that a defensive posture on one front might make it possible to launch an attack on another. The stalemate in the West in 1939 enabled Hitler to invade Poland.

Quester considers the potential objection that offensive advantages are a function of numerical superiority, not the technology of weaponry. He points out that "The significant impact of defensive or offensive technology shows up in the minimum ratios of numerical superiority required for such an offensive. With a basically defensive technology, perhaps a three to one, or five to one superiority, will be required to break through; with a more offensive technology, far less of a superiority may suffice, or no numerical superiority at all may be required, as either side can win if it only beats the other to the offensive."

Quester concludes by noting that many factors can make war more or less likely: "We will need to be certain that technology does not too much favor the offensive. We will also want to be sure that neither side outweighs the other so much in military terms that it can by sheer numbers override any advantages of the defense. But we make our task easier for the second category, as we succeed in the first."

The next section of essays examines the role of offense and defense in World War I. The outbreak of the Great War in July 1914 has inspired much scholarship on offense-defense theory. The general image of states mobilizing and launching offensives in hopes of winning quick victories has contributed much to the development of offense-defense theory. In few other cases does the connection between perceived offense dominance and the onset of war seem to be so strong. At the same time, the apparent dominance of defense in the years of trench warfare on the Western Front suggests that perceptions of offensive advantage were tragically mistaken. The essays in this section explore the consequences of beliefs in offensive advantage, the sources of those beliefs, and some of the reasons why World War I may not support offense-defense theory.

In "The Cult of the Offensive and the Origins of the First World War," Stephen Van Evera develops offense-defense theory and applies it to the outbreak of war in 1914. He argues that Europe's militaries shared a belief in the power of offense on the eve of World War I. In Berlin, Paris, London, St. Petersburg, and even Brussels, generals embraced a "cult of the offensive" that held that offense was superior to defense and that victory depended on offensive action. According to Van Evera, these perceptions of a significant offensive advantage had several consequences, all of which contributed to the outbreak of war in 1914. First, belief in the power of offense made expansionist policies seem possible and attractive. Second, perceived offensive advantages made preemptive war seem more necessary. Third, perceptions of offensive advantage also made preventive war seem imperative. Fourth, states adopted more competitive styles of diplomacy based on brinkmanship and faits accomplis. Fifth, the cult of the offensive led militaries to place a premium on secrecy, leaving civilian leaders in the dark about the war plans of their generals and unaware of the magnitude of the dangers of the July Crisis. In combination, these factors made it very difficult for diplomats to manage the July Crisis. Thus the cult of the offensive directly contributed to the outbreak of war in 1914.

Jack Snyder, in "Civil-Military Relations and the Cult of the Offensive," offers an explanation of why Europe's militaries preferred offensive doctrines. He argues that a crisis of civil-military relations in Germany, France, and Russia was the cause of the cult of the offensive. Offensive military doctrines enabled military organizations to enhance their autonomy, minimize civilian interference, improve their ability to plan, increase their importance in national policy, and receive bigger budgets for larger military forces. By allowing militaries to plan on assuming the initiative, and by making conquest seem feasible, if not easy, and external threats consequently seem great, offensive doctrines served these purposes even though, on the battlefield, they were to prove to be disastrously unsuccessful.

In "1914 Revisited: Allies, Offense, and Instability," Scott Sagan argues against the claim that perceptions of a significant offensive advantage were a major cause of World War I. Whereas Van Evera and Snyder contend that domestic and organizational biases led most European states to exaggerate the power of offensive military action, Sagan suggests that states adopted offensive military strategies to serve their broader political objectives. France and Russia, for example, required offensive military doctrines so that they could assist potentially vulnerable allies. They preferred offense because it was the only means to assist allies, not because they embraced a more general cult of the offensive.

Sagan questions whether the offense-defense balance actually strongly favored defense in 1914. In contrast to writers who claim that defense was dominant, Sagan points out that Germany's Schlieffen Plan almost succeeded in the West and that offensives on the Eastern Front were often successful. If military offensives could succeed or come close to succeeding, there may not have been a significant technologically based defensive advantage in 1914.

Sagan also argues that offensive military postures and a general belief in the efficacy of offensive action were not decisive causes of the instabilities that led to war in July 1914. He suggests that critical military vulnerabilities of the Entente powers and Belgium created military incentives for preemptive attacks. These vulnerabilities, which included the inability of the Russian army to mobilize rapidly and the lack of permanent defenses at the Belgian railway junction of Liège, were more important causes of war than perceptions of an offensive advantage.[21]

In a section of correspondence on "The Origins of Offense and the Consequences of Counterforce," Jack Snyder replies to Scott Sagan's arguments, and Sagan responds.[22] Their exchange reveals the complexities of determining whether offensive military doctrines are stabilizing or destabilizing in a multipolar system with several alliances. Snyder argues that the Schlieffen plan did not come close to success. Although the French forces made many errors in August 1914, they were still able to deploy superior forces at the battle of the Marne and halt the Germans. The advantages of being on the defensive enabled the French to shift forces rapidly to this decisive sector at a time when the German troops were overextended. Snyder points out that Schlieffen himself was aware of the flaws of his plan, but covered them up because he did not want to consider defensive alternatives.

Snyder contends that the need to assist allies did not mean that the European powers required offensive military strategies, except in the case of Russia, which needed to attack Germany to relieve pressure on France. Even in this case, however, the cult of the offensive appears to have driven Russia to launch a premature and disastrous offensive in East Prussia. Germany, on the other hand, did not need an offensive doctrine to assist its ally, Austria-Hungary, because attacking France did little to assist Austria-Hungary. Similarly, France did not need to go on the offensive against Germany to assist Russia. French offensive war plans actually hurt Russia's strategic interests by making a rapid decision on the Western front likely and forcing Russia to mobilize and attack prematurely—policies that made the July Crisis harder to manage and contributed to Russia's military defeats in 1914.

Sagan replies by noting that Russia had strong incentives to adopt an offensive military policy in 1914. It needed to attack Germany to assist France, and it needed to attack Austria-Hungary to prevent that country from sending all of

21. Sagan's essay, as well as the subsequent exchange with Jack Snyder, also raises several arguments about the place of counterforce in U.S. nuclear strategy during the Cold War—issues that now seem to be of largely historical interest.

22. It would have been logical for Stephen Van Evera to respond to Sagan's article, but at the time Van Evera was managing editor of *International Security* and he felt it would be inappropriate for him to contribute a letter to the editor to the journal. See Van Evera, *Causes of War,* pp. 152–160 for a discussion of some of the issues raised by Sagan.

its forces to attack Russia's ally, Serbia. Austria-Hungary also required an offensive capability to attack Russia, thereby relieving Russian pressure on Germany.

Recognizing that the alliance motivations behind French and German choices for offense are complex, Sagan argues that France needed only an offensive option to help protect its Russian ally. The biases of the cult of the offensive explain why France embraced offense far too enthusiastically in 1914 and launched an attack when initially remaining on the defensive might have made more strategic sense. In the German case, Sagan points out that Germany required an offensive capability against Russia to relieve pressure on Austria-Hungary. He also argues that exaggerated faith in the power of the offensive does not explain Germany's decision to launch its offensive against France. German military and civilian leaders opted for the Schlieffen Plan because they felt it was imperative to avoid a war of attrition, a belief that seems warranted by the ultimate course of World War I.

Jonathan Shimshoni offers an important challenge to offense-defense theorists' interpretation of World War I in "Technology, Military Advantage, and World War I: A Case for Military Entrepreneurship." Shimshoni argues that technology does not determine the offense-defense balance. Instead, he suggests that innovative "military entrepreneurship" can use technology to create and manipulate offensive or defensive advantages. There was no technologically determined defensive advantage in 1914, only a failure of the military leadership to devise offensive capabilities that would successfully serve the political goals pursued by their governments. Like Sagan, Shimshoni argues that political and strategic objectives should determine military doctrine. Before 1914, strategic goals dictated offensive doctrines, but the militaries of Europe did not develop offensive operational advantages. Far from being condemned to accept an existing offense-defense balance, military leaders can and should manipulate such advantages through "military entrepreneurship." The course of World War I shows that such innovation was possible. The Germans successfully developed infiltration tactics and the allies learned to mount assaults with tanks. Offensives ultimately ended the war on all fronts.

The final section in this volume offers two important recent explications of offense-defense theory, as well as recent critiques of the theory.[23]

---

23. For an explication and critique of other prominent criticisms of offense-defense theory, see Sean M. Lynn-Jones, "Offense-Defense Theory and Its Critics," *Security Studies*, Vol. 4, No. 4 (Summer 1995), pp. 660–691.

Stephen Van Evera's "Offense, Defense, and the Causes of War," examines how the offense-defense balance affects the probability of war.[24] He defines the offense-defense balance as the feasibility of conquest. The balance is shaped by military factors, geography, domestic social and political factors, and the nature of diplomacy.

Van Evera argues that shifts in the offense-defense balance toward the offense have at least ten effects that make war more likely: (1) empires are easier to conquer; (2) self-defense is more difficult; (3) states become more insecure and thus resist others' expansion more fiercely; (4) first-strike advantages become larger; (5) windows of opportunity and vulnerability open wider; (6) states adopt fait accompli diplomatic tactics; (7) states negotiate less readily and less cooperatively; (8) states maintain greater secrecy in foreign and defense policy; (9) arms racing becomes faster and harder; and (10) as conquest grows easier, states adopt policies such as offensive military doctrines, which make conquest even easier, thereby magnifying all the other effects. These effects emerge whether the offensive advantage is real or only perceived. They operate individually or collectively to make war more likely when conquest is easy.

In Van Evera's version of offense-defense theory, the offense-defense balance is determined by multiple factors. Van Evera argues that military factors, such as technology, doctrine, and force posture and deployments have important effects on the balance. For example, the ability to build impregnable fortifications favors the defense. Motorized armor and the doctrine of blitzkrieg favor the offense. Geography—especially the presence or absence of defensible borders and natural obstacles to conquest—is also important, as is the domestic political and social order of states. Popular regimes, for example, are better able to organize guerrilla resistance to potential conquest, thereby strengthening the defense. Finally, diplomatic factors influence the offense-defense balance: "collective security systems, defensive alliances, and balancing behavior by neutral states" all "impede conquest by adding allies to the defending side."

Van Evera tests offense-defense theory by looking at three cases: Europe since 1789, ancient China during the Spring and Autumn and Warring States eras, and the United States since 1789. Van Evera deduces three prime predictions from the theory: (1) war will be more common when offense is easy or is perceived to be easy; (2) states that have or think they have offensive opportu-

---

24. For a more complete explication of Van Evera's ideas, see his *Causes of War.*

nities will be more likely to initiate wars than other states; and (3) states will be more likely to initiate wars during periods when they have or believe they have offensive opportunities. The theory also predicts that the ten war-causing effects listed above will be present in eras and states where there is a real or perceived offensive advantage. Although each case can test only some of the theory's predictions, taken together they allow a test of most of the predictions.

The cases offer broad support for offense-defense theory. In Europe since 1789, the amount of war tends to correlate loosely with the offense-defense balance and tightly with perceptions of the balance. States that faced real or perceived offensive opportunities and defensive vulnerabilities were most likely to initiate war. In ancient China, Van Evera finds that from 722–221 BCE there was a long-term shift in the offense-defense balance that strengthened the offense. As the theory predicts, war became more common as the offense became more powerful. The history of the United States since 1789 also supports the theory. In general, the relative geographical invulnerability of the United States has enabled it to fight only a few great-power wars. Few of the predicted intervening variables—the processes that cause offensive advantages to lead to war—are present in U.S. diplomatic and military history. The United States has fought few preemptive and preventive wars, has been less prone to acquire colonies than other great powers, has not based its diplomacy on fait accompli tactics, has been less secretive than other great powers, and engaged in arms racing only during the Cold War. The level of U.S. bellicosity also has varied with changes in U.S. perceptions of external threats.

Van Evera concludes that offense-defense theory is a robust theory that can pass some difficult tests. He also contends that it is a useful theory that can be applied to prevent war. Unlike variables such the polarity of the international system, the state of human nature, or the strength of international institutions, the variables identified by offense-defense theory—especially perceptions of the offense-defense balance—can be manipulated by national policies. Van Evera argues that the theory suggests that war can be limited or prevented if states adopt defensive military doctrines and limit offensive military capabilities through arms control. Defensive alliances, such as U.S. security guarantees in Europe and Asia since 1949, also can make conquest harder and war less likely.

In "What Is the Offense-Defense Balance and Can We Measure It?" Charles Glaser and Chaim Kaufmann note that the absence of an agreed definition of the offense-defense balance is a serious shortcoming of offense-defense theory and explicate how they believe this variable should be defined and measured.

They argue that the offense-defense balance should be defined as "the ratio of the cost of the forces that the attacker requires to take territory to the cost of the defender's forces." Glaser and Kaufmann hold that offense-defense theory is only a partial theory of military capabilities, because the offense-defense balance is analytically distinct from two other variables: power, defined as relative resources, and military skill. Each of these three variables might be more important in a given situation. For example, a powerful attacker might prevail over a much weaker defender even if the offense-defense balance favors defense.

Glaser and Kaufmann offer six specifications and assumptions that are necessary to define and measure the offense-defense balance: (1) the need to specify the cost of fighting that the attacker would incur; (2) the need to specify the attacker's territorial goal; (3) defining the offense-defense balance on the basis of the outcomes of wars, not battles; (4) assuming that states act optimally: "within reasonable limits of analysis, states make the best possible decisions for attack or defense, taking into account their own and their opponents' options for strategy and force posture" which means that "the offense-defense balance is the cost ratio of the attacker's best possible offense to the defender's best possible defense"; (5) measuring the balance across specific dyads of states: "the offense-defense balance is well defined only for specific dyads, not for the entire international system" because it "depends on a number of diverse factors—including geography, cumulativity of resources, and nationalism" that are not shared across dyads; and (6) the need to evaluate the "compound balance" between two states.

Glaser and Kaufmann argue for a "broad" approach to measuring the offense-defense balance. All factors that influence the cost-ratio of offense to defense should be used to measure the balance, including technology, geography, the size of forces, the cumulativity of resources, and nationalism. Alliance behavior and first-move advantages should be excluded, however. Glaser and Kaufmann conclude that the offense-defense balance can be measured using the same analytical tools used to perform military net assessments.

In the section of correspondence, "Taking Offense at Offense-Defense Theory," several critics of the preceding two essays have their say, and Van Evera, Glaser, and Kaufmann respond. The ensuing exchange provides an excellent and illuminating overview of some of the central issues in the contemporary debate over offense-defense theory.

James Davis faults Van Evera for broadening the concept of the offense-defense balance by adding diplomatic factors such as collective security systems, defensive alliances, and the balancing tendencies of neutral states to it. He contends that these additions make the theory less parsimonious and that they are variables that the offense-defense balance should attempt to explain, not part of the balance itself. Davis also argues that Van Evera wrongly combines objective and perceptual variants of the offense-defense balance in the same theory. He also argues that Van Evera's methodology is flawed, because Van Evera does not test his theory against plausible contending explanations.

Bernard Finel offers four objections to how Van Evera and Glaser and Kaufmann approach offense-defense theory. First, he argues that Glaser and Kaufmann ignore the complicated interaction effects of different strategies when they call for measuring the offense-defense balance by using the tools of net assessment. In complex war games, a common tool of net assessment of maneuver warfare, outcomes are extremely sensitive to small changes in strategies. Second, Finel argues that Van Evera and Glaser and Kaufmann are wrong to claim that states are more likely to use force when it is easy to seize territory, a central proposition of offense-defense theory. Instead, the rapidity of victory is the crucial determinant in decisions to use force. Rapid victory does not always require seizing and holding territory. Third, Finel contends that Van Evera attempts to aggregate too many variables into his measure of the offense-defense balance, producing a theory that is not parsimonious and that depends on subjective assessments. Many of the variables cited by Van Evera and by Glaser and Kaufmann—mobility, geography, social and political order, diplomatic factors—are difficult to code clearly. It is even more difficult to aggregate them. Finally, Finel argues that offense-defense theory in general lacks parsimony and is unlikely to be useful.

Stacie Goddard argues that neither Van Evera nor Glaser and Kaufmann distinguish the offense-defense balance from the balance of power or military skill, which renders their versions of offense-defense theory tautological and difficult to test. In her view, Van Evera confuses offense-defense balance with the probability of success in war, which may be a function of superior power or other factors unrelated to the offense-defense balance. Like Davis and Finel, Goddard also argues that Van Evera includes too many variables as determinants of the offense-defense balance. Goddard praises Glaser and Kaufmann for their attempt to address methodological shortcomings of earlier definitions of the offense-defense balance, but concludes that their broad definition of the offense-defense balance fails to distinguish the balance from other variables such as power and skill. In particular, she argues that whether states behave in accordance with Glaser and Kaufmann's optimality assumption can only be determined on the basis of war outcomes. She suggests that offense-defense theory may be useful, but it needs a definition of the offense-defense balance that avoids subsuming power and doctrine, as well as far more rigorous empirical tests.

Stephen Van Evera replies to Davis, Finel, and Goddard by making three main arguments. First, he argues that some military technologies and force postures can be characterized as offensive or defensive, rejecting Finel's contention that such characterizations always depend on the context of combat. Secure nuclear deterrents, for example, are fundamentally defensive, as is modern guerilla warfare and the barbed wire, machine guns, and trenches of World War I.

Second, Van Evera argues that offense-defense theory actually is parsimonious, because aggregating lesser-included concepts actually simplifies discussion of security problems and policies. Decisionmakers often focus on whether conquest is easy or hard as they formulate national security policies. Theorists should not hesitate to aggregate factors that influence the ease of conquest, because this corresponds to the way policymakers think. Offense-defense theory also simplifies our understanding of power and war by identifying the taproot cause of many different phenomena.

Third, Van Evera responds to Davis's claim that Van Evera's offense-defense theory is not testable because it conflates objective perceptual variables. Van Evera acknowledges that his article conflated the two sets of predictions because those predictions are essentially the same.

Van Evera concludes with a series of brief replies to criticisms, arguing that victory is necessary to seize territory; historians favor his interpretations, not Finel's; his definition of the offense-defense balance is distinct from the balance of power; there is no methodological problem with using the behavior of states to explain the behavior of states; and others are free to define terms differently in their theories and analyses.

Charles Glaser and Chaim Kaufmann reply to Goddard by arguing that her views on offense-defense theory are not at odds with their own, but her criticisms reflect a misunderstanding of their arguments. They point out that their optimality assumption means only that states choose optimally "within reasonable limits of analysis" not that states choose the best possible force posture, doctrine, and strategy. Thus it is possible to assume optimality for analytical purposes without relying on the evidence provided by war outcomes.

Glaser and Kaufmann reply to Finel's argument about interaction effects by pointing out that many military policy decisions do not depend on interaction effects. Between the two world wars, for example, all navies would have been better off if they had invested less in battleships. In other cases, "interaction" means responding to an adversary's attempt to pursue an optimal policy, which becomes part of the structural constraints that shape a state's own policies. Glaser and Kaufmann recognize that there are some cases in which whether a state's military strategy succeeds or not does indeed depend on guessing its adversary's strategy, just as it does in the "rock-paper-scissors" game. In such cases, the offense-defense balance cannot be measured exactly but spreads out into a "band of uncertainty." In practice such bands are likely to be narrow, because strategic and grand strategic choices, which have the greatest effect on war outcomes, are constrained by geography and resource endowments, rendering strategic choices more like a mutual optimization model than a guessing game.

Glaser and Kaufmann also reject Goddard's argument that their definition of the offense-defense balance includes the balance of power, arguing that some elements of the offense-defense balance that seem to be part of aggregate power are actually determinants of offensive and defensive capabilities. Nationalism is a component of the offense-defense balance, for example, because it makes it easier to defend a national homeland. Nationalism may also increase aggregate power, but it is part of the offense-defense balance because it makes it easier to defend territory.

In response to Finel's argument that the techniques of net assessment may not be adequate for measuring the offense-defense balance, Glaser and Kaufmann say that further research is necessary, but that Finel's pessimism would only be justified if net assessment were never possible. They also reject Finel's argument that their concept of the offense-defense balance assumes that victory depends on conquering territory, arguing that the offense-defense balance reflects the impact of territory only when territory influences war outcomes.

In "Must War Find a Way? A Review Essay," Richard Betts offers a critical review of Stephen Van Evera's *Causes of War: Power and the Roots of Conflict,* a book that presents one of the most prominent and important variants of contemporary offense-defense theory. Betts uses his review essay to criticize Van Evera's book in particular, and offense-defense theory more generally. He makes four main arguments.

First, Betts argues that offense-defense theory "has been overwhelmingly influenced by the frame of reference that developed around nuclear weapons and the application of deterrence theorems to the conventional case of World

War I." This approach has limited the applicability of offense-defense theory to intrastate wars. The influence of nuclear deterrence concepts, which have focused on weapons and military postures as causes of war, has also diverted offense-defense theory away from the substantive causes of war.

Betts faults offense-defense theory for its argument that mutual nuclear deterrence is a situation of defense dominance. He argues that this turns traditional logic on its head by defining the ability to attack with nuclear missiles as a defensive advantage. Earlier definitions of the offense-defense balance of conventional forces held that defense had the advantage when it was operationally difficult to mount a successful offensive. In contrast, mutual nuclear deterrence is a situation in which the political—not military—effect of nuclear weapons makes peace more likely. Thus the U.S.-Soviet nuclear stalemate and the defensive deadlock of 1914–17 do not belong in the same category.

Betts further argues that offense-defense theory has exaggerated the stabilizing effects of nuclear weapons. In his view, Van Evera and other offense-defense theorists fail to consider how stability at the nuclear level may actually make the world safe for conventional war. He argues that Van Evera ignores many arguments against missile defense programs.

In a brief discussion of the outbreak of World War I, Betts points out that some historians do not agree that a "cult of the offensive" caused war in 1914. He criticizes Van Evera for testing offense-defense theory with the case that has inspired much of the theory. Finally, he points out that Van Evera concedes that if the Western allies had developed offensive capabilities in the 1930s, they might have deterred World War II and argues that this indicates that offense-defense theorists have relied too much on the World War I case.

Second, Betts argues that Van Evera "crams so many things into the definition of the offense-defense balance that it becomes a gross megavariable." By including military technology and doctrine, geography, national social structure, diplomatic arrangements, military manpower policies, and other factors in his list of determinants of the offense-defense balance, Van Evera creates a variable that is almost the same as relative power. Betts argues that Van Evera does not explain how to aggregate all these factors to measure the offense-defense balance. He suggests that analysts should rely instead on the distribution of relative power in their attempts to explain the outbreak of wars. Noting that Van Evera argues for his definition of the offense-defense balance on the grounds that it includes variables that can be manipulated by leaders who hope to prevent war, Betts points out that revisionist powers also could manipulate the balance to make conquest easier.

Third, Betts argues that Van Evera—and offense-defense theory more generally—does not consider the substantive stakes of conflict. Whether conquest is easy or hard may be less important than the value states attach to their objectives. In some cases, motivations such honor may lead states to fight unwinnable wars. In others, Betts argues changes in values and interests are more important than the offense-defense balance in decisions to start or avoid war. He thus contends that the Cold War ended, for example, because the Soviet Union abandoned the ideological struggle with the West, not because Gorbachev embraced offense-defense theory. Betts argues that offense-defense theory says little about internal wars, which are about who rules a state, not territory. Offense-defense theory is also unlikely to be relevant to irregular and guerilla warfare.

Finally, Betts faults Van Evera for adopting a normative bias in favor of peace and the status quo. He argues that Van Evera's version of offense-defense theory seems to assume "that war is justified only to defend what one has, not to conquer what others hold." This assumption has much moral appeal, but it fails to recognize that states often take offensive action to change an unjust status quo. Betts suggest that theories should not assume that peace is more natural and just than war.

Keir Lieber's "Grasping the Technological Peace: The Offense-Defense Balance and International Security" presents a deductive and empirical critique of offense-defense theory. Lieber concludes that the technological offense-defense balance, whether real or perceived, has limited effects on political and strategic behavior.

Lieber chooses to test what he calls the "core" version of offense-defense theory. This version holds that the offense-defense balance is primarily determined by technology. The "broad" version, on the other hand, incorporates many additional factors as determinants of the offense-defense balance, including geography, nationalism, regime popularity, the ease of exploiting resources from conquered territory, and others. Lieber argues that the core version is a better candidate for testing because it is more parsimonious and more likely to apply across space and time.

Drawing on the work of offense-defense theorists, Lieber derives two hypotheses from the core version of offense-defense theory: mobility-enhancing technologies favor offense, and firepower-enhancing technologies favor defense. He suggests that the logical case for these hypotheses may be weak. Mobility, for example, may be necessary for a defender that is trying to rush reinforcements to thwart an attack. Nevertheless, Lieber argues that testing these hypotheses constitutes a valid test of offense-defense theory, because proponents of the theory generally accept these hypotheses.

Lieber tests the two hypotheses by examining the impact of four major technological innovations: railroads, the artillery and small arms revolution, tanks, and nuclear weapons. He considers whether the innovation altered military outcomes by making victories more likely for the attacker or the defender, and whether the innovation altered political outcomes by making decisionmakers more likely to initiate war when they perceived an offensive advantage.

Railroads, in Lieber's view, had relatively little impact on military outcomes. The quick and decisive victories by, for example, Prussia in 1864, 1866, and 1870–71 coincided with the spread of railroads in Europe, but were actually the result of Prussia's superior doctrine, organization, and power. In World War I, railroads appear to have contributed to defensive successes by enabling defenders to shift forces to halt attacks. Lieber argues that railroads also did not have the political effect of making war more likely. Between 1850 and 1871, European decisionmakers believed railroads favored the defender, but several wars broke out during this period. After 1871, decisionmakers believed railroads enhanced offense, but this belief coincided with a long era of peace in Europe.

Lieber finds that the technological revolution in small arms and artillery of the late nineteenth and early twentieth century had relatively small military effects. Although the increase in firepower due to improvements in range, accuracy, and rate of fire favored defenders by making massed frontal assaults more difficult, tactical innovations made offensives possible in the Franco-Prussian War, the American Civil War, and World War I. Politically, Lieber argues that European decisionmakers opted for offensive military strategies and war even though they believed that the firepower revolution favored defense, exactly the opposite of what offense-defense theory predicts.

Although many observers argue that the development of the tank enhanced the relative power of offense, Lieber finds that they had indeterminate effects. In World War I, tanks had little impact on operational military outcomes. In World War II, Germany's initial victories with armored offensives reflected German doctrinal and material superiority, not the offensive potential of the tank. When other countries—particularly Russia—became proficient in armored warfare, tanks enhanced the defense. The advent of the tanks also did influence political decisions for war. Lieber argues that Hitler's decision to attack France in 1940, for example, was not based on confidence in the offensive potential of tank forces.

Lieber finds that the nuclear revolution has had mixed effects. Offense-defense theorists generally argue that nuclear weapons enhance defense, because a state with a secure nuclear retaliatory capability becomes unconquerable. Recognizing that no war has occurred between major nuclear powers, Lieber points out that offense-defense theory is undermined by the conflicts between India and Pakistan, the U.S.-Soviet nuclear arms race, and the frequency of U.S. and Soviet military intervention in the third world.

On the basis of the empirical evidence, Lieber concludes that offense-defense theory has exaggerated the extent to which technology—and beliefs about it—shape military and political outcomes.

In "Attack and Conquer? International Anarchy and the Offense-Defense-Deterrence Balance," Karen Ruth Adams presents an empirical test of offense-defense theory. In contrast to offense-defense theorists who have used the theory to explain and predict war, Adams focuses on other predictions of offense-defense theory. She argues that the logic of the theory leads to two broad predictions. First, states are more likely to be conquered in offense-dominant eras. Second, states are more likely to conduct offensive military operations in such eras. Adams holds that "the most direct way" to test hypotheses about the consequences of variations in the offense-defense balance is "to examine the effects of the balance on the incidence of attack and conquest." When the offense-defense balance shifts, these variables are likely to vary more than the incidence of war.

Before testing offense-defense theory, Adams refines the concept of the offense-defense balance by distinguishing between deterrence and defense. Although other offense-defense theorists have argued that nuclear deterrence is equivalent to an extremely large defensive advantage, Adams argues that deterrent operations are different from defensive operations. When deterrence—not defense—is dominant, conquest should be less likely and states will have few incentives to expand. Adams therefore replaces the offense- defense balance with the offense-defense-deterrence balance, which she defines as "the relative efficacy of offense, defense, and deterrence given prevailing conditions."

Adams measures the offense-defense-deterrence balance by examining "the potential lethality, protection, and mobility of state-of-the-art offensive, defensive, and deterrent operations." In contrast to, for example, Van Evera, she treats technology as the sole determinant of the balance. Technology includes the "methods, skills, and tools" that influence how states conduct offensive, defensive, and deterrent operations.

To determine how the offense-defense-deterrence balance has varied since 1800, Adams assesses how technological changes have affected force lethality, protection, and mobility in military operations. After integrating these factors, she codes the 1800–49 period as offense dominant, 1850–1933 as defense dominant, 1933–45 as offense dominant, and 1946–present as deterrence dominant.

Using data on conquest of and attacks by great powers and nuclear states from 1800 to the present, Adams quantitatively tests her hypotheses against alternative explanations of attack and conquest. Her tests find that historical trends in both attack and conquest offer strong support for her hypotheses. She also finds that relative capabilities matter. Great powers are less likely to be conquered and more likely to attack, regardless of the offense-defense-deterrence balance.

Adams argues that her findings have broader implications for international relations theory. With regard to offense-defense theory, she reiterates that "incidence of war is not the best dependent variable to use in testing" theories of offense and defense, that technological factors have significant effects, and that defense and deterrence should be distinguished. Adams also notes that variations in the balance do not explain all of the variation in military doctrine, arms racing, and international cooperation. She also calls for further tests to explain why attack is much more likely than conquest in offense-dominant eras. Factors such as homeland advantage, power balancing, military doctrine, force employment, and attrition may account for this pattern.

More generally, Adams argues that her findings support structural realist theory while casting doubt on offensive realism, defensive realism, and "arguments about the pacific effects of unbalanced power and rising audience costs." Turning to the implications of her findings for contemporary international politics, she argues that the spread of nuclear weapons will accelerate if even one nuclear state pursues a policy of conquest and expansion.

The essays in this volume indicate that offense-defense theory has generated an active research program and some significant theoretical contributions. The intuitive appeal of the various versions of the theory and the existence of historical evidence that suggests that decisionmakers sometimes consider the relative efficacy of offense and defense has stimulated much conceptual innovation and empirical research.

Nevertheless, the debate over offense and defense in international politics suggests that offense-defense theory has some limits. Offense-defense theory—especially when the offense-defense balance is defined narrowly—may not explain as much as its proponents initially hoped and claimed. The narrow, global balance of offensive and defensive military technology may be useful for explaining overall trends in a given international system, such as the likelihood of war, attack, or conquest; whether there are a few large states or many small states; and whether alliances are tight and balancing rapid and vigorous. It may not, however, explain all decisions to initiate war or the outcome of each war. In this sense, offense-defense theory has the same limitations as many other structural theories of international politics.

Some of the variables identified by the theory (e.g., technological factors) may not always be very powerful, compared to others, including relative capabilities and state-level variables. The global technological offense-defense balance may only have a strong effect when it takes on an extreme value, and even then other factors may exert a stronger influence on the foreign policies of individual states.

Defining and measuring the offense-defense balance remains difficult, whether the balance is measured systemically or across particular dyads. The attempts of Van Evera and Glaser and Kaufmann to define and measure the offense-defense balance illustrate that conceptualizing and operationalizing this variable is neither easy nor uncontroversial.

Offense-defense theory also may seem less relevant at a time when traditional patterns of conflict between states and territorial conquest appear to be changing. In an era when security issues often revolve around the "privatization of violence" and the potential use of weapons of mass destruction by terrorist groups, concepts like defense-dominance may not shed light on the most important threats to contemporary global security.[25]

These limits notwithstanding, the essays assembled in this collection, as well as a review of the literature listed as suggestions for further reading at the end of this volume, suggest many potential avenues for future research.

First, other researchers might adapt and apply Van Evera's version of offense-defense theory to other cases. In particular, Van Evera's theory could be extended to cases of internal conflict.[26]

Second, Van Evera's offense-defense theory could be recast as theory of foreign policy. Critics complain that the many elements of Van Evera's version of the offense-defense balance make it difficult to use the balance to make pre-

25. Offense-defense theorists might respond that the 2003 U.S. invasion and occupation of Iraq is a reminder that territorial conquest remains important.
26. See, for example, William Rose, "The Security Dilemma and Ethnic Conflict: Some New Hypotheses," *Security Studies*, Vol. 9, No. 4 (Summer 2000), pp. 1–51.

dictions about the overall pattern of international politics. These problems might be less significant if some or all of the components that Van Evera incorporates into the balance were used to explain variations in the foreign policy of a particular state.

Third, scholars might attempt to develop security dilemma theory instead of offense-defense theory. Although the two theories have much in common and are often conflated, security dilemma theory could be regarded as taking into account the range of political, economic, military, and psychological factors that contribute to the potential for suspicion and hostility in a particular dyadic relationship. Several scholars have adopted this approach.[27]

Fourth, the essay by Glaser and Kaufmann in this volume suggests that offense-defense theory could be integrated into the process of military net assessment, with particular focus on the analysis of the sources of offensive potential.

Fifth, it may be necessary to recast offense-defense theory so that it focuses on other sources of offensive capabilities, such as force employment practices, instead of the technological variables traditionally associated with offense-defense theory.[28] Some observers might question whether this type of approach even qualifies as offense-defense theory, but this classification question is less important than the potential for interesting research.

Sixth, disaggregating elements of offense-defense theory and incorporating these variables into other theories may contribute to theory development. Several important works in international security studies have used concepts borrowed from offense-defense theory, even when they have not used the offense-defense balance as an explanatory variable. Stephen Walt's *The Origins of Alliances,* for example, employs concepts that are central in offense-defense theory—offensive power and geographical proximity—in constructing the broader concept of "threat."[29] T.V. Paul uses the concept of offensive capabilities to explain why states start wars when they are obviously weaker than their adversaries.[30]

Seventh, it may be possible to incorporate offense-defense variables into

27. See Rose, "The Security Dilemma and Ethnic Conflict: Some New Hypotheses"; Barry R. Posen, "The Security Dilemma and Ethnic Conflict," *Survival,* Vol. 35, No. 1 (Spring 1993), pp. 27–57; and Stephen M. Walt, *Revolution and War* (Ithaca, N.Y.: Cornell University Press, 1996).
28. See Biddle, "Rebuilding the Foundations of Offense-Defense Theory."
29. Stephen M. Walt, *The Origins of Alliances* (Ithaca, N.Y.: Cornell University Press, 1987).
30. T.V. Paul, *Asymmetric Conflicts: War Initiation by Weaker Powers* (Cambridge, U.K.: Cambridge University Press, 1984).

formal models. The basic logic of offense-defense theory would appear to lend itself to formal approaches, but they have been rare in the existing offense-defense literature.[31]

Finally, there may be many ways in which exploring the causal logic of offense-defense theory could open up new directions for research. In this volume, the essay by Karen Ruth Adams breaks new ground by reexamining the deductive logic of offense-defense theory and concluding that the dependent variables of attack and conquest deserve more attention. Adams also suggests that states may expand when defense has the advantage if they anticipate future offense-dominance. This observation implies that future research might consider the "intertemporal offense-defense balance" and the extent to which states make decisions on the basis of future expectations of the relative efficacy of offense or defense.[32]

There are almost certainly many other ways in which offense-defense theories, or aspects of them, can inspire innovative new research. In many cases, debates over what the offense-defense balance is and whether it can be measured need not impede further explorations of the role of offense and defense in international politics. Offense-defense theorists need to consider these general issues of definition and measurement, but they also need to focus on identifying variables that appear to have explanatory power, deriving hypotheses, and testing them empirically.

The essays assembled in this volume are only a small portion of the vast literature on offense, defense, and war. The suggestions for further reading at the end of the book include many other books and articles. We nevertheless hope that this collection of essays will help readers to understand the central elements of offense-defense theories, the main themes in the debate over offense and defense in international politics, and potential avenues for further research on these topics.

31. For an important exception, see Robert Powell, "Guns, Butter, and Anarchy," *American Political Science Review*, Vol. 87, No. 1 (March 1993), pp. 115–132.
32. This type of analysis is prominent in research on preventive war and how states respond to anticipated shifts in their relative power. It also has been applied to expectations of future levels of trade. See Dale C. Copeland, "Economic Interdependence and War: A Theory of Trade Expectations," *International Security*, Vol. 20, No. 4 (Spring 1996), pp. 5–41.

Part I:
Foundations of Offense-Defense Theory

Cooperation Under the Security Dilemma

*Robert Jervis**

I. Anarchy and the Security Dilemma

The lack of an international sovereign not only permits wars to occur, but also makes it difficult for states that are satisfied with the status quo to arrive at goals that they recognize as being in their common interest. Because there are no institutions or authorities that can make and enforce international laws, the policies of cooperation that will bring mutual rewards if others cooperate may bring disaster if they do not. Because states are aware of this, anarchy encourages behavior that leaves all concerned worse off than they could be, even in the extreme case in which all states would like to freeze the status quo. This is true of the men in Rousseau's "Stag Hunt." If they cooperate to trap the stag, they will all eat well. But if one person defects to chase a rabbit—which he likes less than stag—none of the others will get anything. Thus, all actors have the same preference order, and there is a solution that gives each his first choice: (1) cooperate and trap the stag (the international analogue being cooperation and disarmament); (2) chase a rabbit while others remain at their posts (maintain a high level of arms while others are disarmed); (3) all chase rabbits (arms competition and high risk of war); and (4) stay at the original position while another chases a rabbit (being disarmed while others are armed).[1] Unless

---

* I am grateful to Robert Art, Bernard Brodie, and Glenn Snyder for comments, and to the Committee on Research of the UCLA Academic Senate for financial support. An earlier version of this essay appeared as Working Paper No. 5, UCLA Program in Arms Control and International Security.

This article was originally printed in *World Politics* (30:2).

---

1. This kind of rank-ordering is not entirely an analyst's invention, as is shown by the following section of a British army memo of 1903 dealing with British and Russian railroad construction near the Persia-Afghanistan border:

> The conditions of the problem may . . . be briefly summarized as follows:
>
> a) If we make a railway to Seistan while Russia remains inactive, we gain a considerable defensive advantage at considerable financial cost;
> b) If Russia makes a railway to Seistan, while we remain inactive, we gain a considerable offensive advantage at considerable financial cost;
> c) If both we and Russia make railways to Seistan, the defensive and offensive advantages may be held to neutralize each other; in other words, we shall have spent a good deal of money and be no better off than we are at present. On the other hand, we shall be no worse off, whereas under alternative (b) we shall be much worse off. Consequently, the theoretical balance of advantage lies with the proposed railway extension from Quetta to Seistan.

W. G. Nicholson, "Memorandum on Seistan and Other Points Raised in the Discussion on the Defence of India," (Committee of Imperial Defence, March 20, 1903). It should be noted that the possibility of neither side building railways was not mentioned, thus strongly biasing the analysis.

each person thinks that the others will cooperate, he himself will not. And why might he fear that any other person would do something that would sacrifice his own first choice? The other might not understand the situation, or might not be able to control his impulses if he saw a rabbit, or might fear that some other member of the group is unreliable. If the person voices any of these suspicions, others are more likely to fear that he will defect, thus making them more likely to defect, thus making it more rational for him to defect. Of course in this simple case—and in many that are more realistic—there are a number of arrangements that could permit cooperation. But the main point remains: although actors may know that they seek a common goal, they may not be able to reach it.

Even when there is a solution that is everyone's first choice, the international case is characterized by three difficulties not present in the Stag Hunt. First, to the incentives to defect given above must be added the potent fear that even if the other state now supports the status quo, it may become dissatisfied later. No matter how much decision makers are committed to the status quo, they cannot bind themselves and their successors to the same path. Minds can be changed, new leaders can come to power, values can shift, new opportunities and dangers can arise.

The second problem arises from a possible solution. In order to protect their possessions, states often seek to control resources or land outside their own territory. Countries that are not self-sufficient must try to assure that the necessary supplies will continue to flow in wartime. This was part of the explanation for Japan's drive into China and Southeast Asia before World War II. If there were an international authority that could guarantee access, this motive for control would disappear. But since there is not, even a state that would prefer the status quo to increasing its area of control may pursue the latter policy.

When there are believed to be tight linkages between domestic and foreign policy or between the domestic politics of two states, the quest for security may drive states to interfere pre-emptively in the domestic politics of others in order to provide an ideological buffer zone. Thus, Metternich's justification for supervising the politics of the Italian states has been summarized as follows:

> Every state is absolutely sovereign in its internal affairs. But this implies that every state must do nothing to interfere in the internal affairs of any other. However, any false or pernicious step taken by any state in its internal affairs may disturb the repose of another state, and this consequent disturbance of another state's repose constitutes an interference in that state's internal affairs. Therefore, every state—or rather, every sovereign of a great power—has the

duty, in the name of the sacred right of independence of every state, to supervise the governments of smaller states and to prevent them from taking false and pernicious steps in their internal affairs.[2]

More frequently, the concern is with direct attack. In order to protect themselves, states seek to control, or at least to neutralize, areas on their borders. But attempts to establish buffer zones can alarm others who have stakes there, who fear that undesirable precedents will be set, or who believe that their own vulnerability will be increased. When buffers are are sought in areas empty of great powers, expansion tends to feed on itself in order to protect what is acquired, as was often noted by those who opposed colonial expansion. Balfour's complaint was typical: "Every time I come to a discussion—at intervals of, say, five years—I find there is a new sphere which we have got to guard, which is supposed to protect the gateways of India. Those gateways are getting further and further away from India, and I do not know how far west they are going to be brought by the General Staff."[3]

Though this process is most clearly visible when it involves territorial expansion, it often operates with the increase of less tangible power and influence. The expansion of power usually brings with it an expansion of responsibilities and commitments; to meet them, still greater power is required. The state will take many positions that are subject to challenge. It will be involved with a wide range of controversial issues unrelated to its core values. And retreats that would be seen as normal if made by a small power would be taken as an index of weakness inviting predation if made by a large one.

The third problem present in international politics but not in the Stag Hung is the security dilemma: many of the means by which a state tries to increase its security decrease the security of others. In domestic society, there are several ways to increase the safety of one's person and property without endangering others. One can move to a safer neighborhood, put bars on the windows, avoid dark streets, and keep a distance from suspicious-looking characters. Of course these measures are not convenient, cheap, or certain of success. But no one save criminals need be alarmed if a person takes them. In international politics, however, one state's gain in security often inadvertently threatens others. In explaining British policy on naval disarmament in the interwar period to the

2. Paul Schroeder, *Metternich's Diplomacy at Its Zenith, 1820–1823* (Westport, Conn.: Greenwood Press 1969), 126.
3. Quoted in Michael Howard, *The Continental Commitment* (Harmondsworth, England: Penguin 1974), 67.

Japanese, Ramsey MacDonald said that "Nobody wanted Japan to be insecure."[4] But the problem was not with British desires, but with the consequences of her policy. In earlier periods, too, Britain had needed a navy large enough to keep the shipping lanes open. But such a navy could not avoid being a menace to any other state with a coast that could be raided, trade that could be interdicted, or colonies that could be isolated. When Germany started building a powerful navy before World War I, Britain objected that it could only be an offensive weapon aimed at her. As Sir Edward Grey, the Foreign Secretary, put it to King Edward VII: "If the German Fleet ever becomes superior to ours, the German Army can conquer this country. There is no corresponding risk of this kind to Germany; for however superior our Fleet was, no naval victory could bring us any nearer to Berlin." The English position was half correct: Germany's navy was an anti-British instrument. But the British often overlooked what the Germans knew full well: "in every quarrel with England, German colonies and trader were . . . hostages for England to take." Thus, whether she intended it or not, the British Navy constituted an important instrument of coercion.[5]

II. *What Makes Cooperation More Likely?*

Given this gloomy picture, the obvious question is, why are we not all dead? Or, to put it less starkly, what kinds of variables ameliorate the impact of anarchy and the security dilemma? The workings of several can be seen in terms of the Stag Hunt or repeated plays of the Prisoner's Dilemma. The Prisoner's Dilemma differs from the Stag Hunt in that there is no solution that is in the best interests of all the participants; there are offensive as well as defensive incentives to defect from the coalition with the others; and, if the game is to be played only once, the only rational response is to defect. But if the game is repeated indefinitely, the latter characteristic no longer holds and we can analyze the game in terms similar to those applied to the Stag Hunt. It would be in the

Prisoner's Dilemma

4. Quoted in Gerald Wheeler, *Prelude to Pearl Harbor* (Columbia: University of Missouri Press 1963), 167.

5. Quoted in Leonard Wainstein, "The Dreadnought Gap," in Robert Art and Kenneth Waltz, eds., *The Use of Force* (Boston: Little, Brown 1971), 155; Raymond Sontag, *European Diplomatic History, 1871–1932* (New York: Appleton-Century-Crofts 1933), 147. The French had made a similar argument 50 years earlier; see James Phinney Baxter III, *The Introduction of the Ironclad Warship* (Cambridge: Harvard University Press 1933), 149. For a more detailed discussion of the security dilemma, see Jervis, *Perception and Misperception in International Politics* (Princeton: Princeton University Press 1976), 62–76.

interest of each actor to have others deprived of the power to defect; each would be willing to sacrifice this ability if others were similarly restrained. But if the others are not, then it is in the actor's interest to retain the power to defect.[6] The game theory matrices for these two situations are given below, with the numbers in the boxes being the order of the actors' preferences.

STAG HUNT

B \ A	COOPERATE	DEFECT
COOPERATE	1, 1	2, 4
DEFECT	4, 2	3, 3

PRISONER'S DILEMMA

B \ A	COOPERATE	DEFECT
COOPERATE	2, 2	1, 4
DEFECT	4, 1	3, 3

We can see the logical possibilities by rephrasing our question: "Given either of the above situations, what makes it more or less likely that the players will cooperate and arrive at CC?" The chances of achieving this outcome will be increased by: (1) anything that increases incentives to cooperate by increasing the gains of mutual cooperation (CC) and/or decreasing the costs the actor will pay if he cooperates and the other does not (CD); (2) anything that decreases the incentives for defecting by decreasing the gains of taking advantage of the other (DC) and/or increasing the costs of mutual noncooperation (DD); (3) anything that increases each side's expectation that the other will cooperate.[7]

THE COSTS OF BEING EXPLOITED (CD)

The fear of being exploited (that is, the cost of CD) most strongly drives the security dilemma; one of the main reasons why international life is not more nasty, brutish, and short is that states are not as vulnerable as men are in a state of nature. People are easy to kill, but as Adam Smith replied to a friend who feared that the Napoleonic Wars would ruin England, "Sir, there is a great deal of ruin in a nation."[8] The easier it is to destroy a state, the greater the reason for

6. Experimental evidence for this proposition is summarized in James Tedeschi, Barry Schlenker, and Thomas Bonoma, *Conflict, Power, and Games* (Chicago: Aldine 1973), 135–41.
7. The results of Prisoner's Dilemma games played in the laboratory support this argument. See Anatol Rapoport and Albert Chammah, *Prisoner's Dilemma* (Ann Arbor: University of Michigan Press 1965), 33–50. Also see Robert Axelrod, *Conflict of Interest* (Chicago: Markham 1970), 60–70.
8. Quoted in Bernard Brodie, *Strategy in the Missile Age* (Princeton: Princeton University Press 1959), 6.

it either to join a larger and more secure unit, or else to be especially suspicious of others, to require a large army, and, if conditions are favorable, to attack at the slightest provocation rather than wait to be attacked. If the failure to eat that day—be it venison or rabbit—means that he will starve, a person is likely to defect in the Stag Hunt even if he really likes venison and has a high level of trust in his colleagues. (Defection is especially likely if the others are also starving or if they know that he is.) By contrast, if the costs of CD are lower, if people are well-fed or states are resilient, they can afford to take a more relaxed view of threats.

A relatively low cost of CD has the effect of transforming the game from one in which both players make their choices simultaneously to one in which an actor can make his choice after the other has moved. He will not have to defect out of fear that the other will, but can wait to see what the other will do. States that can afford to be cheated in a bargain or that cannot be destroyed by a surprise attack can more easily trust others and need not act at the first, and ambiguous, sign of menace. Because they have a margin of time and error, they need not match, or more than match, any others' arms in peacetime. They can mobilize in the prewar period or even at the start of the war itself, and still survive. For example, those who opposed a crash program to develop the H-bomb felt that the U.S. margin of safety was large enough so that even if Russia managed to gain a lead in the race, America would not be endangered. The program's advocates disagreed: "If we let the Russians get the super first, catastrophe becomes all but certain."[9]

When the costs of CD are tolerable, not only is security easier to attain but, what is even more important here, the relatively low level of arms and relatively passive foreign policy that a status-quo power will be able to adopt are less likely to threaten others. Thus it is easier for status-quo states to act on their common interests if they are hard to conquer. All other things being equal, a world of small states will feel the effects of anarchy much more than a world of large ones. Defensible borders, large size, and protection against sudden attack not only aid the state, but facilitate cooperation that can benefit all states.

Of course, if one state gains invulnerability by being more powerful than most others, the problem will remain because its security provides a base from

9. Herbert York, *The Advisors: Oppenheimer, Teller, and the Superbomb* (San Francisco: Freeman, 1976), 56–60.

which it can exploit others. When the price a state will pay for DD is low, it leaves others with few hostages for its good behavior. Others who are more vulnerable will grow apprehensive, which will lead them to acquire more arms and will reduce the chances of cooperation. The best situation is one in which a state will not suffer greatly if others exploit it, for example, by cheating on an arms control agreement (that is, the costs of CD are low); but it will pay a high long-run price if cooperation with the others breaks down—for example, if agreements cease functioning or if there is a long war (that is, the costs of DD are high). The state's invulnerability is then mostly passive; it provides some protection, but it cannot be used to menace others. As we will discuss below, this situation is approximated when it is easier for states to defend themselves than to attack others, or when mutual deterrence obtains because neither side can protect itself.

The differences between highly vulnerable and less vulnerable states are illustrated by the contrasting policies of Britain and Austria after the Napoleonic Wars. Britain's geographic isolation and political stability allowed her to take a fairly relaxed view of disturbances on the Continent. Minor wars and small changes in territory or in the distribution of power did not affect her vital interests. An adversary who was out to overthrow the system could be stopped after he had made his intentions clear. And revolutions within other states were no menace, since they would not set off unrest within England. Austria, surrounded by strong powers, was not so fortunate; her policy had to be more closely attuned to all conflicts. By the time an aggressor-state had clearly shown its colors, Austria would be gravely threatened. And foreign revolutions, be they democratic or nationalistic, would encourage groups in Austria to upset the existing order. So it is not surprising that Metternich propounded the doctrine summarized earlier, which defended Austria's right to interfere in the internal affairs of others, and that British leaders rejected this view. Similarly, Austria wanted the Congress system to be a relatively tight one, regulating most disputes. The British favored a less centralized system. In other words, in order to protect herself, Austria had either to threaten or to harm others, whereas Britain did not. For Austria and her neighbors the security dilemma was acute; for Britain it was not.

The ultimate cost of CD is of course loss of sovereignty. This cost can vary from situation to situation. The lower it is (for instance, because the two states have compatible ideologies, are similar ethnically, have a common culture, or because the citizens of the losing state expect economic benefits), the less the impact of the security dilemma; the greater the costs, the greater the impact of the dilemma. Here is another reason why extreme differences in values and ideologies exacerbate international conflict.

It is through the lowering of the costs of CD that the proposed Rhodesian "safety net"—guaranteeing that whites who leave the country will receive fair payment for their property—would have the paradoxical effect of making it more likely that the whites will stay. This is less puzzling when we see that the whites are in a multi-person Prisoner's Dilemma with each other. Assume that all whites are willing to stay if most of the others stay; but, in the absence of guarantees, if there is going to be a mass exodus, all want to be among the first to leave (because late-leavers will get less for their property and will have more trouble finding a country to take them in). Then the problem is to avoid a self-fulfilling prophecy in which each person rushes to defect because he fears others are going to. In narrowing the gap between the payoff for leaving first (DC) and leaving last (CD) by reducing the cost of the latter, the guarantees make it easier for the whites to cooperate among themselves and stay.

SUBJECTIVE SECURITY DEMANDS. Decision makers act in terms of the vulnerability they feel, which can differ from the actual situation; we must therefore examine the decision makers' subjective security requirements.[10] Two dimensions are involved. First, even if they agree about the objective situation, people can differ about how much security they desire—or, to put it more precisely, about the price they are willing to pay to gain increments of security. The more states value their security above all else (that is, see a prohibitively high cost in CD), the more they are likely to be sensitive to even minimal threats, and to demand high levels of arms. And if arms are positively valued because of pressures from a military-industrial complex, it will be especially hard for status-quo powers to cooperate. By contrast, the security dilemma will not operate as strongly when pressing domestic concerns increase the opportunity costs of armaments. In this case, the net advantage of exploiting the other (DC) will be less, and the costs of arms races (that is, one aspect of DD) will be greater; therefore the state will behave as though it were relatively invulnerable.

10. For the development of the concept of subjective security, see Arnold Wolfers, *Discord and Collaboration* (Baltimore: Johns Hopkins Press 1962), chap. 10. In the present section we assume that the state believes that its security can be best served by increasing its arms; later we will discuss some of the conditions under which this assumption does not hold.

The second aspect of subjective security is the perception of threat (that is, the estimate of whether the other will cooperate).[11] A state that is predisposed to see either a specific other state as an adversary, or others in general as a menace, will react more strongly and more quickly than a state that sees its environment as benign. Indeed, when a state believes that another not only is not likely to be an adversary, but has sufficient interests in common with it to be an ally, then it will actually welcome an increase in the other's power.

British and French foreign policies in the interwar years illustrate these points. After the rise of Hitler, Britain and France felt that increases in each other's arms increased rather than decreased their own security. The differing policies that these states followed toward Germany can be explained by their differences on both dimensions of the variable of subjective security.[12] Throughout the period, France perceived Germany as more of a threat than England did. The British were more optimistic and argued that conciliation could turn Germany into a supporter of the status quo. Furthermore, in the years immediately following World War I, France had been more willing to forego other values in order to increase her security and had therefore followed a more belligerent policy than England, maintaining a larger army and moving quickly to counter German assertiveness. As this example shows, one cannot easily say how much subjective security a state should seek. High security requirements make it very difficult to capitalize on a common interest and run the danger of setting off spirals of arms races and hostility. The French may have paid this price in the 1920's. Low security requirements avoid this trap, but run the risk of having too few arms and of trying to conciliate an aggressor.

One aspect of subjective security related to the predisposition to perceive threat is the state's view of how many enemies it must be prepared to fight. A state can be relaxed about increases in another's arms if it believes that there is a functioning collective security system. The chances of peace are increased in a world in which the prevailing international system is valued in its own right, not only because most states restrain their ambitions and those who do not are deterred (these are the usual claims for a Concert system), but also because of

11. The question of when an actor will see another as a threat is important and understudied. For a valuable treatment (although one marred by serious methodological flaws), see Raymond Cohen, "Threat Perception on International Relations," Ph.D. diss. (Hebrew University 1974). Among the important factors, touched on below, are the lessons from the previous war.
12. Still the best treatment is Arnold Wolfers, *Britain and France Between Two Wars* (New York: Harcourt, Brace 1940).

the decreased chances that the status-quo states will engage in unnecessary conflict out of the quest for security. Indeed, if there were complete faith in collective security, no state would want an army. By contrast, the security dilemma is insoluble when each state fears that many others, far from coming to its aid, are likely to join in any attack. Winston Churchill, as First Lord of the Admiralty, was setting a high security requirement when he noted:

> Besides the Great Powers, there are many small states who are buying or building great ships of war and whose vessels may by purchase, by some diplomatic combinations, or by duress, be brought into the line against us. None of these powers need, like us, navies to defend their actual safety of independence. They build them so as to play a part in world affairs. It is sport to them. It is death to us.[13]

It takes great effort for any one state to be able to protect itself alone against an attack by several neighbors. More importantly, it is next to impossible for all states in the system to have this capability. Thus, a state's expectation that allies will be available and that only a few others will be able to join against it is almost a necessary condition for security requirements to be compatible.

GAINS FROM COOPERATION AND COSTS OF A BREAKDOWN (CC AND DD)

The main costs of a policy of reacting quickly and severely to increases in the other's arms are not the price of one's own arms, but rather the sacrifice of the potential gains from cooperation (CC) and the increase in the dangers of needless arms races and wars (DD). The greater these costs, the greater the incentives to try cooperation and wait for fairly unambiguous evidence before assuming that the other must be checked by force. Wars would be much more frequent—even if the first choice of all states was the status quo—if they were less risky and costly, and if peaceful intercourse did not provide rich benefits. Ethiopia recently asked for guarantees that the Territory of Afars and Issas would not join a hostile alliance against it when it gained independence. A spokesman for the Territory replied that this was not necessary: Ethiopia "already had the best possible guarantee in the railroad" that links the two countries and provides indispensable revenue for the Territory.[14]

The basic points are well known and so we can move to elaboration. First, most statesmen know that to enter a war is to set off a chain of unpredictable and uncontrollable events. Even if everything they see points to a quick vic-

13. Quoted in Peter Gretton, *Former Naval Person* (London: Cassell 1968), 151.
14. Michael Kaufman, "Tension Increases in French Colony," *New York Times*, July 11, 1976.

tory, they are likely to hesitate before all the uncertainties. And if the battlefield often produces startling results, so do the council chambers. The state may be deserted by allies or attacked by neutrals. Or the postwar alignment may rob it of the fruits of victory, as happened to Japan in 1895. Second, the domestic costs of wars must be weighed. Even strong states can be undermined by dissatisfaction with the way the war is run and by the necessary mobilization of men and ideas. Memories of such disruptions were one of the main reasons for the era of relative peace that followed the Napoleonic Wars. Liberal statesmen feared that large armies would lead to despotism; conservative leaders feared that wars would lead to revolution. (The other side of this coin is that when there are domestic consequences of foreign conflict that are positively valued, the net cost of conflict is lowered and cooperation becomes more difficult.) Third—turning to the advantages of cooperation—for states with large and diverse economies the gains from economic exchange are rarely if ever sufficient to prevent war. Norman Angell was wrong about World War I being impossible because of economic ties among the powers; and before World War II, the U.S. was Japan's most important trading partner. Fourth, the gains from cooperation can be increased, not only if each side gets more of the traditional values such as wealth, but also if each comes to value the other's well-being positively. Mutual cooperation will then have a double payoff: in addition to the direct gains, there will be the satisfaction of seeing the other prosper.[15]

While high costs of war and gains from cooperation will ameliorate the impact of the security dilemma, they can create a different problem. If the costs are high enough so that DD is the last choice for both sides, the game will shift to "Chicken." This game differs from the Stag Hunt in that each actor seeks to exploit the other; it differs from Prisoner's Dilemma in that both actors share an interest in avoiding mutual non-cooperation. In Chicken, if you think the other side is going to defect, you have to cooperate because, although being exploited (CD) is bad, it is not as bad as a total breakdown (DD). As the familiar logic of deterrence shows, the actor must then try to convince his adversary that he is going to stand firm (defect) and that the only way the other can avoid disaster is to back down (cooperate). Commitment, the rationality of irrationality, manipulating the communications system, and pretending not to understand the situation, are among the tactics used to reach this goal. The same logic applies when both sides are enjoying great benefits from cooperation.

15. Experimental support for this argument is summarized in Morton Deutsch, *The Resolution of Conflict* (New Haven: Yale University Press 1973), 181–95.

The side that can credibly threaten to disrupt the relationship unless its demands are met can exploit the other. This situation may not be stable, since the frequent use of threats may be incompatible with the maintenance of a cooperative relationship. Still, de Gaulle's successful threats to break up the Common Market unless his partners acceded to his wishes remind us that the shared benefits of cooperation as well as the shared costs of defection can provide the basis for exploitation. Similarly, one reason for the collapse of the Franco-British entente more than a hundred years earlier was that decision makers on both sides felt confident that their own country could safely pursue a policy that was against the other's interest because the other could not afford to destroy the highly valued relationship.[16] Because statesmen realize that the growth of positive interdependence can provide others with new levers of influence over them, they may resist such developments more than would be expected from the theories that stress the advantages of cooperation.

GAINS FROM EXPLOITATION (DC)

Defecting not only avoids the danger that a state will be exploited (CD), but brings positive advantages by exploiting the other (DC). The lower these possible gains, the greater the chances of cooperation. Even a relatively satisfied state can be tempted to expand by the hope of gaining major values. The temptation will be less when the state sees other ways of reaching its goals, and/or places a low value on what exploitation could bring. The gains may be low either because the immediate advantage provided by DC (for example, having more arms than the other side) cannot be translated into a political advantage (for example, gains in territory), or because the political advantage itself is not highly valued. For instance, a state may not seek to annex additional territory because the latter lacks raw materials, is inhabited by people of a different ethnic group, would be costly to garrison, or would be hard to assimilate without disturbing domestic politics and values. A state can reduce the incentives that another state has to attack it, by not being a threat to the latter and by providing goods and services that would be lost were the other to attempt exploitation.

Even where the direct advantages of DC are great, other considerations can reduce the net gain. Victory as well as defeat can set off undesired domestic

16. Roger Bullen, *Palmerston, Guizot, and the Collapse of the Entente Cordiale* (London: Athlone Press 1974), 81, 88, 93, 212. For a different view of this case, see Stanley Mellon, "Entente, Diplomacy, and Fantasy," *Reviews in European History* 11 (September 1976), 376–80.

changes within the state. Exploitation has at times been frowned upon by the international community, thus reducing the prestige of a state that engages in it. Or others might in the future be quicker to see the state as a menace to them, making them more likely to arm, and to oppose it later. Thus, Bismarck's attempts to get other powers to cooperate with him in maintaining the status quo after 1871 were made more difficult by the widely-held mistrust of him that grew out of his earlier aggressions.[17]

THE PROBABILITY THAT THE OTHER WILL COOPERATE

The variables discussed so far influence the payoffs for each of the four possible outcomes. To decide what to do, the state has to go further and calculate the expected value of cooperating or defecting. Because such calculations involve estimating the probability that the other will cooperate, the state will have to judge how the variables discussed so far act on the other. To encourage the other to cooperate, a state may try to manipulate these variables. It can lower the other's incentives to defect by decreasing what it could gain by exploiting the state (DC)—the details would be similar to those discussed in the previous paragraph—and it can raise the costs of deadlock (DD). But if the state cannot make DD the worst outcome for the other, coercion is likely to be ineffective in the short run because the other can respond by refusing to cooperate, and dangerous in the long run because the other is likely to become convinced that the state is aggressive. So the state will have to concentrate on making cooperation more attractive. One way to do this is to decrease the costs the other will pay if it cooperates and the state defects (CD). Thus, the state could try to make the other less vulnerable. It was for this reason that in the late 1950's and early 1960's some American defense analysts argued that it would be good for both sides if the Russians developed hardened missiles. Of course, decreasing the other's vulnerability also decreases the state's ability to coerce it, and opens the possibility that the other will use this protection as a shield behind which to engage in actions inimical to the state. But by sacrificing some ability to harm the other, the state can increase the chances of mutually beneficial cooperation.

The state can also try to increase the gains that will accrue to the other from mutual cooperation (CC). Although the state will of course gain if it receives a

---

17. Similarly, a French diplomat has argued that "the worst result of Louis XIV's abandonment of our traditional policy was the distrust it aroused towards us abroad." Jules Cambon, "The Permanent Bases of French Foreign Policy," *Foreign Affairs,* VIII (January 1930), 179.

share of any new benefits, even an increment that accrues entirely to the other will aid the state by increasing the likelihood that the other will cooperate."[18]

This line of argument can be continued through the infinite regressions that game theory has made familiar. If the other is ready to cooperate when it thinks the state will, the state can increase the chances of CC by showing that it *is* planning to cooperate. Thus the state should understate the gains it would make if it exploited the other (DC) and the costs it would pay if the other exploited it (CD), and stress or exaggerate the gains it would make under mutual cooperation (CC) and the costs it would pay if there is deadlock (DD). The state will also want to convince the other that it thinks that the other is likely to cooperate. If the other believes these things, it will see that the state has strong incentives to cooperate, and so it will cooperate in turn. One point should be emphasized. Because the other, like the state, may be driven to defect by the fear that it will be exploited if it does not, the state should try to reassure it that this will not happen. Thus, when Khrushchev indicated his willingness to withdraw his missiles from Cuba, he simultaneously stressed to Kennedy that "we are of sound mind and understand perfectly well" that Russia could not launch a successful attack against the U.S., and therefore that there was no reason for the U.S. to contemplate a defensive, pre-emptive strike of its own.[19]

There is, however, a danger. If the other thinks that the state has little choice but to cooperate, it can credibly threaten to defect unless the state provides it with additional benefits. Great advantages of mutual cooperation, like high costs of war, provide a lever for competitive bargaining. Furthermore, for a state to stress how much it gains from cooperation may be to imply that it is gaining much more than the other and to suggest that the benefits should be distributed more equitably.

When each side is ready to cooperate if it expects the other to, inspection devices can ameliorate the security dilemma. Of course, even a perfect inspection system cannot guarantee that the other will not later develop aggressive intentions and the military means to act on them. But by relieving immediate worries and providing warning of coming dangers, inspection can meet a

18. This assumes, however, that these benefits to the other will not so improve the other's power position that it will be more able to menace the state in the future.
19. Walter LaFeber, ed., *The Dynamics of World Power; A Documentary History of United States Foreign Policy 1945–1973*, II: *Eastern Europe and the Soviet Union* (New York: Chelsea House in Association with McGraw-Hill 1973), 700.

significant part of the felt need to protect oneself against future threats, and so make current cooperation more feasible. Similar functions are served by breaking up one large transaction into a series of smaller ones.[20] At each transaction each can see whether the other has cooperated; and its losses, if the other defects, will be small. And since what either side would gain by one defection is slight compared to the benefits of continued cooperation, the prospects of cooperation are high. Conflicts and wars among status-quo powers would be much more common were it not for the fact that international politics is usually a series of small transactions.

How a statesman interprets the other's past behavior and how he projects it into the future is influenced by his understanding of the security dilemma and his ability to place himself in the other's shoes. The dilemma will operate much more strongly if statesmen do not understand it, and do not see that their arms—sought only to secure the status quo—may alarm others and that others may arm, not because they are contemplating aggression, but because they fear attack from the first state. These two failures of empathy are linked. A state which thinks that the other knows that it wants only to preserve the status quo and that its arms are meant only for self-preservation will conclude that the other side will react to its arms by increasing its own capability only if it is aggressive itself. Since the other side is not menaced, there is no legitimate reason for it to object to the first state's arms; therefore, objection proves that the other is aggressive. Thus, the following exchange between Senator Tom Connally and Secretary of State Acheson concerning the ratification of the NATO treaty:

Secretary Acheson: [The treaty] is aimed solely at armed aggression.
Senator Connally: In other words, unless a nation . . . contemplates, meditates, or makes plans looking toward aggression or armed attack on another nation, it has no cause to fear this treaty.
Secretary Acheson: That is correct, Senator Connally, and it seems to me that any nation which claims that this treaty is directed against it should be reminded of the Biblical admonition that 'The guilty flee when no man pursueth.'
Senator Connally: That is a very apt illustration.
What I had in mind was, when a State or Nation passes a criminal act, for in-

20. Thomas Schelling, *The Strategy of Conflict* (New York: Oxford University Press 1963), 134–35.

stance, against burglary, nobody but those who are burglars or getting ready to be burglars need have any fear of the Burglary Act. Is that not true?

Secretary Acheson: The only effect [the law] would have [on an innocent person] would be for his protection, perhaps, by deterring someone else. He wouldn't worry about the imposition of the penalties on himself.[21]

The other side of this coin is that part of the explanation for détente is that most American decision makers now realize that it is at least possible that Russia may fear American aggression; many think that this fear accounts for a range of Soviet actions previously seen as indicating Russian aggressiveness. Indeed, even 36 percent of military officers consider the Soviet Union's motivations to be primarily defensive. Less than twenty years earlier, officers had been divided over whether Russia sought world conquest or only expansion.[22]

Statesmen who do not understand the security dilemma will think that the money spent is the only cost of building up their arms. This belief removes one important restraint on arms spending. Furthermore, it is also likely to lead states to set their security requirements too high. Since they do not understand that trying to increase one's security can actually decrease it, they will overestimate the amount of security that is attainable; they will think that when in doubt they can "play it safe" by increasing their arms. Thus it is very likely that two states which support the status quo but do not understand the security dilemma will end up, if not in a war, then at least in a relationship of higher conflict than is required by the objective situation.

The belief that an increase in military strength always leads to an increase in security is often linked to the belief that the only route to security is through military strength. As a consequence, a whole range of meliorative policies will be downgraded. Decision makers who do not believe that adopting a more conciliatory posture, meeting the other's legitimate grievances, or developing mutual gains from cooperation can increase their state's security, will not devote much attention or effort to these possibilities.

On the other hand, a heightened sensitivity to the security dilemma makes it more likely that the state will treat an aggressor as though it were an insecure defender of the status quo. Partly because of their views about the causes of World War I, the British were predisposed to believe that Hitler sought only

21. U.S. Congress, Senate, Committee on Foreign Relations, *Hearings, North Atlantic Treaty,* 81st Cong., 1st sess. (1949), 17.

22. Bruce Russett and Elizabeth Hanson, *Interest and Ideology* (San Francisco: Freeman 1975), 260; Morris Janowitz, *The Professional Soldier* (New York: Free Press 1960), chap. 13.

the rectification of legitimate and limited grievances and that security could best be gained by constructing an equitable international system. As a result they pursued a policy which, although well designed to avoid the danger of creating unnecessary conflict with a status-quo Germany, helped destroy Europe.

GEOGRAPHY, COMMITMENTS, BELIEFS, AND SECURITY THROUGH EXPANSION

A final consideration does not easily fit in the matrix we have been using, although it can be seen as an aspect of vulnerability and of the costs of CD. Situations vary in the ease or difficulty with which all states can simultaneously achieve a high degree of security. The influence of military technology on this variable is the subject of the next section. Here we want to treat the impact of beliefs, geography, and commitments (many of which can be considered to be modifications of geography, since they bind states to defend areas outside their homelands). In the crowded continent of Europe, security requirements were hard to mesh. Being surrounded by powerful states, Germany's problem—or the problem created by Germany—was always great and was even worse when her relations with both France and Russia were bad, such as before World War I. In that case, even a status-quo Germany, if she could not change the political situation, would almost have been forced to adopt something like the Schlieffen Plan. Because she could not hold off both of her enemies, she had to be prepared to defeat one quickly and then deal with the other in a more leisurely fashion. If France or Russia stayed out of a war between the other state and Germany, they would allow Germany to dominate the Continent (even if that was not Germany's aim). They therefore had to deny Germany this ability, thus making Germany less secure. Although Germany's arrogant and erratic behavior, coupled with the desire for an unreasonably high level of security (which amounted to the desire to escape from her geographic plight), compounded the problem, even wise German statesmen would have been hard put to gain a high degree of security without alarming their neighbors.

A similar situation arose for France after World War I. She was committed to protecting her allies in Eastern Europe, a commitment she could meet only by taking the offensive against Germany. But since there was no way to guarantee that France might not later seek expansion, a France that could successfully launch an attack in response to a German move into Eastern Europe would constitute a potential danger to German core values. Similarly, a United States credibly able to threaten retaliation with strategic nuclear weapons if the Soviet Union attacks Western Europe also constitutes a menace, albeit a reduced one, to the Soviet ability to maintain the status quo. The incompatibility of these security requirements is not complete. Herman Kahn is correct in arguing that the United States could have Type II deterrence (the ability to deter a major Soviet provocation) without gaining first-strike capability because the expected Soviet retaliation following an American strike could be great enough to deter the U.S. from attacking unless the U.S. believed it would suffer enormous deprivation (for instance, the loss of Europe) if it did not strike.[23] Similarly, the Franco-German military balance could have been such that France could successfully attack Germany if the latter's armies were embroiled in Eastern Europe, but could not defeat a Germany that was free to devote all her resources to defending herself. But this delicate balance is very hard to achieve, especially because states usually calculate conservatively. Therefore, such a solution is not likely to be available.

For the United States, the problem posed by the need to protect Europe is an exception. Throughout most of its history, this country has been in a much more favorable position: relatively self-sufficient and secure from invasion, it has not only been able to get security relatively cheaply, but by doing so, did not menace others.[24] But ambitions and commitments have changed this situation. After the American conquest of the Philippines, "neither the United States nor Japan could assure protection for their territories by military and naval means without compromising the defenses of the other. This problem would plague American and Japanese statesmen down to 1941."[25] Furthermore, to the extent that Japan could protect herself, she could resist American threats to go to war if Japan did not respect China's independence. These complications were minor compared to those that followed World War II. A world power cannot help but have the ability to harm many others that is out of proportion to the others' ability to harm it.

Britain had been able to gain security without menacing others to a greater degree than the Continental powers, though to a lesser one than the United States. But the acquisition of colonies and a dependence on foreign trade

23. Kahn, *On Thermonuclear War* (Princeton: Princeton University Press 1960), 138–60. It should be noted that the French example is largely hypothetical because France had no intention of fulfilling her obligations once Germany became strong.
24. Wolfers (fn. 9), chap. 15; C. Vann Woodward, "The Age of Reinterpretation," *American Historical Review*, Vol. 67 (October 1960), 1–19.
25. William Braisted, *The United States Navy in the Pacific, 1897–1909* (Austin: University of Texas Press 1958), 240.

sacrificed her relative invulnerability of being an island. Once she took India, she had to consider Russia as a neighbor; the latter was expanding in Central Asia, thus making it much more difficult for both countries to feel secure. The need to maintain reliable sea lanes to India meant that no state could be allowed to menace South Africa and, later, Egypt. But the need to protect these two areas brought new fears, new obligations, and new security requirements that conflicted with those of other European nations. Furthermore, once Britain needed a flow of imports during both peace and wartime, she required a navy that could prevent a blockade. A navy sufficient for that task could not help but be a threat to any other state that had valuable trade.

A related problem is raised by the fact that defending the status quo often means protecting more than territory. Nonterritorial interests, norms, and the structure of the international system must be maintained. If all status-quo powers agree on these values and interpret them in compatible ways, problems will be minimized. But the potential for conflict is great, and the policies followed are likely to exacerbate the security dilemma. The greater the range of interests that have to be protected, the more likely it is that national efforts to maintain the status quo will clash. As a French spokesman put it in 1930: "Security! The term signifies more indeed than the maintenance of a people's homeland, or even of their territories beyond the seas. It also means the maintenance of the world's respect for them, the maintenance of their economic interests, everything in a word, which goes to make up the grandeur, the life itself, of the nation."[26] When security is thought of in this sense, it almost automatically has a competitive connotation. It involves asserting one state's will over others, showing a high degree of leadership if not dominance, and displaying a prickly demeanor. The resulting behavior will almost surely clash with that of others who define their security in the same way.

The problem will be almost insoluble if statesmen believe that their security requires the threatening or attacking of others. "That which stops growing begins to rot," declared a minister to Catherine the Great.[27] More common is the belief that if the other is secure, it will be emboldened to act against one's own state's interests, and the belief that in a war it will not be enough for the state to protect itself: it must be able to take the war to the other's homeland. These

26. Cambon (fn. 17), 185.
27. Quoted in Adam Ulam, *Expansion and Co-Existence* (New York: Praeger 1968), 5. In 1920 the U.S. Navy's General Board similarly declared "A nation must advance or retrocede in world position." Quoted in William Braisted, *The United States Navy in the Pacific, 1909–1922* (Austin: University of Texas Press 1971), 488.

convictions make it very difficult for status-quo states to develop compatible security policies, for they lead the state to conclude that its security requires that others be rendered insecure.

In other cases, "A country engaged in a war of defense might be obliged for strategic reasons to assume the offensive," as a French delegate to an interwar disarmament conference put it.[28] That was the case for France in 1799:

> The Directory's political objectives were essentially defensive, for the French wanted only to protect the Republic from invasion and preserve the security and territory of the satellite regimes in Holland, Switzerland, and Italy. French leaders sought no new conquests; they wanted only to preserve the earlier gains of the Revolution. The Directory believed, however, that only a military offensive could enable the nation to achieve its defensive political objective. By inflicting rapid and decisive defeats upon one or more members of the coalition, the directors hoped to rupture allied unity and force individual powers to seek a separate peace.[29]

It did not matter to the surrounding states that France was not attacking because she was greedy, but because she wanted to be left in peace. Unless there was some way her neighbors could provide France with an alternate route to her goal, France had to go to war.

III. *Offense, Defense, and the Security Dilemma*

Another approach starts with the central point of the security dilemma—that an increase in one state's security decreases the security of others—and examines the conditions under which this proposition holds. Two crucial variables are involved: whether defensive weapons and policies can be distin-guished from offensive ones, and whether the defense or the offense has the advantage. The definitions are not always clear, and many cases are difficult to judge, but these two variables shed a great deal of light on the question of whether status-quo powers will adopt compatible security policies. All the variables discussed so far leave the heart of the problem untouched. But when defensive weapons differ from offensive ones, it is possible for a state to make itself more secure without making others less secure. And when the defense has the advantage over the offense, a large increase in one state's security only slightly decreases

28. Quoted in Marion Boggs, *Attempts to Define and Limit "Aggressive" Armament in Diplomacy and Strategy* (Columbia: University of Missouri Studies, XVI, No. 1, 1941), 41.
29. Steven Ross, *European Diplomatic History, 1789–1815* (Garden City, N.Y.: Doubleday 1969), 194.

the security of the others, and status-quo powers can all enjoy a high level of security and largely escape from the state of nature.

OFFENSE-DEFENSE BALANCE

When we say that the offense has the advantage, we simply mean that it is easier to destroy the other's army and take its territory than it is to defend one's own. When the defense has the advantage, it is easier to protect and to hold than it is to move forward, destroy, and take. If effective defenses can be erected quickly, an attacker may be able to keep territory he has taken in an initial victory. Thus, the dominance of the defense made it very hard for Britain and France to push Germany out of France in World War I. But when superior defenses are difficult for an aggressor to improvise on the battlefield and must be constructed during peacetime, they provide no direct assistance to him.

The security dilemma is at its most vicious when commitments, strategy, or technology dictate that the only route to security lies through expansion. Status-quo powers must then act like aggressors; the fact that they would gladly agree to forego the opportunity for expansion in return for guarantees for their security has no implications for their behavior. Even if expansion is not sought as a goal in itself, there will be quick and drastic changes in the distribution of territory and influence. Conversely, when the defense has the advantage, status-quo states can make themselves more secure without gravely endangering others.[30] Indeed, if the defense has enough of an advantage and if the states are of roughly equal size, not only will the security dilemma cease to inhibit status-quo states from cooperating, but aggression will be next to impossible, thus rendering international anarchy relatively unimportant. If states cannot conquer each other, then the lack of sovereignty, although it presents problems of collective goods in a number of areas, no longer forces states to devote their primary attention to self-preservation. Although, if force were not usable, there would be fewer restraints on the use of nonmilitary instruments, these are rarely powerful enough to threaten the vital interests of a major state.

Two questions of the offense-defense balance can be separated. First, does the state have to spend more or less than one dollar on defensive forces to offset each dollar spent by the other side on forces that could be used to attack? If

30. Thus, when Wolfers (fn. 10), 126, argues that a status-quo state that settles for rough equality of power with its adversary, rather than seeking preponderance, may be able to convince the other to reciprocate by showing that it wants only to protect itself, not menace the other, he assumes that the defense has an advantage.

the state has one dollar to spend on increasing its security, should it put it into offensive or defensive forces? Second, with a given inventory of forces, is it better to attack or to defend? Is there an incentive to strike first or to absorb the other's blow? These two aspects are often linked: if each dollar spent on offense can overcome each dollar spent on defense, and if both sides have the same defense budgets, then both are likely to build offensive forces and find it attractive to attack rather than to wait for the adversary to strike.

These aspects affect the security dilemma in different ways. The first has its greatest impact on arms races. If the defense has the advantage, and if the status-quo powers have reasonable subjective security requirements, they can probably avoid an arms race. Although an increase in one side's arms and security will still decrease the other's security, the former's increase will be larger than the latter's decrease. So if one side increases its arms, the other can bring its security back up to its previous level by adding a smaller amount to its forces. And if the first side reacts to this change, its increase will also be smaller than the stimulus that produced it. Thus a stable equilibrium will be reached. Shifting from dynamics to statics, each side can be quite secure with forces roughly equal to those of the other. Indeed, if the defense is much more potent than the offense, each side can be willing to have forces much smaller than the other's, and can be indifferent to a wide range of the other's defense policies.

The second aspect—whether it is better to attack or to defend—influences short-run stability. When the offense has the advantage, a state's reaction to international tension will increase the chances of war. The incentives for preemption and the "reciprocal fear of surprise attack" in this situation have been made clear by analyses of the dangers that exist when two countries have first-strike capabilities.[31] There is no way for the state to increase its security without menacing, or even attacking, the other. Even Bismarck, who once called preventive war "committing suicide from fear of death," said that "no government, if it regards war as inevitable even if it does not want it, would be so foolish as to leave to the enemy the choice of time and occasion and to wait for the moment which is most convenient for the enemy."[32] In another arena, the same dilemma applies to the policeman in a dark alley confronting a suspected criminal who appears to be holding a weapon. Though racism may indeed be

31. Schelling (fn. 20), chap. 9.
32. Quoted in Fritz Fischer, *War of Illusions* (New York: Norton 1975), 377, 461.

present, the security dilemma can account for many of the tragic shootings of innocent people in the ghettos.

Beliefs about the course of a war in which the offense has the advantage further deepen the security dilemma. When there are incentives to strike first, a successful attack will usually so weaken the other side that victory will be relatively quick, bloodless, and decisive. It is in these periods when conquest is possible and attractive that states consolidate power internally—for instance, by destroying the feudal barons—and expand externally. There are several consequences that decrease the chance of cooperation among status-quo states. First, war will be profitable for the winner. The costs will be low and the benefits high. Of course, losers will suffer; the fear of losing could induce states to try to form stable cooperative arrangements, but the temptation of victory will make this particularly difficult. Second, because wars are expected to be both frequent and short, there will be incentives for high levels of arms, and quick and strong reaction to the other's increases in arms. The state cannot afford to wait until there is unambiguous evidence that the other is building new weapons. Even large states that have faith in their economic strength cannot wait, because the war will be over before their products can reach the army. Third, when wars are quick, states will have to recruit allies in advance.[33] Without the opportunity for bargaining and re-alignments during the opening stages of hostilities, peacetime diplomacy loses a degree of the fluidity that facilitates balance-of-power policies. Because alliances must be secured during peacetime, the international system is more likely to become bipolar. It is hard to say whether war therefore becomes more or less likely, but this bipolarity increases tension between the two camps and makes it harder for status-quo states to gain the benefits of cooperation. Fourth, if wars are frequent, statesmen's perceptual thresholds will be adjusted accordingly and they will be quick to perceive ambiguous evidence as indicating that others are aggressive. Thus, there will be more cases of status-quo powers arming against each other in the incorrect belief that the other is hostile.

When the defense has the advantage, all the foregoing is reversed. The state that fears attack does not pre-empt—since that would be a wasteful use of its military resources—but rather prepares to receive an attack. Doing so does not decrease the security of others, and several states can do it simultaneously; the

33. George Quester, *Offense and Defense in the International System* (New York: John Wiley 1977), 105–06; Sontag (fn. 5), 4–5.

situation will therefore be stable, and status-quo powers will be able to cooperate. When Herman Kahn argues that ultimatums "are vastly too dangerous to give because . . . they are quite likely to touch off a pre-emptive strike,"[34] he incorrectly assumes that it is always advantageous to strike first.

More is involved than short-run dynamics. When the defense is dominant, wars are likely to become stalemates and can be won only at enormous cost. Relatively small and weak states can hold off larger and stronger ones, or can deter attack by raising the costs of conquest to an unacceptable level. States then approach equality in what they can do to each other. Like the .45-caliber pistol in the American West, fortifications were the "great equalizer" in some periods. Changes in the status quo are less frequent and cooperation is more common wherever the security dilemma is thereby reduced.

Many of these arguments can be illustrated by the major powers' policies in the periods preceding the two world wars. Bismarck's wars surprised statesmen by showing that the offense had the advantage, and by being quick, relatively cheap, and quite decisive. Falling into a common error, observers projected this pattern into the future.[35] The resulting expectations had several effects. First, states sought semi-permanent allies. In the early stages of the Franco-Prussian War, Napoleon III had thought that there would be plenty of time to recruit Austria to his side. Now, others were not going to repeat this mistake. Second, defense budgets were high and reacted quite sharply to increases on the other side. It is not surprising that Richardson's theory of arms races fits this period well. Third, most decision makers thought that the next European war would not cost much blood and treasure.[36] That is one reason why war was generally seen as inevitable and why mass opinion was so bellicose. Fourth, once war seemed likely, there were strong pressures to pre-empt. Both sides believed that whoever moved first could penetrate the other deep enough to disrupt mobilization and thus gain an insurmountable advantage.

34. Kahn (fn. 23), 211 (also see 144).
35. For a general discussion of such mistaken learning from the past, see Jervis (fn. 5), chap. 6. The important and still not completely understood question of why this belief formed and was maintained throughout the war is examined in Bernard Brodie, *War and Politics* (New York: Macmillan 1973), 262–70; Brodie, "Technological Change, Strategic Doctrine, and Political Outcomes," in Klaus Knorr, ed., *Historical Dimensions of National Security Problems* (Lawrence: University Press of Kansas 1976), 290–92; and Douglas Porch, "The French Army and the Spirit of the Offensive, 1900–14," in Brian Bond and Ian Roy, eds., *War and Society* (New York: Holmes & Meier 1975), 117–43.
36. Some were not so optimistic. Gray's remark is well-known: "The lamps are going out all over Europe; we shall not see them lit again in our life-time." The German Prime Minister, Bethmann Hollweg, also feared the consequences of the war. But the controlling view was that it would certainly pay for the winner.

(There was no such belief about the use of naval forces. Although Churchill made an ill-advised speech saying that if German ships "do not come out and fight in time of war they will be dug out like rats in a hole,"[37] everyone knew that submarines, mines, and coastal fortifications made this impossible. So at the start of the war each navy prepared to defend itself rather than attack, and the short-run destabilizing forces that launched the armies toward each other did not operate.)[38] Furthermore, each side knew that the other saw the situation the same way, thus increasing the perceived danger that the other would attack, and giving each added reasons to precipitate a war if conditions seemed favorable. In the long and the short run, there were thus both offensive and defensive incentives to strike. This situation casts light on the common question about German motives in 1914: "Did Germany unleash the war deliberately to become a world power or did she support Austria merely to defend a weakening ally," thereby protecting her own position?[39] To some extent, this question is misleading. Because of the perceived advantage of the offense, war was seen as the best route both to gaining expansion and to avoiding drastic loss of influence. There seemed to be no way for Germany merely to retain and safeguard her existing position.

Of course the war showed these beliefs to have been wrong on all points. Trenches and machine guns gave the defense an overwhelming advantage. The fighting became deadlocked and produced horrendous casualties. It made no sense for the combatants to bleed themselves to death. If they had known the power of the defense beforehand, they would have rushed for their own trenches rather than for the enemy's territory. Each side could have done this without increasing the other's incentives to strike. War might have broken out anyway, just as DD is a possible outcome of Chicken, but at least the pressures of time and the fear of allowing the other to get the first blow would not have contributed to this end. And, had both sides known the costs of the war, they would have negotiated much more seriously. The obvious question is why the states did not seek a negotiated settlement as soon as the shape of the war became clear. Schlieffen had said that if his plan failed, peace should be sought.[40]

---

37. Quoted in Martin Gilbert, *Winston S. Churchill*, III, *The Challenge of War, 1914–1916* (Boston: Houghton Mifflin 1971), 84.
38. Quester (fn. 33), 98–99. Robert Art, *The Influence of Foreign Policy on Seapower*, II (Beverly Hills: Sage Professional Papers in International Studies Series, 1973), 14–18, 26–28.
39. Konrad Jarausch, "The Illusion of Limited War: Chancellor Bethmann Hollweg's Calculated Risk, July 1914," *Central European History*, 11 (March 1969), 50.
40. Brodie (fn. 8), 58.

The answer is complex, uncertain, and largely outside of the scope of our concerns. But part of the reason was the hope and sometimes the expectation that breakthroughs could be made and the dominance of the offensive restored. Without that hope, the political and psychological pressures to fight to a decisive victory might have been overcome.

The politics of the interwar period were shaped by the memories of the previous conflict and the belief that any future war would resemble it. Political and military lessons reinforced each other in ameliorating the security dilemma. Because it was believed that the First World War had been a mistake that could have been avoided by skillful conciliation, both Britain and, to a lesser extent, France were highly sensitive to the possibility that interwar Germany was not a real threat to peace, and alert to the danger that reacting quickly and strongly to her arms could create unnecessary conflict. And because Britain and France expected the defense to continue to dominate, they concluded that it was safe to adopt a more relaxed and nonthreatening military posture.[41] Britain also felt less need to maintain tight alliance bonds. The Allies' military posture then constituted only a slight danger to Germany; had the latter been content with the status quo, it would have been easy for both sides to have felt secure behind their lines of fortifications. Of course the Germans were not content, so it is not surprising that they devoted their money and attention to finding ways out of a defense-dominated stalemate. *Blitzkrieg* tactics were necessary if they were to use force to change the status quo.

The initial stages of the war on the Western Front also contrasted with the First World War. Only with the new air arm were there any incentives to strike first, and these forces were too weak to carry out the grandiose plans that had been both dreamed and feared. The armies, still the main instrument, rushed to defensive positions. Perhaps the allies could have successfully attacked while the Germans were occupied in Poland.[42] But belief in the defense was so great that this was never seriously contemplated. Three months after the start of the war, the French Prime Minister summed up the view held by almost

---

41. President Roosevelt and the American delegates to the League of Nations Disarmament Conference maintained that the tank and mobile heavy artillery had reestablished the dominance of the offensive, thus making disarmament more urgent (Boggs, fn. 28, pp. 31, 108), but this was a minority position and may not even have been believed by the Americans. The reduced prestige and influence of the military, and the high pressures to cut government spending throughout this period also contributed to the lowering of defense budgets.

42. Jon Kimche, *The Unfought Battle* (New York: Stein 1968); Nicholas William Bethell, *The War Hitler Won: The Fall of Poland, September 1939* (New York: Holt 1972); Alan Alexandroff and Richard Rosecrance, "Deterrence in 1939," *World Politics*, XXIX (April 1977), 404–24.

everyone but Hitler: on the Western Front there is "deadlock. Two Forces of equal strength and the one that attacks seeing such enormous casualties that it cannot move without endangering the continuation of the war or of the aftermath."[43] The Allies were caught in a dilemma they never fully recognized, let alone solved. On the one hand, they had very high war aims; although unconditional surrender had not yet been adopted, the British had decided from the start that the removal of Hitler was a necessary condition for peace.[44] On the other hand, there were no realistic plans or instruments for allowing the Allies to impose their will on the other side. The British Chief of the Imperial General Staff noted, "The French have no intention of carrying out an offensive for years, if at all"; the British were only slightly bolder.[45] So the Allies looked to a long war that would wear the Germans down, cause civilian suffering through shortages, and eventually undermine Hitler. There was little analysis to support this view—and indeed it probably was not supportable—but as long as the defense was dominant and the numbers on each side relatively equal, what else could the Allies do?

To summarize, the security dilemma was much less powerful after World War I than it had been before. In the later period, the expected power of the defense allowed status-quo states to pursue compatible security policies and avoid arms races. Furthermore, high tension and fear of war did not set off short-run dynamics by which each state, trying to increase its security, inadvertently acted to make war more likely. The expected high costs of war, however, led the Allies to believe that no sane German leader would run the risks entailed in an attempt to dominate the Continent, and discouraged them from risking war themselves.

TECHNOLOGY AND GEOGRAPHY. Technology and geography are the two main factors that determine whether the offense or the defense has the advantage. As Brodie notes, "On the tactical level, as a rule, few physical factors favor the attacker but many favor the defender. The defender usually has the advantage of cover. He characteristically fires from behind some form of shelter while his

43. Roderick Macleod and Denis Kelly, eds., *Time Unguarded: The Ironside Diaries, 1937–1940* (New York: McKay 1962), 173.
44. For a short time, as France was falling, the British Cabinet did discuss reaching a negotiated peace with Hitler. The official history ignores this, but it is covered in P. M. H. Bell, *A Certain Eventuality* (Farnborough, England: Saxon House 1974), 40–48.
45. Macleod and Kelly (fn. 43), 174. In flat contradiction to common sense and almost everything they believed about modern warfare, the Allies planned an expedition to Scandinavia to cut the supply of iron ore to Germany and to aid Finland against the Russians. But the dominant mood was the one described above.

opponent crosses open ground."[46] Anything that increases the amount of ground the attacker has to cross, or impedes his progress across it, or makes him more vulnerable while crossing, increases the advantage accruing to the defense. When states are separated by barriers that produce these effects, the security dilemma is eased, since both can have forces adequate for defense without being able to attack. Impenetrable barriers would actually prevent war; in reality, decision makers have to settle for a good deal less. Buffer zones slow the attacker's progress; they thereby give the defender time to prepare, increase problems of logistics, and reduce the number of soldiers available for the final assault. At the end of the 19th century, Arthur Balfour noted Afghanistan's "non-conducting" qualities. "So long as it possesses few roads, and no railroads, it will be impossible for Russia to make effective use of her great numerical superiority at any point immediately vital to the Empire." The Russians valued buffers for the same reasons; it is not surprising that when Persia was being divided into Russian and British spheres of influence some years later, the Russians sought assurances that the British would refrain from building potentially menacing railroads in their sphere. Indeed, since railroad construction radically altered the abilities of countries to defend themselves and to attack others, many diplomatic notes and much intelligence activity in the late 19th century centered on this subject.[47]

Oceans, large rivers, and mountain ranges serve the same function as buffer zones. Being hard to cross, they allow defense against superior numbers. The defender has merely to stay on his side of the barrier and so can utilize all the men he can bring up to it. The attacker's men, however, can cross only a few at a time, and they are very vulnerable when doing so. If all states were self-sufficient islands, anarchy would be much less of a problem. A small investment in shore defenses and a small army would be sufficient to repel invasion. Only very weak states would be vulnerable, and only very large ones could menace others. As noted above, the United States, and to a lesser extent Great Britain, have partly been able to escape from the state of nature because their geographical positions approximated this ideal.

Although geography cannot be changed to conform to borders, borders can and do change to conform to geography. Borders across which an attack is easy

46. Brodie (fn. 8), 179.

47. Arthur Balfour, "Memorandum," Committee on Imperial Defence, April 30, 1903, pp. 2–3; see the telegrams by Sir Arthur Nicolson, in G. P. Gooch and Harold Temperley, eds., *British Documents on the Origins of the War,* Vol. 4 (London: H.M.S.O. 1929), 429, 524. These barriers do not prevent the passage of long-range aircraft; but even in the air, distance usually aids the defender.

tend to be unstable. States living within them are likely to expand or be absorbed. Frequent wars are almost inevitable since attacking will often seem the best way to protect what one has. This process will stop, or at least slow down, when the state's borders reach—by expansion or contraction—a line of natural obstacles. Security without attack will then be possible. Furthermore, these lines constitute salient solutions to bargaining problems and, to the extent that they are barriers to migration, are likely to divide ethnic groups, thereby raising the costs and lowering the incentives for conquest.

Attachment to one's state and its land reinforce one quasi-geographical aid to the defense. Conquest usually becomes more difficult the deeper the attacker pushes into the other's territory. Nationalism spurs the defenders to fight harder; advancing not only lengthens the attacker's supply lines, but takes him through unfamiliar and often devastated lands that require troops for garrison duty. These stabilizing dynamics will not operate, however, if the defender's war materiel is situated near its borders, or if the people do not care about their state, but only about being on the winning side. In such cases, positive feedback will be at work and initial defeats will be insurmountable.[48]

Imitating geography, men have tried to create barriers. Treaties may provide for demilitarized zones on both sides of the border, although such zones will rarely be deep enough to provide more than warning. Even this was not possible in Europe, but the Russians adopted a gauge for their railroads that was broader than that of the neighboring states, thereby complicating the logistics problems of any attacker—including Russia.

Perhaps the most ambitious and at least temporarily successful attempts to construct a system that would aid the defenses of both sides were the interwar naval treaties, as they affected Japanese-American relations. As mentioned earlier, the problem was that the United States could not defend the Philippines without denying Japan the ability to protect her home islands.[49] (In 1941 this dilemma became insoluble when Japan sought to extend her control to Malaya and the Dutch East Indies. If the Philippines had been invulnerable, they could have provided a secure base from which the U.S. could interdict Japanese shipping between the homeland and the areas she was trying to conquer.) In the 1920's and early 1930's each side would have been willing to grant the other

---

48. See, for example, the discussion of warfare among Chinese warlords in Hsi-Sheng Chi, "The Chinese Warlord System as an International System," in Morton Kaplan, ed., *New Approaches to Inter.ational Relations* (New York: St. Martin's 1968), 405–25.

49. Some American decision makers, including military officers, thought that the best way out of the dilemma was to abandon the Philippines.

security for its possessions in return for a reciprocal grant, and the Washington Naval Conference agreements were designed to approach this goal. As a Japanese diplomat later put it, their country's "fundamental principle" was to have "a strength insufficient for attack and adequate for defense."[50] Thus, Japan agreed in 1922 to accept a navy only three-fifths as large as that of the United States, and the U.S. agreed not to fortify its Pacific islands.[51] (Japan had earlier been forced to agree not to fortify the islands she had taken from Germany in World War I.) Japan's navy would not be large enough to defeat America's anywhere other than close to the home islands. Although the Japanese could still take the Philippines, not only would they be unable to move farther, but they might be weakened enough by their efforts to be vulnerable to counterattack. Japan, however, gained security. An American attack was rendered more difficult because the American bases were unprotected and because, until 1930, Japan was allowed unlimited numbers of cruisers, destroyers, and submarines that could weaken the American fleet as it made its way across the ocean.[52]

The other major determinant of the offense-defense balance is technology. When weapons are highly vulnerable, they must be employed before they are attacked. Others can remain quite invulnerable in their bases. The former characteristics are embodied in unprotected missiles and many kinds of bombers. (It should be noted that it is not vulnerability *per se* that is crucial, but the location of the vulnerability. Bombers and missiles that are easy to destroy only after having been launched toward their targets do not create destabilizing dynamics.) Incentives to strike first are usually absent for naval forces that are threatened by a naval attack. Like missiles in hardened silos, they are usually well protected when in their bases. Both sides can then simultaneously be prepared to defend themselves successfully.

In ground warfare under some conditions, forts, trenches, and small groups of men in prepared positions can hold off large numbers of attackers. Less frequently, a few attackers can storm the defenses. By and large, it is a contest between fortifications and supporting light weapons on the one hand, and mobility and heavier weapons that clear the way for the attack on the other. As

50. Quoted in Elting Morrison, *Turmoil and Tradition: A Study of the Life and Times of Henry L. Stimson* (Boston: Houghton Mifflin 1960), 326.
51. The U.S. "refused to consider limitations on Hawaiian defenses, since these works posed no threat to Japan." Braisted (fn. 27), 612.
52. That is part of the reason why the Japanese admirals strongly objected when the civilian leaders decided to accept a seven-to-ten ratio in lighter craft in 1930. Stephen Pelz, *Race to Pearl Harbor* (Cambridge: Harvard University Press 1974), 3.

the erroneous views held before the two world wars show, there is no simple way to determine which is dominant. "[T]hese oscillations are not smooth and predictable like those of a swinging pendulum. They are uneven in both extent and time. Some occur in the course of a single battle or campaign, others in the course of a war, still others during a series of wars." Longer-term oscillations can also be detected:

> The early Gothic age, from the twelfth to the late thirteenth century, with its wonderful cathedrals and fortified places, was a period during which the attackers in Europe generally met serious and increasing difficulties, because the improvement in the strength of fortresses outran the advance in the power of destruction. Later, with the spread of firearms at the end of the fifteenth century, old fortresses lost their power to resist. An age ensued during which the offense possessed, apart from short-term setbacks, new advantages. Then, during the seventeenth century, especially after about 1660, and until at least the outbreak of the War of the Austrian Succession in 1740, the defense regained much of the ground it had lost since the great medieval fortresses had proved unable to meet the bombardment of the new and more numerous artillery.[53]

Another scholar has continued the argument: "The offensive gained an advantage with new forms of heavy mobile artillery in the nineteenth century, but the stalemate of World War I created the impression that the defense again had an advantage; the German invasion in World War II, however, indicated the offensive superiority of highly mechanized armies in the field."[54]

The situation today with respect to conventional weapons is unclear. Until recently it was believed that tanks and tactical air power gave the attacker an advantage. The initial analyses of the 1973 Arab-Israeli war indicated that new anti-tank and anti-aircraft weapons have restored the primacy of the defense. These weapons are cheap, easy to use, and can destroy a high proportion of the attacking vehicles and planes that are sighted. It then would make sense for a status-quo power to buy lots of $20,000 missiles rather than buy a few half-million dollar tanks and multi-million dollar fighter-bombers. Defense would be possible even against a large and well-equipped force; states that care primarily about self-protection would not need to engage in arms races. But fur-

---

53. John Nef, *War and Human Progress* (New York: Norton 1963), 185. Also see *ibid.*, 237, 242–43, and 323; C. W. Oman, *The Art of War in the Middle Ages* (Ithaca, N.Y.: Cornell University Press 1953), 70–72; John Beeler, *Warfare in Feudal Europe, 730–1200* (Ithaca, N.Y.: Cornell University Press 1971), 212–14; Michael Howard, *War in European History* (London: Oxford University Press 1976), 33–37.

54. Quincy Wright, *A Study of War* (abridged ed.; Chicago: University of Chicago Press 1964), 142. Also see 63–70, 74–75. There are important exceptions to these generalizations—the American Civil War, for instance, falls in the middle of the period Wright says is dominated by the offense.

ther examinations of the new technologies and the history of the October War cast doubt on these optimistic conclusions and leave us unable to render any firm judgment.[55]

Concerning nuclear weapons, it is generally agreed that defense is impossible—a triumph not of the offense, but of deterrence. Attack makes no sense, not because it can be beaten off, but because the attacker will be destroyed in turn. In terms of the questions under consideration here, the result is the equivalent of the primacy of the defense. First, security is relatively cheap. Less than one percent of the G.N.P. is devoted to deterring a direct attack on the United States; most of it is spent on acquiring redundant systems to provide a lot of insurance against the worst conceivable contingencies. Second, both sides can simultaneously gain security in the form of second-strike capability. Third, and related to the foregoing, second-strike capability can be maintained in the face of wide variations in the other side's military posture. There is no purely military reason why each side has to react quickly and strongly to the other's increases in arms. Any spending that the other devotes to trying to achieve first-strike capability can be neutralized by the state's spending much smaller sums on protecting its second-strike capability. Fourth, there are no incentives to strike first in a crisis.

Important problems remain, of course. Both sides have interests that go well beyond defense of the homeland. The protection of these interests creates conflicts even if neither side desires expansion. Furthermore, the shift from defense to deterrence has greatly increased the importance and perceptions of resolve. Security now rests on each side's belief that the other would prefer to run high risks of total destruction rather than sacrifice its vital interests. Aspects of the security dilemma thus appear in a new form. Are weapons procurements used as an index of resolve? Must they be so used? If one side fails to respond to the other's buildup, will it appear weak and thereby invite predation? Can both sides simultaneously have images of high resolve or is there a zero-sum element involved? Although these problems are real, they are not as severe as those in the prenuclear era: there are many indices of resolve, and states do not so much judge images of resolve in the abstract as ask how likely it is that the other will stand firm in a particular dispute. Since states are most

---

55. Geoffrey Kemp, Robert Pfaltzgraff, and Uri Ra'anan, eds., *The Other Arms Race* (Lexington, Mass.: D. C. Heath 1975); James Foster, "The Future of Conventional Arms Control," *Policy Sciences,* No. 8 (Spring 1977), 1–19.

likely to stand firm on matters which concern them most, it is quite possible for both to demonstrate their resolve to protect their own security simultaneously.

OFFENSE-DEFENSE DIFFERENTIATION

The other major variable that affects how strongly the security dilemma operates is whether weapons and policies that protect the state also provide the capability for attack. If they do not, the basic postulate of the security dilemma no longer applies. A state can increase its own security without decreasing that of others. The advantage of the defense can only ameliorate the security dilemma. A differentiation between offensive and defensive stances comes close to abolishing it. Such differentiation does not mean, however, that all security problems will be abolished. If the offense has the advantage, conquest and aggression will still be possible. And if the offense's advantage is great enough, status-quo powers may find it too expensive to protect themselves by defensive forces and decide to procure offensive weapons even though this will menace others. Furthermore, states will still have to worry that even if the other's military posture shows that it is peaceful now, it may develop aggressive intentions in the future.

Assuming that the defense is at least as potent as the offense, the differentiation between them allows status-quo states to behave in ways that are clearly different from those of aggressors. Three beneficial consequences follow. First, status-quo powers can identify each other, thus laying the foundations for cooperation. Conflicts growing out of the mistaken belief that the other side is expansionist will be less frequent. Second, status-quo states will obtain advance warning when others plan aggression. Before a state can attack, it has to develop and deploy offensive weapons. If procurement of these weapons cannot be disguised and takes a fair amount of time, as it almost always does, a status-quo state will have the time to take countermeasures. It need not maintain a high level of defensive arms as long as its potential adversaries are adopting a peaceful posture. (Although being so armed should not, with the one important exception noted below, alarm other status-quo powers.) States do, in fact, pay special attention to actions that they believe would not be taken by a status-quo state because they feel that states exhibiting such behavior are aggressive. Thus the seizure or development of transportation facilities will alarm others more if these facilities have no commercial value, and therefore can only be wanted for military reasons. In 1906, the British rejected a Russian protest about their activities in a district of Persia by claiming that this area was "only of [strategic] importance [to the Russians] if they wished to attack the Indian frontier, or to put pressure upon us by making us think that they intend to attack it."[56]

The same inferences are drawn when a state acquires more weapons than observers feel are needed for defense. Thus, the Japanese spokesman at the 1930 London naval conference said that his country was alarmed by the American refusal to give Japan a 70 percent ratio (in place of a 60 percent ratio) in heavy cruisers: "As long as America held that ten percent advantage, it was possible for her to attack. So when America insisted on sixty percent instead of seventy percent, the idea would exist that they were trying to keep that possibility, and the Japanese people could not accept that."[57] Similarly, when Mussolini told Chamberlain in January 1939 that Hitler's arms program was motivated by defensive considerations, the Prime Minister replied that "German military forces were now so strong as to make it impossible for any Power or combination of Powers to attack her successfully. She could not want any further armaments for defensive purposes; what then did she want them for?"[58]

Of course these inferences can be wrong—as they are especially likely to be because states underestimate the degree to which they menace others.[59] And when they are wrong, the security dilemma is deepened. Because the state thinks it has received notice that the other is aggressive, its own arms building will be less restrained and the chances of cooperation will be decreased. But the dangers of incorrect inferences should not obscure the main point: when offensive and defensive postures are different, much of the uncertainty about the other's intentions that contributes to the security dilemma is removed.

The third beneficial consequence of a difference between offensive and defensive weapons is that if all states support the status quo, an obvious arms control agreement is a ban on weapons that are useful for attacking. As President Roosevelt put it in his message to the Geneva Disarmament Conference in 1933: "If all nations will agree wholly to eliminate from possession and use the weapons which make possible a successful attack, defenses automatically will

56. Richard Challener, *Admirals, Generals, and American Foreign Policy, 1898–1914* (Princeton: Princeton University Press 1973), 273; Grey to Nicolson, in Gooch and Temperley (fn. 47), 414.
57. Quoted in James Crowley, *Japan's Quest for Autonomy* (Princeton: Princeton University Press 1966), 49. American naval officers agreed with the Japanese that a ten-to-six ratio would endanger Japan's supremacy in her home waters.
58. E. L. Woodward and R. Butler, eds., *Documents on British Foreign Policy, 1919–1939*, Third series, III (London: H.M.S.O. 1950), 526.
59. Jervis (fn. 5), 69–72, 352–55.

become impregnable, and the frontiers and independence of every nation will become secure."[60] The fact that such treaties have been rare—the Washington naval agreements discussed above and the anti-ABM treaty can be cited as examples—shows either that states are not always willing to guarantee the security of others, or that it is hard to distinguish offensive from defensive weapons.

Is such a distinction possible? Salvador de Madariaga, the Spanish statesman active in the disarmament negotiations of the interwar years, thought not: "A weapon is either offensive or defensive according to which end of it you are looking at." The French Foreign Minister agreed (although French policy did not always follow this view): "Every arm can be employed offensively or defensively in turn. . . . The only way to discover whether arms are intended for purely defensive purposes or are held in a spirit of aggression is in all cases to enquire into the intentions of the country concerned." Some evidence for the validity of this argument is provided by the fact that much time in these unsuccessful negotiations was devoted to separating offensive from defensive weapons. Indeed, no simple and unambiguous definition is possible and in many cases no judgment can be reached. Before the American entry into World War I, Woodrow Wilson wanted to arm merchantmen only with guns in the back of the ship so they could not initiate a fight, but this expedient cannot be applied to more common forms of armaments.[61]

There are several problems. Even when a differentiation is possible, a status-quo power will want offensive arms under any of three conditions. (1) If the offense has a great advantage over the defense, protection through defensive forces will be too expensive. (2) Status-quo states may need offensive weapons to regain territory lost in the opening stages of a war. It might be possible, however, for a state to wait to procure these weapons until war seems likely, and they might be needed only in relatively small numbers, unless the aggressor was able to construct strong defenses quickly in the occupied areas. (3) The state may feel that it must be prepared to take the offensive either because the other side will make peace only if it loses territory or because the state has commitments to attack if the other makes war on a third party. As noted above, status-quo states with extensive commitments are often forced to behave like aggressors. Even when they lack such commitments, status-quo states must

60. Quoted in Merze Tate, *The United States and Armaments* (Cambridge: Harvard University Press 1948), 108.
61. Boggs (fn. 28), 15, 40.

worry about the possibility that if they are able to hold off an attack, they will still not be able to end the war unless they move into the other's territory to damage its military forces and inflict pain. Many American naval officers after the Civil War, for example, believed that "only by destroying the commerce of the opponent could the United States bring him to terms."[62]

A further complication is introduced by the fact that aggressors as well as status-quo powers require defensive forces as a prelude to acquiring offensive ones, to protect one frontier while attacking another, or for insurance in case the war goes badly. Criminals as well as policemen can use bulletproof vests. Hitler as well as Maginot built a line of forts. Indeed, Churchill reports that in 1936 the German Foreign Minister said: "As soon as our fortifications are constructed [on our western borders] and the countries in Central Europe realize that France cannot enter German territory, all these countries will begin to feel very differently about their foreign policies, and a new constellation will develop."[63] So a state may not necessarily be reassured if its neighbor constructs strong defenses.

More central difficulties are created by the fact that whether a weapon is offensive or defensive often depends on the particular situation—for instance, the geographical setting and the way in which the weapon is used. "Tanks . . . spearheaded the fateful German thrust through the Ardennes in 1940, but if the French had disposed of a properly concentrated armored reserve, it would have provided the best means for their cutting off the penetration and turning into a disaster for the Germans what became instead an overwhelming victory."[64] Anti-aircraft weapons seem obviously defensive—to be used, they must wait for the other side to come to them. But the Egyptian attack on Israel in 1973 would have been impossible without effective air defenses that covered the battlefield. Nevertheless, some distinctions are possible. Sir John Simon, then the British Foreign Secretary, in response to the views cited earlier, stated that just because a fine line could not be drawn, "that was no reason for saying that there were not stretches of territory on either side which all practical men and women knew to be well on this or that side of the line." Although there are almost no weapons and

62. Kenneth Hagan, *American Gunboat Diplomacy and the Old Navy, 1877–1889* (Westport, Conn.: Greenwood Press 1973), 20.
63. Winston Churchill, *The Gathering Storm* (Boston: Houghton 1948), 206.
64. Brodie, *War and Politics* (fn. 35), 325.

strategies that are useful only for attacking, there are some that are almost exclusively defensive. Aggressors could want them for protection, but a state that relied mostly on them could not menace others. More frequently, we cannot "determine the absolute character of a weapon, but [we can] make a comparison . . . [and] discover whether or not the offensive potentialities predominate, whether a weapon is more useful in attack or in defense."[65]

The essence of defense is keeping the other side out of your territory. A purely defensive weapon is one that can do this without being able to penetrate the enemy's land. Thus a committee of military experts in an interwar disarmament conference declared that armaments "incapable of mobility by means of self-contained power," or movable only after long delay, were "only capable of being used for the defense of a State's territory."[66] The most obvious examples are fortifications. They can shelter attacking forces, especially when they are built right along the frontier,[67] but they cannot occupy enemy territory. A state with only a strong line of forts, fixed guns, and a small army to man them would not be much of a menace. Anything else that can serve only as a barrier against attacking troops is similarly defensive. In this category are systems that provide warning of an attack, the Russian's adoption of a different railroad gauge, and nuclear land mines that can seal off invasion routes.

If total immobility clearly defines a system that is defensive only, limited mobility is unfortunately ambiguous. As noted above, short-range fighter aircraft and anti-aircraft missiles can be used to cover an attack. And, unlike forts, they can advance with the troops. Still, their inability to reach deep into enemy territory does make them more useful for the defense than for the offense. Thus, the United States and Israel would have been more alarmed in the early 1970's had the Russians provided the Egyptians with long-range instead of short-range aircraft. Naval forces are particularly difficult to classify in these terms, but those that are very short-legged can be used only for coastal defense.

Any forces that for various reasons fight well only when on their own soil in effect lack mobility and therefore are defensive. The most extreme example

65. Boggs (fn. 28), 42, 83. For a good argument about the possible differentiation between offensive and defensive weapons in the 1930's, see Basil Liddell Hart, "Aggression and the Problem of Weapons," *English Review*, Vol. 55 (July 1932), 71–78.
66. Quoted in Boggs (fn. 28), 39.
67. On these grounds, the Germans claimed in 1932 that the French forts were offensive (*ibid.*, 49). Similarly, fortified forward naval bases can be necessary for launching an attack; see Braisted (fn. 27), 643.

would be passive resistance. Noncooperation can thwart an aggressor, but it is very hard for large numbers of people to cross the border and stage a sit-in on another's territory. Morocco's recent march on the Spanish Sahara approached this tactic, but its success depended on special circumstances. Similarly, guerrilla warfare is defensive to the extent to which it requires civilian support that is likely to be forthcoming only in opposition to a foreign invasion. Indeed, if guerrilla warfare were easily exportable and if it took ten defenders to destroy each guerrilla, then this weapon would not only be one which could be used as easily to attack the other's territory as to defend one's own, but one in which the offense had the advantage: so the security dilemma would operate especially strongly.

If guerrillas are unable to fight on foreign soil, other kinds of armies may be unwilling to do so. An army imbued with the idea that only defensive wars were just would fight less effectively, if at all, if the goal were conquest. Citizen militias may lack both the ability and the will for aggression. The weapons employed, the short term of service, the time required for mobilization, and the spirit of repelling attacks on the homeland, all lend themselves much more to defense than to attacks on foreign territory.[68]

Less idealistic motives can produce the same result. A leading student of medieval warfare has described the armies of that period as follows: "Assembled with difficulty, insubordinate, unable to maneuver, ready to melt away from its standard the moment that its short period of service was over, a feudal force presented an assemblage of unsoldierlike qualities such as have seldom been known to coexist. Primarily intended to defend its own borders from the Magyar, the Northman, or the Saracen . . . , the institution was utterly unadapted to take the offensive."[69] Some political groupings can be similarly described. International coalitions are more readily held together by fear than by hope of gain. Thus Castlereagh was not being entirely self-serving when in 1816 he argued that the Quadruple Alliance "could only have owed its origin to a sense of common danger; in its very nature it must be conservative; it cannot threaten either the security or the liberties of other States."[70] It is no accident that most of the major campaigns of expansion have been waged by one domi-

68. The French made this argument in the interwar period; see Richard Challener, *The French Theory of the Nation in Arms* (New York: Columbia University Press 1955), 181–82. The Germans disagreed; see Boggs (fn. 28), 44–45.
69. Oman (fn. 53), 57–58.
70. Quoted in Charles Webster, *The Foreign Policy of Castlereagh*, II, *1815–1822* (London: G. Bell and Sons 1963), 510.

nant nation (for example, Napoleon's France and Hitler's Germany), and that coalitions among relative equals are usually found defending the status quo. Most gains from conquest are too uncertain and raise too many questions of future squabbles among the victors to hold an alliance together for long. Although defensive coalitions are by no means easy to maintain—conflicting national objectives and the free-rider problem partly explain why three of them dissolved before Napoleon was defeated—the common interest of seeing that no state dominates provides a strong incentive for solidarity.

Weapons that are particularly effective in reducing fortifications and barriers are of great value to the offense. This is not to deny that a defensive power will want some of those weapons if the other side has them: Brodie is certainly correct to argue that while their tanks allowed the Germans to conquer France, properly used French tanks could have halted the attack. But France would not have needed these weapons if Germany had not acquired them, whereas even if France had no tanks, Germany could not have foregone them since they provided the only chance of breaking through the French lines. Mobile heavy artillery is, similarly, especially useful in destroying fortifications. The defender, while needing artillery to fight off attacking troops or to counterattack, can usually use lighter guns since they do not need to penetrate such massive obstacles. So it is not surprising that one of the few things that most nations at the interwar disarmament conferences were able to agree on was that heavy tanks and mobile heavy guns were particularly valuable to a state planning an attack.[71]

Weapons and strategies that depend for their effectiveness on surprise are almost always offensive. That fact was recognized by some of the delegates to the interwar disarmament conferences and is the principle behind the common national ban on concealed weapons. An earlier representative of this widespread view was the mid-19th-century Philadelphia newspaper that argued: "As a measure of defense, knives, dirks, and sword canes are entirely useless. They are fit only for attack, and all such attacks are of murderous character. Whoever carries such a weapon has prepared himself for homicide."[72]

It is, of course, not always possible to distinguish between forces that are most effective for holding territory and forces optimally designed for taking it. Such a distinction could not have been made for the strategies and weapons in

71. Boggs (fn. 28), 14–15, 47–48, 60.
72. Quoted in Philip Jordan, *Frontier Law and Order* (Lincoln: University of Nebraska Press 1970), 7; also see 16–17.

Europe during most of the period between the Franco-Prussian War and World War I. Neither naval forces nor tactical air forces can be readily classified in these terms. But the point here is that when such a distinction is possible, the central characteristic of the security dilemma no longer holds, and one of the most troublesome consequences of anarchy is removed.

OFFENSE-DEFENSE DIFFERENTIATION AND STRATEGIC NUCLEAR WEAPONS. In the interwar period, most statesmen held the reasonable position that weapons that threatened civilians were offensive.[73] But when neither side can protect its civilians, a counter-city posture is defensive because the state can credibly threaten to retaliate only in response to an attack on itself or its closest allies. The costs of this strike are so high that the state could not threaten to use it for the less-than-vital interest of compelling the other to abandon an established position.

In the context of deterrence, offensive weapons are those that provide defense. In the now familiar reversal of common sense, the state that could take its population out of hostage, either by active or passive defense or by destroying the other's strategic weapons on the ground, would be able to alter the status quo. The desire to prevent such a situation was one of the rationales for the anti-ABM agreements; it explains why some arms controllers opposed building ABM's to protect cities, but favored sites that covered ICBM fields. Similarly, many analysts want to limit warhead accuracy and favor multiple re-entry vehicles (MRV's), but oppose multiple independently targetable re-entry vehicles (MIRV's). The former are more useful than single warheads for penetrating city defenses, and ensure that the state has a second-strike capability. MIRV's enhance counterforce capabilities. Some arms controllers argue that this is also true of cruise missiles, and therefore do not want them to be deployed either. There is some evidence that the Russians are not satisfied with deterrence and are seeking to regain the capability for defense. Such an effort, even if not inspired by aggressive designs, would create a severe security dilemma.

What is most important for the argument here is that land-based ICBM's are both offensive and defensive, but when both sides rely on Polaris-type systems (SLBM's), offense and defense use different weapons. ICBM's can be used either to destroy the other's cities in retaliation or to initiate hostilities by attacking the other's strategic missiles. Some measures—for instance, hardening of missile sites and warning systems—are purely defensive, since they do not make a first strike easier. Others are predominantly offensive—for instance,

---

73. Boggs (fn. 28), 20, 28.

passive or active city defenses, and highly accurate warheads. But ICBM's themselves are useful for both purposes. And because states seek a high level of insurance, the desire for protection as well as the contemplation of a counterforce strike can explain the acquisition of extremely large numbers of missiles. So it is very difficult to infer the other's intentions from its military posture. Each side's efforts to increase its own security by procuring more missiles decreases, to an extent determined by the relative efficacy of the offense and the defense, the other side's security. That is not the case when both sides use SLBM's. The point is not that sea-based systems are less vulnerable than land-based ones (this bears on the offense-defense ratio) but that SLBM's are defensive, retaliatory weapons. First, they are probably not accurate enough to destroy many military targets.[74] Second, and more important, SLBM's are not the main instrument of attack against other SLBM's. The hardest problem confronting a state that wants to take its cities out of hostage is to locate the other's SLBM's, a job that requires not SLBM's but anti-submarine weapons. A state might use SLBM's to attack the other's submarines (although other weapons would probably be more efficient), but without anti-submarine warfare (ASW) capability the task cannot be performed. A status-quo state that wanted to forego offensive capability could simply forego ASW research and procurement.

There are two difficulties with this argument, however. First, since the state's SLBM's are potentially threatened by the other's ASW capabilities, the state may want to pursue ASW research in order to know what the other might be able to do and to design defenses. Unless it does this, it cannot be confident that its submarines are safe. Second, because some submarines are designed to attack surface ships, not launch missiles, ASW forces have missions other than taking cities out of hostage. Some U.S. officials plan for a long war in Europe which would require keeping the sea lanes open against Russian submarines. Designing an ASW force and strategy that would meet this threat without endangering Soviet SLBM's would be difficult but not impossible, since the two missions are somewhat different.[75] Furthermore, the Russians do not need ASW forces to combat submarines carrying out conventional missions; it might be in America's interest to sacrifice the ability to meet a threat that is not likely to materialize in order to reassure the Russians that we are not menacing their retaliatory capability.

74. See, however, Desmond Ball, "The Counterforce Potential of American SLBM Systems," *Journal of Peace Research*, XIV (No. 1, 1977), 23–40.
75. Richard Garwin, "Anti-Submarine Warfare and National Security," *Scientific American*, Vol. 227 (July 1972), 14–25.

When both sides rely on ICBM's, one side's missiles can attack the other's, and so the state cannot be indifferent to the other's building program. But because one side's SLBM's do not menace the other's, each side can build as many as it wants and the other need not respond. Each side's decision on the size of its force depends on technical questions, its judgment about how much destruction is enough to deter, and the amount of insurance it is willing to pay for—and these considerations are independent of the size of the other's strategic force. Thus the crucial nexus in the arms race is severed.

Here two objections not only can be raised but have been, by those who feel that even if American second-strike capability is in no danger, the United States must respond to a Soviet buildup. First, the relative numbers of missiles and warheads may be used as an index of each side's power and will. Even if there is no military need to increase American arms as the Russians increase theirs, a failure to respond may lead third parties to think that the U.S. has abandoned the competition with the U.S.S.R. and is no longer willing to pay the price of world leadership. Furthermore, if either side believes that nuclear "superiority" matters, then, through the bargaining logic, it will matter. The side with "superiority" will be more likely to stand firm in a confrontation if it thinks its "stronger" military position helps it, or if it thinks that the other thinks its own "weaker" military position is a handicap. To allow the other side to have more SLBM's—even if one's own second-strike capability is unimpaired—will give the other an advantage that can be translated into political gains.

The second objection is that superiority *does* matter, and not only because of mistaken beliefs. If nuclear weapons are used in an all-or-none fashion, then all that is needed is second-strike capability. But limited, gradual, and controlled strikes are possible. If the other side has superiority, it can reduce the state's forces by a slow-motion war of attrition. For the state to strike at the other's cities would invite retaliation; for it to reply with a limited counterforce attack would further deplete its supply of missiles. Alternatively, the other could employ demonstration attacks—such as taking out an isolated military base or exploding a warhead high over a city—in order to demonstrate its resolve. In either of these scenarios, the state will suffer unless it matches the other's arms posture.[76]

These two objections, if valid, mean that even with SLBM's one cannot distinguish offensive from defensive strategic nuclear weapons. Compellence

76. The latter scenario, however, does not require that the state closely match the number of missiles the other deploys.

may be more difficult than deterrence,[77] but if decision makers believe that numbers of missiles or of warheads influence outcomes, or if these weapons can be used in limited manner, then the posture and policy that would be needed for self-protection is similar to that useful for aggression. If the second objection has merit, security would require the ability to hit selected targets on the other side, enough ammunition to wage a controlled counterforce war, and the willingness to absorb limited countervalue strikes. Secretary Schlesinger was correct in arguing that this capability would not constitute a first-strike capability. But because the "Schlesinger Doctrine" could be used not only to cope with a parallel Russian policy, but also to support an American attempt to change the status quo, the new American stance would decrease Russian security. Even if the U.S.S.R. were reassured that the present U.S. Government lacked the desire or courage to do this, there could be no guarantee that future governments would not use the new instruments for expansion. Once we move away from the simple idea that nuclear weapons can only be used for all-out strikes, half the advantage of having both sides rely on a sea-based force would disappear because of the lack of an offensive-defensive differentiation. To the extent that military policy affects political relations, it would be harder for the United States and the Soviet Union to cooperate even if both supported the status quo.

Although a full exploration of these questions is beyond the scope of this paper, it should be noted that the objections rest on decision makers' beliefs—beliefs, furthermore, that can be strongly influenced by American policy and American statements. The perceptions of third nations of whether the details of the nuclear balance affect political conflicts—and, to a lesser extent, Russian beliefs about whether superiority is meaningful—are largely derived from the American strategic debate. If most American spokesmen were to take the position that a secure second-strike capability was sufficient and that increments over that (short of a first-strike capability) would only be a waste of money, it is doubtful whether America's allies or the neutrals would judge the superpowers' useful military might or political will by the size of their stockpiles. Although the Russians stress war-fighting ability, they have not contended that marginal increases in strategic forces bring political gains; any attempt to do so could be rendered less effective by an American assertion that this is nonsense.

77. Thomas Schelling, *Arms and Influence* (New Haven: Yale University Press 1966), 69–78. Schelling's arguments are not entirely convincing, however. For further discussion, see Jervis, "Deterrence Theory Re-Visited," Working Paper No. 14, UCLA Program in Arms Control and International Security.

The bargaining advantages of possessing nuclear "superiority" work best when both sides acknowledge them. If the "weaker" side convinces the other that it does not believe there is any meaningful difference in strength, then the "stronger" side cannot safely stand firm because there is no increased chance that the other will back down.

This kind of argument applies at least as strongly to the second objection. Neither side can employ limited nuclear options unless it is quite confident that the other accepts the rules of the game. For if the other believes that nuclear war cannot be controlled, it will either refrain from responding—which would be fine—or launch all-out retaliation. Although a state might be ready to engage in limited nuclear war without acknowledging this possibility—and indeed, that would be a reasonable policy for the United States—it is not likely that the other would have sufficient faith in that prospect to initiate limited strikes unless the state had openly avowed its willingness to fight this kind of war. So the United States, by patiently and consistently explaining that it considers such ideas to be mad and that any nuclear wars will inevitably get out of control, could gain a large measure of protection against the danger that the Soviet Union might seek to employ a "Schlesinger Doctrine" against an America that lacked the military ability or political will to respond in kind. Such a position is made more convincing by the inherent implausibility of the arguments for the possibility of a limited nuclear war.

In summary, as long as states believe that all that is needed is second-strike capability, then the differentiation between offensive and defensive forces that is provided by reliance on SLBM's allows each side to increase its security without menacing the other, permits some inferences about intentions to be drawn from military posture, and removes the main incentive for status-quo powers to engage in arms races.

IV. Four Worlds

The two variables we have been discussing—whether the offense or the defense has the advantage, and whether offensive postures can be distinguished from defensive ones—can be combined to yield four possible worlds.

The first world is the worst for status-quo states. There is no way to get security without menacing others, and security through defense is terribly difficult to obtain. Because offensive and defensive postures are the same, status-quo states acquire the same kind of arms that are sought by aggressors. And because the offense has the advantage over the defense, attacking is the best

4 WORLDS

	OFFENSE HAS THE ADVANTAGE	DEFENSE HAS THE ADVANTAGE
OFFENSIVE POSTURE NOT DISTINGUISHABLE FROM DEFENSIVE ONE	1 Doubly dangerous	2 Security dilemma, but security requirements may be compatible.
OFFENSIVE POSTURE DISTINGUISHABLE FROM DEFENSIVE ONE	3 No security dilemma, but aggression possible. Status-quo states can follow different policy than aggressors. Warning given.	4 Doubly stable

route to protecting what you have; status-quo states will therefore behave like aggressors. The situation will be unstable. Arms races are likely. Incentives to strike first will turn crises into wars. Decisive victories and conquests will be common. States will grow and shrink rapidly, and it will be hard for any state to maintain its size and influence without trying to increase them. Cooperation among status-quo powers will be extremely hard to achieve.

There are no cases that totally fit this picture, but it bears more than a passing resemblance to Europe before World War I. Britain and Germany, although in many respects natural allies, ended up as enemies. Of course much of the explanation lies in Germany's ill-chosen policy. And from the perspective of our theory, the powers' ability to avoid war in a series of earlier crises cannot be easily explained. Nevertheless, much of the behavior in this period was the product of technology and beliefs that magnified the security dilemma. Decision makers thought that the offense had a big advantage and saw little difference between offensive and defensive military postures. The era was characterized by arms races. And once war seemed likely, mobilization races created powerful incentives to strike first.

In the nuclear era, the first world would be one in which each side relied on vulnerable weapons that were aimed at similar forces and each side understood the situation. In this case, the incentives to strike first would be very high—so high that status-quo powers as well as aggressors would be sorely tempted to pre-empt. And since the forces could be used to change the status quo as well as to preserve it, there would be no way for both sides to increase their security simultaneously. Now the familiar logic of deterrence leads both sides to see the dangers in this world. Indeed, the new understanding of this situation was one reason why vulnerable bombers and missiles were replaced. Ironically, the 1950's would have been more hazardous if the decision makers had been aware of the dangers of their posture and had therefore felt greater pressure to strike first. This situation could be recreated if both sides were to rely on MIRVed ICBM's.

In the second world, the security dilemma operates because offensive and defensive postures cannot be distinguished; but it does not operate as strongly as in the first world because the defense has the advantage, and so an increment in one side's strength increases its security more than it decreases the other's. So, if both sides have reasonable subjective security requirements, are of roughly equal power, and the variables discussed earlier are favorable, it is quite likely that status-quo states can adopt compatible security policies. Although a state will not be able to judge the other's intentions from the kinds of weapons it procures, the level of arms spending will give important evidence. Of course a state that seeks a high level of arms might be not an aggressor but merely an insecure state, which if conciliated will reduce its arms, and if confronted will reply in kind. To assume that the apparently excessive level of arms indicates aggressiveness could therefore lead to a response that would deepen the dilemma and create needless conflict. But empathy and skillful statesmanship can reduce this danger. Furthermore, the advantageous position of the defense means that a status-quo state can often maintain a high degree of security with a level of arms lower than that of its expected adversary. Such a state demonstrates that it lacks the ability or desire to alter the status quo, at least at the present time. The strength of the defense also allows states to react slowly and with restraint when they fear that others are menacing them. So, although status-quo powers will to some extent be threatening to others, that extent will be limited.

This world is the one that comes closest to matching most periods in history. Attacking is usually harder than defending because of the strength of fortifications and obstacles. But purely defensive postures are rarely possible because fortifications are usually supplemented by armies and mobile guns which can support an attack. In the nuclear era, this world would be one in which both sides relied on relatively invulnerable ICBM's and believed that limited nuclear war was impossible. Assuming no MIRV's, it would take more than one attacking missile to destroy one of the adversary's. Pre-emption is therefore unattractive. If both sides have large inventories, they can ignore all but drastic increases on the other side. A world of either ICBM's or SLBM's in which both sides adopted the "Schlesinger Doctrine" would probably fit in this category too. The means of preserving the status quo would also be the means of changing it, as we discussed earlier. And the defense usually would have the advantage, because compellence is more difficult than deterrence. Although a state might succeed in changing the status quo on issues that matter much more to it than to others, status-quo powers could deter major provocations under most circumstances.

In the third world there may be no security dilemma, but there are security problems. Because states can procure defensive systems that do not threaten others, the dilemma need not operate. But because the offense has the advantage, aggression is possible, and perhaps easy. If the offense has enough of an advantage, even a status-quo state may take the initiative rather than risk being attacked and defeated. If the offense has less of an advantage, stability and cooperation are likely. because the status-quo states will procure defensive forces. They need not react to others who are similarly armed, but can wait for the warning they would receive if others started to deploy offensive weapons. But each state will have to watch the others carefully, and there is room for false suspicions. The costliness of the defense and the allure of the offense can lead to unnecessary mistrust, hostility, and war, unless some of the variables discussed earlier are operating to restrain defection.

A hypothetical nuclear world that would fit this description would be one in which both sides relied on SLBM's, but in which ASW techniques were very effective. Offense and defense would be different, but the former would have the advantage. This situation is not likely to occur; but if it did, a status-quo state could show its lack of desire to exploit the other by refraining from threatening its submarines. The desire to have more protecting you than merely the other side's fear of retaliation is a strong one, however, and a state that knows that it would not expand even if its cities were safe is likely to believe that the other would not feel threatened by its ASW program. It is easy to see how such a world could become unstable, and how spirals of tensions and conflict could develop.

The fourth world is doubly safe. The differentiation between offensive and defensive systems permits a way out of the security dilemma; the advantage of the defense disposes of the problems discussed in the previous paragraphs. There is no reason for a status-quo power to be tempted to procure offensive forces, and aggressors give notice of their intentions by the posture they adopt. Indeed, if the advantage of the defense is great enough, there are no security problems. The loss of the ultimate form of the power to alter the status quo would allow greater scope for the exercise of nonmilitary means and probably would tend to freeze the distribution of values.

This world would have existed in the first decade of the 20th century if the decision makers had understood the available technology. In that case, the European powers would have followed different policies both in the long run and in the summer of 1914. Even Germany, facing powerful enemies on both sides, could have made herself secure by developing strong defenses. France could also have made her frontier almost impregnable. Furthermore, when crises arose, no one would have had incentives to strike first. There would have been no competitive mobilization races reducing the time available for negotiations.

In the nuclear era, this world would be one in which the superpowers relied on SLBM's, ASW technology was not up to its task, and limited nuclear options were not taken seriously. We have discussed this situation earlier; here we need only add that, even if our analysis is correct and even if the policies and postures of both sides were to move in this direction, the problem of violence below the nuclear threshold would remain. On issues other than defense of the homeland, there would still be security dilemmas and security problems. But the world would nevertheless be safer than it has usually been.

Offense and Defense in the International System

George H. Quester

The interaction of "offense" and "defense" is a topic much discussed in the analysis of warfare. But semantic confusions about this interaction have caused altogether too many commentators in recent years to avoid using these terms.

We have seen a general endorsement of "defensive" intentions, perhaps out of respect for mutual spheres of autonomy in international and even personal interactions; no one should thus "offend" by pushing into someone else's sphere. President Kennedy won additional support in his confrontation with Khrushchev over Russian missiles in Cuba, by labeling the missiles as offensive rather than defensive. The United States maintains a Defense Department today, instead of a War Department, and Japan has Self-Defense Forces, rather than Armed Forces. Nations might thus wish to seem on the defensive even when they were marching their troops forward, if only because most neutral spectators will be more sympathetic when this image can be maintained. If everyone wants to be "defending," and no one wants to be "offensive," do we then have no choice but to discard the distinction altogether, as a tool of analysis in international affairs?

We argue here that the distinction has far more value than this, precisely when we focus on defensive or offensive capability, quite apart from intention. In sports, everyone wants to win, but there are times when the best one can do is to keep the score as it is (e.g., when the other team is at bat in baseball, when the other team has the ball in football). In war, there may similarly be times when military technology shows us that the best we can do is to keep the "score" (in terms of acres of territory held) the same; for example, when the fortifications on each side are so powerful that it would be foolish for either army to storm out into no-man's-land to try to push the line forward. At other times, the environment may suggest great advantages for launching an attack.

This is primarily a book, then, about military capability. Can we identify the technical, political, and social factors that make it advantageous to strike out offensively at the enemy, rather than to sit in prepared defensive positions waiting for him to strike? And what then are the military and political consequences of such an offensive or defensive preponderance? Whether troops are in fact marching forward, or standing in their bunkers, is importantly a function of the incentives generated by the shape and weapons of the battlefield.

This essay includes parts of chapters 1 and 18 of his *Offense and Defense in the International System* (Indianapolis: Wiley, 1977).

What if one can kill more enemy soldiers by attacking than one loses of his own infantry in the process? Offenses are favored, and are far more likely, when nature structures the casualty exchange-rates this way. What if the reverse is instead true, if more soldiers or airplanes of the attacking force will be destroyed than of the force being attacked? Here a natural incentive appears for remaining on the defensive.

Much, but not all, of war, is thus driven by such calculations of the relative attrition of the two forces, by the "counterforce" implications of battles. The terms defense and offense must also be applied to some "countervalue" aspects of warfare (where soldiers inflict pain on populations, rather than disabling other soldiers), but the most historically significant interaction here rests with the counterforce calculations. Battle begins when an army elects to cross the demarcation line and assault another army. If each side stayed on its own side of the line ("stayed on the defensive"), battles and wars would not get past the stage of "phony wars" (the label the press bestowed on the French and German maneuvers at the outbreak of World War II in 1939, when each army kept largely to its own territory).

If a single bomber airplane can use gravity for the destruction of many hostile airplanes that it catches on an airfield below, this is a technological development that favors the offensive. If a well-planned minefield imposes heavy casualties on an army trying to attack a fortified machine-gun nest, this is a technology that discourages taking the offensive, and reinforces the defense.

Some weapons might have helped offensive as well as defensive operations, if only they could have been moved along with an advancing army to help it win the battles; yet the inability of such weapons to move consigns them to helping only the defense. Minefields are an example of this, as are fixed artillery positions along coasts, or fixed antiaircraft guns. Since it cannot be decided that these weapons will be brought to the enemy, the enemy must decide to come to the weapon. The weapons become supremely defensive, whenever a foreign decision to violate frontiers is indispensable to providing them with a target.

Mobility thus generally supports the offensive. First, one can invade with impunity if one can bring along all the "comforts of home," all of one's most deadly vehicles of destruction. Second, the ability to move may allow an attacking force to exploit various weak spots or blind spots of the force that is standing in place. Many an army might be defeated, even by a numerically inferior force, if its food or gasoline supplies could be suddenly assaulted by a roving enemy attack. Armies facing north may be thrown into disarray if they are suddenly attacked from the south by an opposing force that rolls around their flank. Armies suddenly cut off from home, even large armies, may be demoralized as a result. Airplanes are typically vulnerable to enemy airplanes that have gotten above them.

Third, the ability to move allows an attacking force to group itself and regroup itself, to assemble temporary numerical superiorities as it pleases, when it decides to begin battles. Victory often goes to the side with the largest force in any particular battle. In force-attrition terms, the weaker side may suffer much the heaviest casualties (the casualties perhaps being proportional to the inverse of the square of the force-ratios, as conjectured by the mathematician F. W. Lanchester).[1] It may make sense, therefore, to take the offensive, when this allows one to choose the time and place of battle, and to choose the odds. When armies cannot move quickly enough to allow this kind of strategic exercise, the defensive again looks more favorable, and fewer battles get fought.

Another characteristic of weapons has great significance for the power of the offense. If a weapon can be potent only for temporary durations, it favors taking the initiative, rather than waiting until some time when the weapon will have lost its impact, when the enemy's similar weapon will have grown to full strength. Various examples can be offered. Airplanes cannot stay aloft indefinitely. We might thus be under strong pressure to strike before the enemy air force gets into the air and before we must again land to refuel. Populations cannot be totally mobilized indefinitely, if the civilian economy is to continue to function. We might thus again strike while our army is mobilized, and while the enemy's battalions are not, to avoid his waiting to strike when we again have had to be demobilized to return to civilian occupations. The ability of man to fly (temporarily), or to be totally mobilized into armed forces (temporarily), thus may favor the offense, and threaten peace.

By contrast, any weapon that relates to peculiarities of terrain will be supportive of the defense. Perhaps some kinds of light firearms are really useful only in ambushes; hence, residents who have lived in a region for a long time and know its highways and byways will be advantaged by such arms, as opposed to an invading force offensively pushing into the region. Peculiarities such as marshland, mountains or jungle, or even our contemporary urban sprawl, will similarly favor the defense, if only because the army defending such a region is likely to be more at home with its peculiarities than the alien army trying to break in. Americans may thus be handicapped in the jungles of

1. The original exposition of this can be found in F. W. Lanchester, *Aircraft in Warfare* (London: Constable and Co., 1916), reprinted in James R. Newman (ed.), *The World of Mathematics* (New York: Simon and Schuster, 1956), Vol. IV.

Southeast Asia, while Bedouins were disadvantaged in 1948 within the streets of Jerusalem, and Burgundians would be outfought by the Swiss once the battle shifted into the Swiss mountains. If jungles or mountains or any of the other peculiarities listed interfere with mobility this is, of course an additional bolster for the defense.

In our discussions of defensive or offensive inclinations above, we refer to the effectiveness of weapons when directed against other weapons and the men who fire them; this can be labeled as the "counterforce" effect. If such weapons are permanently fixed, the defense is bolstered; if they are mobile, the offense is favored. But weapons can also be effective in killing civilians and imposing pain and destruction generally, and we must also therefore consider what is sometimes labeled the "countervalue" effect.[2]

The very process of fighting a war almost always imposes some costs on the combattants. Only the more psychologically disturbed persons on this earth will ever welcome such carnage and destruction for its own sake. Most will regard these costs of war as bearable only if they are coupled to the fruits of victory, or if they stave off the even worse costs of defeat. Perhaps it will be possible someday to fight a war with no permanent damage whatsoever to the soldiers fighting it, or to the land over which it is fought. Conceivably it would be no more destructive than an American game of football. For the moment, the costs of war per se remain far higher than this and, indeed, they have been higher for as long as we have known war in the civilized world.

Yet, given that the costs of war are serious, these costs have nonetheless gone up and down as the technology and social structure of our world has evolved. There have been periods in which most costs of this kind were shunted to the losing side, thus simply giving the possible victors all the more incentive to press on with their military initiatives. At other times, however, such costs have been borne more equally by winners and losers, perhaps because the battlefield took a most devastating toll of the young men on both sides, or because the destructiveness of war could not be hemmed in on some battlefield to spare the "home front."

If the battlefield costs promised to be too great, neither side might thus wish to commence hostilities, even when either might win by taking the offensive.

---

2. For a clear and extensive analysis of the interaction of "deterring" countervalue retaliation and defensive strengths, see Glenn Snyder, *Deterrence and Defense* (Princeton: Princeton University Press, 1961). Also see Thomas C. Schelling, *The Strategy of Conflict* (Cambridge, Mass.: Harvard University Press, 1960; New York: Oxford University Press, 1963), and Thomas C. Schelling, *Arms and Influence* (New Haven, Conn.: Yale University Press, 1966), paperback.

Moving one step further along, we today have air forces and nuclear weapons which may allow a totally defeated nation to impose the most horrendous destruction on the winning side's cities and people, in a vindictive last-gasp of retaliation. This also can deter the pursuit of victory, and prevent war.

A weapon that can level a city might thus be considered as "offensive" in that it surely "changes the score," just as an ABM weapon that kept the city intact might thus be viewed as "defensive." But when leveling a city is not related to incapacitating the enemy's armed forces and weapons, a very different kind of logic and motivation have come into effect. It is easy to explain why we would want to shoot enemy soldiers, since otherwise they may shoot us. It is far less obvious why we would ever naturally want to kill the enemy's civilians. The logics of retaliation, and threats of retaliation (i. e., deterrence), are real enough, but they require a more complicated and deeper explanation.

In a strict sense, common English usage thus may confuse strategic understanding by labeling weapons aimed at civilians as "offensive," alongside weapons aimed at other weapons. For most of our discussion here, the phrase "offensive" will instead refer to technology and techniques that reward counter*force* initiatives, destroying more weapons on the other side than are lost in the attack and incapacitating more soldiers than are lost. It is this kind of "offensive" that encourages war, and it takes a "defensive" that reverses such counterforce exchange rates to discourage military operations. The capability for a "counter*value* offensive," by contrast, seems to support peace, while "population defense" may make war more likely. (The problem becomes all the more complicated, of course, if the counter*force* offensive is ever so effective on "first strike" as to preclude all counter*value* retaliation.)

Yet these, with occasional exceptions, are primarily conceptual and strategic complications of the middle of the twentieth century. For much of history, the deliberately countervalue attack could have been dismissed as a separate strategic option; to reach and kill the other side's civilians, one had to incapacitate and defeat his armed forces first, since they stood in the way.

Wars are no picnic, and only madmen begin them lightheartedly. Yet when victors in war have to risk little or no retaliation, statesmen may often enough want to embark on the offensive, wherever victory goes with the initiative, so that a potential enemy cannot claim the initiative and victory for itself. Likelihoods of war are thus clearly influenced by how effective the offensive weapons seems to be, as compared with the defensive, and by how much the rival nations invest in each. If both sides are primed to reap advantages by pushing into each other's territory, war may be extremely likely whenever political crisis erupts. If the defense holds the advantage, by contrast, each side in a crisis will probably wait a little longer, in hopes that the other will foolishly take the offensive.

The importance of defensive advantage for peace is hardly a recent discovery. For example, there is a very nicely articulated discussion of this relationship in the classic writings of Karl von Clausewitz:

> The third cause which catches hold, like a ratchet wheel in machinery, from time to time producing a complete standstill, is the greater strength of the defensive form. A may feel too weak to attack B, from which it does not follow that B is strong enough for an attack on A. The addition of strength, which the defensive gives is not merely lost by assuming the offensive, but also passes to the enemy just as, figuratively expressed, the difference of $a + b$ and $a - b$ is equal to $2b$. Therefore it may so happen that both parties, at one and the same time, not only feel themselves too weak to attack, but also are so in reality.
>
> Thus even in the midst of the act of War itself, anxious sagacity and the apprehension of too great danger find vantage ground, by means of which they can exert their power, and tame the elementary impetuosity of War.[3]

Clausewitz, defining all-out combat as a hypothetically normal pattern for war, then posed the question of why it so often did not happen that way. Yet history often enough has seen two armies on opposite sides of a river, each hoping that the other would be foolish enough to wade through on the attack, exposing itself to the fire of prepared positions on dry ground. If the generals on each side are wise, peace will de facto persist.

Moreover, such an acknowledgement of the power of the defensive is found in the writings of the Chinese military commentator Sun Tzu, 500 years before the birth of Christ. In commenting on weaknesses and strengths, Sun notes a number of substantial advantages and strengths of the defensive.

1. Generally, he who occupies the field of battle first and awaits his enemy is at ease; he who comes later to the scene and rushes into the fight is weary.
2. And therefore those skilled in war bring the enemy to the field of battle and are not brought there by him.[4]

Defensive strengths carried to their logical extreme might make wars meaningless, as "declarations of war" would always be followed by "phony wars"; offensive strengths carried to their extreme might make wars inevitable, wars that lead to conquest. Paradoxically, one might then find a form of "peace" at

3. Karl von Clausewitz, *On War* (London: Pelican, 1968).
4. Sun Tzu, *The Art of War* Oxford: Oxford University Press, 1963).

either extreme, peace however based on enormously different political structures.

There is thus another significant consequence for whether the defense or offense is more effective. The strength of the defensive position, for example the Swiss position in the Alps, is conducive to political independence, as small units can defy the much larger military forces of any empire aspiring to control them. Although offensive weaponry is conducive to war instead of peace, it is also conducive to a final political decision, as wars are pushed through to total conquest. Perhaps real peace must be based on world government. If so, weapons systems that were enormously effective for a brief period of offensive campaigns across borders could produce the war that finally produced lasting peace. We may condemn Louis XIV, Napoleon, and Hitler for beginning wars, and also for the kinds of regimes they would have imposed on England and all the rest of the world. Yet if these men were reaching for the kinds of world empire that Rome once achieved, they were in the process also of striving for a world that would have ceased to be "international," as the last offensive succeeds and ends all offensives. The strength of the defense was not obvious enough to dissuade such men at the outset. It was strong enough in the end to prolong the military confrontations to which we are so accustomed.

Perhaps the worst situation of all is thus some blend of offensive and defensive that prolongs wars, that is, offensive prospects just bright enough to tempt each side to initiate hostilities, and defensive prospects bright enough to lead all sides to resist in hopes that defeat may yet be avoided, that victory may yet be won.

Yet if men cared only about peace, there would long ago have been a rush to submit to a single dominant sovereignty which by its very monopoly of military force made war impossible. Men and nations would have given up their linguistic and cultural autonomy, their economic home rule and privileged affluences, and their political and social peculiarities. What are the terms "sovereignty" and "independence," after all, but euphemisms for "international anarchy" and the "risk of war"? Men, in short, would have abandoned their defensive redoubts and submitted to world government, if peace was really valued above all. It is not so valued.[5]

Since the defenses of the nation are not abandoned, the risk of war emerges whenever offensive campaigns become tempting. If offensive force could replace the international arena with a world empire, much might be different.

5. For a deeper and broader analysis, see Kenneth N. Waltz, *Man, the State, and War: A Theoretical Analysis* (New York: Columbia University Press, 1964).

As things stand, however, offensive force merely tends to replace peace with war.

But how much of even this "normal" incidence of war should be blamed on the attractiveness of the offense? Isn't it possible that men may be hostile enough to want to fight and kill each other even when technology had offered no incentives to the offensive? If two nations are intent on occupying exactly the same real estate, their relationship may indeed be hostile beyond all remedy, Israel versus the Arab states being a major example for the present. We might also perhaps conceive of someone like Hitler who enjoyed cruelty and violence for its own sake, thus making war very likely, even if the offense is not greatly advantaged.

Or men might instead simply be so stupid as to expect victory when their military ventures were sure to lead to defeat; when both sides to a war each expect to defeat the other on the battlefield, one of them at least should have studied the problem more carefully. It might also be that each side is counting on the lack of resolve of the other, in effect expecting the other to give up first as in a tug-of-war game, or the game of "chicken." Fear of an approaching disaster such as World War III, or fatigue at the continuing costs of a World War I or a Vietnam War, could indeed cause either side to surrender the points at issue; yet if each side expects the other to surrender tomorrow, they may each continue the war today, and one of them must be in error.[6] Finally, one can have wars of simple mutual misunderstanding, wherein each side mistakenly assumes hostility in the other, and then takes precautionary moves that seem to confirm all the suspicions.

We thus have listed at least five distinct variables affecting the likelihood of war:

1. The extent of political disagreement between the states.
2. Mistaken assumptions about who would win a war (i.e., both sides may expect to have better soldiers).
3. Mistaken assessments of mutual resolve (i.e., each side under-rates the other's willingness to make sacrifices in an endurance contest).
4. Mistaken assumptions of hostility.
5. A state of military technique that rewards taking the offense.

Any one of these variables can improve or worsen the chances for peace. We contend here that the last one, the topic of this book, is now the most worthy of

6. Illustrations and analysis of these prolongations of war can be found in Fred C. Ikle, *Every War Must End* (New York: Columbia University Press, 1971).

attention, because it may indeed be more manageable than the others, because it in recent times has received less analysis than the other variables, and because it may greatly increase or decrease the problems caused by the others.

Much of war is caused neither by blatant hostility nor by elementary stupidity. The crucial fact is that peace needs to be assured even for times when statesmen are not sure of their facts, when they are necessarily guessing as to what their state's policies should be. What is wrong about the offensive technology, under conditions of uncertainty, of even moderate misinformation or hostility, is that the attack is encouraged in the manner of "shoot first, ask questions later." Offensive operations then get launched even where states had defensive intentions.

As long as we are consigned to a world of separate states, therefore, defensive developments in military practice will be more welcome than offensive ones, simply because they make war less likely among such states. The tone of this book indeed generally reflects an approval for defense, and a sense of tragedy when offensive opportunities appear. There are a few exceptional points in history where a slightly stronger offensive bias might have produced world government once and for all; perhaps this offered an opportunity for the union of all mankind. Yet a near miss for world empire may simply produce the casualties of the Napoleonic Wars, and not enough positive results to match this sacrifice.

This is thus an account of how much military possibilities have determined the shape of the international system and the likelihoods of peace or war. For much of the analysis, our attention is directed toward the real impact of weapons, presuming that the generals and prime ministers have often judged these impacts correctly. At points, however, the contrast will have to be drawn between the real and expected impact, especially when technology begins to come along so rapidly that predictions of offensive or defensive superiorities become prone to error. How tragic that World War I was to be launched on the illusion of offensive advantage, and then was to be prolonged by the reality of defensive advantage! What statesmen believe about war and weapons is crucial; what wars and weapons are really like will also be crucial when wars get fought, and whenever predictions about war become at all astute.

In places the analysis in this book will seem to have wandered perilously close to a discussion of states' intentions rather than capabilities. Yet this will hopefully only be true when the social environment of these intentions also reshapes capabilities. If rulers of the eighteenth century did not wish to seize a province, unless its material property were kept unspoiled in the process, this forced them to invest in quartermaster systems for their armies, and thus encumbered what otherwise might have been more of an offensive capability. Wars as a result were less likely, not because anyone was much less interested in conquest, but because the preconditions for attractive conquest made armies slower moving, and slower moving armies are better at defense than offense. Similar results may arise if it turns out to be true that Mao's China wants Marxist revolutions to take place all over the globe, but only if they have some genuinely indigenous aspects.

By contrast, the stress on social discipline and civic virtues in Rome may or may not have made Romans more expansion-minded. But it clearly made them more expansion-capable, since the disciplined Legions that could march to battles at great distances made offensives altogether more practicable than in the past.

Offense and Defense: Some Conclusions

Throughout this book there has been a tone of approval for the defense, based on the enhanced likelihood of peace where defenses seem more formidable than offensive weapons, and based also on the separate sovereign autonomies such defenses may maintain. Offenses produce war and/or empire; defenses support independence and peace. But all of this requires qualification.

If defenses allowed the degree of autonomous fractionation that applied in feudal Europe, for example, we would hardly applaud, for the damage done to commerce and culture and the quality of life in such a degree of local autonomy is too high a price to pay for "home rule." Western culture oscillates between extolling the dignity of the individual and praising community spirit, between endorsements of national independence and international integration. If this means anything, it suggests that we disapprove of either extreme.

Defenses have also often enough asserted themselves to prolong wars after they were begun, after each side's optimistic expectations of offensive victory were undone. Rulers and peoples may wish to persevere in a war, once it has begun, even when all initial expectations seem to have been disproved. Here mankind would be much better off if the offensive expectations were proved correct. Peace is preferable to a six-day war; but a six-day war is preferable to a six-year war. If the tank had only been perfected by either side in 1915, a great number of lives might have been saved.

Why are wars likely to be prolonged by defensive strengths, even when convincing defenses might have kept them from being started in the first place? As noted in our discussion of World War I, governments may be reluctant to admit that such wars were a mistaken investment, based on some foolish confidence in the offensive. Where governments must be "reelected" in one fashion or another, there may be a terrible temptation to press ahead, "throwing good money after bad." This kind of pathology, based on voter assumption that incumbents should be blamed for bad news and praised for good, is of course not limited to war decisions. If some great public works scheme, for example the Albany Mall, becomes enormously more costly than had been expected, governments may yet press on with it rather than "admitting defeat."

When two governments are locked into such a psychology, wars may be very difficult to terminate once they have been begun, even when it has become clear that the military technology indeed will not favor the initiative. Each side now considers itself in an endurance contest, betting that the other side will more want peace, will more be willing to surrender. This can explain the prolongation of the Vietnam War as easily as the prolongation of World War I. The crucial miscommunication is now almost exactly the opposite of what was involved in the outbreak of wars of apparent offensive advantage. Where the first-strike seems attractive, each side may be driven to exaggerate the adversary's willingness to exploit such a first-strike option, and thus feel driven to preempt that adversary; on "worst case" assumptions, each side has to exaggerate the opponent's hostility. After the war reaches stalemate, however, each side will be underrating the opponent's hostility, exaggerating his peace-mindedness.

One may find wars of attrition in history that were deliberately begun as such right from the outset, rather than appearing only as the inadvertent outcome of stalemated offensives. These would therefore be wars that cannot be remedied by eliminating the initial temptations of the attack. For the motivational reasons noted, however, such endurance contests are more likely when each side has already made a heavy initial investment in the outcome of the contest, a heavy initial investment very possibly based on uncertain but promising options for an offensive.

When wars are stalemated but not terminated, the contest of comparative endurance and attrition thus becomes a counter-value exercise. Each side hampers the other's trade; each side ties up the other's human and economic resources. Each side imposes military costs on the other, and on itself, by occasionally trying the offensive again. It is sad but almost inevitably true that any perfect defense will not be fully acknowledged in its time. Because the attack is thought of as heroic, because "imaginative" thinkers are drawn to the offense, military commanders will often enough still try to break through defensive barriers, even when stalemate is real. Since each side may be equally prone to this temptation, the costs of prolonging the war are thus expanded by the "bill" for this kind of probing.

It is probably true in parallel that no perfect offensive is ever fully acknowledged in its own time either. Everything, as Clausewitz noted, is a little uncertain. The pure "prisoner's dilemma" situation, when each side knows for certain that offensives and double crosses are profitable, knowing for certain that the adversary will think the same way, is thus not so common. Offensive capabilities more often provoke the initiation of wars in a realm of uncertainty, where the offensive looks good but might fail, where the other side looks menacing, but might not really have intended to attack (given the risks of fail-tire, given the battlefield costs that are almost inevitable in even the most skillfully executed preemptive offensive first-strike campaigns).

If a war is already underway, therefore, we might root for success for the offensive, just to get the war over with (that is, as long as the final offense does not produce victory for an enemy that will conquer *us*). If the world is at peace, however, we might more normally root for strength in the defense.

Even a "six-day" war is very costly, as compared with peace. Wars might be based on mutual hate, or just on mutual misunderstanding; the participants may never know for sure. At least, we know that weapons which encourage sitting still are generally preferable whenever political antagonisms are uncertain in their intensity. If our generals tell us to "wait and see," to wait for the other side to make the mistake of attacking, our statesmen may in the intervening days discover that the political issues were not so serious as to merit war in any event. A military system that demands haste has quite the opposite effect. Generals may be given to exaggerating their own military prowess and underestimating that of their opponents. If this is true, wars might break out simply because both sides confidently expected to win. If the defense is rated as generally superior to the offense, such error nonetheless is less likely to do damage; if the reverse is true, errors of this kind are all the more likely to produce disastrous wars.

One must be careful here to assess the situation for an entire front in comparing offense and defense. If a defensive advantage is installed for every mile of an international boundary, it could presumably reduce the likelihood of any serious warfare. If it applied along only a part, however, it might simply make warfare easier on some other front. Louis XIV used Vauban's genius not only to defend France, but to free forces to conquer neighboring territories. The apparent stalemate of defense on the western front in 1939 allowed Hitler more securely to send his armies into Poland. The erection of an air defense screen for Egypt in 1973 made it easier for Egyptian forces to try to storm across the Suez Canal again. An effective air defense system for North Vietnam would have discouraged air warfare, but might have emboldened Hanoi in its support for guerrilla offensives on the ground.

Also, as we have previously pointed out, one must always worry today about what is being "defended." If the vulnerability of our population to murderous attack is all that holds us back from launching a World War III, then a protection of that population will have an impact injurious to peace.

Most observers of military events will still view the "military balance" as a comparison of rival military forces rather than a comparison of offense with defense. Their most pressing question may still normally be "who is the stronger, who will win if war comes?," rather than "does the technology favor caution or preemption?" The intent of this book is to suggest a priority for the latter kind of question.

Yet the traditional analyst could counter that offensives in history have occurred often enough because, and only because, one side was superior to the other in numbers of troops or in firepower. Isn't the attack normally predicated on such a superiority? Isn't it an imbalance of forces between two adversaries that normally explains the beginnings of a war?

Obviously a vast superiority in force strength can give an aggressor an offensive option, almost no matter what the particular technology of weaponry may be. Could ten Swiss soldiers defend their country against 10,000 Austrians? Of course not. The significant impact of defensive or offensive technology shows up in the minimum ratios of numerical superiority required for such an offensive. With a basically defensive technology, perhaps a three to one, or five to one, superiority will be required to break through; with a more offensive technology, far less of a superiority may suffice, or no numerical superiority at all may be required, as either side can win if it only beats the other to the offensive. The last situation is the most horrendous for peace, because it creates the kinds of mutually reinforcing fears that leave little stability for the prevention of war.

The overlap of capabilities are shown in Figures L and M, where the horizontal scale shows the ratio of A's forces to B's, and each side's capabilities and incentives for striking at the other are annotated. In the case of a defensive technology (Figure L), it can never be true that both sides have an incentive to strike simultaneously; the "mirror—image" misperception thus becomes much less likely as the cause of mutually unwanted war.

Offensive opportunities	Side A	Yes	No	No
	Side B	No	No	Yes
	Forces	*A* > *B*	*A* = *B*	*B* > *A*

Figure L. With defensive technology.

In the case of an offensive technology (Figure M), however, there is no force-ratio that does not leave at least one side seeing a military advantage to attacking; moreover, as noted, a range of force-ratios exists that could put both sides into a race to the offensive. This, politically, is the situation that should at highest priority be avoided.

But what about all these cases where at least one side indeed has the ability to take the offensive? As noted, this can occur even if casualty exchange-rates favor the defense, when one side substantially outnumbers the other in troops, so that it could afford to bear the disadvantage of being on the offensive. It could occur when both sides are ensconced in castles, but only one possesses seige equipment. If one side can take the offensive with prospects of success, when the other side cannot, isn't this just as destabilizing and undesirable as the confrontations in which both sides might successfully strike first in quest of victory?

This question has allegedly arisen at various stages of international military confrontations, for example when the United States alone possessed the atomic bomb from 1945 to 1949, or when either of the major powers have been momentarily ahead in strategic nuclear weapons numbers. The USSR was wrongly suspected of being ahead during the "bomber gap" of 1955 to 1957, and the "missile gap" of 1957 to 1961. The United States may in truth have been substantially in front for the years 1949 to 1954 and from 1960 to 1966. A similar logical and strategic question arises with reference to guerrilla war tactics, which may favor a taking of the offensive for Marxist forces in various places around the world, but might not facilitate a simultaneous offensive in the opposite direction.

An offensive capability for one side only is indeed not as destabilizing as when the attack seems feasible for both sides; but it is still more threatening

Offensive opportunities	Side A	Yes	Yes	No
	Side B	No	Yes	Yes
	Forces	$A > B$	$A = B$	$B > A$

Figure M. With offensive technology.

than when such opportunities are available to neither. When blue alone can attack with any hope of success, blue at least for the moment need not brood about the possibility that red will be seeking to preempt this attack, or that blue must be quick to preempt red's assault. Blue's attack might still come, for fear that the offensive opportunity will not remain at hand forever, as the technology continually changes, and political arrangements change. But the urgency is less when it is simply the march of technology that is being raced, instead of a symmetrically conceived enemy attack.

The preservation of peace entails the mobilization of many talents, and the monitoring of many variables. Peace is furthered when nations are relatively content with the status quo, as compared with alternatives that might be accomplished by resort to arms; peace is just as obviously furthered when nations agree about what is desirable in the future, about what should be reformed or replaced in the present. Political and moral theorists have devoted more than a little effort to defining and encouraging the background factors that might thus dispose our world away from war.

But where the political realm succeeds, the military realm may yet create sizable obstacles to peace. And where the political realm fails, the military confrontation may yet serve to moderate conflict. In the military realm, the assurance of peace will at least require concern for both the senses of "balance" that are contrasted immediately above. We will need to be certain that technology does not too much favor the offensive. We will also want to be sure that neither side outweighs the other so much in military terms that it can by sheer numbers override any advantages of the defense. But we make our task easier for the second category, as we succeed in the first.

Part II:
Offense, Defense, and the First World War

The Cult of the Offensive and the Origins of the First World War

Stephen Van Evera

During the decades before the First World War a phenomenon which may be called a "cult of the offensive" swept through Europe. Militaries glorified the offensive and adopted offensive military doctrines, while civilian elites and publics assumed that the offense had the advantage in warfare, and that offensive solutions to security problems were the most effective.

This article will argue that the cult of the offensive was a principal cause of the First World War, creating or magnifying many of the dangers which historians blame for causing the July crisis and rendering it uncontrollable. The following section will first outline the growth of the cult of the offensive in Europe in the years before the war, and then sketch the consequences which international relations theory suggests should follow from it. The second section will outline consequences which the cult produced in 1914, and the final section will suggest conclusions and implications for current American policy.

The Cult of the Offensive and International Relations Theory

THE GROWTH OF THE CULT

The gulf between myth and the realities of warfare has never been greater than in the years before World War I. Despite the large and growing advantage which defenders gained against attackers as a result of the invention of rifled and repeating small arms, the machine gun, barbed wire, and the development of railroads, Europeans increasingly believed that attackers would hold the advantage on the battlefield, and that wars would be short and "decisive"—a "brief storm," in the words of the German Chancellor,

---

I would like to thank Jack Snyder, Richard Ned Lebow, Barry Posen, Marc Trachtenberg, and Stephen Walt for their thoughtful comments on earlier drafts of this paper.

Stephen Van Evera is a Research Fellow at the Center for Science and International Affairs, Harvard University.

Bethmann Hollweg.[1] They largely overlooked the lessons of the American Civil War, the Russo–Turkish War of 1877–78, the Boer War, and the Russo–Japanese War, which had demonstrated the power of the new defensive technologies. Instead, Europeans embraced a set of political and military myths which obscured both the defender's advantages and the obstacles an aggressor would confront. This mindset helped to mold the offensive military doctrines which every European power adopted during the period 1892–1913.[2]

In Germany, the military glorified the offense in strident terms, and inculcated German society with similar views. General Alfred von Schlieffen, author of the 1914 German war plan, declared that "Attack is the best defense," while the popular publicist Friedrich von Bernhardi proclaimed that "the offensive mode of action is by far superior to the defensive mode," and that "the superiority of offensive warfare under modern conditions is greater than formerly."[3] German Chief of Staff General Helmuth von Moltke also endorsed "the principle that the offensive is the best defense," while General August von Keim, founder of the Army League, argued that "Germany ought to be armed for attack," since "the offensive is the only way of insuring victory."[4] These assumptions guided the Schlieffen Plan, which envisaged rapid and decisive attacks on Belgium, France, and Russia.

1. Quoted in L.L. Farrar, Jr., "The Short War Illusion: The Syndrome of German Strategy, August–December 1914," *Militaergeschictliche Mitteilungen*, No. 2 (1972), p. 40.
2. On the origins of the cult of the offensive, see Jack Lewis Snyder, "Defending the Offensive: Biases in French, German, and Russian War Planning, 1870–1914" (Ph.D. dissertation, Columbia University, 1981), forthcoming as a book from Cornell University Press in 1984; Snyder's essay in this issue; and my "Causes of War" (Ph.D. dissertation, University of California, Berkeley, 1984), chapter 7. On the failure of Europeans to learn defensive lessons from the wars of 1860–1914, see Jay Luvaas, *The Military Legacy of the Civil War: The European Inheritance* (Chicago: University of Chicago Press, 1959); and T.H.E. Travers, "Technology, Tactics, and Morale: Jean de Bloch, the Boer War, and British Military Theory, 1900–1914," *Journal of Modern History*, Vol. 51 (June 1979), pp. 264–286. Also relevant is Bernard Brodie, *Strategy in the Missile Age* (Princeton: Princeton University Press, 1965), pp. 42–52.

A related work which explores the sources of offensive and defensive doctrines before World War II is Barry R. Posen, *The Sources of Military Doctrine: France, Britain, and Germany Between the World Wars* (Ithaca: Cornell University Press, 1984), pp. 47–51, 67–74, and passim.
3. Gerhard Ritter, *The Schlieffen Plan: Critique of a Myth*, trans. Andrew and Eva Wilson, with a Foreword by B.H. Liddell Hart (London: Oswald Wolff, 1958; reprint ed., Westport, Conn.: Greenwood Press, 1979), p. 100; and Friedrich von Bernhardi, *How Germany Makes War* (New York: George H. Doran Co., 1914), pp. 153, 155.
4. Imanuel Geiss, ed., *July 1914: The Outbreak of the First World War: Selected Documents* (New York: W.W. Norton, 1967), p. 357; and Wallace Notestein and Elmer E. Stoll, eds., *Conquest and Kultur: Aims of the Germans in Their Own Words* (Washington, D.C.: U.S. Government Printing Office, 1917), p. 43. Similar ideas developed in the German navy; see Holger H. Herwig, *Politics*

In France, the army became "Obsessed with the virtues of the offensive," in the words of B.H. Liddell Hart, an obsession which also spread to French civilians.[5] The French army, declared Chief of Staff Joffre, "no longer knows any other law than the offensive. . . . Any other conception ought to be rejected as contrary to the very nature of war,"[6] while the President of the French Republic, Clément Fallières, announced that "The offensive alone is suited to the temperament of French soldiers. . . . We are determined to march straight against the enemy without hesitation."[7] Emile Driant, a member of the French chamber of deputies, summarized the common view: "The first great battle will decide the whole war, and wars will be short. The idea of the offense must penetrate the spirit of our nation."[8] French military doctrine reflected these offensive biases.[9] In Marshall Foch's words, the French army adopted "a single formula for success, a single combat doctrine, namely, the decisive power of offensive action undertaken with the resolute determination to march on the enemy, reach and destroy him."[10]

Other European states displayed milder symptoms of the same virus. The British military resolutely rejected defensive strategies despite their experience in the Boer War which demonstrated the power of entrenched defenders against exposed attackers. General W.G. Knox wrote, "The defensive is never an acceptable role to the Briton, and he makes little or no study of it," and General R.C.B. Haking argued that the offensive "will win as sure as there is a sun in the heavens."[11] The Russian Minister of War, General V.A. Sukhomlinov, observed that Russia's enemies were directing their armies "towards guaranteeing the possibility of dealing rapid and decisive blows.

of *Frustration: The United States in German Naval Planning, 1889–1941* (Boston: Little, Brown & Co., 1976), pp. 42–66.
5. B.H. Liddell Hart, *Through the Fog of War* (New York: Random House, 1938), p. 57.
6. In 1912, quoted in John Ellis, *The Social History of the Machine Gun* (New York: Pantheon, 1975), pp. 53–54.
7. Barbara Tuchman, *The Guns of August* (New York: Dell, 1962), p. 51.
8. In 1912, quoted in John M. Cairns, "International Politics and the Military Mind: The Case of the French Republic, 1911–1914," *The Journal of Modern History*, Vol. 25, No. 3 (September 1953), p. 282.
9. On the offensive in French prewar thought, see B.H. Liddell Hart, "French Military Ideas before the First World War," in Martin Gilbert, ed., *A Century of Conflict, 1850–1950* (London: Hamilton Hamish, 1966), pp. 135–148.
10. Richard D. Challener, *The French Theory of the Nation in Arms, 1866–1939* (New York: Columbia University Press, 1955), p. 81. Likewise, Joffre later explained that Plan XVII, his battle plan for 1914, was less a plan for battle than merely a plan of "concentration. . . . I adopted no preconceived idea, other than a full determination to take the offensive with all my forces assembled." Theodore Ropp, *War in the Modern World*, rev. ed. (New York: Collier, 1962), p. 229.
11. In 1913 and 1914, quoted in Travers, "Technology, Tactics, and Morale," p. 275.

. . . We also must follow this example."[12] Even in Belgium the offensive found proponents: under the influence of French ideas, some Belgian officers favored an offensive strategy, proposing the remarkable argument that "To ensure against our being ignored it was essential that we should attack," and declaring that "We must hit them where it hurts."[13]

Mythical or mystical arguments obscured the technical dominion of the defense, giving this faith in the offense aspects of a cult, or a mystique, as Marshall Joffre remarked in his memoirs.[14] For instance, Foch mistakenly argued that the machine gun actually strengthened the offense: "Any improvement of firearms is ultimately bound to add strength to the offensive. . . . Nothing is easier than to give a mathematical demonstration of that truth." If two thousand men attacked one thousand, each man in both groups firing his rifle once a minute, he explained, the "balance in favor of the attack" was one thousand bullets per minute. But if both sides could fire ten times per minute, the "balance in favor of the attacker" would increase to ten thousand, giving the attack the overall advantage.[15] With equally forced logic, Bernhardi wrote that the larger the army the longer defensive measures would take to execute, owing to "the difficulty of moving masses"; hence, he argued, as armies grew, so would the relative power of the offense.[16]

British and French officers suggested that superior morale on the attacking side could overcome superior defensive firepower, and that this superiority in morale could be achieved simply by assuming the role of attacker, since offense was a morale-building activity. One French officer contended that "the offensive doubles the energy of the troops" and "concentrates the thoughts of the commander on a single objective,"[17] while British officers declared that "Modern [war] conditions have enormously increased the value of moral quality," and "the moral attributes [are] the primary causes of all great success."[18] In short, mind would prevail over matter; morale would triumph over machine guns.

---

12. In 1909, quoted in D.C.B. Lieven, *Russia and the Origins of the First World War* (New York: St. Martin's Press, 1983), p. 113.
13. See Tuchman, *Guns of August,* pp. 127–131.
14. Marshall Joffre, *Mémoires du Maréchel Joffre* (Paris: Librarie Plon, 1932), p. 33. Joffre speaks of "le culte de l'offensive" and "d'une 'mystique de l'offensive'" of "le caractère un peu irraisonné."
15. Ropp, *War in the Modern World,* p. 218.
16. Ibid., p. 203. See also Bernhardi, *How Germany Makes War,* p. 154.
17. Captain Georges Gilbert, quoted in Snyder, "Defending the Offensive," pp. 80–81.
18. The *Field Service Regulations* of 1909 and Colonel Kiggell, quoted in Travers, "Technology, Tactics, and Morale," pp. 273, 276–277.
Even when European officers recognized the new tactical power of the defense, they often

Europeans also tended to discount the power of political factors which would favor defenders. Many Germans believed that "bandwagoning" with a powerful state rather than "balancing" against it was the guiding principle in international alliance-formation.[19] Aggressors would gather momentum as they gained power, because opponents would be intimidated into acquiescence and neutrals would rally to the stronger side. Such thinking led German Chancellor Bethmann Hollweg to hope that "Germany's growing strength . . . might force England to realize that [the balance of power] principle had become untenable and impracticable and to opt for a peaceful settlement with Germany,"[20] and German Secretary of State Gottlieb von Jagow to forecast British neutrality in a future European war: "We have not built our fleet in vain," and "people in England will seriously ask themselves whether it will be just that simple and without danger to play the role of France's guardian angel against us."[21] German leaders also thought they might frighten Belgium into surrender: during the July crisis Moltke was "counting on the possibility of being able to come to an understanding [with Belgium] when the Belgian Government realizes the seriousness of the situation."[22] This ill-founded belief in bandwagoning reinforced the general belief that conquest was relatively easy.

The belief in easy conquest eventually pervaded public images of international politics, manifesting itself most prominently in the widespread application of Darwinist notions to international affairs. In this image, states competed in a decisive struggle for survival which weeded out the weak and ended in the triumph of stronger states and races—an image which assumed a powerful offense. "In the struggle between nationalities," wrote former

---

resisted the conclusion that the defender would also hold the strategic advantage. Thus Bernhardi wrote that while "the defense as a form of fighting is stronger than the attack," it remained true that "in the conduct of war as a whole the offensive mode of action is by far superior to the defensive mode, especially under modern conditions." Bernhardi, *How Germany Makes War,* p. 155. See also Snyder, "Defending the Offensive," pp. 152–154, 253–254; and Travers, "Technology, Tactics, and Morale," passim.

19. On these concepts, see Kenneth N. Waltz, *Theory of International Politics* (Reading, Mass.: Addison–Wesley, 1979), pp. 125–127; and Stephen M. Walt, "The Origins of Alliances" (Ph.D. dissertation, University of California, Berkeley, 1983).

20. December 2, 1914, quoted in Fritz Fischer, *War of Illusions: German Policies from 1911 to 1914,* trans. Marian Jackson, with a Foreword by Alan Bullock (New York: W.W. Norton, 1975), p. 69.

21. February 1914, quoted in Geiss, *July 1914,* p. 25. For more examples, see Fischer, *War of Illusions,* pp. 133, 227; and Wayne C. Thompson, *In the Eye of the Storm: Kurt Riezler and the Crises of Modern Germany* (Iowa City: University of Iowa Press, 1980), p. 120.

22. August 3, quoted in Bernadotte E. Schmitt, *The Coming of the War: 1914,* 2 vols. (New York: Charles Scribner's Sons, 1930), Vol. 2, p. 390n.

German Chancellor Bernhard von Bülow, "one nation is the hammer and the other the anvil; one is the victor and the other the vanquished. . . . it is a law of life and development in history that where two national civilisations meet they fight for ascendancy."[23] A writer in the London *Saturday Review* portrayed the Anglo–German competition as "the first great racial struggle of the future: here are two growing nations pressing against each other . . . all over the world. One or the other has to go; one or the other will go."[24] This Darwinist foreign policy thought reflected and rested upon the implicit assumption that the offense was strong, since "grow or die" dynamics would be impeded in a defense-dominant world where growth could be stopped and death prevented by self-defense.

CONSEQUENCES OF OFFENSE-DOMINANCE

Recent theoretical writing in international relations emphasizes the dangers that arise when the offense is strong relative to the defense.[25] If the theory outlined in these writings is valid, it follows that the cult of the offensive was a reason for the outbreak of the war.

Five major dangers relevant to the 1914 case may develop when the offense is strong, according to this recent writing. First, states adopt more aggressive

23. Prince Bernhard von Bülow, *Imperial Germany*, trans. Marie A. Lewenz (New York: Dodd, Mead & Co., 1915), p. 291. On international social Darwinism, see also H.W. Koch, "Social Imperialism as a Factor in the 'New Imperialism,'" in H.W. Koch, ed., *The Origins of the First World War* (London: Macmillan, 1972), pp. 329–354.

24. Joachim Remak, *The Origins of World War I, 1871–1914* (Hinsdale, Ill.: Dryden Press, 1967), p. 85. Likewise the British Colonial Secretary, Joseph Chamberlain, declared that "the tendency of the time is to throw all power into the hands of the greater empires," while the "minor kingdoms" seemed "destined to fall into a secondary and subordinate place. . . ." In 1897, quoted in Fischer, *War of Illusions*, p. 35.

25. See Robert Jervis's pathbreaking article, "Cooperation under the Security Dilemma," *World Politics*, Vol. 30, No. 2 (January 1978), pp. 167–214; and Chapter 3 of my "Causes of War." Also relevant are George H. Quester, *Offense and Defense in the International System* (New York: John Wiley & Sons, 1977); John Herz, "Idealist Internationalism and the Security Dilemma," *World Politics*, Vol. 2, No. 2 (January 1950), pp. 157, 163; and Herbert Butterfield, *History and Human Relations* (London: Collins, 1950), pp. 19–20. Applications and elaborations include: Shai Feldman, *Israeli Nuclear Deterrence* (New York: Columbia University Press, 1982); idem, "Superpower Security Guarantees in the 1980's," in *Third World Conflict and International Security, Part II*, Adelphi Paper No. 167 (London: International Institute for Strategic Studies, 1981), pp. 34–44; Barry R. Posen, "Inadvertent Nuclear War? Escalation and NATO's Northern Flank," *International Security*, Vol. 7, No. 2 (Fall 1982), pp. 28–54; Jack Lewis Snyder, "Perceptions of the Security Dilemma in 1914," in Robert Jervis and Richard Ned Lebow, eds., *Perceptions and Deterrence*, forthcoming in 1985; and Kenneth N. Waltz, *The Spread of Nuclear Weapons: More May Be Better*, Adelphi Paper No. 171 (London: International Institute for Strategic Studies, 1981). Of related interest is John J. Mearsheimer, *Conventional Deterrence* (Ithaca: Cornell University Press, 1983).

foreign policies, both to exploit new opportunities and to avert new dangers which appear when the offense is strong. Expansion is more tempting, because the cost of aggression declines when the offense has the advantage. States are also driven to expand by the need to control assets and create the conditions they require to secure themselves against aggressors, because security becomes a scarcer asset. Alliances widen and tighten as states grow more dependent on one another for security, a circumstance which fosters the spreading of local conflicts. Moreover, each state is more likely to be menaced by aggressive neighbors who are governed by the same logic, creating an even more competitive atmosphere and giving states further reason to seek security in alliances and expansion.

Second, the size of the advantage accruing to the side mobilizing or striking first increases, raising the risk of preemptive war.[26] When the offense is strong, smaller shifts in ratios of forces between states create greater shifts in their relative capacity to conquer territory. As a result states have greater incentive to mobilize first or strike first, if they can change the force ratio in their favor by doing so. This incentive leads states to mobilize or attack to

---

26. In a "preemptive" war, either side gains by moving first; hence, one side moves to exploit the advantage of moving first, or to prevent the other side from doing so. By contrast, in a "preventive" war, one side foresees an adverse shift in the balance of power, and attacks to avoid a more difficult fight later.

"Moving first" in a preemptive war can consist of striking first *or mobilizing* first, if mobilization sets in train events which cause war, as in 1914. Thus a war is preemptive if statesmen attack because they believe that it pays to strike first; or if they mobilize because they believe that it pays to mobilize first, even if they do not also believe that it pays to strike first, if mobilizations open "windows" which spur attacks for "preventive" reasons, or if they produce other effects which cause war. Under such circumstances war is caused by preemptive actions which are not acts of war, but which are their equivalent since they produce conditions which cause war.

A preemptive war could also involve an attack by one side and mobilization by the other—for instance, one side might mobilize to forestall an attack, or might attack to forestall a mobilization, as the Germans apparently attacked Liège to forestall Belgian preparations to defend it (see below). Thus four classes of preemption are possible: an attack to forestall an attack, an attack to forestall a mobilization, a mobilization to forestall an attack, or a mobilization to forestall a mobilization (such as the Russian mobilizations in 1914).

The size of the incentive to preempt is a function of three factors: the degree of secrecy with which each side could mobilize its forces or mount an attack; the change in the ratio of forces which a secret mobilization or attack would produce; and the size and value of the additional territory which this changed ratio would allow the attacker to conquer or defend. If secret action is impossible, or if it would not change force ratios in favor of the side moving first, or if changes in force ratios would not change relative ability to conquer territory, then there is no first-strike or first-mobilization advantage. Otherwise, states have some inducement to move first.

On preemption, see Thomas C. Schelling, *Arms and Influence* (New Haven: Yale University Press, 1966), pp. 221–259; and idem, *Strategy of Conflict* (New York: Oxford University Press, 1963), pp. 207–254.

seize the initiative or deny it to adversaries, and to conceal plans, demands, and grievances to avoid setting off such a strike by their enemies, with deleterious effects on diplomacy.

Third, "windows" of opportunity and vulnerability open wider, forcing faster diplomacy and raising the risk of preventive war. Since smaller shifts in force ratios have larger effects on relative capacity to conquer territory, smaller prospective shifts in force ratios cause greater hope and alarm, open bigger windows of opportunity and vulnerability, and enhance the attractiveness of exploiting a window by launching a preventive attack.

Fourth, states adopt more competitive styles of diplomacy—brinkmanship and presenting opponents with *faits accomplis,* for instance—since the gains promised by such tactics can more easily justify the risks they entail. At the same time, however, the risks of adopting such strategies also increase, because they tend to threaten the vital interests of other states more directly. Because the security of states is more precarious and more tightly interdependent, threatening actions force stronger and faster reactions, and the political ripple effects of *faits accomplis* are larger and harder to control.

Fifth, states enforce tighter political and military secrecy, since national security is threatened more directly if enemies win the contest for information. As with all security assets, the marginal utility of information is magnified when the offense is strong; hence states compete harder to gain the advantage and avoid the disadvantage of disclosure, leading states to conceal their political and military planning and decision-making more carefully.

The following section suggests that many of the proximate causes of the war of 1914 represent various guises of these consequences of offense-dominance: either they were generated or exacerbated by the assumption that the offense was strong, or their effects were rendered more dangerous by this assumption. These causes include: German and Austrian expansionism; the belief that the side which mobilized or struck first would have the advantage; the German and Austrian belief that they faced "windows of vulnerability"; the nature and inflexibility of the Russian and German war plans and the tight nature of the European alliance system, both of which spread the war from the Balkans to the rest of Europe; the imperative that "mobilization meant war" for Germany; the failure of Britain to take effective measures to deter Germany; the uncommon number of blunders and mistakes committed by statesmen during the July crisis; and the ability of the Central powers to evade blame for the war. Without the cult of the offensive these problems probably would have been less acute, and their effects would have posed smaller risks. Thus the cult of the offensive was a mainspring driving many of the mechanisms which brought about the First World War.

The Cult of the Offensive and the Causes of the War

GERMAN EXPANSION AND ENTENTE RESISTANCE

Before 1914 Germany sought a wider sphere of influence or empire, and the war grew largely from the political collision between expansionist Germany and a resistant Europe. Germans differed on whether their empire should be formal or informal, whether they should seek it in Europe or overseas, and whether they should try to acquire it peacefully or by violence, but a broad consensus favored expansion of some kind. The logic behind this expansionism, in turn, rested on two widespread beliefs which reflected the cult of the offensive: first, that German security required a wider empire; and second, that such an empire was readily attainable, either by coercion or conquest. Thus German expansionism reflected the assumption that conquest would be easy both for Germany and for its enemies.

Prewar statements by German leaders and intellectuals reflected a pervasive belief that German independence was threatened unless Germany won changes in the status quo. Kaiser Wilhelm foresaw a "battle of Germans against the Russo–Gauls for their very existence," which would decide "the existence or non-existence of the Germanic race in Europe,"[27] declaring: "The question for Germany is to be or not to be."[28] His Chancellor, Bethmann Hollweg, wondered aloud if there were any purpose in planting new trees at his estate at Hohenfinow, near Berlin, since "in a few years the Russians would be here anyway."[29] The historian Heinrich von Treitschke forecast that "in the long run the small states of central Europe can not maintain themselves,"[30] while other Germans warned, "If Germany does not rule the world . . . it will disappear from the map; it is a question of either or," and "Germany will be a world power or nothing."[31] Similarly, German military officers predicted that "without colonial possessions [Germany] will suffocate in her small territory or else will be crushed by the great world powers" and

---

27. In 1912, quoted in Thompson, *Eye of the Storm*, p. 42.
28. In 1912, quoted in Fischer, *War of Illusions*, p. 161.
29. V.R. Berghahn, *Germany and the Approach of War in 1914* (London: Macmillan, 1973), p. 186.
30. In 1897, quoted in Notestein and Stoll, *Conquest and Kultur*, p. 21.
31. Houston Chamberlain and Ernest Hasse, quoted in Fischer, *War of Illusions*, pp. 30, 36.

foresaw a "supreme struggle, in which the existence of Germany will be at stake. . . ."[32]

Germans also widely believed that expansion could solve their insecurity: "Room; they must make room. The western and southern Slavs—or we! . . . Only by growth can a people save itself."[33] German expansionists complained that German borders were constricted and indefensible, picturing a Germany "badly protected by its unfavorable geographic frontiers. . . ."[34] Expansion was the suggested remedy: "Our frontiers are too narrow. We must become land-hungry, must acquire new regions for settlement. . . ."[35] Expanded borders would provide more defensible frontiers and new areas for settlement and economic growth, which in turn would strengthen the German race against its competitors: "the continental expansion of German territory [and] the multiplication on the continent of the German peasantry . . . would form a sure barrier against the advance of our enemies. . . ."[36] Such utterances came chiefly from the hawkish end of the German political spectrum, but they reflected widely held assumptions.

Many Germans also failed to see the military and political obstacles to expansion. The Kaiser told departing troops in early August, "You will be home before the leaves have fallen from the trees,"[37] and one of his generals predicted that the German army would sweep through Europe like a bus full of tourists: "In two weeks we shall defeat France, then we shall turn round, defeat Russia and then we shall march to the Balkans and establish order there."[38] During the July crisis a British observer noted the mood of "supreme confidence" in Berlin military circles, and a German observer reported that the German General Staff "looks ahead to war with France with great confidence, expects to defeat France within four weeks. . . ."[39] While some

---

32. *Nauticus,* in 1900, quoted in Berghahn, *Germany and the Approach of War in 1914,* p. 29; and Colmar von der Goltz, quoted in Notestein and Stoll, *Conquest and Kultur,* p. 119.
33. Otto Richard Tannenberg, in 1911, quoted in Notestein and Stoll, *Conquest and Kultur,* p. 53.
34. Crown Prince Wilhelm, in 1913, quoted in ibid., p. 44. Likewise Walter Rathenau complained of German "frontiers which are too long and devoid of natural protection, surrounded and hemmed in by rivals, with a short coastline. . . ." In July 1914, quoted in Fischer, *War of Illusions,* p. 450.
35. Hermann Vietinghoff-Scheel, in 1912, quoted in William Archer, ed., *501 Gems of German Thought* (London: T. Fisher Unwin, 1916), p. 46.
36. Albrecht Wirth, in 1901, quoted in Notestein and Stoll, *Conquest and Kultur,* p. 52.
37. Quoted in Tuchman, *Guns of August,* p. 142.
38. Von Loebell, quoted in Fischer, *War of Illusions,* p. 543.
39. The English Military Attaché, quoted in Luigi Albertini, *The Origins of the War of 1914,* 3 vols., trans. and ed. Isabella M. Massey (London: Oxford University Press, 1952–57; reprint ed.,

German military planners recognized the tactical advantage which defenders would hold on the battlefield, most German officers and civilians believed they could win a spectacular, decisive victory if they struck at the right moment.

Bandwagon logic fed hopes that British and Belgian opposition to German expansion could be overcome. General Moltke believed that "Britain is peace loving" because in an Anglo–German war "Britain will lose its domination at sea which will pass forever to America"[40]; hence Britain would be intimidated into neutrality. Furthermore, he warned the Belgians, "Small countries, such as Belgium, would be well advised to rally to the side of the strong if they wished to retain their independence," expecting Belgium to follow this advice if Germany applied enough pressure.[41]

Victory, moreover, would be decisive and final. In Bülow's words, a defeat could render Russia "incapable of attacking us for at least a generation" and "unable to stand up for twenty-five years," leaving it "lastingly weakened,"[42] while Bernhardi proposed that France "must be annihilated once and for all as a great power."[43]

Thus, as Robert Jervis notes: "Because of the perceived advantage of the offense, war was seen as the best route both to gaining expansion and to avoiding drastic loss of influence. There seemed to be no way for Germany merely to retain and safeguard her existing position."[44] The presumed power of the offense made empire appear both feasible and necessary. Had Germans recognized the real power of the defense, the notion of gaining wider empire would have lost both its urgency and its plausibility.

Security was not Germany's only concern, nor was it always a genuine one. In Germany, as elsewhere, security sometimes served as a pretext for expansion undertaken for other reasons. Thus proponents of the "social imperialism" theory of German expansion note that German elites endorsed imperialism, often using security arguments, partly to strengthen their do-

---

Westport, Conn.: Greenwood Press, 1980), Vol. 3, p. 171; and Lerchenfeld, the Bavarian ambassador in Berlin, quoted in Fischer, *War of Illusions*, p. 503.
40. In 1913, quoted in Fischer, *War of Illusions*, p. 227.
41. In 1913, quoted in Albertini, *Origins of the War*, Vol. 3, p. 441. See also Bernhardi's dismissal of the balance of power, in Friedrich von Bernhardi, *Germany and the Next War*, trans. Allen H. Powles (New York: Longmans, Green & Co., 1914), p. 21.
42. In 1887, quoted in Fischer, *War of Illusions*, p. 45.
43. In 1911, quoted in Tuchman, *Guns of August*, p. 26.
44. Jervis, "Cooperation under the Security Dilemma," p. 191.

mestic political and social position.[45] Likewise, spokesmen for the German military establishment exaggerated the threat to Germany and the benefits of empire for organizationally self-serving reasons. Indeed, members of the German elite sometimes privately acknowledged that Germany was under less threat than the public was being told. For example, the Secretary of State in the Foreign Office, Kiderlen-Wächter, admitted, "If we do not conjure up a war into being, no one else certainly will do so," since "The Republican government of France is certainly peace-minded. The British do not want war. They will never give cause for it. . . ."[46]

Nevertheless, the German public believed that German security was precarious, and security arguments formed the core of the public case for expansion. Moreover, these arguments proved persuasive, and the chauvinist public climate which they created enabled the elite to pursue expansion, whatever elite motivation might actually have been. Indeed, some members of the German government eventually felt pushed into reckless action by an extreme chauvinist public opinion which they felt powerless to resist. Admiral von Müller later explained that Germany pursued a bellicose policy during the July crisis because "The government, already weakened by domestic disunity, found itself inevitably under pressure from a great part of the German people which had been whipped into a high-grade chauvinism by Navalists and Pan-Germans."[47] Bethmann Hollweg felt his hands tied by an expansionist public climate: "With these idiots [the Pan-Germans] one cannot conduct a foreign policy—on the contrary. Together with other factors they will eventually make any reasonable course impossible for us."[48] Thus the search for security was a fundamental cause of German conduct, whether or not the elite was motivated by security concerns, because the elite was

45. Examples are: Arno Mayer, "Domestic Causes of the First World War," in Leonard Krieger and Fritz Stern, eds., *The Responsibility of Power* (New York: Macmillan, 1968), pp. 286–300; Berghahn, *Germany and the Approach of War;* Fischer, *War of Illusions,* pp. 257–258; and Imanuel Geiss, *German Foreign Policy, 1871–1914* (Boston: Routledge & Kegan Paul, 1976). A criticism is Marc Trachtenberg, "The Social Interpretation of Foreign Policy," *Review of Politics,* Vol. 40, No. 3 (July 1978), pp. 341–350.
46. In 1910, quoted in Geiss, *German Foreign Policy,* p. 126.
47. Admiral von Müller, quoted in Fritz Stern, *The Failure of Illiberalism* (London: Allen & Unwin, 1972), p. 94.
48. In 1909, quoted in Konrad H. Jarausch, *The Enigmatic Chancellor: Bethmann Hollweg and the Hubris of Imperial Germany* (New Haven: Yale University Press, 1973), p. 119. See also ibid., p. 152; and Geiss, *German Foreign Policy,* pp. 135–137. As Jules Cambon, French ambassador to Germany, perceptively remarked: "It is false that in Germany the nation is peaceful and the government bellicose—the exact opposite is true." In 1911, quoted in Jarausch, *Enigmatic Chancellor,* p. 125.

allowed or even compelled to adopt expansionist policies by a German public which found security arguments persuasive.

The same mixture of insecurity and perceived opportunity stiffened resistance to German expansion and fuelled a milder expansionism elsewhere in Europe, intensifying the conflict between Germany and its neighbors. In France the nationalist revival and French endorsement of a firm Russian policy in the Balkans were inspired partly by a growing fear of the German threat after 1911,[49] partly by an associated concern that Austrian expansion in the Balkans could shift the European balance of power in favor of the Central Powers and thereby threaten French security, and partly by belief that a war could create opportunities for French expansion. The stiffer French "new attitude" on Balkan questions in 1912 was ascribed to the French belief that "a territorial acquisition on the part of Austria would affect the general balance of power in Europe and as a result touch the particular interests of France"—a belief which assumed that the power balance was relatively precarious, which in turn assumed a world of relatively strong offense.[50] At the same time some Frenchmen looked forward to "a beautiful war which will deliver all the captives of Germanism,"[51] inspired by a faith in the power of the offensive that was typified by the enthusiasm of Joffre's deputy, General de Castelnau: "Give me 700,000 men and I will conquer Europe!"[52]

Russian policy in the Balkans was driven both by fear that Austrian expansion could threaten Russian security and by hopes that Russia could destroy its enemies if war developed under the right conditions. Sazonov saw a German–Austrian Balkan program to "deliver the Slavonic East, bound hand and foot, into the power of Austria–Hungary," followed by the German seizure of Constantinople, which would gravely threaten Russian security by placing all of Southern Russia at the mercy of German power.[53] Eventually a "German Khalifate" would be established, "extending from the banks of the Rhine to the mouth of the Tigris and Euphrates," which would reduce

---

49. See Eugen Weber, *The Nationalist Revival in France, 1905–1914* (Berkeley and Los Angeles: University of California Press, 1968), passim; and Snyder, "Defending the Offensive," pp. 32–33.
50. By the Russian ambassador to Paris, A.P. Izvolsky, quoted in Schmitt, *Coming of the War,* Vol. 1, p. 21.
51. *La France Militaire,* in 1913, quoted in Weber, *Nationalist Revival in France,* p. 127.
52. In 1913, quoted in L.C.F. Turner, *Origins of the First World War* (London: Edward Arnold, 1970), p. 53.
53. Serge Sazonov, *Fateful Years, 1909–1916* (London: Jonathan Cape, 1928), p. 179. See also Schmitt, *Coming of the War,* Vol. 1, p. 87.

"Russia to a pitiful dependence upon the arbitrary will of the Central Powers."[54] At the same time some Russians believed these threats could be addressed by offensive action: Russian leaders spoke of the day when "the moment for the downfall of Austria–Hungary arrives,"[55] and the occasion when "The Austro-Hungarian ulcer, which today is not yet so ripe as the Turkish, may be cut up."[56] Russian military officers contended that "the Austrian army represents a serious force. . . . But on the occasion of the first great defeats all of this multi-national and artificially united mass ought to disintegrate."[57]

In short, the belief that conquest was easy and security scarce was an important source of German–Entente conflict. Without it, both sides could have adopted less aggressive and more accommodative policies.

THE INCENTIVE TO PREEMPT

American strategists have long assumed that World War I was a preemptive war, but they have not clarified whether or how this was true.[58] Hence two questions should be resolved to assess the consequences of the cult of the offensive: did the states of Europe perceive an incentive to move first in 1914, which helped spur them to mobilize or attack? If so, did the cult of the offensive help to give rise to this perception?

The question of whether the war was preemptive reduces to the question of why five principal actions in the July crisis were taken. These actions are: the Russian preliminary mobilization ordered on July 25–26; the partial Russian mobilization against Austria–Hungary ordered on July 29; the Russian

54. Sazonov, *Fateful Years*, pp. 191, 204.
55. Izvolsky, in 1909, quoted in Schmitt, *Coming of the War*, Vol. 1, p. 129.
56. Sazonov, in 1913, quoted in ibid., p. 135.
57. *Sbornik glavnogo upravleniia general'nogo shtaba*, the secret magazine of the Russian general staff, in 1913, quoted in William C. Fuller, "The Russian Empire and Its Potential Enemies" (manuscript, 1980), p. 21.

British resistance was also driven by security concerns: during the July crisis the London *Times* warned that "the ruin of France or the Low Countries would be the prelude to our own," while other interventionists warned that Antwerp in German hands would be a "pistol pointed at the heart of England," and that the German threat to France and the Low Countries created "a deadly peril for ourselves." The *Times* on August 4, quoted in Geoffrey Marcus, *Before the Lamps Went Out* (Boston: Little, Brown, 1965), p. 305; and the *Pall Mall Gazette* and James Gavin, on July 29 and August 2, quoted in ibid., pp. 243, 268.

58. Suggesting that World War I was preemptive are: Herman Kahn, *On Thermonuclear War*, 2nd ed. (New York: The Free Press, 1969), pp. 359–362; Schelling, *Arms and Influence*, pp. 223–224; Jervis, "Cooperation under the Security Dilemma," pp. 191–192; Quester, *Offense and Defense*, pp. 110–111; Richard Ned Lebow, *Between Peace and War: The Nature of International Crisis* (Baltimore, Md.: The Johns Hopkins University Press, 1981), pp. 238–242.

full mobilization ordered on July 30; French preliminary mobilization measures ordered during July 25–30; and the German attack on the Belgian fortress at Liège at the beginning of the war. The war was preemptive if Russia and France mobilized preemptively, since these mobilizations spurred German and Austrian mobilization, opening windows which helped cause war. Thus while the mobilizations were not acts of war, they caused effects which caused war. The war was also preemptive if Germany struck Liège preemptively, since the imperative to strike Liège was one reason why "mobilization meant war" to Germany.

The motives for these acts cannot be determined with finality; testimony by the actors is spotty and other direct evidence is scarce. Instead, motives must be surmised from preexisting beliefs, deduced from circumstances, and inferred from clues which may by themselves be inconclusive. However, three pieces of evidence suggest that important preemptive incentives existed, and helped to shape conduct. First, most European leaders apparently believed that mobilization by either side which was not answered within a very few days, or even hours, could affect the outcome of the war. This judgment is reflected both in the length of time which officials assumed would constitute a militarily significant delay between mobilization and offsetting counter-mobilization, and in the severity of the consequences which they assumed would follow if they mobilized later than their opponents.

Second, many officials apparently assumed that significant mobilization measures and preparations to attack could be kept secret for a brief but significant period. Since most officials also believed that a brief unanswered mobilization could be decisive, they concluded that the side which mobilized first would have the upper hand.

Third, governments carried out some of their mobilization measures in secrecy, suggesting that they believed secret measures were feasible and worthwhile.

THE PERCEIVED SIGNIFICANCE OF SHORT DELAYS. Before and during the July crisis European leaders used language suggesting that they believed a lead in ordering mobilization of roughly one to three days would be significant. In Austria, General Conrad believed that "every day was of far-reaching importance," since "any delay might leave the [Austrian] forces now assembling in Galicia open to being struck by the full weight of a Russian offensive in the midst of their deployment."[59] In France, Marshall Joffre warned the

59. July 29, quoted in Albertini, *Origins*, Vol. 2, p. 670.

French cabinet that "any delay of twenty-four hours in calling up our reservists" once German preparations began would cost France "ten to twelve miles for each day of delay; in other words, the initial abandonment of much of our territory."[60] In Britain, one official believed that France "cannot possibly delay her own mobilization for even the fraction of a day" once Germany began to mobilize.[61]

In Germany, one analyst wrote that "A delay of a single day . . . can scarcely ever be rectified."[62] Likewise Moltke, on receiving reports of preparations in France and Russia during the July crisis, warned that "the military situation is becoming from day to day more unfavorable for us," and would "lead to fateful consequences for us" if Germany did not respond.[63] On July 30 he encouraged Austria to mobilize, warning that "every hour of delay makes the situation worse, for Russia gains a start."[64] On August 1, the Prussian ministry of war was reportedly "very indignant over the day lost for the mobilization" by the German failure to mobilize on July 30.[65] The German press drove home the point that if mobilization by the adversary went unanswered even briefly, the result could be fatal, one German newspaper warning that "Every delay [in mobilizing] would cost us an endless amount of blood" if Germany's enemies gained the initiative; hence "it would be disastrous if we let ourselves be moved by words not to carry on our preparations so quickly. . . ."[66]

---

60. July 29, from Marshall Joffre, *The Personal Memoirs of Marshall Joffre*, 2 vols., trans. T. Bentley Mott (New York: Harper & Brothers, 1932), Vol. 1, p. 125.
61. Eyre Crowe, on July 27, quoted in Geiss, *July 1914*, p. 251.
62. Kraft zu Hohenlohe-Ingelfingen, in 1898, quoted in Ropp, *War in the Modern World*, p. 203.
63. To Bethmann Hollweg, on July 29, quoted in Geiss, *July 1914*, p. 284.
64. Quoted in Schmitt, *Coming of the War*, Vol. 2, p. 196.
65. Ibid., p. 265n.
66. The *Reinisch-Westfälische Zeitung*, July 31, quoted in Jonathan French Scott, *The Five Weeks* (New York: John Day Co., 1927), p. 146.

Likewise after the war General von Kluck, who commanded the right wing of the German army in the march on Paris, claimed that if the German army had been mobilized and deployed "three days earlier, a more sweeping victory and decisive result would probably have been gained" against France, and Admiral Tirpitz complained that German diplomats had given Britain and Belgium several crucial days warning of the German attack on July 29, which "had an extraordinarily unfavorable influence on the whole course of the war." A delay of "only a few days" in the preparation of the British expeditionary force "might have been of the greatest importance to us." Schmitt, *Coming of the War*, Vol. 2, p. 148n.; and Albertini, *Origins*, Vol. 3, p. 242n.

A more relaxed opinion was expressed by the Prussian war minister, General Falkenhayn, who seemed to feel that it would be acceptable if German mobilization "follows two or three days later than the Russian and Austrian," since it "will still be completed more quickly than theirs." Schmitt, *Coming of the War*, Vol. 2, p. 147. However, he also expressed himself in favor

Thus time was measured in small units: "three days," "day to day," "a single day," "the fraction of a day," or even "every hour." Moreover, the consequences of conceding the initiative to the adversary were thought to be extreme. The Russian Minister of Agriculture, Alexander Krivoshein, warned that if Russia delayed its mobilization "we should be marching toward a certain catastrophe,"[67] and General Janushkevich warned the Russian foreign minister that "we were in danger of losing [the war] before we had time to unsheath our sword" by failing to mobilize promptly against Germany.[68] General Joffre feared that France would find itself "in an irreparable state of inferiority" if it were outstripped by German mobilization.[69] And in Germany, officials foresaw dire consequences if Germany conceded the initiative either in the East or the West. Bethmann Hollweg explained to one of his ambassadors that if German mobilization failed to keep pace with the Russian, Germany would suffer large territorial losses: "East Prussia, West Prussia, and perhaps also Posen and Silesia [would be] at the mercy of the Russians."[70] Such inaction would be "a crime against the safety of our fatherland."[71]

Germans also placed a high value on gaining the initiative at Liège, since Liège controlled a vital Belgian railroad junction, and German forces could not seize Liège with its tunnels and bridges intact unless they surprised the Belgians. As Moltke wrote before the war, the advance through Belgium "will hardly be possible unless Liège is in our hands . . . the possession of Liège is the *sine qua non* of our advance." But seizing Liège would require "meticulous preparation and surprise" and "is only possible if the attack is made at once, before the areas between the forts are fortified," "immediately" after the declaration of war.[72] In short, the entire German war plan would be ruined if Germany allowed Belgium to prepare the defense of Liège.

This belief that brief unanswered preparations and actions could be decisive reflected the implicit assumption that the offense had the advantage. Late mobilization would cost Germany control of East and West Prussia only

---

of preemption at other junctures. See ibid., p. 297; and Berghahn, *Germany and the Approach of War*, p. 203.

67. To Sazonov, July 30, quoted in Geiss, *July 1914*, p. 311.
68. To Sazonov, July 30, quoted in Albertini, *Origins*, Vol. 2, p. 566.
69. August 1, Poincaré reporting Joffre's view, quoted in Albertini, *Origins*, Vol. 3, p. 100.
70. August 1, quoted in Schmitt, *Coming of the War*, Vol. 2, p. 264.
71. August 1, quoted in Albertini, *Origins*, Vol. 3, p. 167.
72. Ritter, *The Schlieffen Plan*, p. 166. On the Liège attack, see also Snyder, "Defending the Offensive," pp. 203, 285–287.

if Russian offensive power were strong, and German defensive power were weak; mobilizing late could only be a "crime against the safety" of Germany if numerically superior enemies could destroy it; lateness could only confront Russia with "certain catastrophe" or leave it in danger of "losing before we have time to unsheath our sword" if Germany could develop a powerful offensive with the material advantage it would gain by preparing first; and lateness could only condemn France to "irreparable inferiority" if small material inferiority translated into large territorial losses. Had statesmen understood that in reality the defense had the advantage, they also would have known that the possession of the initiative could not be decisive, and could have conceded it more easily.

WAS SECRET PREPARATION BELIEVED FEASIBLE? The belief that delay could be fatal would have created no impulse to go first had European leaders believed that they could detect and offset their opponents' preparations immediately. However, many officials believed that secret action for a short time was possible. Russian officials apparently lacked confidence in their own ability to detect German or Austrian mobilization, and their decisions to mobilize seem to have been motivated partly by the desire to forestall surprise preparation by their adversaries. Sazonov reportedly requested full mobilization on July 30 partly from fear that otherwise Germany would "gain time to complete her preparations in secret."[73] Sazonov offers confirmation in his memoirs, explaining that he had advised mobilization believing that "The perfection of the German military organization made it possible by means of personal notices to the reservists to accomplish a great part of the work quietly." Germany could then "complete the mobilization in a very short time. This circumstance gave a tremendous advantage to Germany, but we could counteract it to a certain extent by taking measures for our own mobilization in good time."[74]

Similar reasoning contributed to the Russian decision to mobilize against Austria on July 29. Sazonov explains that the mobilization was undertaken in part "so as to avoid the danger of being taken unawares by the Austrian

73. Paleologue's diary, quoted in Albertini, *Origins*, Vol. 2, p. 619.
74. Sazonov, *Fateful Years*, pp. 202–203. The memorandum of the day of the Russian foreign ministry for July 29 records that Russian officials had considered whether Germany seriously sought peace, or whether its diplomacy "was only intended to lull us to sleep and so to postpone the Russian mobilization and thus gain time wherein to make corresponding preparations." Quoted in Geiss, *July 1914*, pp. 296–297.

preparations."[75] Moreover, recent experience had fuelled Russian fears of an Austrian surprise: during the Balkan crisis of 1912, the Russian army had been horrified to discover that Austria had secretly mobilized in Galicia, without detection by Russian intelligence; and this experience resolved the Russian command not to be caught napping again. In one observer's opinion, "the experience of 1912 . . . was not without influence as regards Russia's unwillingness to put off her mobilization in the July days of 1914."[76]

Top Russian officials also apparently believed that Russia could itself mobilize secretly, and some historians ascribe the Russian decision to mobilize partly to this erroneous belief. Luigi Albertini writes that Sazonov did not realize that the mobilization order would be posted publicly and that, accordingly, he "thought Russia could mobilize without Germany's knowing of it immediately."[77] Albertini reports that the German ambassador caused "real stupefaction" by appearing at the Russian ministry for foreign affairs with a red mobilization poster on the morning of mobilization,[78] and concludes that the "belief that it was possible to proceed to general mobilization without making it public may well have made Sazonov more inclined to order it."[79]

Contemporary accounts confirm that the Russian leadership believed in their own ability to mobilize in secret. The memorandum of the Russian Ministry for Foreign Affairs records that Sazonov sought to "proceed to the general mobilization as far as possible secretly and without making any public announcement concerning it," in order "To avoid rendering more acute our relations with Germany."[80] And in informing his government of Russian preliminary mobilization measures which began on July 26, the French ambassador indicated Russian hopes that they could maintain secrecy: "Secret preparations will, however, commence already today,"[81] and "the military districts of Warsaw, Vilna and St. Petersburg are secretly making preparations."[82] His telegram informing Paris of Russian general mobilization ex-

75. Sazonov, *Fateful Years*, p. 188.
76. A.M. Zayonchovsky, quoted in Lieven, *Russia and the Origins of the First World War*, p. 149.
77. Albertini, *Origins*, Vol. 2, p. 624.
78. Ibid., quoting Taube who quoted Nolde.
79. Ibid., p. 573. See also p. 584, suggesting that "Sazonov was such a greenhorn in military matters as to imagine the thing could be done, and was only convinced of the contrary when on 31 July he saw the red notices, calling up reservists, posted up in the streets of St. Petersburg." This point "provides the key to many mysteries" (p. 624).
80. For July 31, in Geiss, *July 1914*, p. 326.
81. Paleologue, July 25, in Albertini, *Origins*, Vol. 2, p. 591.
82. Paleologue, July 26, in ibid., p. 592.

plained that "the Russian government has decided to proceed secretly to the first measures of general mobilization."[83]

Like their Russian counterparts, top French officials also apparently feared that Germany might mobilize in secret, which spurred the French to their own measures. Thus during the July crisis General Joffre spoke of "the concealments [of mobilization] which are possible in Germany,"[84] and referred to "information from excellent sources [which] led us to fear that on the Russian front a sort of secret mobilization was taking place [in Germany]."[85] In his memoirs, Joffre quotes a German military planning document acquired by the French government before the July crisis, which he apparently took to indicate German capabilities, and which suggested that Germany could take "quiet measures . . . in preparation for mobilization," including "a discreet assembly of complementary personnel and materiel" which would "assure us advantages very difficult for other armies to realize in the same degree."[86] The French ambassador to Berlin, Jules Cambon, also apparently believed that Germany could conduct preliminary mobilization measures in secret, became persuaded during the July crisis that it had in fact done this, and so informed Paris: "In view of German habits, [preliminary measures] can be taken without exciting the population or causing indiscretions to be committed. . . ."[87] For their part the Germans apparently did not believe that they or their enemies could mobilize secretly, but they did speak in terms suggesting that Germany could surprise the Belgians: German planners referred to the "*coup de main*" at Liège and the need for "meticulous preparation and surprise."[88]

To sum up, then, French policymakers feared that Germany could mobilize secretly; Russians feared secret mobilization by Germany or Austria, and hoped Russian mobilization could be secret; while Central Powers planners

---

83. Ibid., p. 620.
84. August 1, quoted in Joffre, *Personal Memoirs*, p. 128.
85. July 29, quoted in ibid., p. 120.
86. Ibid., p. 127.
87. Cambon dispatch to Paris, July 21, quoted in ibid., p. 119. Joffre records that Cambon's telegram, which mysteriously did not arrive in Paris until July 28, convinced him that "for seven days at least the Germans had been putting into effect the plan devised for periods of political tension and that our normal methods of investigation had not revealed this fact to us. Our adversaries could thus reach a condition of mobilization that was almost complete," reflecting Joffre's assumption that secret German measures were possible.
88. Moltke, quoted in Ritter, *The Schlieffen Plan*, p. 166.

saw less possibility for preemptive mobilization by either side, but hoped to mount a surprise attack on Belgium.[89]

DID STATESMEN ACT SECRETLY? During the July crisis European statesmen sometimes informed their opponents before they took military measures, but on other occasions they acted secretly, suggesting that they believed the initiative was both attainable and worth attaining, and indicating that the desire to seize the initiative may have entered into their decisions to mobilize. German leaders warned the French of their preliminary measures taken on July 29,[90] and their pre-mobilization and mobilization measures taken on July 31;[91] and they openly warned the Russians on July 29 that they would mobilize if Russia conducted a partial mobilization.[92] Russia openly warned Austria on July 27 that it would mobilize if Austria crossed the Serbian frontier,[93] and then on July 28 and July 29 openly announced to Germany and Austria its partial mobilization of July 29,[94] and France delayed full mobilization until after Germany had taken the onus on itself by issuing ultimata to Russia and France. However, Russia, France, and Germany tried

89. During the July crisis, adversaries actually detected signs of most major secret mobilization activity in roughly 6–18 hours, and took responsive decisions in 1–2 days. Accordingly, the maximum "first mobilization advantage" which a state could gain by forestalling an adversary who otherwise would have begun mobilizing first was roughly 2–4 days. Orders for Russian preliminary mobilization measures were issued in sequential telegrams transmitted between 4:00 p.m. on July 25 and 3:26 a.m. on July 26; Berlin received its first reports of these measures early on July 26; and at 4:00 p.m. on July 27 the German intelligence board concluded that Russian premobilization had in fact begun, for a lag of roughly one and one-half to two days between the issuance of orders and their definite detection. Sidney B. Fay, *The Origins of the World War*, 2 vols., 2nd ed. rev. (New York: Free Press, 1966), Vol. 2, pp. 310–315; and Ulrich Trumpener, "War Premeditated? German Intelligence Operations in July 1914," *Central European History*, Vol. 9 (1976), pp. 67–70. Full Russian mobilization was ordered at 6:00 p.m. on July 30, first rumors reached Berlin very late on July 30, more definite but inconclusive information was received around 7:00 a.m. July 31, reliable confirmation was received at 11:45 a.m., and German preliminary mobilization was ordered at 1:00 p.m., for a lag of roughly 20 hours. Fay, *Origins of the World War*, Vol. 2, p. 473; Schmitt, *Coming of the War*, Vol. 2, pp. 211–212, 262–265; and Trumpener, "War Premeditated?," pp. 80–83. French preliminary measures were begun on July 25, expanded on July 26, further expanded on July 27, and remained substantially undetected on July 28. Secondary sources do not clarify when Germany detected French preliminary measures, but it seems that German discovery lagged roughly two days behind French actions. Schmitt, *Coming of the War*, Vol. 2, pp. 17–19; Joffre, *Personal Memoirs*, pp. 115–118; and Trumpener, "War Premeditated?," pp. 71–73. As for Liège, it was not captured as quickly as German planners had hoped, but was not properly defended when the Germans arrived, and was taken in time to allow the advance into France.

90. Albertini, *Origins*, Vol. 2, p. 491.

91. Schmitt, *Coming of the War*, Vol. 2, pp. 267–268.

92. Ibid., p. 105.

93. Albertini, *Origins*, Vol. 2, p. 529.

94. Ibid., pp. 549, 551; and Geiss, *July 1914*, pp. 262, 278, 299.

to conceal four of the five major preemptive actions of the crisis: the Russians hid both their preliminary measures of July 25–26 and their general mobilization of July 30, the French attempted to conceal their preliminary mobilization measures of July 25–29, and the Germans took great care to conceal their planned *coup de main* against Liège. Thus states sometimes conceded the initiative, but sought it at critical junctures.

Overall, evidence suggests that European leaders saw some advantage to moving first in 1914: the lags which they believed significant lay in the same range as the lags they believed they could gain or forestall by mobilizing first. These perceptions probably helped spur French and Russian decisions to mobilize, which in turn helped set in train the German mobilization, which in turn meant war partly because the Germans were determined to preempt Liège. Hence the war was in some modest measure preemptive.

If so, the cult of the offensive bears some responsibility. Without it, statesmen would not have thought that secret mobilization or preemptive attack could be decisive. The cult was not the sole cause of the perceived incentive to preempt; rather, three causes acted together, the others being the belief that mobilization could briefly be conducted secretly, and the systems of reserve manpower mobilization which enabled armies to multiply their strength in two weeks. The cult had its effect by magnifying the importance of these other factors in the minds of statesmen, which magnified the incentive to preempt which these factors caused them to perceive. The danger that Germany might gain time to complete preparations in secret could only alarm France and Russia if Germany could follow up these preparations with an effective offensive; otherwise, early secret mobilization could *not* give "a tremendous advantage" to Germany, and such a prospect would not require a forestalling response. Sazonov could have been tempted to mobilize secretly only if early Russian mobilization would forestall important German gains, or could provide important gains for Russia, as could only have happened if the offense were powerful.

"WINDOWS" AND PREVENTIVE WAR

Germany and Austria pursued bellicose policies in 1914 partly to shut the looming "windows" of vulnerability which they envisioned lying ahead, and partly to exploit the brief window of opportunity which they thought the summer crisis opened. This window logic, in turn, grew partly from the cult of the offensive, since it depended upon the implicit assumption that the offense was strong. The shifts in the relative sizes of armies, economies, and alliances which fascinated and frightened statesmen in 1914 could have cast such a long shadow only in a world where material advantage promised decisive results in warfare, as it could only in an offense-dominant world.

The official communications of German leaders are filled with warnings that German power was in relative decline, and that Germany was doomed unless it took drastic action—such as provoking and winning a great crisis which would shatter the Entente, or directly instigating a "great liquidation" (as one general put it).[95] German officials repeatedly warned that Russian military power would expand rapidly between 1914 and 1917, as Russia carried out its 1913–1914 Great Program, and that in the long run Russian power would further outstrip German power because Russian resources were greater.[96] In German eyes this threat forced Germany to act. Secretary of State Jagow summarized a view common in Germany in a telegram to one of his ambassadors just before the July crisis broke:

> Russia will be ready to fight in a few years. Then she will crush us by the number of her soldiers; then she will have built her Baltic fleet and her strategic railways. Our group in the meantime will have become steadily weaker. . . . I do not desire a preventive war, but if the conflict should offer itself, we ought not to shirk it.[97]

Similarly, shortly before Sarajevo the Kaiser reportedly believed that "the big Russian railway constructions were . . . preparations for a great war which could start in 1916" and wondered "whether it might not be better to attack than to wait."[98] At about the same time Chancellor Bethmann Hollweg declared bleakly, "The future belongs to Russia which grows and grows and becomes an even greater nightmare to us,"[99] warning that "After the completion of their strategic railroads in Poland our position [will be] untenable."[100] During the war, Bethmann confessed that the "window" argument

95. Von Plessen, quoted in Isabell V. Hull, *The Entourage of Kaiser Wilhelm II, 1888–1918* (New York: Cambridge University Press, 1982), p. 261. Thus Bethmann summarized German thinking when he suggested on July 8 that the Sarajevo assassination provided an opportunity either for a war which "we have the prospect of winning" or a crisis in which "we still certainly have the prospect of maneuvering the Entente apart. . . ." Thompson, *In the Eye of the Storm*, p. 75.

96. The Russian program planned a 40 percent increase in the size of the peacetime Russian army and a 29 percent increase in the number of officers over four years. Lieven, *Russia & the Origins of the First World War*, p. 111.

97. July 18, quoted in Schmitt, *Coming of the War*, Vol. 1, p. 321.

98. June 21, quoted in Fischer, *War of Illusions*, p. 471, quoting Max Warburg.

99. July 7, quoted in ibid., p. 224, quoting Riezler.

100. July 7, quoted in Jarausch, "The Illusion of Limited War," p. 57. Likewise on July 20, he expressed terror at Russia's "growing demands and colossal explosive power. In a few years

had driven German policy in 1914: "Lord yes, in a certain sense it was a preventive war," motivated by "the constant threat of attack, the greater likelihood of its inevitability in the future, and by the military's claim: today war is still possible without defeat, but not in two years!"[101]

Window logic was especially prevalent among the German military officers, many of whom openly argued for preventive war during the years before the July crisis. General Moltke declared, "I believe a war to be unavoidable and: the sooner the better" at the infamous "war council" of December 8, 1912,[102] and he expressed similar views to his Austrian counterpart, General Conrad, in May 1914: "to wait any longer meant a diminishing of our chances; as far as manpower is concerned, one cannot enter into a competition with Russia,"[103] and "We [the German Army] are ready, the sooner the better for us."[104] During the July crisis Moltke remarked that "we shall never hit it again so well as we do now with France's and Russia's expansion of their armies incomplete," and argued that "the singularly favorable situation be exploited for military action."[105] After the war Jagow recalled a conversation with Moltke in May 1914, in which Moltke had spelled out his reasoning:

> In two–three years Russia would have completed her armaments. The military superiority of our enemies would then be so great that he did not know how we could overcome them. Today we would still be a match for them. In his opinion there was no alternative to making preventive war in order to defeat the enemy while we still had a chance of victory. The Chief of General Staff therefore proposed that I should conduct a policy with the aim of provoking a war in the near future.[106]

Other members of the German military shared Moltke's views, pressing for preventive war because "conditions and prospects would never become

---

she would be supreme—and Germany her first lonely victim." Quoted in Lebow, *Between Peace and War*, p. 258n.

101. Jarausch, "The Illusion of Limited War," p. 48. Likewise Friedrich Thimme quoted Bethmann during the war: "He also admits that our military are quite convinced that they could still be victorious in the war, but that in a few years time, say in 1916 after the completion of Russia's railway network, they could not. This, of course, also affected the way in which the Serbian question was dealt with." Quoted in Volker R. Berghahn and Martin Kitchen, eds., *Germany in the Age of Total War* (Totowa, N.J.: Barnes and Noble, 1981), p. 45.

102. Fischer, *War of Illusions*, p. 162.

103. Berghahn, *Germany and the Approach of War*, p. 171.

104. Geiss, *German Foreign Policy*, p. 149.

105. Berghahn, *Germany and the Approach of War*, p. 203.

106. Quoted in J.C.G. Röhl, ed., *From Bismarck to Hitler: The Problem of Continuity in German History* (London: Longman, 1970), p. 70.

better."[107] General Gebstattel recorded the mood of the German leadership on the eve of the war: "Chances better than in two or three years hence and the General Staff is reported to be confidently awaiting events."[108] The Berlin *Post,* a newspaper which often reflected the views of the General Staff, saw a window in 1914: "at the moment the state of things is favorable for us. France is not yet ready for war. England has internal and colonial difficulties, and Russia recoils from the conflict because she fears revolution at home. Ought we to wait until our adversaries are ready?" It concluded that Germany should "prepare for the inevitable war with energy and foresight" and "begin it under the most favorable conditions."[109]

German leaders also saw a tactical window of opportunity in the political constellation of July 1914, encouraging them to shut their strategic window of vulnerability. In German eyes, the Sarajevo assassination created favorable conditions for a confrontation, since it guaranteed that Austria would join Germany against Russia and France (as it might not if war broke out over a colonial conflict or a dispute in Western Europe), and it provided the Central powers with a plausible excuse, which raised hopes that Britain might remain neutral. On July 8, Bethmann Hollweg reportedly remarked, "If war comes from the east so that we have to fight for Austria–Hungary and not Austria–Hungary for us, we have a chance of winning."[110] Likewise, the German ambassador to Rome reportedly believed on July 27 that "the present moment is extraordinarily favorable to Germany,"[111] and the German ambassador to London even warned the British Prime Minister that "there was some feeling in Germany . . . that trouble was bound to come and therefore it would be better not to restrain Austria and let trouble come now, rather than later."[112]

The window logic reflected in these statements is a key to German conduct in 1914: whether the Germans were aggressive or restrained depended on

---

107. Leuckart's summary of the views of the General Staff, quoted in Geiss, *July 1914,* p. 69. For more on advocacy of preventive war by the German army, see Martin Kitchen, *The German Officer Corps, 1890–1914* (Oxford: Clarendon Press, 1968), pp. 96–114; and Hull, *Entourage of Kaiser Wilhelm II,* pp. 236–265.
108. August 2, quoted in Fischer, *War of Illusions,* p. 403.
109. February 24, 1914, in Schmitt, *Coming of the War,* Vol. 1, p. 100n.; and Fischer, *War of Illusions,* pp. 371–272.
110. Jarausch, "Illusion of Limited War," p. 58. Earlier Bülow had explained why the Agadir crisis was an unsuitable occasion for war in similar terms: "In 1911 the situation was much worse. The complication would have begun with Britain; France would have remained passive, it would have forced us to attack and then there would have been no *causus foederis* for Austria . . . whereas Russia was obliged to join in." In 1912, quoted in Fischer, *War of Illusions,* p. 85.
111. Schmitt, *Coming of the War,* Vol. 2, p. 66n.
112. Ibid., Vol. 1, p. 324, quoting Lichnowsky, on July 6.

whether at a given moment they thought windows were open or closed. Germany courted war on the Balkan question after Sarajevo because window logic led German leaders to conclude that war could not be much worse than peace, and might even be better, if Germany could provoke the right war under the right conditions against the right opponents. German leaders probably preferred the status quo to a world war against the entire Entente, but evidence suggests that they also preferred a continental war against France and Russia to the status quo—as long as Austria joined the war, and as long as they could also find a suitable pretext which they could use to persuade the German public that Germany fought for a just cause. This, in turn, required that Germany engineer a war which engaged Austrian interests, and in which Germany could cast itself as the attacked, in order to involve the Austrian army, to persuade Britain to remain neutral, and to win German public support. These window considerations help explain both the German decision to force the Balkan crisis to a head and German efforts to defuse the crisis after it realized that it had failed to gain British neutrality. The German peace efforts after July 29 probably represent a belated effort to reverse course after it became clear that the July crisis was not such an opportune war window after all.

Window logic also helped to persuade Austria to play the provocateur for Germany. Like their German counterparts, many Austrian officials believed that the relative strength of the central powers was declining, and saw in Sarajevo a rare opportunity to halt this decline by force. Thus the Austrian War Minister, General Krobatin, argued in early July that "it would be better to go to war immediately, rather than at some later period, because the balance of power must in the course of time change to our disadvantage," while the Austrian Foreign Minister, Count Berchtold, favored action because "our situation must become more precarious as time goes on,"[113] warning that unless Austria destroyed the Serbian army in 1914, it would face "another attack [by] Serbia in much more unfavorable conditions" in two or three years.[114] Likewise, the Austrian foreign ministry reportedly believed that, "if Russia would not permit the localization of the conflict with Serbia, the present moment was more favorable for a reckoning than a later one would be";[115] General Conrad believed, "If it comes to war with Russia—as

113. July 7, quoted in Geiss, *July 1914,* pp. 81, 84.
114. July 31, quoted in Schmitt, *Coming of the War,* Vol. 2, p. 218.
115. Ibid., Vol. 1, p. 372, quoting Baron von Tucher on July 18.

it must some day—today is as good as any other day";[116] and the Austrian ambassador to Italy believed an Austro–Serbian war would be "a piece of real good fortune," since "for the Triple Alliance the present moment is more favorable than another later."[117]

Thus the First World War was in part a "preventive" war, launched by the Central powers in the belief that they were saving themselves from a worse fate in later years. The cult of the offensive bears some responsibility for that belief, for in a defense-dominated world the windows which underlie the logic of preventive war are shrunken in size, as the balance of power grows less elastic to the relative sizes of armies and economies; and windows cannot be shut as easily by military action. Only in a world taken by the cult of the offensive could the window logic which governed German and Austrian conduct have proved so persuasive: Germans could only have feared that an unchecked Russia could eventually "crush us by the numbers of her soldiers," or have seen a "singularly favorable situation" in 1914 which could be "exploited by military action" if material superiority would endow the German and Russian armies with the ability to conduct decisive offensive operations against one another. Moltke claimed he saw "no alternative to making preventive war," but had he believed that the defense dominated, better alternatives would have been obvious.

The cult of the offensive also helped cause the arms race before 1914 which engendered the uneven rates of military growth that gave rise to visions of windows. The German army buildup after 1912 was justified by security arguments: Bethmann Hollweg proclaimed, "For Germany, in the heart of Europe, with open boundaries on all sides, a strong army is the most secure guarantee of peace," while the Kaiser wrote that Germany needed "More ships and soldiers . . . because our existence is at stake."[118] This buildup provoked an even larger Russian and French buildup, which created the windows which alarmed Germany in 1914.[119] Thus the cult both magnified

WWI Preventive War

---

116. In October 1913, quoted in Gerhard Ritter, *The Sword and the Scepter: The Problem of Militarism in Germany*, 4 vols., trans. Heinz Norden (Coral Gables, Fla.: University of Miami Press, 1969–73), Vol. 2, p. 234. Likewise the *Militärisch Rundschau* argued for provoking war: "Since we shall have to accept the contest some day, let us provoke it at once." On July 15, 1914, quoted in Schmitt, *Coming of the War*, Vol. 1, p. 367. For more on preventive war and the Austrian army, see Ritter, *Sword and the Scepter*, Vol. 2, pp. 227–239.
117. Count Merey, July 29, quoted in Albertini, *Origins*, Vol. 2, p. 383.
118. Both in 1912, quoted in Jarausch, *Enigmatic Chancellor*, p. 95; and Fischer, *War of Illusions*, p. 165.
119. On the motives for the Russian buildup, see P.A. Zhilin, "Bol'shaia programma po usileniiu russkoi armii," *Voenno-istoricheskii zhurnal*, No. 7 (July 1974), pp. 90–97.

the importance of fluctuations in ratios of forces and helped to fuel the arms race which fostered them.

THE SCOPE AND INFLEXIBILITY OF MOBILIZATION PLANS

The spreading of World War I outward from the Balkans is often ascribed to the scope and rigidity of the Russian and German plans for mobilization, which required that Russia must also mobilize armies against Germany when it mobilized against Austria–Hungary, and that Germany also attack France and Belgium if it fought Russia. Barbara Tuchman writes that Europe was swept into war by "the pull of military schedules," and recalls Moltke's famous answer when the Kaiser asked if the German armies could be mobilized to the East: "Your Majesty, it cannot be done. The deployment of millions cannot be improvised. If Your Majesty insists on leading the whole army to the East it will not be an army ready for battle but a disorganized mob of armed men with no arrangements for supply."[120] Likewise, Herman Kahn notes the "rigid war plan[s]" of 1914, which "were literally cast in concrete,"[121] and David Ziegler notes the influence of military "planning in advance," which left "no time to improvise."[122]

The scope and character of these plans in turn reflected the assumption that the offense was strong. In an offense-dominant world Russia would have been prudent to mobilize against Germany if it mobilized against Austria–Hungary; and Germany probably would have been prudent to attack Belgium and France at the start of any Russo–German war. Thus the troublesome railroad schedules of 1914 reflected the offense-dominant world in which the schedulers believed they lived. Had they known that the defense was powerful, they would have been drawn towards flexible plans for limited deployment on single frontiers; and had such planning prevailed, the war might have been confined to Eastern Europe or the Balkans.

Moreover, the "inflexibility" of the war plans may have reflected the same offensive assumptions which determined their shape. Russian and German soldiers understandably developed only options which they believed prudent to exercise, while omitting plans which they believed would be dangerous to implement. These judgments in turn reflected their own and their adver-

120. Tuchman, *Guns of August*, pp. 92, 99.
121. Kahn, *On Thermonuclear War*, pp. 359, 362.
122. David W. Ziegler, *War, Peace and International Politics* (Boston: Little, Brown, 1977), p. 25.

saries' offensive ideas. Options were few because these offensive ideas seemed to narrow the range of prudent choice.

Lastly, the assumption of offense-dominance gave preset plans greater influence over the conduct of the July crisis, by raising the cost of improvisation if statesmen insisted on adjusting plans at the last minute. Russian statesmen were told that an improvised partial mobilization would place Russia in a "extremely dangerous situation,"[123] and German civilians were warned against improvisation in similar terms. This in turn reflected the size of the "windows" which improvised partial mobilizations would open for the adversary on the frontier which the partial mobilization left unguarded, which in turn reflected the assumption that the offense was strong (since if defenses were strong a bungled mobilization would create less opportunity for others to exploit). Thus the cult of the offensive gave planners greater power to bind statesmen to the plans they had prepared.

RUSSIAN MOBILIZATION PLANS. On July 28, 1914, Russian leaders announced that partial Russian mobilization against Austria would be ordered on July 29. They took this step to address threats emanating from Austria, acting partly to lend emphasis to their warnings to Austria that Russia would fight if Serbia were invaded, partly to offset Austrian mobilization against Serbia, and partly to offset or forestall Austrian mobilization measures which they believed were taking place or which they feared might eventually take place against Russia in Galicia.[124] However, after this announcement was made, Russian military officers advised their civilian superiors that no plans for partial mobilization existed, that such a mobilization would be a "pure improvisation," as General Denikin later wrote, and that sowing confusion in the Russian railway timetables would impede Russia's ability to mobilize later on its northern frontier. General Sukhomlinov warned the Czar that "much time would be necessary in which to re-establish the normal conditions for any further mobilization" following a partial mobilization, and General Yanushkevich flatly told Sazonov that general mobilization "could not be put into operation" once partial mobilization began.[125] Thus Russian lead-

---

123. By Generals Yanushkevich and Sukhomlinov, according to Sazonov, quoted in Albertini, *Origins*, Vol. 2, p. 566. See also M.F. Schilling, "Introduction," in *How the War Began*, trans. W. Cyprian Bridge, with a Foreword by S.D. Sazonov (London: Allen & Unwin, 1925), pp. 16, 63.
124. On the Russian decision, see Schmitt, *Coming of the War*, Vol. 2, pp. 85–87, 94–101; and Albertini, *Origins*, Vol. 2, pp. 539–561.
125. Anton I. Denikin, *The Career of a Tsarist Officer: Memoirs, 1872–1916*, trans. Margaret Patoski (Minneapolis: University of Minnesota Press, 1975), p. 222; Albertini, *Origins*, Vol. 2, p. 559; Schilling, *How the War Began*, p. 16.

ers were forced to choose between full mobilization or complete retreat, choosing full mobilization on July 30.

The cult of the offensive set the stage for this decision by buttressing Russian military calculations that full mobilization was safer than partial. We have little direct evidence explaining why Russian officers had prepared no plan for partial mobilization, but we can deduce their reasoning from their opinions on related subjects. These suggest that Russian officers believed that Germany would attack Russia if Russia fought Austria, and that the side mobilizing first would have the upper hand in a Russo–German war (as I have outlined above). Accordingly, it followed logically that Russia should launch any war with Austria by preempting Germany.

Russian leaders had three principal reasons to fear that Germany would not stand aside in an Austro–Russian conflict. First, the Russians were aware of the international Social Darwinism then sweeping Germany, and the expansionist attitude toward Russia which this worldview engendered. One Russian diplomat wrote that Germany was "beating all records of militarism" and "The Germans are not . . . wholly without the thought of removing from Russia at least part of the Baltic coastline in order to place us in the position of a second Serbia" in the course of a campaign for "German hegemony on the continent."[126] Russian military officers monitored the bellicose talk across the border with alarm, one intelligence report warning: "In Germany at present, the task of gradually accustoming the army and the population to the thought of the inevitability of conflict with Russia has begun," noting the regular public lectures which were then being delivered in Germany to foster war sentiment.[127]

Second, the Russians were aware of German alarm about windows and the talk of preventive war which this alarm engendered in Germany. Accordingly, Russian leaders expected that Germany might seize the excuse offered by a Balkan war to mount a preventive strike against Russia, especially since a war arising from the Balkans was a "best case" scenario for Germany, involving Austria on the side of Germany as it did. Thus General Yanushkevich explained Russia's decision to mobilize against Germany in 1914: "We knew well that Germany was ready for war, that she was longing

126. G.N. Trubetskoy, in 1909, quoted in Lieven, *Russia & the Origins of the First World War*, p. 96.
127. The Kiev District Staff, February 23, 1914, quoted in Fuller, "The Russian Empire and Its Potential Enemies," p. 17.

for it at that moment, because our big armaments program was not yet completed . . . and because our war potential was not as great as it might be." Accordingly, Russia had to expect war with Germany: "We knew that war was inevitable, not only against Austria, but also against Germany. For this reason partial mobilization against Austria alone, which would have left our front towards Germany open . . . might have brought about a disaster, a terrible disaster."[128] In short, Russia had to strike to preempt a German preventive strike against Russia.

Third, the Russians knew that the Germans believed that German and Austrian security were closely linked. Germany would therefore feel compelled to intervene in any Austro–Russian war, because a Russian victory against Austria would threaten German safety. German leaders had widely advertised this intention: for instance, Bethmann Hollweg had warned the Reichstag in 1912 that if the Austrians "while asserting their interests should against all expectations be attacked by a third party, then we would have to come resolutely to their aid. And then we would fight for the maintenance of our own position in Europe and in defense of our future and security."[129] And in fact this was precisely what happened in 1914: Germany apparently decided to attack on learning of Russian *partial* mobilization, before Russian full mobilization was known in Germany.[130] This suggests that the role of "inflexible" Russian plans in causing the war is overblown—Russian full mobilization was sufficient but not necessary to cause the war; but it also helps explain why these plans were drawn as they were, and supports the view that some of the logic behind them was correct, given the German state of mind with which Russia had to contend.

---

128. Albertini, *Origins*, Vol. 2, p. 559. See also Fuller, "The Russian Empire and Its Potential Enemies," p. 16.

129. In 1912, quoted in Stern, *Failure of Illiberalism*, p. 84. Likewise the Kaiser explained that security requirements compelled Germany to defend Austria: "If we are forced to take up arms it will be to help *Austria*, not only to defend ourselves against Russia but against the Slavs in general and to remain *German*. . . ." In 1912, quoted in Fischer, *War of Illusions*, pp. 190–191, emphasis in original. The German White Book also reflected this thinking, declaring that the "subjugation of all the Slavs under Russian sceptre" would render the "position of the Teutonic race in Central Europe untenable." August 3, 1914, quoted in Geiss, *German Foreign Policy*, p. 172.

130. See Schmitt, *Coming of the War*, Vol. 2, pp. 198–199; and Albertini, *Origins*, Vol. 3, pp. 7, 17–27; also Vol. 2, p. 485n. As Jagow plainly told the Russians on July 29: "If once you mobilize against Austria, then you will also take serious measures against us. . . . We are compelled to proclaim mobilization against Russia. . . ." Schmitt, *Coming of the War*, Vol. 2, p. 140.

In sum, Russians had to fear that expansionist, preventive, and alliance concerns might induce Germany to attack, which in turn reflected the German assumption that the offense was strong. The Russian belief that it paid to mobilize first reflected the effects of the same assumption in Russia. Had Europe known that the defense dominated, Russians would have had less reason to fear that an Austro–Russian war would spark a German attack, since the logic of expansionism and preventive war would presumably have been weaker in Germany, and Germany could more easily have tolerated some reduction in Austrian power without feeling that German safety was also threatened. At the same time, Russian soldiers would presumably have been slower to assume that they could improve their position in a Russo–German war by mobilizing preemptively. In short, the logic of general mobilization in Russia largely reflected and depended upon conclusions deduced from the cult of the offensive, or from its various manifestations. Without the cult of the offensive, a partial southern mobilization would have been the better option for Russia.

It also seems probable that the same logic helped persuade the Russian General Staff to eschew planning for a partial mobilization. If circumstances argued against a partial mobilization, they also argued against planning for one, since this would raise the risk that Russian civilians might actually implement the plan. This interpretation fits with suggestions that Russian officers exaggerated the difficulties of partial mobilization in their representations to Russian civilians.[131] If Russian soldiers left a partial mobilization option undeveloped because they believed that it would be dangerous to exercise, it follows that they also would emphasize the difficulty of improvising a southern option, since they also opposed it on other grounds.

GERMAN MOBILIZATION PLANS. The Schlieffen Plan was a disastrous scheme which only approached success because the French war plan was equally foolish: had the French army stood on the defensive instead of lunging into Alsace–Lorraine, it would have smashed the German army at the French frontier. Yet General Schlieffen's plan was a sensible response to the offense-

131. See L.C.F. Turner, "The Russian Mobilization in 1914," *Journal of Contemporary History*, Vol. 3, No. 1 (January 1968), pp. 72–74. But see also Lieven, *Russia and the Origins of the First World War*, pp. 148–150.

Likewise, German soldiers exaggerated the difficulties of adapting to eastward mobilization, as many observers note, e.g., Tuchman, *Guns of August*, p. 100, and Lebow, *Between Peace and War*, p. 236.

dominant world imagined by many Germans. The plan was flawed because it grew from a fundamentally flawed image of warfare.

In retrospect, Germany should have retained the later war plan of the elder Moltke (Chief of Staff from 1857 to 1888), who would have conducted a limited offensive in the east against Russia while standing on the defensive in the west.[132] However, several considerations pushed German planners instead toward Schlieffen's grandiose scheme, which envisioned a quick victory against Belgium and France, followed by an offensive against Russia.

First, German planners assumed that France would attack Germany if Germany fought Russia, leaving Germany no option for a one-front war. By tying down German troops in Poland, an eastern war would create a yawning window of opportunity for France to recover its lost territories, and a decisive German victory over Russia would threaten French security by leaving France to face Germany alone. For these reasons they believed that France would be both too tempted and too threatened to stand aside. Bernhardi, among others, pointed out "the standing danger that France will attack us on a favorable occasion, as soon as we find ourselves involved in complications elsewhere."[133] The German declaration of war against France explained that France might suddenly attack from behind if Germany fought Russia; hence, "Germany cannot leave to France the choice of the moment" at which to attack.[134]

Second, German planners assumed that "window" considerations required a German offensive against either France or Russia at the outset of any war against the Entente. German armies could mobilize faster than the combined Entente armies; hence, the ratio of forces would most favor Germany at the beginning of the war. Therefore, Germany would do best to force an early decision, which in turn required that it assume the offensive, since otherwise its enemies would play a waiting game. As one observer explained, Germany

---

132. Assessing the Schlieffen Plan are Ritter, *The Schlieffen Plan,* and Snyder, "Defending the Offensive," pp. 189–294.

133. Bernhardi, quoted in Anon., *Germany's War Mania* (London: A.W. Shaw, 1914), p. 161.

134. Albertini, *Origins,* Vol. 3, p. 194. Moreover, these fears reflected views found in France. When Poincaré was asked on July 29 if he believed war could be avoided, he reportedly replied: "It would be a great pity. We should never again find conditions better." Albertini, *Origins,* Vol. 3, p. 82n. Likewise, in 1912 the French General Staff concluded that a general war arising from the Balkans would leave Germany "at the mercy of the Entente" because Austrian forces would be diverted against Serbia, and "the Triple Entente would have the best chances of success and might gain a victory which would enable the map of Europe to be redrawn." Turner, *Origins,* p. 36. See also the opinions of Izvolsky and Bertie in Schmitt, *Coming of the War,* Vol. 1, pp. 20–21, and Vol. 2, p. 349n.

"has the speed and Russia has the numbers, and the safety of the German Empire forbade that Germany should allow Russia time to bring up masses of troops from all parts of her wide dominions."[135] Germans believed that the window created by these differential mobilization rates was big, in turn, because they believed that both Germany and its enemies could mount a decisive offensive against the other with a small margin of superiority. If Germany struck at the right time, it could win easily—Germans hoped for victory in several weeks, as noted above—while if it waited it was doomed by Entente numerical superiority, which German defenses would be too weak to resist.

Third, German planners believed that an offensive against France would net them more than an offensive against Russia, which explains the western bias of the Schlieffen Plan. France could be attacked more easily than Russia, because French forces and resources lay within closer reach of German power; hence, as Moltke wrote before the war, "A speedy decision may be hoped for [against France], while an offensive against Russia would be an interminable affair."[136] Moreover, France was the more dangerous opponent not to attack, because it could take the offensive against Germany more quickly than Russia, and could threaten more important German territories if Germany left its frontier unguarded. Thus Moltke explained that they struck westward because "Germany could not afford to expose herself to the danger of attack by strong French forces in the direction of the Lower Rhine," and Wegerer wrote later that the German strike was compelled by the need to protect the German industrial region from French attack.[137] In German eyes these considerations made it too dangerous to stand on the defensive in the West in hopes that war with France could be avoided.

Finally, German planners believed that Britain would not have time to bring decisive power to bear on the continent before the German army overran France. Accordingly, they discounted the British opposition which their attack on France and Belgium would elicit: Schlieffen declared that if the British army landed, it would be "securely billeted" at Antwerp or "arrested" by the German armies,[138] while Moltke said he hoped that it would

---

135. Goschen, in Schmitt, *Coming of the War,* Vol. 2, p. 321.
136. Moltke, in General Ludendorff, *The General Staff and its Problems,* trans. F.A. Holt (New York: E.P. Dutton, n.d.), Vol. 1, p. 61.
137. Geiss, *July 1914,* p. 357; and Alfred von Wegerer, *A Refutation of the Versailles War Guilt Thesis,* trans. Edwin H. Zeydel (New York: Alfred A. Knopf, 1930), p. 310.
138. Ritter, *Schlieffen Plan,* pp. 71, 161–162; and Geiss, *German Foreign Policy,* p. 101. See also Ritter, *Schlieffen Plan,* p. 161. But see also Moltke quoted in Turner, *Origins of the World War,* p. 64.

land so that the German army "could take care of it."[139] In accordance with their "bandwagon" worldview, German leaders also hoped that German power might cow Britain into neutrality; or that Britain might hesitate before entering the war, and then might quit in discouragement once the French were beaten—Schlieffen expected that, "If the battle [in France] goes in favor of the Germans, the English are likely to abandon their enterprise as hopeless"—which led them to further discount the extra political costs of attacking westward.[140]

Given these four assumptions, an attack westward, even one through Belgium which provoked British intervention, was the most sensible thing for Germany to do. Each assumption, in turn, was a manifestation of the belief that the offense was strong. Thus while the Schlieffen Plan has been widely criticized for its political and military naiveté, it would have been a prudent plan had Germans actually lived in the offense-dominant world they imagined. Under these circumstances quick mobilization would have in fact given them a chance to win a decisive victory during their window of opportunity, and if they had failed to exploit this window by attacking, they would eventually have lost; the risk of standing on the defense in the West in hopes that France would not fight would have been too great; and the invasion of France and Belgium would have been worth the price, because British power probably could not have affected the outcome of the war.

Thus the belief in the power of the offense was the linchpin which held Schlieffen's logic together, and the main criticisms which can be levelled at the German war plan flow from the falsehood of this belief. German interests would have been better served by a limited, flexible, east-only plan which conformed to the defensive realities of 1914. Moreover, had Germany adopted such a plan, the First World War might well have been confined to Eastern Europe, never becoming a world war.

"MOBILIZATION MEANS WAR"

"Mobilization meant war" in 1914 because mobilization meant war to Germany: the German war plan mandated that special units of the German standing army would attack Belgium and Luxemburg immediately after mobilization was ordered, and long before it was completed. (In fact Germany

139. Ritter, *Sword and the Scepter*, Vol. 2, p. 157.
140. Ritter, *The Schlieffen Plan*, p. 163. See also Bethmann Hollweg, quoted in Fischer, *War of Illusions*, pp. 169, 186–187.

invaded Luxemburg on August 1, the same day on which it ordered full mobilization.) Thus Germany had no pure "mobilization" plan, but rather had a "mobilization and attack" plan under which mobilizing and attacking would be undertaken simultaneously. As a result, Europe would cascade into war if any European state mobilized in a manner which eventually forced German mobilization.

This melding of mobilization and attack in Germany reflected two decisions to which I have already alluded. First, Germans believed that they would lose their chance for victory and create a grave danger for themselves if they gave the Entente time to mobilize its superior numbers. In German eyes, German defenses would be too weak to defeat this superiority. As one German apologist later argued, "Germany could never with success have warded off numerically far superior opponents by means of a defensive war against a mobilized Europe" had it mobilized and stood in place. Hence it was "essential for the Central Powers to begin hostilities as soon as possible" following mobilization.[141] Likewise, during the July crisis, Jagow explained that Germany must attack in response to Russian mobilization because "we are obliged to act as fast as possible before Russia has the time to mobilize her army."[142]

Second, the German war plan depended on the quick seizure of Liège. Germany could only secure Liège quickly if German troops arrived before Belgium prepared its defense, and this in turn depended on achieving surprise against Belgium. Accordingly, German military planners enshrouded the planned Liège attack in such dark secrecy that Bethmann Hollweg, Admiral Tirpitz, and possibly even the Kaiser were unaware of it.[143] They also felt compelled to strike as soon as mobilization was authorized, both because Belgium would strengthen the defenses of Liège as a normal part of the Belgian mobilization which German mobilization would engender, and because otherwise Belgium eventually might divine German intentions towards Liège and focus upon preparing its defense and destroying the critical bridges and tunnels which it controlled.

---

141. Von Wegerer, *Refutation*, pp. 307–309.
142. August 4, quoted in Alfred Vagts, *Defense and Diplomacy* (New York: Kings Crown Press, 1956), p. 306. Likewise Bethmann Hollweg explained that, if Russia mobilized, "we could hardly sit and talk any longer because we have to strike immediately in order to have any chance of winning at all." Fischer, *War of Illusions*, p. 484.
143. Albertini, *Origins*, Vol. 2, p. 581; Vol. 3, pp. 195, 250, 391; Ritter, *Sword and the Scepter*, Vol. 2, p. 266; and Fay, *Origins*, Vol. 1, pp. 41–42.

Both of these decisions in turn reflected German faith in the power of the offense, and were not appropriate to a defense-dominant world. Had Germans recognized the actual power of the defense, they might have recognized that neither Germany nor its enemies could win decisively even by exploiting a fleeting material advantage, and decided instead to mobilize without attacking. The tactical windows that drove Germany to strike in 1914 were a mirage, as events demonstrated during 1914–1918, and Germans would have known this in advance had they understood the power of the defense. Likewise, the Liège *coup de main* was an artifact of Schlieffen's offensive plan; if the Germans had stuck with the elder Moltke's plan, they could have abandoned both the Liège attack and the compulsion to strike quickly which it helped to engender.

BRINKMANSHIP AND FAITS ACCOMPLIS

Two *faits accomplis* by the Central powers set the stage for the outbreak of the war: the Austrian ultimatum to Serbia on July 23, and the Austrian declaration of war against Serbia on July 28. The Central powers also planned to follow these with a third *fait accompli*, by quickly smashing Serbia on the battlefield before the Entente could intervene. These plans and actions reflected the German strategy for the crisis: "*fait accompli* and then friendly towards the Entente, the shock can be endured," as Kurt Riezler had summarized.[144]

This *fait accompli* strategy deprived German leaders of warning that their actions would plunge Germany into a world war, by depriving the Entente of the chance to warn Germany that it would respond if Austria attacked Serbia. It also deprived diplomats of the chance to resolve the Austro–Serbian dispute in a manner acceptable to Russia. Whether this affected the outcome of the crisis depends on German intentions—if Germany sought a pretext for a world war, then this missed opportunity had no importance, but if it preferred the status quo to world war, as I believe it narrowly did, then the decision to adopt *fait accompli* tactics was a crucial step on the road to war.

144. July 8, quoted in John A. Moses, *The Politics of Illusion: The Fischer Controversy in German Historiography* (London: George Prior, 1975), p. 39. Austria declared war on Serbia, as one German diplomat explained, "in order to forestall any attempt at mediation" by the Entente; and the rapid occupation of Serbia was intended to "confront the world with a '*fait accompli*.'" Tschirschky, in Schmitt, *Coming of the War,* Vol. 2, p. 5; and Jagow, in Albertini, *Origins,* Vol. 2, p. 344; see also pp. 453–460.

Had Germany not done so, it might have recognized where its policies led before it took irrevocable steps, and have drawn back.

The influence of the cult of the offensive is seen both in the German adoption of this *fait accompli* strategy and in the disastrous scope of the results which followed in its train. Some Germans, such as Kurt Riezler, apparently favored brinkmanship and *fait accompli* diplomacy as a means of peaceful expansion.[145] Others probably saw it as a means to provoke a continental war. In either case it reflected a German willingness to trade peace for territory, which reflected German expansionism—which in turn reflected security concerns fuelled by the cult of the offensive. Even those who saw *faits accomplis* as tools of peaceful imperialism recognized their risks, believing that necessity justified the risk. Thus Riezler saw the world in Darwinistic terms: "each people wants to grow, expand, dominate and subjugate others without end . . . until the world has become an organic unity under [single] domination."[146] *Faits accomplis* were dangerous tools whose adoption reflected the dangerous circumstances which Germans believed they faced.

The cult of the offensive also stiffened the resistance of the Entente to the Austro-German *fait accompli,* by magnifying the dangers they believed it posed to their own security.[147] Thus Russian leaders believed that Russian security would be directly jeopardized if Austria crushed Serbia, because they valued the power which Serbia added to their alliance, and because they feared a domino effect, running to Constantinople and beyond, if Serbia were overrun. Sazonov believed that Serbian and Bulgarian military power was a vital Russian resource, "five hundred thousand bayonets to guard the Balkans" which "would bar the road forever to German penetration, Austrian invasion."[148] If this asset were lost, Russia's defense of its own territories would be jeopardized by the German approach to Constantinople: Sazonov warned the Czar, "First Serbia would be gobbled up; then will come Bulgaria's turn, and then we shall have her on the Black Sea." This would be

145. On Riezler's thought, see Moses, *Politics of Illusion,* pp. 27–44; and Thompson, *In the Eye of the Storm.*
146. Quoted in Moses, *Politics of Illusion,* pp. 28, 31. Likewise during the war Riezler wrote that unless Germany gained a wider sphere of influence in Europe "we will in the long run be crushed between the great world empires . . . Russia and England." Thompson, *In the Eye of the Storm,* p. 107.
147. I am grateful to Jack Snyder for this and related observations.
148. Schmitt, *Coming of the War,* Vol. 1, p. 131n. See also Lieven, *Russia and the Origins of the First World War,* pp. 40–41, 99–100, 147.

"the death-warrant of Russia" since in such an event "the whole of southern Russia would be subject to [Germany]."[149]

Similar views could be found in France. During the July crisis one French observer warned that French and Serbian security were closely intertwined, and the demise of Serbia would directly threaten French security:

> To do away with Serbia means to double the strength which Austria can send against Russia: to double Austro–Hungarian resistance to the Russian Army means to enable Germany to send some more army corps against France. For every Serbian soldier killed by a bullet on the Morava one more Prussian soldier can be sent to the Moselle. . . . It is for us to grasp this truth and draw the consequences from it before disaster overtakes Serbia.[150]

These considerations helped spur the Russian and French decisions to begin military preparations on July 25, which set in train a further sequence of events: German preliminary preparations, which were detected and exaggerated by French and Russian officials, spurring them on to further measures, which helped spur the Germans to their decision to mobilize on July 30. The effects of the original *fait accompli* rippled outward in ever-wider circles, because the reactions of each state perturbed the safety of others—forcing them to react or preempt, and ultimately forcing Germany to launch a world war which even it preferred to avoid.

Had Europe known that, in reality, the defense dominated, these dynamics might have been dampened: the compulsion to resort to *faits accomplis*, the scope of the dangers they raised for others, and the rippling effects engendered by others' reactions all would have been lessened. States still might have acted as they did, but they would have been less pressured in this direction.

PROBLEMS OF ALLIANCES: UNCONDITIONALITY AND AMBIGUITY

Two aspects of the European alliance system fostered the outbreak of World War I and helped spread the war. First, both alliances had an unconditional, offensive character—allies supported one another unreservedly, regardless of whether their behavior was defensive or provocative. As a result a local war would tend to spread throughout Europe. And second, German leaders were not convinced that Britain would fight as an Entente member, which

149. Fay, *Origins*, Vol. 2, p. 300; Sazonov, *Fateful Years*, p. 179; Schmitt, *Coming of the War*, Vol. 1, p. 87.
150. J. Herbette, July 29, in Albertini, *Origins*, Vol. 2, p. 596.

encouraged Germany to confront the Entente. In both cases the cult of the offensive contributed to the problem.

UNCONDITIONAL ("TIGHT") ALLIANCES. Many scholars contend that the mere existence of the Triple Alliance and the Triple Entente caused and spread the war. Sidney Fay concluded, "The greatest single underlying cause of the War was the system of secret alliance," and Raymond Aron argued that the division of Europe into two camps "made it inevitable that any conflict involving two great powers would bring general war."[151] But the problem with the alliances of 1914 lay less with their existence than with their nature. A network of defensive alliances, such as Bismarck's alliances of the 1880s, would have lowered the risk of war by facing aggressors with many enemies, and by making status quo powers secure in the knowledge that they had many allies. Wars also would have tended to remain localized, because the allies of an aggressor would have stood aside from any war that aggressor had provoked. Thus the unconditional nature of alliances rather than their mere existence was the true source of their danger in 1914.

The Austro–German alliance was offensive chiefly and simply because its members had compatible aggressive aims. Moreover, German and Russian mobilization plans left their neighbors no choice but to behave as allies by putting them all under threat of attack. But the Entente also operated more unconditionally, or "tightly," because Britain and France failed to restrain Russia from undertaking mobilization measures during the July crisis. This was a failure in alliance diplomacy, which in turn reflected constraints imposed upon the Western allies by the offensive assumptions and preparations with which they had to work.

First, they were hamstrung by the offensive nature of Russian military doctrine, which left them unable to demand that Russia confine itself to defensive preparations. All Russian preparations were inherently offensive, because Russian war plans were offensive. This put Russia's allies in an "all or nothing" situation—either they could demand that Russia stand unprepared, or they could consent to provocative preparations. Thus the British ambassador to St. Petersburg warned that Britain faced a painful decision, to "choose between giving Russia our active support or renouncing her friendship."[152] Had Russia confined itself to preparing its own defense, it

---

151. Fay, *Origins*, Vol. 1, p. 34; and Raymond Aron, *The Century of Total War* (Boston: Beacon Press, 1955), p. 15.
152. Buchanan, in Fay, *Origins*, Vol. 2, p. 379.

would have sacrificed its Balkan interests by leaving Austria free to attack Serbia, and this it would have been very reluctant to do. However, the British government was probably willing to sacrifice Russia's Balkan interests to preserve peace;[153] what Britain was unable to do was to frame a request to Russia which would achieve this, because there was no obvious class of defensive activity that it could demand. Edward Grey, the British Foreign Secretary, wrote later:

> I felt impatient at the suggestion that it was for me to influence or restrain Russia. I could do nothing but express pious hopes in general terms to Sazonov. If I were to address a direct request to him that Russia should not mobilize, I knew his reply: Germany was much more ready for war than Russia; it was a tremendous risk for Russia to delay her mobilization. . . . I did most honestly feel that neither Russian nor French mobilization was an unreasonable or unnecessary precaution.[154]

One sees in this statement a losing struggle to cope with the absence of defensive options. Russia was threatened, and must mobilize. How could Britain object?

Britain and France were also constrained by their dependence upon the strength and unity of the Entente for their own security, which limited their ability to make demands on Russia. Because they feared they might fracture the Entente if they pressed Russia too hard, they tempered their demands to preserve the alliance. Thus Poincaré wrote later that France had been forced to reconcile its efforts to restrain Russia with the need to preserve the Franco–Russian alliance, "the break up of which would leave us in isolation at the mercy of our rivals."[155] Likewise Winston Churchill recalled that "the one thing [the Entente states] would not do was repudiate each other. To do this might avert the war for the time being. It would leave each of them to face the next crisis alone. They did not dare to separate."[156] These fears were probably overdrawn, since Russia had no other option than alliance with the other Entente states, but apparently they affected French and British behavior.[157] This in turn reflected the assumption in France and Britain that the security of the Entente members was closely interdependent.

---

153. See Geiss, *July 1914*, p. 176; and Albertini, *Origins*, Vol. 2, p. 295.
154. Albertini, *Origins*, Vol. 2, p. 518.
155. Ibid., p. 605.
156. Winston Churchill, *The Unknown War* (New York: Charles Scribner's Sons, 1931), p. 103.
157. Thus Grey later wrote that he had feared a "diplomatic triumph on the German side and humiliation on the other as would smash the Entente, and if it did not break the Franco–Russian

French leaders also felt forced in their own interests to aid Russia if Russia embroiled itself with Germany, because French security depended on the maintenance of Russian power. This in turn undermined the French ability to credibly threaten to discipline a provocative Russia. Thus the British ambassador to Paris reflected French views when he cabled that he could not imagine that France would remain quiescent during a Russo–German war, because "If [the] French undertook to remain so, the Germans would first attack [the] Russians and, if they defeated them, they would then turn round on the French."[158] This prospect delimited French power to restrain Russian conduct.

Third, British leaders were unaware that German mobilization meant war, hence that peace required Britain to restrain Russia from mobilizing first, as well as attacking. As a result, they took a more relaxed view of Russian mobilization than they otherwise might, while frittering away their energies on schemes to preserve peace which assumed that war could be averted even after the mobilizations began.[159] This British ignorance reflected German failure to explain clearly to the Entente that mobilization did indeed mean war—German leaders had many opportunities during the July crisis to make this plain, but did not do so.[160] We can only guess why Germany was silent, but German desire to avoid throwing a spotlight on the Liège operation probably played a part, leading German soldiers to conceal the plan from German civilians, which led German civilians to conceal the political implications of the plan from the rest of Europe.[161] Thus preemptive planning threw a shroud of secrecy over military matters, which obscured the mechanism that would unleash the war and rendered British statesmen less able

---

alliance, would leave it without spirit, a spineless and helpless thing." Likewise during July 1914 Harold Nicolson wrote: "Our attitude during the crisis will be regarded by Russia as a test and we must be careful not to alienate her." Schmitt, *Coming of the War*, Vol. 2, pp. 38, 258.

158. Bertie, on August 1, in Schmitt, *Coming of the War*, Vol. 2, p. 349n.

159. Geiss, *July 1914*, pp. 198, 212–213, 250–251; Albertini, *Origins*, Vol. 2, pp. 330–336.

160. See Albertini, *Origins*, Vol. 2, pp. 479–481; Vol. 3, pp. 41–43, 61–65. Albertini writes that European leaders "had no knowledge of what mobilization actually was . . . what consequences it brought with it, to what risks it exposed the peace of Europe. They looked on it as a measure costly, it is true, but to which recourse might be had without necessarily implying that war would follow." This reflected German policy: Bethmann's ultimatum to Russia "entirely omitted to explain that for Germany to mobilize meant to begin war," and Sazonov gathered "the distinct impression that German mobilization was not equivalent to war" from his exchanges with German officials. Vol. 2, p. 479; Vol. 3, pp. 41–43.

161. Kautsky and Albertini suggest that the German deception was intended to lull the Russians into military inaction, but it seems more likely that they sought to lull the Belgians. Albertini, *Origins*, Vol. 3, p. 43.

to wield British power effectively for peace by obscuring what it was that Britain had to do.

Lastly, the nature of German war plans empowered Russia to involve France, and probably Britain also, in war, since Germany would be likely to start any eastern war by attacking westward, as Russian planners were aware. Hence France and Britain would probably have to fight for Russia even if they preferred to stand aside, because German planners assumed that France would fight eventually and planned accordingly, and the plans they drew would threaten vital British interests. We have no direct evidence that Russian policies were emboldened by these considerations, but it would be surprising if they never occurred to Russian leaders.

These dynamics reflected the general tendency of alliances toward tightness and offensiveness in an offense-dominant world. Had Europe known that the defense had the advantage, the British and French could have more easily afforded to discipline Russia in the interest of peace, and this might have affected Russian calculations. Had Russia had a defensive military strategy, its allies could more easily and legitimately have asked it to confine itself to defensive preparations. Had British leaders better understood German war plans, they might have known to focus their efforts on preventing Russian mobilization. And had German plans been different, Russian leaders would have been more uncertain that Germany would entangle the Western powers in eastern wars, and perhaps proceeded more cautiously.

The importance of the failure of the Western powers to restrain Russia can be exaggerated, since Russia was not the chief provocateur in the July crisis. Moreover, too much can be made of factors which hamstrung French restraint of Russia, since French desire to prevent war was tepid at best, so French inaction probably owed as much to indifference as inability. Nevertheless, Russian mobilization was an important step toward a war which Britain, if not France, urgently wanted to prevent; hence, to that extent, the alliance dynamics which allowed it helped bring on the war.

THE AMBIGUITY OF BRITISH POLICY. The British government is often accused of causing the war by failing to warn Germany that Britain would fight. Thus Albertini concludes that "to act as Grey did was to allow the catastrophe to happen,"[162] and Germans themselves later argued that the British had led them on, the Kaiser complaining of "the grossest deception" by the British.[163]

162. Ibid., Vol. 2, p. 644.
163. Ibid., p. 517. See also Tirpitz, quoted in ibid., Vol. 3, p. 189.

The British government indeed failed to convey a clear threat to the Germans until after the crisis was out of control, and the Germans apparently were misled by this. Jagow declared on July 26 that "we are sure of England's neutrality," while during the war the Kaiser wailed, "If only someone had told me beforehand that England would take up arms against us!"[164] However, this failure was not entirely the fault of British leaders; it also reflected their circumstances. First, they apparently felt hamstrung by the lack of a defensive policy option. Grey voiced fear that if he stood too firmly with France and Russia, they would grow too demanding, while Germany would feel threatened, and "Such a menace would but stiffen her attitude."[165]

Second, British leaders were unaware of the nature of the German policy to which they were forced to react until very late, which left them little time in which to choose and explain their response. Lulled by the Austro–German *fait accompli* strategy, they were unaware until July 23 that a crisis was upon them. On July 6, Arthur Nicolson, undersecretary of the British foreign office, cheerfully declared, "We have no very urgent and pressing question to preoccupy us in the rest of Europe."[166] They also were apparently unaware that a continental war would begin with a complete German conquest of Belgium, thanks to the dark secrecy surrounding the Liège operation. Britain doubtless would have joined the war even if Germany had not invaded Belgium, but the Belgian invasion provoked a powerful emotional response in Britain which spurred a quick decision on August 4. This reaction suggests that the British decision would have been clearer to the British, hence to the Germans, had the nature of the German operation been known in advance.

Thus the British failure to warn Germany was due as much to German secrecy as to British indecision. Albertini's condemnation of Grey seems unfair: governments cannot easily take national decisions for war in less than a week in response to an uncertain provocation. The ambiguity of British policy should be recognized as an artifact of the secret styles of the Central powers, which reflected the competitive politics and preemptive military doctrines of the times.

WHY SO MANY "BLUNDERS"?

Historians often ascribe the outbreak of the war to the blunders of a mediocre European leadership. Barbara Tuchman describes the Russian Czar as having

164. Ibid., Vol. 2, p. 429; and Tuchman, *Guns of August*, p. 143. See also Albertini, *Origins*, Vol. 2, pp. 514–527, 643–650; and Jarausch, "Illusion of Limited War."
165. Albertini, *Origins*, Vol. 2, p. 631; and Schmitt, *Coming of the War*, Vol. 2, p. 90.
166. Schmitt, *Coming of the War*, Vol. 1, pp. 417–418.

"a mind so shallow as to be all surface," and Albertini refers to the "untrained, incapable, dull-witted Bethmann-Hollweg," the "mediocrity of all the personages" in the German government, and the "short-sighted and unenlightened" Austrians. Ludwig Reiners devotes a chapter to "Berchtold's Blunders"; Michael Howard notes the "bland ignorance among national leaders" of defense matters; and Oron Hale claims that "the men who directed international affairs in 1914 were at the lowest level of competence and ability in several decades."[167]

Statesmen often did act on false premises or fail to anticipate the consequences of their actions during the July crisis. For instance, Russian leaders were initially unaware that a partial mobilization would impede a later general mobilization;[168] they probably exaggerated the military importance of mobilizing against Austria quickly;[169] they falsely believed Germany would acquiesce to their partial mobilization; they probably exaggerated the significance of the Austrian bombardment of Belgrade;[170] they falsely believed a general Russian mobilization could be concealed from Germany; and they mobilized without fully realizing that for Germany "mobilization meant war."[171]

German leaders encouraged Russia to believe that Germany would tolerate a partial Russian mobilization, and failed to explain to Entente statesmen that mobilization meant war, leading British and Russian leaders to assume that it did not.[172] They also badly misread European political sentiment, hoping that Italy, Sweden, Rumania, and even Japan would fight with the Central powers, and that Britain and Belgium would stand aside.[173] For their part, Britain and Italy failed to warn Germany of their policies; and Britain acquiesced to Russian mobilization, apparently without realizing that Russian mobilization meant German mobilization, which meant war. Finally, intelli-

---

167. Tuchman, *Guns of August*, p. 78; Albertini, *Origins*, Vol. 2, pp. 389, 436; Vol. 3, p. 253; Ludwig Reiners, *The Lamps Went Out in Europe* (New York: Pantheon, 1955), pp. 112–122; Howard quoted in Schelling, *Arms and Influence*, p. 243; and Oron J. Hale, *The Great Illusion: 1900–1914* (New York: Harper & Row, 1971), p. 285.
168. Albertini, *Origins*, Vol. 2, pp. 295–296.
169. See Turner, *Origins of the First World War*, pp. 92–93; Albertini, *Origins*, Vol. 2, p. 409; Vol. 3, pp. 230–231; but see also Lieven, *Russia and the Origins of the First World War*, pp. 148–149.
170. Reiners, *Lamps Went Out in Europe*, p. 135; and Albertini, *Origins*, Vol. 2, p. 553.
171. Albertini, *Origins*, Vol. 2, p. 574, 579–581; Vol. 3, pp. 56, 60–65.
172. Ibid., Vol. 2, pp. 332, 479–482, 485, 499–500, 550; Vol. 3, pp. 41–43, 61–65; Geiss, *July 1914*, pp. 245, 253, 266.
173. See Albertini, *Origins*, Vol. 2, pp. 334, 673, 678; Vol. 3, p. 233; Geiss, *July 1914*, pp. 226, 255, 302, 350–353; Schmitt, *Coming of the War*, Vol. 1, pp. 72–74, 322; Vol. 2, pp. 52–55, 149, 390n. Also relevant is Albertini, *Origins*, Vol. 2, pp. 308–309, 480, 541.

gence mistakes on both sides made matters worse. Russian leaders exaggerated German and Austrian mobilization measures, some German reports exaggerated Russian mobilizations, and French officials exaggerated German measures, which helped spur both sides to take further measures.[174]

What explains this plethora of blunders and accidents? Perhaps Europe was unlucky in the leaders it drew, but conditions in 1914 also made mistakes easy to make and hard to undo. Because secrecy was tight and *faits accomplis* were the fashion, facts were hard to acquire. Because windows were large and preemption was tempting, mistakes provoked rapid, dramatic reactions that quickly made the mistake irreversible. Statesmen seem like blunderers in retrospect partly because the international situation in 1914 was especially demanding and unforgiving of error. Historians castigate Grey for failing to rapidly take drastic national decisions under confusing and unexpected circumstances in the absence of domestic political consensus, and criticize Sazonov for his shaky grasp of military details on July 28 which no Russian civilian had had in mind five days earlier. The standard implicit in these criticisms is too stiff—statecraft seldom achieves such speed and precision. The blame for 1914 lies less with the statesmen of the times than with the conditions of the times and the severe demands these placed on statesmen.

BLAMECASTING

The explosive conditions created by the cult of the offensive made it easier for Germany to spark war without being blamed, by enabling that country to provoke its enemies to take defensive or preemptive steps which confused the question of responsibility for the war. German advocates of preventive war believed that Germany had to avoid blame for its outbreak, to preserve British neutrality and German public support for the war. Moreover, they seemed confident that the onus for war *could* be substantially shifted onto their opponents. Thus Moltke counselled war but warned that "the attack must be started by the Slavs,"[175] Bethmann Hollweg decreed that "we must

174. See generally Lebow, *Between Peace and War*, pp. 238–242; and Albertini, *Origins*, Vol. 3, pp. 67–68. For details on Russia see Albertini, *Origins*, Vol. 2, pp. 499, 545–546, 549, 566–567, 570–571, 576; Schmitt, *Coming of the War*, Vol. 2, pp. 97–98, 238, 244n.; Schilling, *How the War Began*, pp. 61–62; and Sazonov, *Fateful Years*, pp. 193, 199–200, 202–203. For details on France, see Joffre, *Personal Memoirs*, Vol. 1, pp. 117–128; and Albertini, *Origins*, Vol. 2, p. 647; Vol. 3, p. 67. On Germany see Trumpener, "War Premeditated?," pp. 73–74; Albertini, *Origins*, Vol. 2, pp. 529, 560, 637; Vol. 3, pp. 2–3, 6–9; and Geiss, *July 1914*, pp. 291–294.
175. In 1913, in Albertini, *Origins*, Vol. 2, p. 486.

give the impression of being forced into war,"[176] and Admiral von Müller summarized German policy during the July crisis as being to "keep quiet, letting Russia put herself in the wrong, but then not shying away from war."[177] "It is very important that we should appear to have been provoked" in a war arising from the Balkans, wrote Jagow, for "then—but probably only then—Britain can remain neutral."[178] And as the war broke out, von Müller wrote, "The mood is brilliant. The government has succeeded very well in making us appear as the attacked."[179]

These and other statements suggest an official German hope that German responsibility could be concealed. Moreover, whatever the source of this confidence, it had a sound basis in prevailing military conditions, which blurred the distinction between offensive and defensive conduct, and forced such quick reactions to provocation that the question of "who started it?" could later be obscured. Indeed, the German "innocence campaign" during and after the war succeeded for many years partly because the war developed from a rapid and complex chemistry of provocation and response which could easily be misconstrued by a willful propagandist or a gullible historian.[180] Defenders seemed like aggressors to the untrained eye, because all defended quickly and aggressively. Jack Snyder rightly points out elsewhere in this issue that German war plans were poorly adapted for the strategy of brinkmanship and peaceful expansion which many Germans pursued until 1914, but prevailing European military arrangements and beliefs also facilitated the deceptions in which advocates of preventive war believed Germany had to engage.

---

176. On July 27, 1914, in Fischer, *War of Illusions*, p. 486.
177. On July 27, in J.C.G. Röhl, "Admiral von Müller and the Approach of War, 1911–1914," *Historical Journal*, Vol. 12, No. 4 (1969), p. 669. In the same spirit, Bernhardi (who hoped for Russian rather than British neutrality) wrote before the war that the task of German diplomacy was to spur a French attack, continuing: "[W]e must not hope to bring about this attack by waiting passively. Neither France nor Russia nor England need to attack in order to further their interests. . . . [Rather] we must initiate an active policy which, without attacking France, will so prejudice her interests or those of England that both these States would feel themselves compelled to attack us. Opportunities for such procedures are offered both in Africa and in Europe. . . ." Bernhardi, *Germany and the Next War*, p. 280.
178. In 1913, in Fischer, *War of Illusions*, p. 212.
179. Röhl, "Admiral von Müller," p. 670.
180. On this innocence campaign, see Imanuel Geiss, "The Outbreak of the First World War and German War Aims," in Walter Laqueur and George L. Mosse, eds., *1914: The Coming of the First World War* (New York: Harper and Row, 1966), pp. 71–78.

Conclusion

The cult of the offensive was a major underlying cause of the war of 1914, feeding or magnifying a wide range of secondary dangers which helped pull the world to war. The causes of the war are often catalogued as an unrelated grab-bag of misfortunes which unluckily arose at the same time; but many shared a common source in the cult of the offensive, and should be recognized as its symptoms and artifacts rather than as isolated phenomena.

The consequences of the cult of the offensive are illuminated by imagining the politics of 1914 had European leaders recognized the actual power of the defense. German expansionists then would have met stronger arguments that empire was needless and impossible, and Germany could have more easily let the Russian military buildup run its course, knowing that German defenses could still withstand Russian attack. All European states would have been less tempted to mobilize first, and each could have tolerated more preparations by adversaries before mobilizing themselves, so the spiral of mobilization and counter-mobilization would have operated more slowly, if at all. If armies mobilized, they might have rushed to defend their own trenches and fortifications, instead of crossing frontiers, divorcing mobilization from war. Mobilizations could more easily have been confined to single frontiers, localizing the crisis. Britain could more easily have warned the Germans and restrained the Russians, and all statesmen could more easily have recovered and reversed mistakes made in haste or on false information. Thus the logic that led Germany to provoke the 1914 crisis would have been undermined, and the chain reaction by which the war spread outward from the Balkans would have been very improbable. In all likelihood, the Austro–Serbian conflict would have been a minor and soon-forgotten disturbance on the periphery of European politics.

This conclusion does not depend upon how one resolves the "Fischer controversy" over German prewar aims; while the outcome of the Fischer debate affects the *way* in which the cult caused the war, it does not affect the importance which the cult should be assigned. If one accepts the Fischer–Geiss–Röhl view that German aims were very aggressive, then one emphasizes the role of the cult in feeding German expansionism, German window thinking, and the German ability to catalyze a war while concealing responsibility for it by provoking a preemption by Germany's adversaries. If one believes that Germany was less aggressive, then one focuses on the role of the incentive to preempt in spurring the Russian and French decisions to mobilize, the nature of Russian and German mobilization plans, the British failure to restrain Russia and warn Germany, the scope and irreversibility of the effects of the Austro–German *fait accompli*, and the various other blunders of statesmen.[181] The cult of the offensive would play a different role in the history as taught by these two schools, but a central role in both.

The 1914 case thus supports Robert Jervis and other theorists who propose that an offense-dominant world is more dangerous, and warns both superpowers against the offensive ideas which many military planners in both countries favor. Offensive doctrines have long been dogma in the Soviet military establishment, and they are gaining adherents in the United States as well. This is seen in the declining popularity of the nuclear strategy of "assured destruction" and the growing fashionability of "counterforce" nuclear strategies,[182] which are essentially offensive in nature.[183]

The 1914 case bears directly on the debate about these counterforce strategies, warning that the dangers of counterforce include but also extend far beyond the well-known problems of "crisis instability" and preemptive war. If the superpowers achieved disarming counterforce capabilities, or if they believed they had done so, the entire political universe would be disturbed. The logic of self-protection in a counterforce world would compel much of the same behavior and produce the same phenomena that drove the world to war in 1914—dark political and military secrecy, intense competition for resources and allies, yawning windows of opportunity and vulnerability, intense arms-racing, and offensive and preemptive war plans of great scope and violence. Smaller political and military mistakes would have larger and less reversible consequences. Crises would be harder to control, since military

181. A useful review of the debate about German aims is Moses, *Politics of Illusion*.
182. On the growth of offensive ideas under the Reagan Administration, see Barry R. Posen and Stephen Van Evera, "Defense Policy and the Reagan Administration: Departure from Containment," *International Security*, Vol. 8, No. 1 (Summer 1983), pp. 24–30. On counterforce strategies, a recent critical essay is Robert Jervis, *The Illogic of American Nuclear Strategy* (Ithaca: Cornell University Press, 1984).
183. "Counterforce" forces include forces which could preemptively destroy opposing nuclear forces before they are launched, forces which could destroy retaliating warheads in flight towards the attacker's cities, and forces which could limit the damage which retaliating warheads could inflict on the attacker's society if they arrived. Hence, "counterforce" weapons and programs include highly accurate ICBMs and SLBMs (which could destroy opposing ICBMs) *and* air defense against bombers, ballistic missile defense for cities, and civil defense. Seemingly "defensive" programs such as the Reagan Administration's ballistic missile defense ("Star Wars") program and parallel Soviet ballistic missile defense programs are in fact *offensive* under the inverted logic of a MAD world. See Posen and Van Evera, "Defense Policy and the Reagan Administration," pp. 24–25.

alerts would open and close larger windows, defensive military preparations would carry larger offensive implications, and smaller provocations could spur preemptive attack. Arms control would be harder to achieve, since secrecy would impede verification and treaties which met the security requirements of both sides would be harder to frame, which would circumscribe the ability of statesmen to escape this frightful world by agreement.

"Assured destruction" leaves much to be desired as a nuclear strategy, and the world of "mutual assured destruction" ("MAD") which it fosters leaves much to be desired as well. But 1914 warns that we tamper with MAD at our peril: any exit from MAD to a counterforce world would create a much more dangerous arrangement, whose outlines we glimpsed in the First World War.

Civil-Military Relations and the Cult of the Offensive, 1914 and 1984

Jack Snyder

Military technology should have made the European strategic balance in July 1914 a model of stability, but offensive military strategies defied those technological realities, trapping European statesmen in a war-causing spiral of insecurity and instability. As the Boer and Russo–Japanese Wars had foreshadowed and the Great War itself confirmed, prevailing weaponry and means of transport strongly favored the defender. Tactically, withering firepower gave a huge advantage to entrenched defenders; strategically, defenders operating on their own territory could use railroads to outmaneuver marching invaders. Despite these inexorable constraints, each of the major continental powers began the war with an offensive campaign. These war plans and the offensive doctrines behind them were in themselves an important and perhaps decisive cause of the war. Security, not conquest, was the principal criterion used by the designers of the plans, but their net effect was to reduce everyone's security and to convince at least some states that only preventive aggression could ensure their survival.

Even if the outbreak of war is taken as a given, the offensive plans must still be judged disasters. Each offensive failed to achieve its ambitious goals and, in doing so, created major disadvantages for the state that launched it. Germany's invasion of Belgium and France ensured that Britain would join the opposing coalition and implement a blockade. The miscarriage of France's ill-conceived frontal attack almost provided the margin of help that the Schlieffen Plan needed. Though the worst was averted by a last-minute railway maneuver, the Germans nonetheless occupied a key portion of France's industrial northeast, making a settlement based on the status quo ante impossible to negotiate. Meanwhile, in East Prussia the annihilation of an over-extended Russian invasion force squandered troops that might have

---

Robert Jervis, William McNeill, Cynthia Roberts, and Stephen Van Evera provided helpful comments on this paper, which draws heavily on the author's forthcoming book, *The Ideology of the Offensive: Military Decision Making and the Disasters of 1914* (Ithaca, N.Y.: Cornell University Press, 1984).

Jack Snyder is an Assistant Professor in the Political Science Department, Columbia University.

been decisive if used to reinforce the undermanned advance into Austria. In each case, a defensive or more limited offensive strategy would have left the state in a more favorable strategic position.

None of these disasters was unpredictable or unpredicted. It was not only seers like Ivan Bloch who anticipated the stalemated positional warfare. General Staff strategists themselves, in their more lucid moments, foresaw these outcomes with astonishing accuracy. Schlieffen directed a war game in which he defeated his own plan with precisely the railway maneuver that Joffre employed to prevail on the Marne. In another German war game, which actually fell into Russian hands, Schlieffen used the advantage of railway mobility to defeat piecemeal the two prongs of a Russian advance around the Masurian Lakes—precisely the maneuver that led to the encirclement of Sazonov's Second Army at Tannenberg in August 1914. This is not to say that European war planners fully appreciated the overwhelming advantages of the defender; partly they underrated those advantages, partly they defied them. The point is that our own 20/20 hindsight is not qualitatively different from the understanding that was achievable by the historical protagonists.[1]

Why then were these self-defeating, war-causing strategies adopted? Although the particulars varied from country to country, in each case strategic policymaking was skewed by a pathological pattern of civil-military relations that allowed or encouraged the military to use wartime operational strategy to solve its institutional problems. When strategy went awry, it was because a penchant for offense helped the military organization to preserve its autonomy, prestige, and traditions, to simplify its institutional routines, or to resolve a dispute within the organization. As further discussion will show, it was not just a quirk of fate that offensive strategies served these functions. On balance, offense tends to suit the needs of military organizations better than defense does, and militaries normally exhibit at least a moderate preference for offensive strategies and doctrines for that reason. What was special about the period before World War I was that the state of civil-military relations in each of the major powers tended to exacerbate that normal offensive bias, either because the lack of civilian control allowed it to grow

1. Gerhard Ritter, *The Schlieffen Plan* (New York: Praeger, 1958), p. 60, note 34; A.A. Polivanov, *Voennoe delo*, No. 14 (1920), p. 421, quoted in Jack Snyder, *The Ideology of the Offensive: Military Decision Making and the Disasters of 1914* (Ithaca: Cornell University Press, 1984), chapter 7.

unchecked or because an abnormal degree of civil-military conflict heightened the need for a self-protective ideology.

In part, then, the "cult of the offensive" of 1914 reflected the endemic preference of military organizations for offensive strategies, but it also reflected particular circumstances that liberated or intensified that preference. The nature and timing of these catalytic circumstances, though all rooted in problems of civil-military relations, were different in each country. Indeed, if war had broken out as late as 1910, the Russian and French armies would both have fought quite defensively.[2]

Germany was the first European power to commit itself to a wildly overambitious offensive strategy, moving steadily in this direction from 1891 when Schlieffen became the Chief of the General Staff. The root of this pathology was the complete absence of civilian control over plans and doctrine, which provided no check on the natural tendency of mature military organizations to institutionalize and dogmatize doctrines that support the organizational goals of prestige, autonomy, and the elimination of novelty and uncertainty. Often, as in this case, it is offense that serves these interests best.[3]

France moved in 1911 from a cautious counteroffensive strategy towards the reckless frontal assault prescribed by the *offensive à outrance*. The roots of this doctrine can also be traced to a problem in civil-military relations. The French officer corps had always been wary of the Third Republic's inclination towards shorter and shorter terms of military service, which threatened the professional character and traditions of their organization. Touting the offense was a way to contain this threat, since everyone agreed that an army based on reservists and short-service conscripts would be good only for defense. The Dreyfus Affair and the radical military reforms that followed it heightened the officer corps' need for a self-protective ideology that would justify the essence and defend the autonomy of their organization. The extreme doctrine of the *offensive à outrance* served precisely this function, helping to discredit the defensive, reservist-based plans of the politicized

---

2. One reason that the war did not happen until 1914 was that Russian offensive power did not seriously threaten Germany until about that year. In this sense, the fact that all the powers had offensive strategies in the year the war broke out is to be explained more by their strategies' interactive consequences than by their common origins.

3. Snyder, *Ideology of the Offensive,* chapters 1, 4, and 5. I have profited greatly from the works of Barry Posen, *The Sources of Military Doctrine* (Ithaca: Cornell University Press, 1984), and Stephen Van Evera, "Causes of War" (Ph.D. dissertation, University of California, Berkeley, 1984), who advance similar arguments.

"republican" officers who ran the French military under civilian tutelage until the Agadir crisis of 1911. Given a freer rein in the harsher international climate, General Joffre and the Young Turks around him used the offensive doctrine to help justify a lengthening of the term of service and to reemphasize the value of a more highly professionalized army.[4]

Russia's drift towards increasingly overcommitted offensive plans between 1912 and 1914 was also abetted by the condition of civil-military relations. The problem in this case was the existence of two powerful veto groups within the military, one in the General Staff that favored an offensive against Germany and another centered on the Kiev military district that wanted to attack Austria. Forces were insufficient to carry out both missions, but there was no strong, centralized civilian authority who could or would enforce a rational priority commensurate with Russian means. Lacking firm civilian direction, the two military factions log-rolled the issue, each getting to implement its preferred offensive but with insufficient troops.[5]

It might be argued that these pathologies of civil-military relations are unique to the historical setting of this period. Civilians may have been ignorant of military affairs in a way that has been unequaled before or since. The transition in this period of the officer corps from an aristocratic caste to a specialized profession may have produced a uniquely unfavorable combination of the ill effects of both. Finally, social changes associated with rapid industrialization and urbanization may have provided a uniquely explosive setting for civil-military relations, as class conflicts reinforced civil-military conflicts.[6] Even if this is true, however, the same general patterns may persist but with lesser intensity, and understanding the circumstances that provoke more intense manifestations may help to forestall their recurrence.

Such a recurrence, whether intense or mild, is not a farfetched scenario. As in 1914, today's military technologies favor the defender of the status quo, but the superpowers are adopting offensive counterforce strategies in defiance of these technological constraints. Like machine guns and railroads, survivable nuclear weapons render trivial the marginal advantages to be gained by striking first. In the view of some, this stabilizing effect even neutralizes whatever first-strike advantages may exist at the conventional level, since the fear of uncontrollable escalation will restrain even the first

4. Snyder, *Ideology of the Offensive,* chapters 2 and 3. See also Samuel Williamson, *The Politics of Grand Strategy* (Cambridge: Harvard University Press, 1969).
5. Snyder, *Ideology of the Offensive,* chapters 6 and 7. See also A.M. Zaionchkovskii, *Podgotovka Rossii k imperialisticheskoi voine* (Moscow: Gosvoenizdat, 1926).
6. Van Evera, "Causes of War," chapter 7, explores these questions briefly.

steps in that direction. Since the would-be aggressor has the "last clear chance" to avoid disaster and normally cares less about the outcome than the defender does, mutual assured destruction works strongly for stability and the defense of the status quo. In this way, the absolute power to inflict punishment eases the security dilemma. All states possessing survivable second-strike forces can be simultaneously secure.[7]

Even those who are not entirely satisfied by the foregoing line of argument—and I include myself among them—must nevertheless admit the restraining effect that the irrevocable power to punish has had on international politics. Caveats aside, the prevailing military technology tends to work for stability, yet the strategic plans and doctrines of both superpowers have in important ways defied and undermined that basic reality. As in 1914, the danger today is that war will occur because of an erroneous belief that a disarming, offensive blow is feasible and necessary to ensure the attacker's security.

In order to understand the forces that are eroding the stability of the strategic balance in our own era, it may be helpful to reflect on the causes and consequences of the "cult of the offensive" of 1914. In proceeding towards this goal, I will discuss, first, how offensive strategies promoted war in 1914 and, second, why each of the major continental powers developed offensive military strategies. Germany will receive special attention because the Schlieffen Plan was the mainspring tightening the European security dilemma in 1914, because the lessons of the German experience can be more broadly generalized than those of the other cases, and because of the need to correct the widespread view that Germany's military strategy was determined by its revisionist diplomatic aims. After examining the domestic sources of military strategy in Germany, France, and Russia, I will discuss the effect of each state's policies on the civil-military relations and strategies of its neighbors. A concluding section will venture some possible applications of these findings to the study of contemporary Soviet military doctrine.

How Offense Promoted War

Conventional wisdom holds that World War I was caused in part by runaway offensive war plans, but historians and political scientists have been remark-

---

7. The best and most recent expression of this view is Robert Jervis, *The Illogic of American Nuclear Strategy* (Ithaca: Cornell University Press, 1984).

ably imprecise in reconstructing the logic of this process. Their vagueness has allowed critics of arms controllers' obsession with strategic instability to deny that the war resulted from "the reciprocal fear of surprise attack" or from any other by-product of offensive strategy.[8] Stephen Van Evera's contribution to this issue takes a major step towards identifying the manifold ways in which offensive strategies and doctrines promoted war in 1914. I would add only two points to his compelling argument. The first identifies some remaining puzzles about the perception of first-strike advantage in 1914; the second elaborates on Germany's incentive for preventive attack as the decisive way in which offensive military strategy led Europe towards war.

Van Evera cites statements and behavior indicating that European military and political decision-makers believed that the first army to mobilize and strike would gain a significant advantage. Fearing that their own preparations were lagging (or hoping to get a jump on the opponent), authorities in all of the countries felt pressed to take military measures that cut short the process of diplomacy, which might have converged on the solution of a "halt in Belgrade" if given more time. What is lacking in this story is a clear explanation of how the maximum gain or loss of two days could decisively affect the outcome of the campaign.

Planning documents suggest that no one believed that a two-day edge would allow a disarming surprise attack. Planners in all countries guarded against preemptive attacks on troops disembarking at railheads by concentrating their forces out of reach of such a blow. The only initial operation that depended on this kind of preemptive strike against unprepared forces was the German *coup de main* against the Belgian transport bottleneck of Liège. As the July crisis developed, the German General Staff was caused some anxiety by the progress of Belgian preparations to defend Liège, which jeopardized the smooth implementation of the Schlieffen Plan, but Moltke's attitude was not decisively influenced by this incentive to preempt.[9] In any event, it was Russia that mobilized first, and there is little to suggest that preemption was decisive in this case either. Prewar planning documents and

8. Even the usually crystal-clear Thomas Schelling is a bit murky on this point. See his *Arms and Influence* (New Haven: Yale University Press, 1966), pp. 221–225. For a critic, see Stephen Peter Rosen, "Nuclear Arms and Strategic Defense," *Washington Quarterly*, Vol. 4, No. 2 (Spring 1981), pp. 83–84.
9. Ulrich Trumpener, "War Premeditated? German Intelligence Operations in July 1914," *Central European History*, Vol. 9, No. 1 (March 1976), p. 80.

staff exercises show that the Russians worried about being preempted, but took sufficient precautions against it. They also indicate that preemption was not particularly feared if Austria was embroiled in the Balkans—precisely the conditions that obtained in July 1914. On the offensive side, however, the incentive to strike first might have been an important factor. Van Evera points out that the difference between the best case (mobilizing first) and the worst case (mobilizing second) was probably a net gain of four days (two gained plus two not lost). Given the Russians' aim of putting pressure on Germany's rear before the campaign in France was decided, four days was not a negligible consideration. To save just two days, the Russians were willing to begin their advance without waiting for the formation of their supply echelons. Thus, time pressure imposed by military exigencies may explain the haste of the crucial Russian mobilization. It should be stressed, however, that it was neither "the reciprocal fear of surprise attack" nor the chance of preempting the opponent's unalerted forces that produced this pressure. Rather, it was the desire to close Germany's window of opportunity against France that gave Russia an incentive to strike first.[10]

A second elaboration of Van Evera's argument, which will be crucial for understanding the following sections of this paper, is that offensive plans not only reflected the belief that states are vulnerable and conquest is easy; they actually caused the states adopting them to *be* vulnerable and consequently fearful. Even the Fischer school, which emphasizes Germany's "grasping for 'World Power'" as the primary cause of the war, admits that Germany's decision to provoke a conflict in 1914 was also due to the huge Russian army increases then in progress, which would have left Germany at Russia's mercy upon their completion in 1917.[11] This impending vulnerabil-

10. Russia, 10-i otdel General'nogo shtaba RKKA, *Vostochnoprusskaia operatsiia: sbornik dokumentov* (Moscow: Gosvoenizdat, 1939), especially p. 62, which reproduces a Russian General Staff intelligence estimate dated March 1, 1914. Van Evera's quotations suggest that decision-makers in all countries exhibited more concern about being preempted than seems warranted by actual circumstances. One explanation may be that the military oversold this danger as a way of guarding against the risk of excessive civilian foot-dragging, which was clearly a concern among the French military, at least. Another possibility is that there was a disconnect between the operational level of analysis, where it was obvious that no one could disrupt his opponent's concentration, and the more abstract level of doctrine, where the intangible benefits of "seizing the initiative" were nonetheless considered important. See Snyder, *Ideology of the Offensive*, chapters 2 and 3.

11. The Germans saw the planned 40 percent increase in the size of the Russian standing army as a threat to Germany's physical survival, not just a barrier foreclosing opportunities to expand. This is expressed most clearly in the fear that the power shift would allow Russia to force a

ity, though real enough, was largely a function of the Schlieffen Plan, which had to strip the eastern front in order to amass the forces needed to deal with the strategic conundrums and additional opponents created by the march through Belgium. If the Germans had used a positional defense on the short Franco–German border to achieve economies of force, they could have handled even the enlarged Russian contingents planned for 1917.[12]

In these ways, offensive strategies helped to cause the war and ensured that, when war occurred, it would be a world war. Prevailing technologies should have made the world of 1914 an arms controllers' dream; instead, military planners created a nightmare of strategic instability.

Germany: Uncontrolled Military or Militarized Civilians?

The offensive character of German war planning in the years before World War I was primarily an expression of the professional interests and outlook of the General Staff. Civilian foreign policy aims and attitudes about international politics were at most a permissive cause of the Schlieffen Plan. On balance, the General Staff's all-or-nothing war plan was more a hindrance than a help in implementing the diplomats' strategy of brinkmanship. The reason that the military was allowed to indulge its strategic preferences was not so much that the civilians agreed with them; rather, it was because war planning was considered to be within the autonomous purview of the General Staff. Military preferences were never decisive on questions of the use of force, however, since this was not considered their legitimate sphere. But indirectly, war plans trapped the diplomats by handing them a blunt instrument suitable for massive preventive war, but ill-designed for controlled coercion. The military's unchecked preference for an unlimited offensive strategy and the mismatch between German military and diplomatic strategy were important causes of strategic instability rooted in the problem of civil-military relations. This section will trace those roots and point out some implications relevant to contemporary questions.

The Schlieffen Plan embodied all of the desiderata commonly found in field manuals and treatises on strategy written by military officers: it was an

---

revision of the status quo in the Balkans, leading to Austria's collapse. See especially Fritz Fischer, *War of Illusions: German Policies from 1911 to 1914* (New York: W.W. Norton, 1975; German edition 1969), pp. 377–379, 427.

12. This is argued in Snyder, *Ideology of the Offensive*, chapter 4.

offensive campaign, designed to seize the initiative, to exploit fleeting opportunities, and to achieve a decisive victory by the rapid annihilation of the opponents' military forces. War was to be an "instrument of politics," not in the sense that political ends would restrain and shape military means, but along lines that the General Staff found more congenial: war would solve the tangle of political problems that the diplomats could not solve for themselves. "The complete defeat of the enemy always serves politics," argued General Colmar von der Goltz in his influential book, *The Nation in Arms*. "Observance of this principle not only grants the greatest measure of freedom in the political sphere but also gives widest scope to the proper use of resources in war."[13]

To do this, Schlieffen sought to capitalize on the relatively slow mobilization of the Russian army, which could not bring its full weight to bear until the second month of the campaign. Schlieffen reasoned that he had to use this "window of opportunity" to decisively alter the balance of forces in Germany's favor. Drawing on precedents provided by Moltke's campaigns of 1866 and 1870 as well as his later plans for a two-front war, Schlieffen saw that a rapid decision could be achieved only by deploying the bulk of the German army on one front in order to carry out a grandiose encirclement maneuver. France had to be the first victim, because the Russians might spoil the encirclement by retreating into their vast spaces. With Paris at risk, the French would have to stand and fight. By 1897, Schlieffen had concluded that this scheme could not succeed without traversing Belgium, since the Franco–German frontier in Alsace–Lorraine was too narrow and too easily defended to permit a decisive maneuver. In the mature conception of 1905, most of the German army (including some units that did not yet exist) would march for three or four weeks through Belgium and northern France, encircling and destroying the French army, and then board trains for the eastern front to reinforce the few divisions left to cover East Prussia.

Even Schlieffen was aware that his plan was "an enterprise for which we are too weak."[14] He and his successor, the younger Moltke, understood most of the pitfalls of this maneuver quite well: the gratuitous provocation of new enemies, the logistical nightmares, the possibility of a rapid French rede-

---

13. Gerhard Ritter, *The Sword and the Scepter: The Problem of Militarism in Germany* (Coral Gables: University of Miami Press, 1969; German edition 1954), Vol. 1, p. 196, citing *Das Volk in Waffen* (5th ed., 1889), p. 129.
14. Ritter, *Schlieffen Plan*, p. 66.

ployment to nullify the German flank maneuver, the numerical insufficiency of the Germany army, the tendency of the attacker's strength to wane with every step forward and the defender's to grow, and the lack of time to finish with France before Russia would attack. The General Staff clung to this plan not because they were blind to its faults, but because they thought all the alternatives were worse. To mollify Austria in 1912, they went through the motions of gaming out a mirror-image of the Schlieffen Plan pointed towards the east, concluding that the French would defeat the weak forces left in the Rhineland long before a decision could be reached against Russia.[15] What the General Staff refused to consider seriously after 1890 was the possibility of an equal division of their forces between west and east, allowing a stable defensive against France and a limited offensive with Austria against Russia. (This was the combination that Germany used successfully in 1915 and that the elder Moltke had resigned himself to in the 1890s.)

Around the turn of the century, the General Staff played some war games based on a defensive in the west. These led to the embarrassing conclusion that the French would have great difficulty overwhelming even a modest defensive force. In future years, when games with this premise were played, the German defenders were allotted fewer forces, while Belgians and Dutch were arbitrarily added to the attacking force. Stacking the deck against the defensive appeared not only in war-gaming but also in Schlieffen's abstract expostulations of doctrine. Even some German critics caught him applying a double standard, arbitrarily granting the attacker advantages in mobility, whereas the reality should have been quite the opposite.[16]

In short, German war planning, especially after 1890, showed a strong bias in favor of offensive schemes for decisive victory and against defensive or more limited offensive schemes, even though the latter had a greater prospect of success. This bias cannot be explained away by the argument that Germany would have been at an economic disadvantage in a long war against Russia and hence had to gamble everything on a quick victory. As the actual war showed, this was untrue. More important, Schlieffen hit upon economic rationalizations for his war plan only after it had already been in place for years. Moreover, he actively discouraged serious analysis of wartime economics, deciding *a priori* that the only good war was a short war and that

15. Louis Garros, "Préludes aux invasions de la Belgique," *Revue historique de l'armée* (March 1949), pp. 37–38; French archival documents cited in Snyder, *Ideology of the Offensive*, chapter 4.
16. Friedrich von Bernhardi, *On War of Today* (London: Rees, 1912), Vol. 1, p. 44.

the only way to end a war quickly was to disarm the opponent decisively.[17] These conclusions were not in themselves unreasonable, but Schlieffen reached them before he did his analysis and then arranged the evidence in order to justify his preferred strategy.

The explanation for the General Staff's bias in favor of offensive strategy is rooted in the organizational interests and parochial outlook of the professional military. The Germans' pursuit of a strategy for a short, offensive, decisive war despite its operational infeasibility is simply an extreme case of an endemic bias of military organizations. Militaries do not always exhibit a blind preference for the offensive, of course. The lessons of 1914–1918 had a tempering effect on the offensive inclinations of European militaries, for example.[18] Still, exceptions and questionable cases notwithstanding, initial research indicates that militaries habitually prefer offensive strategies, even though everyone from Clausewitz to Trevor Dupuy has proved that the defender enjoys a net operational advantage.[19]

EXPLAINING THE OFFENSIVE BIAS

Several explanations for this offensive bias have been advanced. A number of them are consistent with the evidence provided by the German case. A particularly important explanation stems from the division of labor and the narrow focus of attention that necessarily follows from it. The professional training and duties of the soldier force him to focus on threats to his state's security and on the conflictual side of international relations. Necessarily preoccupied with the prospect of armed conflict, he sees war as a pervasive aspect of international life. Focusing on the role of military means in ensuring the security of the state, he forgets that other means can also be used towards that end. For these reasons, the military professional tends to hold a simplified, zero-sum view of international politics and the nature of war, in which wars are seen as difficult to avoid and almost impossible to limit.

17. Lothar Burchardt, *Friedenswirtschaft und Kriegsvorsorge: Deutschlands wirtschaftliche Rüstungsbestrebungen vor 1914* (Boppard am Rhein: Boldt, 1968), pp. 15, 163–164.
18. However, this effect should not be overdrawn. Barry Posen, *Sources of Military Doctrine*, has recently demonstrated that the French collapse in 1940 was due not to a Maginot Line mentality but to the overcommitment of forces to the offensive campaign in Belgium.
19. Possible biases in civilian views on offense and defense have not been studied systematically. For Trevor Dupuy's attempts to analyze quantitatively offensive and defensive operations in World War II, see his *Numbers, Predictions and War* (New York: Bobbs-Merrill, 1979), chapter 7, and other publications of his "HERO" project.

When the hostility of others is taken for granted, prudential calculations are slanted in favor of preventive wars and preemptive strikes. Indeed, as German military officers were fond of arguing, the proper role of diplomacy in a Hobbesian world is to create favorable conditions for launching preventive war. A preventive grand strategy requires an offensive operational doctrine. Defensive plans and doctrines will be considered only after all conceivable offensive schemes have been decisively discredited. Under uncertainty, such discrediting will be difficult, so offensive plans and doctrines will frequently be adopted even if offense is not easier than defense in the operational sense.

The assumption of extreme hostility also favors the notion that decisive, offensive operations are always needed to end wars. If the conflict of interest between the parties is seen as limited, then a decisive victory may not be needed to end the fighting on mutually acceptable terms. In fact, denying the opponent his objectives by means of a successful defense may suffice. However, when the opponent is believed to be extremely hostile, disarming him completely may seem to be the only way to induce him to break off his attacks. For this reason, offensive doctrines and plans are needed, even if defense is easier operationally.

Kenneth Waltz argues that states are socialized to the implications of international anarchy.[20] Because of their professional preoccupations military professionals become "oversocialized." Seeing war more likely than it really is, they increase its likelihood by adopting offensive plans and buying offensive forces. In this way, the perception that war is inevitable becomes a self-fulfilling prophecy.

A second explanation emphasizes the need of large, complex organizations to operate in a predictable, structured environment. Organizations like to work according to a plan that ties together the standard operating procedures of all the subunits into a prepackaged script. So that they can stick to this script at all costs, organizations try to dominate their environment rather than react to it. Reacting to unpredictable circumstances means throwing out the plan, improvising, and perhaps even deviating from standard operating procedures. As Barry Posen points out, "taking the offensive, exercising the initiative, is a way of structuring the battle."[21] Defense, in contrast, is more reactive, less structured, and harder to plan. Van Evera argues that the

---

20. Kenneth N. Waltz, *Theory of International Politics* (Reading, Mass.: Addison-Wesley, 1979).
21. Posen, *Sources of Military Doctrine*, chapter 2.

military will prefer a task that is easier to plan even if it is more difficult to execute successfully.[22] In Russia, for example, regional staffs complained that the General Staff's defensive war plan of 1910 left their own local planning problem too unstructured. They clamored for an offensive plan with specified lines of advance, and in 1912 they got it.[23]

The German military's bias for the offensive may have derived in part from this desire to structure the environment, but evidence on this point is mixed. The elder Moltke developed clockwork mobilization and rail transport plans leading to offensive operations, but he scoffed at the idea that a campaign plan could be mapped out step-by-step from the initial deployment through to the crowning encirclement battle. For him, strategy remained "a system of *ad hoc* expedients . . . , the development of an original idea in accordance with continually changing circumstances."[24] This attitude may help to explain his willingness to entertain defensive alternatives when his preferred offensive schemes began to look too unpromising. The Schlieffen Plan, in contrast, was a caricature of the link between rigid planning and an unvarying commitment to the offensive. Even here, however, there is some evidence that fits poorly with the hypothesis that militaries prefer offense because it allows them to fight according to their plans and standard operating procedures. Wilhelm Groener, the General Staff officer in charge of working out the logistical preparations for the Schlieffen Plan, recognized full well that the taut, ambitious nature of the plan would make it impossible to adhere to normal, methodical supply procedures. Among officers responsible for logistics, "the feeling of responsibility must be so great that in difficult circumstances people free themselves from procedural hindrances and take the responsibility for acting in accordance with common sense."[25] Nonetheless, it is difficult to ignore the argument ubiquitously advanced by European military writers that defense leads to uncertainty, confusion, passivity, and incoherent action, whereas offense focuses the efforts of the army and the mind of the commander on a single, unwavering goal. Even when they understood the uncertainties and improvisations required by offensive operations, as Groener did, they may still have feared the uncertainties of the defensive more. An offensive plan at least gives the illusion of certainty.

---

22. Van Evera, "Causes of War," chapter 7.
23. Zaionchkovskii, *Podgotovka Rossii k imperialisticheskoi voine*, pp. 244, 277.
24. Quoted by Hajo Holborn, "Moltke and Schlieffen," in Edward M. Earle, ed., *Makers of Modern Strategy* (Princeton: Princeton University Press, 1971), p. 180.
25. Papers of Wilhelm Groener, U.S. National Archives, roll 18, piece 168, p. 5.

Another possibility, however, is that this argument for the offensive was used to justify a doctrine that was preferred primarily on other grounds. French military publicists invoked such reasoning more frequently, for example, during periods of greater threat to traditional military institutions.[26]

Other explanations for the offensive bias are rooted even more directly in the parochial interests of the military, including the autonomy, prestige, size, and wealth of the organization.[27] The German case shows the function of the offensive strategy as a means towards the goal of operational autonomy. The elder Moltke succinctly stated the universal wish of military commanders: "The politician should fall silent the moment that mobilization begins."[28] This is least likely to happen in the case of limited or defensive wars, where the whole point of fighting is to negotiate a diplomatic solution. Political considerations—and hence politicians—have to figure in operational decisions. The operational autonomy of the military is most likely to be allowed when the operational goal is to disarm the adversary quickly and decisively by offensive means. For this reason, the military will seek to force doctrine and planning into this mold.

The prestige, self-image, and material health of military institutions will prosper if the military can convince civilians and themselves that wars can be short, decisive, and socially beneficial. One of the attractions of decisive, offensive strategies is that they hold out the promise of a demonstrable return on the nation's investment in military capability. Von der Goltz, for example, pushed the view that "modern wars have become the nation's way of doing business"—a perspective that made sense only if wars were short, cheap, and hence offensive.[29] The German people were relatively easy to convince of this, because of the powerful example provided by the short, offensive, nation-building wars of 1866 and 1870, which cut through political fetters and turned the officer corps into demigods. This historical backdrop gave the General Staff a mantel of unquestioned authority and legitimacy in operational questions; it also gave them a reputation to live up to. Later, when technological and strategic circumstances challenged the viability of their

---

26. See the argument in Snyder, *Ideology of the Offensive*, chapter 3, citing especially Georges Gilbert, *Essais de critique militaire* (Paris: Librairie de la Nouvelle Revue, 1890), pp. 43, 47–48.
27. Posen and Van Evera, in analyzing organizational interests in this way, have drawn on the categories laid out by Morton Halperin, *Bureaucratic Politics and Foreign Policy* (Washington: Brookings, 1974), chapter 3.
28. Quoted by Bernard Brodie, *War and Politics* (New York: Macmillan, 1973), p. 11.
29. Quoted by Van Evera from Ferdinand Foch, *The Principles of War* (New York: Fly, 1918), p. 37.

formula for a short, victorious war, General Staff officers like Schlieffen found it difficult to part with the offensive strategic formulae that had served their state and organization so effectively. As Posen puts it, offense makes soldiers "specialists in victory," defense makes them "specialists in attrition," and in our own era mutual assured destruction makes them "specialists in slaughter."[30]

THE EVOLUTION OF GERMAN WAR PLANNING

The foregoing arguments could, for the most part, explain the offensive bias of the military in many countries and in many eras. What remains to be explained is why this offensive bias became so dogmatic and extreme in Germany before 1914. The evolution of the General Staff's strategic thinking from 1870 to 1914 suggests that a tendency towards doctrinal dogmatism and extremism may be inherent in mature military organizations that develop under conditions of near-absolute autonomy in doctrinal questions. This evolution, which occurred in three stages, may be typical of the maturation of uncontrolled, self-evaluating organizations and consequently may highlight the conditions in which doctrinal extremism might recur in our own era.[31]

The first stage was dominated by the elder Moltke, who established the basis tenets of the organizational ideology of the German General Staff. These were the inevitability and productive nature of war, the indispensability of preventive war, and the need for an operational strategy that could provide rapid, decisive victories. Moltke was the creator, not a captive of his doctrines and did not implement them in the manner of a narrow technician. He was willing to think in political terms and to make his opinion heard in political matters. This practice had its good and bad sides. On one hand, it allowed him to consider war plans that gave diplomacy some role in ending the war; on the other, it spurred him to lobby for preventive war against France in 1868 and against Russia in 1887. Moltke thought he understood what international politics was all about, but he understood it in a military way. In judging the opportune moment for war, Moltke looked exclusively at military factors, whereas Bismarck focused primarily on preparing domestic and foreign opinion for the conflict.[32]

30. Posen, *Sources of Military Doctrine.*
31. Van Evera uses the concept of the self-evaluating organization, drawing on the work of James Q. Wilson.
32. Ritter, *Sword and Scepter,* Vol. 1, pp. 217–218, 245.

Schlieffen, the key figure in the second stage of the General Staff's development, was much more of a technocrat than Moltke. Not a founder, he was a systematizer and routinizer. Schlieffen dogmatized Moltke's strategic precepts in a way that served the mature institution's need for a simple, standardized doctrine to facilitate the training of young officers and the operational planning of the General Staff. In implementing this more dogmatic doctrine, Schlieffen and his colleagues lacked Moltke's ability to criticize fundamental assumptions and tailor doctrine to variations in circumstances. Thus, Moltke observed the defender's increasing advantages and decided reluctantly that the day of the rapid, decisive victory was probably gone, anticipating that "two armies prepared for battle will stand opposite each other, neither wishing to begin battle."[33] Schlieffen witnessed even further developments in this direction in the Russo–Japanese War, but concluded only that the attacker had to redouble his efforts. "The armament of the army has changed," he recognized, "but the fundamental laws of combat remain the same, and one of these laws is that one cannot defeat the enemy without attacking."[34]

Seeing himself as primarily a technician, Schlieffen gave political considerations a lesser place in his work than had Moltke. Again, this had both good and bad consequences. On one hand, Schlieffen never lobbied for preventive war in the way Moltke and Waldersee had, thinking such decisions were not his to make. When asked, of course, he was not reluctant to tell the political authorities that the time was propitious, as he did in 1905. On the other hand, Schlieffen had a more zero-sum, apolitical view of the conduct of warfare than did the elder Moltke. Consequently, his war plans excluded any notion of political limitations on the conduct of war or diplomatic means to end it.[35]

Contrasting the problems of civilian control of the military in stages one and two, we see that the founders' generation, being more "political," chal-

33. Helmuth von Moltke, *Die Deutschen Aufmarschpläne, 1871–1890*, Ferdinand von Schmerfeld, ed. (Berlin: Mittler, 1929), p. 122ff.

34. The quotation is from an 1893 comment on an operational exercise, quoted by O. von Zoellner, "Schlieffens Vermächtnis," *Militärwissenschaftliche Rundschau* (Sonderheft, 1938), p. 18, but identical sentiments are expressed in Schlieffen's "Krieg in der Gegenwart," *Deutsche Revue* (1909).

35. Brodie, *War and Politics*, p. 58, reports a perhaps apocryphal statement by Schlieffen that if his plan failed to achieve decisive results, then Germany should negotiate an end to the war. Even if he did say this, the possibility of negotiations had no effect on his war planning, in contrast to that of the elder Moltke.

lenges the political elite on questions of the use of force, but as if in compensation, is more capable of self-evaluation and self-control in its war planning. The technocratic generation, however, is less assertive politically but also less capable of exercising political judgment in its own work. The founders' assertiveness is the more dramatic challenge to political control, but as the German case shows, Bismarck was able to turn back the military's direct lobbying for preventive war, which was outside of the military's legitimate purview even by the Second Reich's skewed standards of civil-military relations. Much more damaging in the long run was Schlieffen's unobtrusive militarism, which created the conditions for a preventive war much more surely than Moltke's overt efforts did.

A third stage, which was just developing on the eve of World War I, combined the worst features of the two previous periods. Exemplary figures in this final stage were Erich Ludendorff and Wilhelm Groener, products of a thoroughgoing socialization to the organizational ideology of the German General Staff. Groener, describing his own war college training, makes it clear that not only operational principles but also a militaristic philosophy of life were standard fare in the school's curriculum. These future functionaries and leaders of the General Staff were getting an intensive course in the same kind of propaganda that the Army and Navy Leagues were providing the general public. They came out of this training believing in the philosophy of total war, demanding army increases that their elders were reluctant to pursue and fearing that "weaklings" like Bethmann Hollweg would throw away the army's glorious victories.[36]

An organizational explanation for this third stage would point to the self-amplifying effects of the organizational ideology in a mature, self-evaluating unit. An alternative explanation also seems plausible, however. Geoff Eley, in his study of right-wing radical nationalism in Wilhelmine Germany, argues that emerging counterelites used national populist causes and institutions like the Navy and Army Leagues as weapons aimed at the political monopoly retained by the more cautious traditional elite, who were vulnerable to criticism on jingoistic issues.[37] This pattern fits the cases of Groener and Ludendorff, who were middle-class officers seeking the final transformation of the

36. Helmut Haeussler, *General William Groener and the Imperial German Army* (Madison: The State Historical Society of Wisconsin, 1962), p. 72.
37. Geoff Eley, *Reshaping the German Right: Radical Nationalism and Political Change after Bismarck* (New Haven: Yale University Press, 1980).

old Prussian army into a mass organ of total war, which would provide upward mobility for their own kind. German War Ministers, speaking for conservative elements in the army and the state, had traditionally resisted large increases in the size of the army, which would bring more bourgeois officers into the mess and working-class soldiers into the ranks; it would also cost so much that the Junkers' privileged tax status would be brought into question. This alternative explanation makes it difficult to know whether organizational ideologies really tend toward self-amplification or whether extremist variants only occur from some particular motivation, as the French case suggests.

THE MISMATCH BETWEEN MILITARY STRATEGY AND DIPLOMACY

It is sometimes thought that Germany required an unlimited, offensive military strategy because German civilian elites were hell-bent on overturning the continental balance of power as a first step in their drive for "World Power." In this view, the Schlieffen Plan was simply the tool needed to achieve this high-risk, high-payoff goal, around which a national consensus of both military and civilian elites had formed.[38] There are several problems with this view. The first is that the civilians made virtually no input into the strategic planning process. Contrary to the unsupported assertions of some historians, the shift from Moltke's plan for a limited offensive against Russia to Schlieffen's plan for a more decisive blow aimed at France had nothing to do with the fall of Bismarck or the "New Course" in foreign policy. Rather, Schlieffen saw it as a technical change, stemming from an improved Russian ability to defend their forward theater in Poland. Nor was Schlieffen chosen to head the General Staff because of the strategy he preferred. Schlieffen had simply been the next in line as deputy chief under Waldersee, who was fired primarily because he dared to criticize the Kaiser's tactical decisions in a mock battle.[39] Later, when Reich Chancellor von Bülow learned of Schlieffen's intention to violate Belgian neutrality, his reaction was: "if the Chief of Staff, especially a strategic authority such as Schlieffen, believes such a measure to be necessary, then it is the obligation of diplomacy to adjust to it and prepare for it in every possible way."[40] In 1912 Foreign Secretary von

---

38. See, for example, L.L. Farrar, Jr., *Arrogance and Anxiety* (Iowa City: University of Iowa Press, 1981), pp. 23–24.
39. Ritter, *Schlieffen Plan*, pp. 17–37; Norman Rich and M.H. Fisher, eds., *The Holstein Papers* (Cambridge: Cambridge University Press, 1963), Vol. 3, pp. 347, note 1, and 352–353.
40. Ritter, *Schlieffen Plan*, pp. 91–92.

Jagow urged a reevaluation of the need to cross Belgian territory, but a memo from the younger Moltke ended the matter.[41] In short, the civilians knew what Schlieffen was planning to do, but they were relatively passive bystanders in part because military strategy was not in their sphere of competence and legitimate authority, and perhaps also because they were quite happy with the notion that the war could be won quickly and decisively. This optimism alleviated their fear that a long war would mean the destruction of existing social and economic institutions, no matter who won it. The decisive victory promised by the Schlieffen Plan may have also appealed to civilian elites concerned about the need for spectacular successes as a payoff for the masses' enthusiastic participation in the war. Trying to justify the initial war plan from the retrospective vantage point of 1919, Bethmann Hollweg argued that "offense in the East and defense in the West would have implied that we expected at best a draw. With such a slogan no army and no nation could be led into a struggle for their existence."[42] Still, this is a long way from the totally unfounded notion that Holstein and Schlieffen cooked up the Schlieffen Plan expressly for the purpose of bullying France over the Morocco issue and preparing the way for "Welt Politik."[43] The Schlieffen Plan had some appeal for German civilian elites, but the diplomats may have had serious reservations about it, as the Jagow episode suggests. Mostly, the civilians passively accepted whatever operational plan the military deemed necessary.

If German diplomats had devised a military strategy on their own, it is by no means certain that they would have come up with anything like the Schlieffen Plan. This all-or-nothing operational scheme fit poorly with the diplomatic strategy of expansion by means of brinkmanship and controlled, coercive pressure, which they pursued until 1914. In 1905, for example, it is clear that Bülow, Holstein, and Wilhelm II had no inclination to risk a world war over the question of Morocco.

"The originators of *Weltpolitik* looked forward to a series of small-scale, marginal foreign policy successes," says historian David Kaiser, "not to a major war."[44] Self-deterred by the unlimited character of the Schlieffen Plan,

41. Fischer, *War of Illusions*, p. 390.
42. Konrad Jarausch, *The Enigmatic Chancellor* (New Haven: Yale University Press, 1973), p. 195.
43. This is implied by Martin Kitchen, *The German Officer Corps* (Oxford: Oxford University Press, 1968), p. 104, and Imanuel Geiss, *German Foreign Policy, 1871–1914* (London: Routledge & Kegan Paul, 1976), pp. 101–103.
44. David E. Kaiser, "Germany and the Origins of the First World War," *Journal of Modern History*, Vol. 55, No. 3 (September 1983), p. 448.

they had few military tools that they could use to demonstrate resolve in a competition in risk-taking. The navy offered a means for the limited, demonstrative use of force, namely the dispatch of the gunboat *Panther* to the Moroccan port of Agadir, but the army was an inflexible tool. At one point in the crisis, Schlieffen told Bülow that the French were calling up reservists on the frontier. If this continued, Germany would have to respond, setting off a process that the Germans feared would be uncontrollable.[45] Thus, the German military posture and war plan served mainly to deter the German diplomats, who did not want a major war even though Schlieffen told them the time was favorable. They needed limited options, suitable for coercive diplomacy, not unlimited options, suitable for preventive war. With the Schlieffen Plan, they could not even respond to the opponent's precautionary moves without setting off a landslide toward total war.

This mismatch between military and diplomatic strategy dogged German policy down through 1914. Bethmann Hollweg described his strategy in 1912 as one of controlled coercion, sometimes asserting German demands, sometimes lulling and mollifying opponents to control the risk of war. "On all fronts we must drive forward quietly and patiently," he explained, "without having to risk our existence."[46] Bethmann's personal secretary, Kurt Riezler, explained this strategy of calculated risk in a 1914 volume, *Grundzüge der Weltpolitik*. A kind of cross between Thomas Schelling and Norman Angell, Riezler explained that wars were too costly to actually fight in the modern, interdependent, capitalist world. Nonetheless, states can still use the threat of war to gain unilateral advantages, forcing the opponent to calculate whether costs, benefits, and the probability of success warrant resorting to force. His calculations can be affected in several ways. Arms-racing can be used, *á la* Samuel Huntington, as a substitute for war—that is, a bloodless way to show the opponent that he would surely lose if it came to a fight. Brinkmanship and bluffing can be used to demonstrate resolve; *faits accomplis* and salami tactics can be used to shift the onus for starting the undesired war onto the opponent. But, Riezler warns, this strategy will not work if one is greedy and impatient to overturn the balance of power. Opponents will fight if they sense that their vital interests are at stake. Consequently, "victory

---

45. Holstein to Radolin, June 28, 1905, in *Holstein Papers*, Vol. 4, p. 347.
46. Jarausch, *Enigmatic Chancellor*, pp. 110–111.

belongs to the steady, tenacious, and gradual achievement of small successes . . . without provocation."[47]

Although this may have been a fair approximation of Bethmann's thinking in 1912, the theory of the calculated risk had undergone a major transformation by July 1914. By that time, Bethmann wanted a major diplomatic or military victory and was willing to risk a continental war—perhaps even a world war—to achieve it. *Fait accompli* and onus-shifting were still part of the strategy, but with a goal of keeping Britain out of the war and gaining the support of German socialists, not with a goal of avoiding war altogether.

The Schlieffen Plan played an important role in the transformation of Bethmann's strategy and in its failure to keep Britain neutral in the July crisis. Riezler's diary shows Bethmann's obsession in July 1914 with Germany's need for a dramatic victory to forestall the impending period of vulnerability that the Russian army increases and the possible collapse of Austria-Hungary would bring on.[48] As I argued earlier, the Schlieffen Plan only increased Germany's vulnerability to the Russian buildup, stripping the eastern front and squandering forces in the vain attempt to knock France out of the war. In this sense, it was the Schlieffen Plan that led Bethmann to transform the calculated-risk theory from a cautious tool of coercive diplomacy into a blind hope of gaining a major victory without incurring an unwanted world war.

Just as the Schlieffen Plan made trouble for Bethmann's diplomacy, so too German brinkmanship made trouble for the Schlieffen Plan. The Russian army increases, provoked by German belligerence in the 1909 Bosnian crisis and Austrian coercion of the Serbs in 1912, made the German war plan untenable.[49] The arms-racing produced by this aggressive diplomacy was not a "substitute for war"; rather, it created a window of vulnerability that helped to cause the war. Thus, Riezler (and Bethmann) failed to consider how easily a diplomatic strategy of calculated brinkmanship could set off a chain of uncontrollable consequences in a world of military instability.

---

47. Andreas Hillgruber, *Germany and the Two World Wars* (Cambridge: Harvard University Press, 1981), pp. 22–24; J.J. Ruedorffer (pseud. for Kurt Riezler), *Grundzüge der Weltpolitik in der Gegenwart* (Berlin: Deutsche Verlags-Anstalt, 1914), especially pp. 214–232; quotation from Jarausch, *Enigmatic Chancellor*, pp. 143–144.
48. Jarausch, *Enigmatic Chancellor*, p. 157.
49. P.A. Zhilin, "Bol'shaia programma po usileniiu russkoi armii," *Voenno-istoricheskii zhurnal*, No. 7 (July 1974), pp. 90–97, shows the connection between the 1913 increases and the Balkan crisis of 1912. He also shows that this project, with its emphasis on increasing the standing army and providing rail lines to speed its concentration, was directly connected to the offensive character of Russia's increasingly overcommitted, standing-start, short-war campaign plan.

Even the transformed version of the calculated-risk theory, implemented in July 1914, was ill-served by the Schlieffen Plan. If Bethmann had had eastern-oriented or otherwise limited military options, all sorts of possibilities would have been available for defending Austria, bloodying the Russians, driving a wedge between Paris and St. Petersburg, and keeping Britain neutral. In contrast, the Schlieffen Plan cut short any chance for coercive diplomacy and ensured that Britain would fight. In short, under Bethmann as well as Bülow, the Schlieffen Plan was hardly an appropriate tool underwriting the brinkmanship and expansionist aims of the civilian elite. Rather, the plan was the product of military organizational interests and misconceptions that reduced international politics to a series of preventive wars. The consequences of the all-or-nothing war plan were, first, to reduce the coercive bargaining leverage available to German diplomats, and second, to ensnare German diplomacy in a security dilemma that forced the abandonment of the strategy of controlled risks. Devised by military officers who wanted a tool appropriate for preventive war, the Schlieffen Plan trapped Germany in a situation where preventive war seemed like the only safe option.

In summary, three generalizations emerge from the German case. First, military organizations tend to exhibit a bias in favor of offensive strategies, which promote organizational prestige and autonomy, facilitate planning and adherence to standard operating procedures, and follow logically from the officer corps' zero-sum view of international politics. Second, this bias will be particularly extreme in mature organizations which have developed institutional ideologies and operational doctrines with little civilian oversight. Finally, the destabilizing consequences of an inflexible, offensive military strategy are compounded when it is mismatched with a diplomatic strategy based on the assumption that risks can be calculated and controlled through the skillful fine-tuning of threats.

France: Civil-Military Truce and Conflict

France before the Dreyfus Affair exemplifies the healthiest pattern of civil-military relations among the European states, but after Dreyfus, the most destructive. In the former period civilian defense experts who understood and respected the military contained the latent conflict between the professional army and republican politicians by striking a bargain that satisfied the main concerns of both sides. In this setting, the use of operational doctrine as a weapon of institutional defense was minimal, so plans and doctrine were a moderate combination of offense and defense. After the Dreyfus watershed, the truce broke. Politicians set out to "republicanize" the army, and the officer corps responded by developing the doctrine of *offensive à outrance*, which helped to reverse the slide towards a military system based overwhelmingly on reservists and capable only of defensive operations.[50]

The French army had always coexisted uneasily with the Third Republic. Especially in the early years, most officers were Bonapartist or monarchist in their political sentiments, and Radical politicians somewhat unjustifiably feared a military coup against Parliament in support of President MacMahon, a former Marshal. The military had its own fears, which were considerably more justified. Responding to constituent demands, republican politicians gradually worked to reduce the length of military service from seven to three years and to break down the quasi-monastic barriers insulating the regiment from secular, democratic trends in French society at large. Military professionals, while not averse to all reform, rightly feared a slippery slope towards a virtual militia system, in which the professional standing army would degenerate into a school for the superficial, short-term training of France's decidedly unmilitary youth. War college professors and military publicists like Georges Gilbert, responding to this danger, began by the 1880s to promote an offensive operational doctrine, which they claimed could only be implemented by well-trained, active-duty troops.[51]

This explosive situation was well managed by nationalist republican leaders like Léon Gambetta, leader of the French national resistance in the second phase of the Franco–Prussian War, and especially Charles de Freycinet, organizer of Gambetta's improvised popular armies. As War Minister in the 1880s and 1890s, Freycinet defused military fears and won their acceptance of the three-year service. He backed the military on questions of matériel, autonomy in matters of military justice, and selection of commanders on the basis of professional competence rather than political acceptability. At the same time, he pressed for more extensive use of the large pool of reservist manpower that was being created by the three-year conscription system, and the military was reasonably accommodating. In this context of moderate civil-military relations, war plans and doctrine were also moderate. Henri Bonnal's

---

50. Presenting somewhat contrasting views of French civil-military relations during this period are Douglas Porch, *March to the Marne: The French Army, 1871–1914* (Cambridge: Cambridge University Press, 1981) and David B. Ralston, *The Army of the Republic: The Place of the Military in the Political Evolution of France, 1871–1914* (Cambridge: M.I.T. Press, 1967).
51. See, for example, Gilbert, *Essais*, p. 271.

"defensive-offensive" school was the Establishment doctrine, reflected in the cautious, counteroffensive war plans of that era.[52]

Freycinet and other republican statesmen of the militant neo-Jacobin variety cherished the army as the instrument of revanche and as a truly popular institution, with roots in the *levée en masse* of the Wars of the Revolution. Though he wanted to democratize the army, Freycinet also cared about its fighting strength and morale, unlike many later politicians who were concerned only to ease their constituents' civic obligations. His own moderate policies, respectful of military sensitivities but insistent on key questions of civilian control, elicited a moderate response from military elites, whose propensity to develop a self-protective organizational ideology was thus held in check.

The deepening of the Dreyfus crisis in 1898 rekindled old fears on both sides and destroyed the system of mutual respect and reassurance constructed by Freycinet. The military's persistence in a blatant miscarriage of justice against a Jewish General Staff officer accused of espionage confirmed the republicans' view of the army as a state within the state, subject to no law but the reactionary principles of unthinking obedience and blind loyalty. When conservatives and monarchists rallied to the military's side, it made the officer corps appear (undeservedly) to be the spearhead of a movement to overthrow the Republic. Likewise, attacks by the Dreyfusards confirmed the worst fears of the military. Irresponsible Radicals were demanding to meddle in the army's internal affairs, impeaching the integrity of future wartime commanders, and undermining morale. Regardless of Dreyfus's guilt or innocence, the honor of the military had to be defended for the sake of national security.

The upshot of the affair was a leftward realignment of French politics. The new Radical government appointed as War Minister a young reformist general, Louis André, with instructions to "republicanize" the army. André, aided by an intelligence network of Masonic Lodges, politicized promotions and war college admissions, curtailed officers' perquisites and disciplinary powers, and forced Catholic officers to participate in inventorying church property. In 1905, the term of conscription was reduced to two years, with reservists intended to play a more prominent role in war plans, field exercises, and the daily life of the regiment.

52. Charles de Freycinet, *Souvenirs, 1878–1893* (New York: Da Capo, 1973).

In this hostile environment, a number of officers—especially the group of "Young Turks" around Colonel Loyzeaux de Grandmaison—began to reemphasize in extreme form the organizational ideology propounded earlier by Gilbert. Its elements read like a list of the errors of Plan 17: *offensive à outrance*, mystical belief in group *élan* achieved by long service together, denigration of reservists, and disdain for reactive war plans driven by intelligence estimates. Aided by the Agadir Crisis of 1911, General Joffre and other senior figures seeking a reassertion of professional military values used the Young Turks' doctrine to scuttle the reformist plans of the "republican" commander in chief, Victor Michel, and to hound him from office. Michel, correctly anticipating the Germans' use of reserve corps in the opening battles and the consequent extension of their right wing across northern Belgium, had sought to meet this threat by a cordon defense, making intensive use of French reservists. Even middle-of-the-road officers considered ruinous the organizational changes needed to implement this scheme. It was no coincidence that Grandmaison's operational doctrine provided a tool for attacking Michel's ideas point-by-point, without having to admit too blatantly that it was the institutional implications of Michel's reservist-based plan that were its most objectionable aspect.[53] Having served to oust Michel in 1911, the Grandmaison doctrine also played a role (along with the trumped-up scenario of a German standing-start attack) in justifying a return to the three-year term of service in 1913. The problem was that this ideology, so useful as a tool for institutional defense, became internalized by the French General Staff, who based Plan 17 on its profoundly erroneous tenets.

Obviously, there is much that is idiosyncratic in the story of the *offensive à outrance*. The overlapping of social and civil-military cleavages, which produced an unusually intense threat to the "organizational essence" and autonomy of the French army, may have no close analog in the contemporary era. At a higher level of abstraction, however, a broadly applicable hypothesis may nonetheless be gleaned from the French experience. That is, doctrinal bias is likely to become more extreme whenever strategic doctrine can be used an an ideological weapon to protect the military organization from threats to its institutional interests. Under such circumstances, doctrine be-

53. An internal General Staff document that was highly critical of Michel's scheme stated: "It is necessary only to remark that this mixed force would require very profound changes in our regulations, our habits, our tactical rules, and the organization of our staffs." Cited in Snyder, *Ideology of the Offensive*, chapter 3.

comes unhinged from strategic reality and responds primarily to the more pressing requirements of domestic and intragovernmental politics.

Russia: Institutional Pluralism and Strategic Overcommitment

Between 1910 and 1912, Russia changed from an extremely cautious defensive war plan to an overcommitted double offensive against both Germany and Austria. The general direction of this change can be easily explained in terms of rational strategic calculations. Russia's military power had increased relative to Germany's, making an offensive more feasible, and the tightening of alliances made it more obvious that Germany would deploy the bulk of its army against France in the first phase of the fighting, regardless of the political circumstances giving rise to the conflict. Russian war planners consequently had a strong incentive to invade Germany or Austria during the "window of opportunity" provided by the Schlieffen Plan. Attacking East Prussia would put pressure on Germany's rear, thus helping France to survive the onslaught; attacking the Austrian army in Galicia might decisively shift the balance of power by knocking Germany's ally out of the war, while eliminating opposition to Russian imperial aims in Turkey and the Balkans.[54]

What is harder to explain is the decision to invade both Germany and Austria, which ensured that neither effort would have sufficient forces to achieve its objectives. At a superficial level the explanation for this failure to set priorities is simple enough: General Yuri Danilov and the General Staff in St. Petersburg wanted to use the bulk of Russia's forces to attack Germany, while defending against Austria; General Mikhail Alekseev and other regional commanders wanted to attack Austria, leaving a weak defensive screen facing East Prussia. Each faction had powerful political connections and good arguments. No higher arbiter could or would choose between the contradictory schemes, so a *de facto* compromise allowed each to pursue its preferred offensive with insufficient forces. At this level, we have a familiar tale of bureaucratic politics producing an overcommitted, Christmas-tree "resultant."[55]

---

54. Apart from Zaionchkovskii, the most interesting work on Russian strategy is V.A. Emets, *Ocherki vneshnei politiki Rossii v period pervoi mirovoi voiny: vzaimootnosheniia Rossii s soiuznikami po voprosam vedeniia voiny* (Moscow: Nauka, 1977).
55. On the characteristics of compromised policy, see Warner Schilling, "The Politics of National Defense: Fiscal 1950," in Schilling et al., *Strategy, Politics, and Defense Budgets* (New York: Columbia University Press, 1962), pp. 217–218.

At a deeper level, however, several puzzles remain. One is that "where you sat" bureaucratically was only superficially related to "where you stood" on the question of strategy. Alekseev was the Chief-of-Staff-designate of the Austrian front, so had an interest in making his turf the scene of the main action. But Alekseev had always preferred an Austria-first strategy, even when he had been posted to the General Staff in St. Petersburg. Similarly, Danilov served under General Zhilinskii, the Chief of Staff who negotiated a tightening of military cooperation with France after 1911, so his bureaucratic perspective might explain his adoption of the Germany-first strategy that France preferred. But Danilov's plans had always given priority to the German front, even in 1908–1910 when he doubted the reliability and value of France as an ally.[56] Thus, this link between bureaucratic position and preferred strategy was mostly spurious.

Bureaucratic position does explain why Alekseev's plan attracted wide support among military district chiefs of staff, however. These regional planners viewed the coming war as a problem of battlefield operations, not grand strategy. Alekseev's scheme was popular with them, because it proposed clear lines of advance across open terrain. Danilov's plans, in contrast, were a source of frustration for the commanders who would have to implement them. His defensive 1910 plan perplexed them, because it offered no clear objectives.[57] His 1913 plan for an invasion of East Prussia entailed all sorts of operational difficulties that local commanders would have to overcome: inordinate time pressure, the division of the attacking force by the Masurian Lakes, and the defenders' one-sided advantages in rail lines, roads, fortifications, and river barriers.

Nonetheless, the main differences between Danilov and Alekseev were intellectual, not bureaucratic.[58] Danilov was fundamentally pessimistic about Russia's ability to compete with modern, efficient Germany. He considered Russia too weak to indulge in imperial dreams, whether against Austria or Turkey, arguing that national survival required an absolute priority be given to containing the German danger. In 1910, this pessimism was expressed in his ultra-defensive plan, based on the fear that Russia would have to face Germany virtually alone. By 1913–1914, Danilov's pessimism took a different form. The improved military balance, the tighter alliance with France after

---

56. Zaionchkovskii, *Podgotovka Rossii k imperialisticheskoi voine*, pp. 184–190.
57. Ibid., pp. 206–207.
58. See Schilling, "Politics of National Defense," for this distinction.

Agadir, and telling criticism from Alekseev convinced Danilov that a porcupine strategy was infeasible politically and undesirable strategically. Now his nightmare was that France would succumb in a few weeks, once again leaving backward Russia to face Germany virtually alone. To prevent this, he planned a hasty attack into East Prussia, designed to draw German forces away from the decisive battle in France.

Alekseev was more optimistic about Russian prospects, supporting imperial adventures in Asia and anticipating that a "sharp rap" would cause Austria to collapse. Opponents of Danilov's Germany-first strategy also tended to argue that a German victory against France would be Pyrrhic. Germany would emerge from the contest bloodied and lacking the strength or inclination for a second round against Russia. A Russo–German condominium would ensue, paving the way for Russian hegemony over the Turkish Straits and in the Balkans.[59]

Available evidence is insufficient to explain satisfactorily the sources of these differing views. Personality differences may explain Danilov's extreme pessimism and Alekseev's relative optimism, but this begs the question of why each man was able to gain support for his view. What evidence exists points to idiosyncratic explanations: Danilov's plan got support from Zhilinskii (it fit the agreements he made with Joffre), the commander-designate of the East Prussian front (it gave him more troops), and the General Staff apparatus (a military elite disdainful of and pessimistic about the rabble who would implement their plans). Alekseev won support from operational commanders and probably from Grand Duke Nikolai Nikolaevitch, the future commander-in-chief and a quintessential optimist about Russian capabilities and ambitions. The War Minister, the Czar, and the political parties seem to have played little role in strategic planning, leaving the intramilitary factions to logroll their own disputes.[60]

Perhaps the most important question is why the outcome of the logrolling was not to scale down the aims of both offensives to fit the diminished forces available to each. In particular, why did Danilov insist on an early-start, two-pincer advance into East Prussia, when the weakness of each pincer made them both vulnerable to piecemeal destruction? Why not wait a few days

59. *Documents diplomatiques français (1871–1914)*, Series 2, Vol. XII, p. 695, and other sources cited in Snyder, *Ideology of the Offensive*, chapter 7.
60. Norman Stone, *The Eastern Front, 1914–1917* (New York: Scribner's, 1975), chapter 1, presents some speculations about factional alignments, but evidence is inconclusive in this area.

until each pincer could be reinforced by late-arriving units, or why not advance only on one side of the lakes? The answer seems to lie in Danilov's extreme fears about the viability of the French and his consequent conviction that Russian survival depended on early and substantial pressure on the German rear. This task was a necessity, given his outlook, something that had to be attempted whether available forces were adequate or not. Trapped by his pessimism about Russia's prospects in the long run, Danilov's only way out was through unwarranted optimism about operational prospects in the short run. Like most cornered decision-makers, Danilov saw the "necessary" as possible.

This is an important theme in the German case as well. Schlieffen and the younger Moltke demonstrated an ability to be ruthlessly realistic about the shortcomings of their operational plans, but realism was suppressed when it would call into question their fundamental beliefs and values. Schlieffen's qualms about his war plan's feasibility pervade early drafts, but disappear later on, without analytical justification. He entertained doubts as long as he thought they would lead to improvements, but once he saw that no further tinkering would resolve the plan's remaining contradictions, he swept them under the rug. The younger Moltke did the same thing, resorting to blithe optimism only on make-or-break issues, like the seizure of Liège, where a realistic assessment of the risks would have spotlighted the dubiousness of *any* strategy for rapid, decisive victory. Rather than totally rethink their strategic assumptions, which were all bound up with fundamental interests and even personal characteristics, all of these strategists chose to see the "necessary" as possible.[61]

Two hypotheses emerge from the Russian case. The first points to bureaucratic logrolling as a factor that is likely to exacerbate the normal offensive bias of military organizations. In the absence of a powerful central authority, two factions or suborganizations will each pursue its own preferred offensive despite a dramatic deficit of available forces. Thus, offensives that are moderately ambitious when considered separately become extremely overcommitted under the pressure of scarce resources and the need to logroll with

61. Groener, writing in the journal *Wissen und Wehr* in 1927, p. 532, admitted that it had been mere "luck" that an "extremely important" tunnel east of Liège was captured intact by the Germans in August 1914. Ritter, *Schlieffen Plan*, p. 166, documents Moltke's uncharacteristic optimism about quickly seizing Liège and avoiding the development of a monumental logistical bottleneck there. In the event, the Belgians actually ordered the destruction of their bridges and rail net, but the orders were not implemented systematically.

other factions competing for their allocation. The German case showed how the lack of civilian control can produce doctrinal extremism when the military is united; the Russian case shows how lack of civilian control can also lead to extreme offensives when the military is divided.

The second hypothesis, which is supported by the findings of cognitive theory, is that military decision-makers will tend to overestimate the feasibility of an operational plan if a realistic assessment would require forsaking fundamental beliefs or values.[62] Whenever offensive doctrines are inextricably tied to the autonomy, "essence," or basic worldview of the military, the cognitive need to see the offensive as possible will be strong.

External Influences on Strategy and Civil-Military Relations

The offensive strategies of 1914 were largely domestic in origin, rooted in bureaucratic, sociopolitical, and psychological causes. To some extent, however, external influences exacerbated—and occasionally diminished—these offensive biases. Although these external factors were usually secondary, they are particularly interesting for their lessons about sources of leverage over the destabilizing policies of one's opponents. The most important of these lessons—and the one stressed by Van Evera elsewhere in this issue—is that offense tends to promote offense and defense tends to promote defense in the international system.

One way that offense was exported from one state to another was by means of military writings. The French discovered Clausewitz in the 1880s, reading misinterpretations of him by contemporary German militarists who focused narrowly on his concept of the "decisive battle." At the same time, reading the retrograde Russian tactician Dragomirov reinforced their homegrown overemphasis on the connection between the offensive and morale. Russian writings later reimported these ideas under the label of *offensive à outrance,* while borrowing from Germany the short-war doctrine. Each of Europe's militaries cited the others in parroting the standard lessons drawn from the Russo–Japanese War: offense was becoming tactically more difficult but was still advantageous strategically. None of this shuffling and sharing of rationales for offense was the initial cause of anyone's offensive bias. Everyone was exporting offense to everyone else; no one was just receiving.

---

62. Irving Janis and Leon Mann, *Decision Making* (New York: Free Press, 1977).

Its main effect was mutual reinforcement. The military could believe (and argue to others) that offense must be advantageous, since everyone else said so, and that the prevalence of offensive doctrines was somebody else's fault.[63]

The main vehicle for exporting offensive strategies was through aggressive policies, not offensive ideas. The aggressive diplomacy and offensive war plans of one state frequently encouraged offensive strategies in neighboring states both directly, by changing their strategic situation, and indirectly, by changing their pattern of civil-military relations. German belligerence in the Agadir crisis of 1911 led French civilians to conclude that war was likely and that they had better start appeasing their own military by giving them leaders in which they would have confidence. This led directly to Michel's fall and the rise of Joffre, Castelnau, and the proponents of the *offensive à outrance*. German belligerence in the Bosnian crisis of 1908–1909 had a similar, if less direct effect on Russia. It convinced Alekseev that a limited war against Austria alone would be impossible, and it put everyone in a receptive mood when the French urged the tightening of the alliance in 1911.[64] Before Bosnia, people sometimes thought in terms of a strategic modus vivendi with Germany; afterwards, they thought in terms of a breathing spell while gaining strength for the final confrontation. Combined with the Russians' growing realization of the probable character of the German war plan, this led inexorably to the conclusions that war was coming, that it could not be limited, and that an unbridled offensive was required to exploit the window of opportunity provided by the Schlieffen Plan's westward orientation. Caught in this logic, Russian civilians who sought limited options in July 1914 were easily refuted by Danilov and the military. Completing the spiral, the huge Russian arms increases provoked by German belligerence allowed the younger Moltke to argue persuasively that Germany should seek a pretext for preventive war before those increases reached fruition in 1917. This recommendation was persuasive only in the context of the Schlieffen Plan, which made Germany look weaker than it really was by creating needless enemies and wasting troops on an impossible task. Without the Schlieffen Plan, Germany would not have been vulnerable in 1917.

In short, the European militaries cannot be blamed for the belligerent diplomacy that set the ball rolling towards World War I. Once the process began, however, their penchant for offense and their quickness to view war

63. Snyder, *Ideology of the Offensive*, chapters 2 and 3.
64. Ibid., chapter 7, citing Zaionchkovskii, pp. 103, 350, and other sources.

as inevitable created a slide towards war that the diplomats did not foresee.[65] The best place to intervene to stop the destabilizing spiral of exported offense was, of course, at the beginning. If German statesmen had had a theory of civil-military relations and of the security dilemma to help them calculate risks more accurately, their choice of a diplomatic strategy might have been different.

If offense gets exported when states adopt aggressive policies, it also gets exported when states try to defend themselves in ways that are indistinguishable from preparations for aggression.[66] In the 1880s, the Russians improved their railroads in Poland and increased the number of troops there in peacetime, primarily in order to decrease their vulnerability to German attack in the early weeks of a war. The German General Staff saw these measures as a sign that a Russian attack was imminent, so counseled launching a preventive strike before Russian preparations proceeded further. Bismarck thought otherwise, so the incident did not end in the same way as the superficially similar 1914 case. Several factors may account for the difference: Bismarck's greater power over the military, his lack of interest in expansion for its own sake, and the absence of political conditions that would make war seem inevitable to anyone but a General Staff officer. Perhaps the most important difference, however, was that in 1914 the younger Moltke was anticipating a future of extreme vulnerability, whereas in 1887 the elder Moltke was anticipating a future of strategic stalemate. Moltke, planning for a defense in the west in any event, believed that the Germans could in the worst case hold out for 30 years if France and Russia forced war upon them.[67]

Although states can provoke offensive responses by seeming too aggressive, they can also invite offensive predation by seeming too weak. German hopes for a rapid victory, whether expressed in the eastward plan of the 1880s or the westward Schlieffen Plan, always rested on the slowness of Russia's mobilization. Likewise, Germany's weakness on the eastern front, artificially created by the Schlieffen Plan, promoted the development of offensive plans in Russia. Finally, Belgian weakness allowed the Germans to

---

65. Isabel V. Hull, *The Entourage of Kaiser Wilhelm II, 1888–1918* (New York: Cambridge University Press, 1982), discusses the effect on the Kaiser of his military aides' incessant warnings that war was inevitable.
66. Robert Jervis, "Cooperation under the Security Dilemma," *World Politics*, Vol. 31, No. 2 (January 1978), pp. 199–210.
67. Barbara Tuchman, *The Guns of August* (1962; rpt., New York: Dell, 1971), p. 38; see also *Aufmarschpläne*, pp. 150–156, for Moltke's last war plan of February 1888.

retain their illusions about decisive victory by providing an apparent point of entry into the French keep.

States who want to export defense, then, should try to appear neither weak nor aggressive. The French achieved this in the early 1880s, when a force posture heavy on fortifications made them an unpromising target and an ineffective aggressor. In the short run, this only redirected Moltke's offensive toward a more vulnerable target, Russia. But by 1888–1890, when Russia too had strengthened its fortifications and its defensive posture in Poland generally, Moltke was stymied and became very pessimistic about offensive operations. Schlieffen, however, was harder to discourage. When attacking Russia became unpromising, he simply redirected his attention towards France, pursuing the least unpromising offensive option. For hard core cases like Schlieffen, one wonders whether any strategy of non-provocative defense, no matter how effective and non-threatening, could induce abandoning the offensive.

Soviet Strategy and Civil-Military Relations

In 1914, flawed civil-military relations exacerbated and liberated the military's endemic bias for offensive strategies, creating strategic instability despite military technologies that aided the defender of the status quo. Some of the factors that produced this outcome may have been peculiar to that historical epoch. The full professionalization of military staffs had been a relatively recent development, for example, and both civilians and military were still groping for a satisfactory *modus vivendi*. After the First World War, military purveyors of the "cult of the offensive" were fairly well chastened except in Japan, where the phenomenon was recapitulated. Our own era has seen nothing this extreme, but more moderate versions of the military's offensive bias are arguably still with us. It will be worthwhile, therefore, to reiterate the kinds of conditions that have intensified this bias in the past in order to assess the likelihood of their recurrence.

First, offensive bias is exacerbated when civilian control is weak. In Germany before 1914, a long period of military autonomy in strategic planning allowed the dogmatization of an offensive doctrine, rooted in the parochial interests and outlook of the General Staff. In Russia, the absence of firm, unified civilian control fostered logrolling between two military factions, compounding the offensive preferences exhibited by each. Second, offensive bias grows more extreme when operational doctrine is used as a weapon in civil-military disputes about domestic politics, institutional arrangements, or other nonstrategic issues. The French *offensive à outrance,* often dismissed as some mystical aberration, is best explained in these terms.

Once it appears, an acute offensive bias tends to be self-replicating and resistant to disconfirming evidence. Offensive doctrinal writings are readily transmitted across international boundaries. More important, offensive strategies tend to spread in a chain reaction, since one state's offensive tends to create impending dangers or fleeting opportunities for other states, who must adopt their own offensives to forestall or exploit them. Finally, hard operational evidence of the infeasibility of an offensive strategy will be rationalized away when the offensive is closely linked to the organization's "essence," autonomy, or fundamental ideology.

I believe that these findings, derived from the World War I cases, resonate strongly with the development of Soviet nuclear strategy and with certain patterns in the U.S.–Soviet strategic relationship. At a time when current events are stimulating considerable interest in the state of civil-military relations in the Soviet Union, the following thoughts are offered not as answers but as questions that researchers may find worth considering.

Soviet military doctrine, as depicted by conventional wisdom, embodies all of the desiderata typically expressed in professional military writings throughout the developed world since Napoleon. Like Schlieffen's doctrine, it stresses offense, the initiative, and decisive results through the annihilation of the opponent's ability to resist. It is suspicious of political limitations on violence based on mutual restraint, especially in nuclear matters. Both in style and substance, Sidorenko reads like a throwback to the military writers of the Second Reich, warning that "a forest which has not been completely cut down grows up again."[68] The similarity is not accidental. Not only does offense serve some of the same institutional functions for the Soviet military as it did for the German General Staff, but Soviet doctrine is to some degree their lineal descendant. "In our military schools," a 1937 Pravda editorial averred, "we study Clausewitz, Moltke, Schlieffen, and Ludendorff."[69] Soviet nuclear doctrine also parallels pre-1914 German strategy in that both cut against the grain of the prevailing technology. The Soviets have never been

---

68. Quoted by Benjamin Lambeth, "Selective Nuclear Options and Soviet Strategy," in Johan Holst and Uwe Nerlich, *Beyond Nuclear Deterrence* (New York: Crane, Russak, 1977), p. 92.
69. Raymond Garthoff, *Soviet Military Doctrine* (Glencoe, Ill.: Free Press, 1953), p. 56.

in a position to achieve anything but disaster by seizing the initiative and striving for decisive results; neither was Schlieffen.

There are also parallels in the political and historical circumstances that permitted the development of these doctrines. The Soviet victories in World War II, like the German victories in 1866 and 1870, were nation-building and regime-legitimating enterprises that lent prestige and authority to the military profession, notwithstanding Stalin's attempt to check it. This did not produce a man on horseback in either country, nor did it allow the military to usurp authority on questions of the use of force. But in both cases the military retained a monopoly of military operational expertise and was either never challenged or eventually prevailed in practical doctrinal disputes. In the German case, at least, it was military autonomy on questions of operational plans and doctrine that made war more likely; direct lobbying for preventive strikes caused less trouble because it was clearly illegitimate.

While many accounts of the origins of Soviet nuclear strategy acknowledge the effect of the professional military perspective, they often lay more stress on civilian sources of offensive, warfighting doctrines: for example, Marxism–Leninism, expansionist foreign policy goals, and historical experiences making Russia a "militarized society." Political leaders, in this view, promote or at least accept the military's warfighting doctrine because it serves their foreign policy goals and/or reflects a shared view of international politics as a zero-sum struggle. Thus, Lenin is quoted as favoring a preemptive first strike, Frunze as linking offense to the proletarian spirit. The military principle of annihilation of the opposing armed force is equated with the Leninist credo of *kto kogo*.[70]

Although this view may capture part of the truth, it fails to account for recurrent statements by Soviet political leaders implying that nuclear war is unwinnable, that meaningful damage limitation cannot be achieved through superior warfighting capabilities, and that open-ended expenditures on strategic programs are wasteful and perhaps pointless. These themes have been voiced in the context of budgetary disputes (not just for public relations purposes) by Malenkov, Khrushchev, Brezhnev, and Ustinov. To varying degrees, all of these civilian leaders have chafed at the cost of open-ended warfighting programs and against the redundant offensive capabilities de-

70. Herbert Dinerstein, *War and the Soviet Union* (New York: Praeger, 1962), pp. 210–211; Garthoff, *Soviet Military Doctrine*, pp. 65, 149.

manded by each of several military suborganizations. McNamara discovered in the United States that the doctrine of mutual assured destruction, with its emphasis on the irrelevance of marginal advantages and the infeasibility of counterforce damage-limitation strategies, had great utility in budgetary debates. Likewise, recent discussions in the Soviet Union on the feasibility of victory seem to be connected with the question of how much is enough. Setting aside certain problems of nuance and interpretation, a case can be made that the civilian leadership, speaking through Defense Minister Ustinov, has been using strategic doctrine to justify slowing down the growth of military spending. In the context of arguments about whether the Reagan strategic buildup will really make the Soviet Union more vulnerable, Ustinov has quite clearly laid out the argument that neither superpower can expect to gain anything by striking first, since both have survivable retaliatory forces and launch-on-warning capabilities. Thus, Ustinov has been stressing that the importance of surprise is diminishing and that "preemptive nuclear strikes are alien to Soviet military doctrine." Ogarkov, the Chief of the General Staff, has been arguing the opposite on all counts: the U.S. buildup is truly threatening, the international scene is akin to the 1930s, the surprise factor is growing in importance, damage limitation is possible (though "victory" is problematic), and consequently the Soviet Union must spare no expense in preparing to defend itself.[71]

This is somewhat reminiscent of the French case in World War I, in which civilians and the military were using doctrinal arguments as weapons in disputes on other issues. Two related dangers arise in such situations. The first is that doctrinal argumentation and belief, responding to political and organizational necessity, lose their anchoring in strategic realities and become dogmatic and extremist. The second is that a spiral dynamic in the political dispute may carry doctrine along with it. That is, the harder each side fights to prevail on budgetary or organizational questions, the more absolute and unyielding their doctrinal justifications will become. In this regard, it would be interesting to see whether the periods in which Soviet military spokesmen

71. Citations to the main statements by Ogarkov and Ustinov can be found in Dan L. Strode and Rebecca V. Strode, "Diplomacy and Defense in Soviet National Security Policy," *International Security*, Vol. 8, No. 2 (Fall 1983), pp. 91–116. Quotation from William Garner, *Soviet Threat Perceptions of NATO's Eurostrategic Missiles* (Paris: Atlantic Institute for International Affairs, 1983), p. 69, citing *Pravda*, July 25, 1981. I have benefitted from discussions of the Ogarkov and Ustinov statements with Lawrence Caldwell, Stephen Coffey, Clifford Kupchan, and Cynthia Roberts, who advanced a variety of interpretations not necessarily similar to my own.

were arguing hardest that "victory is possible" coincided with periods of sharp budgetary disputes.

Even if some of the above is true, the pattern may be a weak one in comparison with the French case. Ustinov is more like Freycinet than André, and marginal budgetary issues do not carry the same emotional freight as the threats to organizational "essence" mounted in the Dreyfus aftermath. Still, if we consider that the Soviet case couples some of the autonomy problems of the German case with some of the motivational problems of the French case, a volatile mixture may be developing.

Another civil-military question is whether Soviet military doctrine is mismatched with Soviet diplomacy. On the surface, it may seem that the awe-inspiring Soviet military machine and its intimidating offensive doctrine are apt instruments for supporting a policy of diplomatic extortion. It may, however, pose the same problem for Soviet statesmen that the Schlieffen Plan did for Bülow and Bethmann. Soviet leaders may be self-deterred by the all-or-nothing character of their military options.[72] Alternatively, if the Soviets try to press ahead with a diplomacy based on the "Bolshevik operational code" principles of controlled pressure, limited probes, and controlled, calculated risks, they may find themselves trapped by military options that create risks which cannot be controlled.

These problems may not arise, however, since the Soviets seem to have turned away from Khrushchev's brinkmanship diplomacy. In the Brezhnev era, Soviet doctrine on the political utility of nuclear forces stressed its role as an umbrella deterring intervention against "progressive" political change.[73] Insofar as limited options and "salami tactics" are more clearly indispensable for compellent than for deterrent strategies, this would help to solve the Soviet diplomats' mismatch problem. The "last clear chance" to avoid disaster would be shifted onto the United States. This solution to the diplomats'

---

72. Increased Soviet attention to the "conventional option" since the late 1960s would seem to have mitigated this problem, but in fact it may have compounded it. Military interest in preparing for a conventional phase and acquiring capabilities for escalation dominance in the theater may derive more from obvious organizational motives than from a fundamental change in the military's mind-set of "inflexible over-response." In Soviet thinking, limitations seem to be based less on mutual restraint than on NATO's willingness to see its theater nuclear forces destroyed during the conventional phase. This raises the nightmarish possibility that the Soviet leadership could embark on war thinking that it had a conventional option, whereas in fact unrestrained conventional operations and preemptive incentives at the theater nuclear level would lead to rapid escalation.

73. Coit Blacker, "The Kremlin and Detente: Soviet Conceptions, Hopes, and Expectations," in Alexander George, ed., *Managing U.S.–Soviet Rivalry* (Boulder: Westview, 1983), pp. 122–123.

problem might cause problems for the military's budget rationale, however, since strategic parity should be sufficient to carry out a strictly deterrent function.

The German case suggests that extremism in strategic thinking may depend a great deal on institutionalization and dogmatization of doctrine in the mature military organization. If Roman Kolkowicz's "traditionalists" are equated with the Moltke generation and his "modernist" technocrats with the Schlieffen generation, do we find a parallel in the dogmatization of doctrine? Benjamin Lambeth argues that Soviet doctrine is quite flexible and creative, but so was Schlieffen on questions of how to implement his strategic tenets under changing conditions.[74] Creativity within the paradigm of decisive, offensive operations may coexist with utter rigidity towards options that would require a change in the basic paradigm. For example, the Soviet ground forces adapted creatively to improvements in precision-guided munitions (PGMs) that seemed to threaten the viability of their offensive doctrine; they did not consider, however, that PGMs might offer an opportunity to give up their fundamentally offensive orientation. As for the third phase of organizational evolution, are there any parallels to Ludendorff or Groener among younger Soviet officers? Are they forging links to Russian nationalists, whose social base Alexander Yanov describes in ways that are strongly reminiscent of Eley's account of the ultranationalist German right?[75]

Any discussion of the extremist potential of Soviet strategy must consider the strong reality constraint imposed by the mutual-assured-destruction relationship. Despite the reckless rhetoric of some junior officers, it seems clear that when the head of the Strategic Rocket Forces said in 1967 that "a sudden preemptive strike cannot give [the aggressor] a decisive advantage," he knew that launch-on-warning and the hardening of silos made this true for both sides.[76] And today Ogarkov does not deny that a scot-free victory is impossible. But despite this, the theme of damage limitation remains strong in Soviet military thinking, and we should remember those World War I strategists who saw the "necessary" as possible, no matter how realistically they did their operational calculations.

---

74. Lambeth, "Selective Nuclear Options"; Kolkowicz, *The Soviet Military and the Communist Party* (Princeton: Princeton University Press, 1967).
75. Alexander Yanov, *Detente after Brezhnev* (Berkeley: Institute of International Studies, University of California, 1977).
76. Garner, *Soviet Threat Perceptions*, p. 69.

Finally, how have the policies of the United States affected the development of civil-military relations and strategic doctrine in the U.S.S.R.? Some analysts argue that the Ogarkov–Ustinov debates ended in May 1983 with Ustinov's capitulation, at least on the level of rhetoric. Although leadership politics may have been a factor, a more important reason may have been the Reagan "Star Wars" speech and the Reagan defense program generally.[77] Echoing the developments in France in 1911, rising levels of external threat may have helped the military to win the doctrinal argument and achieve its institutional aims in the underlying issues tied to the doctrinal dispute. This episode may also be seen as the latest round of a process of exporting and re-importing warfighting strategies. The impact of Soviet counterforce doctrines on the American strategic debate in the 1970s is obvious; now the fruits of our conversion are perhaps being harvested by Ogarkov in Soviet debates on military budgets and operational policies.

Whatever the precise reality of current civil-military relations in the Soviet Union, patterns revealed by the World War I cases suggest that the Soviet Union manifests several "risk factors" that could produce an extreme variant of the military's endemic offensive bias. The historical parallel further suggests that the actions of rival states can play an important role in determining how these latent risks unfold. Aggressive policies were liable to touch off these latent dangers, but vulnerability also tended to encourage the opponent to adopt an offensive strategy. Postures that were both invulnerable and non-provocative got the best results, but even these did not always dissuade dogmatic adherents to the "cult of the offensive." Although Soviet persistence in working the problems of conventional and nuclear offensives does recall the dogged single-mindedness of a Schlieffen, nuclear weapons pose a powerful reality constraint for which no true counterpart existed in 1914. Consequently, if the twin dangers of provocation and vulnerability are avoided, there should be every hope of keeping Soviet "risk factors" under control. The current drift of the strategic competition, however, makes that not a small "if."

77. Setting these debates into the context of U.S.–Soviet relations are Lawrence T. Caldwell and Robert Legvold, "Reagan Through Soviet Eyes," *Foreign Policy*, No. 52 (Fall 1983), pp. 3–21.

1914 Revisited

Allies, Offense, and Instability

Scott D. Sagan

The origins of the First World War continue to be of great interest today because there are a number of striking similarities between the events of 1914 and contemporary fears about paths by which a nuclear war could begin. July 1914 was a brinksmanship crisis, resulting in a war that everyone was willing to risk but that no one truly wanted. During the crisis, the political leaderships' understanding of military operations and control over critical war preparations were often tenuous at best. In 1914, the perceived incentives to strike first, once war was considered likely, were great, and the rapidity and inflexibility of offensive war plans limited the time available to diplomats searching for an acceptable political solution to the crisis. In a world in which the possibility of massive nuclear retaliation has made the deliberate, premeditated initiation of nuclear war unlikely, there is widespread concern that a repetition of the Sarajevo scenario may occur: an apparently insignificant incident sparking—through a dangerous mixture of miscalculations, inadvertent escalation, and loss of control over events—a tragic and unintended war.[1] Indeed, for a student of the July crisis, even specific phrases in the current nuclear debate can be haunting: what former Secretary of Defense Harold Brown meant to be a comforting metaphor, that the Soviet Union would never risk its society on "a cosmic throw of the dice," is less reassuring to those who recall German Chancellor Theobald von Bethmann-Hollweg's statement, "If the iron dice are now to be rolled, may God help us," made just hours before Germany declared war against Russia on August 1, 1914.[2]

I would like to thank Robert Art, Stanley Hoffmann, Jack Levy, and Edward Rhodes for their helpful comments on earlier drafts of this paper.

Scott D. Sagan is a Lecturer in the Government Department, Harvard University.

1. Recent discussions of nuclear strategy that utilize the 1914 analogy include Graham T. Allison, Albert Carnesale, and Joseph S. Nye, Jr., eds., *Hawks, Doves, and Owls* (New York: Norton, 1985), especially pp. 210–217; Paul J. Bracken, *The Command and Control of Nuclear Forces* (New Haven: Yale University Press, 1983), pp. 222–223, 239–240; and Miles Kahler, "Rumors of War: The 1914 Analogy," *Foreign Affairs*, Vol. 58, No. 2 (Winter 1979–80).
2. Harold Brown, *Department of Defense Annual Report for Fiscal Year 1979* (Washington, D.C.: U.S. Government Printing Office, 1978), p. 63; and Karl Kautsky, ed., *Outbreak of the World War:*

Prior to 1914, the general staffs of each of the European great powers had designed elaborate and inflexible offensive war plans, which were implemented in a series of mobilizations and countermobilizations at the end of the July crisis. In August, all the continental powers took the offensive: the Germans attacked across Belgium and Luxembourg into France; the French army launched a massive assault against German positions in Alsace–Lorraine; and the Russian army, although not yet fully mobilized, immediately began simultaneous offenses against Germany and Austria–Hungary. In retrospect, the war plans of the great powers had disastrous political and military consequences. The negative political consequences were seen at the cabinet meetings during the July crisis, for the pressures to begin mobilization and launch offensives promptly, according to the military timetables, contributed greatly to the dynamic of escalation and the political leaderships' loss of freedom of action. In Berlin, for example, as Bethmann–Hollweg frankly admitted to the Prussian Ministry of State, once the Russians began to mobilize, "control had been lost and the stone had started rolling."[3] The military consequences were seen on the battlefield. Each of the major offensive campaigns was checked or repulsed with enormous costs: some 900,000 men were missing, taken prisoner, wounded, or dead by the end of 1914.[4]

Historians and political scientists have long sought to understand why the great powers all had offensive military doctrines when the military technology of 1914—barbed wire, machine guns, and railroads—appears to have favored the defense. The popular explanation is that European soldiers and statesmen blithely ignored the demonstrations of defensive firepower in the American Civil War and the Russo–Japanese War and simply believed that the next European war would be like the last (an offensive victory as in the Franco–Prussian war), but that has never been satisfactory. For, in fact, numerous European military observers were in the United States from 1861–65 and in Manchuria during the 1904–5 conflict. Observer reports were widely distributed, the German, French, and British armies sponsored multivolume official histories of the Russo–Japanese War, and throughout the period prior to 1914 prolonged and heated debates raged in European military journals

---

German Documents (New York: Oxford University Press, 1924), No. 553, p. 441. (Hereinafter, *German Documents*.)
3. *German Documents*, No. 456, p. 382.
4. The estimate is from Michael Howard, "Men Against Fire: The Doctrine of the Offensive in 1914," in Peter Paret, ed., *The Makers of Modern Strategy from Machiavelli to the Nuclear Age* (Princeton: Princeton University Press, 1986), p. 510.

about the relative effectiveness of offensive and defensive tactics and strategies.[5]

Recent scholarship has suggested a new explanation which, using organization theory, emphasizes the degree to which the organizational interests of the professional military are advanced by offensive military doctrines, regardless of whether offensives are recommended by perceived national interests or prevailing technology.[6] Jack Snyder and Stephen Van Evera have found, in the 1914 case, an extreme example of this phenomenon: the "cult of the offensive."[7] This new explanation for the origins of the First World War is becoming widely accepted, and no one has challenged its validity.[8]

This essay reviews the organizational theory arguments and their "cult of the offensive" application to the events of 1914. It concludes that this approach seriously misrepresents the *causes* of the offensive doctrines of 1914 and, therefore, the underlying causes of the war. By focusing on the organizational interests of the professional military, the "cult of the offensive" theory has overlooked the more fundamental causes of the World War I offensive doctrines: the political objectives and alliance commitments of the great powers.

This essay also argues that the "cult of the offensive" theory misrepresents the *consequences* of the 1914 offensive military doctrines. Although offensive

---

5. These writings are reviewed in ibid.; and Michael Howard, "Men Against Fire: Expectations of War in 1914," in Steven E. Miller, ed., *Military Strategy and the Origins of the First World War: An International Security Reader* (Princeton: Princeton University Press, 1985), pp. 41–57. Also see T.H.E. Travers, "The Offensive and the Problem of Innovation in British Military Thought 1870–1915," *Journal of Contemporary History*, Vol. 13 (1978), pp. 531–553; Travers, "Technology, Tactics, and Morale: Jean de Bloch, The Boer War, and British Military Theory, 1900–1914," *Journal of Modern History*, Vol. 51, No. 2 (June 1979), pp. 264–286; and Jay Luvaas, *The Military Legacy of the Civil War: The European Inheritance* (Chicago: University of Chicago Press, 1959).
6. See Barry R. Posen, *The Sources of Military Doctrine: France, Britain and Germany Between the World Wars* (Ithaca: Cornell University Press, 1984); and Stephen Van Evera, "Causes of War" (Ph.D. dissertation, University of California, Berkeley, 1984), especially chapter 7.
7. Jack Snyder, "Civil–Military Relations and the Cult of the Offensive, 1914 and 1984" and Stephen Van Evera, "The Cult of the Offensive and the Origins of the First World War," both in Miller, *Military Strategy and the Origins of the First World War*. Also see Snyder, *The Ideology of the Offensive: Military Decision Making and the Disasters of 1914* (Ithaca: Cornell University Press, 1984); and Van Evera, "Why Cooperation Failed in 1914," *World Politics*, Vol. 38, No. 1 (October 1985), pp. 97–98.
8. See Allison et al., *Hawks, Doves, and Owls*, p. 212; Robert Axelrod and Robert O. Keohane, "Achieving Cooperation under Anarchy: Strategies and Institutions," *World Politics*, Vol. 38, No. 1 (October 1985), pp. 230–231; Richard Ned Lebow, "The Soviet Offensive in Europe: The Schlieffen Plan Revisited?," *International Security*, Vol. 9, No. 4 (Spring 1985), pp. 52–53, 68–69; Jack S. Levy, "Organizational Routines and the Causes of War," *International Studies Quarterly*, in press; and Steven E. Miller, "Introduction: The Great War and the Nuclear Age," in Miller, *Military Strategy and the Origins of the First World War*, p. 3.

military doctrines were necessary, they were not sufficient to cause the strategic instability witnessed in the July crisis; the critical preemptive incentives and pressures to move quickly felt by the German General Staff would not have been so strong without specific military vulnerabilities of the Entente powers and Belgium. In addition, even given the German offensive doctrine and war plans, it appears likely that Berlin would have been deterred in 1914 if the British government had issued a clear and credible threat to intervene in a continental war early in the July crisis. Furthermore, while the "cult of the offensive" theory correctly identifies the problem of offensive instability during the July crisis, it ignores the critical strategic dangers that would have resulted if European statesmen had adopted purely defensive strategies in 1914. These conclusions are, finally, of more than historical interest, for they suggest that the explicit lessons that the "cult of the offensive" theorists offer for contemporary American deterrent strategy are quite misleading.

Offenses, Military Biases, and the Security Dilemma

International relations theory posits the existence of a common security dilemma between sovereign states and has stressed the pernicious impact of offensive military forces and doctrines in exacerbating the problem.[9] The security dilemma exists when actions taken by one state solely for the purposes of increasing its own security simultaneously threaten another state, decreasing its security. This dilemma can be vicious "even in the extreme case in which all states would like to freeze the status quo," when two conditions exist: first, when defensive weapons and strategies cannot be distinguished from offensive ones and, second, when offensive military operations are considered easier than defensive operations.[10] Such conditions are said to produce a number of dangers. When offense is easier, or when one cannot differentiate between offenses and defenses, "unnecessary" arms races are made more likely. Under such conditions, the incentives to launch preventive wars are increased whenever the balance of power is shifting in favor of an adversary. Likewise, when war is considered likely, preemptive

9. Robert Jervis, "Cooperation under the Security Dilemma," *World Politics*, Vol. 30, No. 2 (January 1978), pp. 167–214. Also see George H. Quester, *Offense and Defense in the International System* (New York: John Wiley and Sons, 1977); Van Evera, "Causes of War," chapter 3; and Jack S. Levy, "The Offensive/Defensive Balance of Military Technology: A Theoretical and Historical Analysis," *International Studies Quarterly*, Vol. 28 (1984), pp. 219–238.
10. Jervis, "Cooperation under the Security Dilemma," pp. 167, 186–187.

incentives are increased to the degree that striking the first offensive blow is considered advantageous compared to waiting to be attacked.

Earlier work on the subject focused largely on the effect of military technology on the offensive/defensive balance, and many arms control negotiations have sought to promote "stability" by identifying and limiting offensive weapons and promoting defensive ones.[11] This approach's assumption, that status quo powers will pursue defensive, "stabilizing" military capabilities if possible, is challenged by the new "cult of the offensive" literature which emphasizes that military organizations display a strong preference for offensive forces and doctrines, even if the predominant military technology favors defense.

Five related and reinforcing explanations are offered for the military's bias in favor of the offense. First, offensive doctrines enhance the power and size of military organizations.[12] Offenses are usually technologically more complex and quantitatively more demanding than defenses. They often require larger forces, longer range weapons, and more extensive logistic capabilities. Since military organizations, like other organizations, seek to enhance their own size and wealth, as a rule, they will prefer offenses. Second, offensive doctrines tend to promote military autonomy. As Jack Snyder explains, "The operational autonomy of the military is most likely to be allowed when the operational goal is to disarm the adversary quickly and decisively by offensive means."[13] Not only are defensive operations, because they tend to be less complex, easier for civilian leaders to understand, but defenses also can lead to prolonged conflict on one's own soil, increasing the likelihood of civilian interference. In addition, offensive doctrines may require professional armies, rather than more "civilianized" conscripted armies. Third, offenses enhance the prestige and self-image of military officers. Defensive operations are often seen as passive, less challenging, and less glorious. As Barry Posen puts it, offenses can make soldiers "specialists in victory," while defenses merely turn them into "specialists in attrition."[14]

---

11. Ibid., pp. 186–214. For related works see Quester, *Offense and Defense in the International System*; Marion W. Boggs, *Attempts to Define and Limit "Aggressive" Armament in Diplomacy and Strategy* (Columbia, Mo.: University of Missouri, 1941); and B.H. Liddell Hart, "Aggression and the Problem of Weapons," *The English Review*, July 1932, pp. 71–78.
12. Posen, *Sources of Military Doctrine*, p. 49.
13. Snyder, "Civil–Military Relations and the Cult of the Offensive," p. 121; see also Posen, *Sources of Military Doctrine*, pp. 49–50.
14. Posen, *Sources of Military Doctrine*, p. 49. A recent examination of the causes of the persistence of offensive tactics on the part of the Confederacy emphasizes the influence of Southern culture's romantic notions of soldierly honor. See Grady McWhiney and Perry D. Jamieson, *Attack and*

The fourth explanation offered for the military's offensive bias is that offenses structure military campaigns in favorable ways. Taking the initiative helps ensure that your standard scenario and operations plans, instead of the enemy's, dominate at least in the initial battles of a war.[15] This advantage of offensive operations has been repeated in statements of the "principle of the initiative" in military manuals at least since Jomini's maxim of 1807:

> The general who takes the initiative knows what he is going to do; he conceals his movements, surprises and crushes an extremity or weak point. The general who waits is beaten at one point before he learns of the attack.[16]

The fifth explanation emphasizes the effect that military officers' training and duties have on their beliefs about the need for decisive military operations that tend to be offensive in nature. Officers, it is argued, are necessarily preoccupied with the possibility of war; they tend to see the adversary as extremely hostile and war as a natural, indeed often an inevitable, part of international politics. Such beliefs lead them to favor preventive wars or preemptive strikes when necessary and decisive operations when possible. "Seeing war more likely than it really is," Snyder concludes, "[military professionals] increase its likelihood by adopting offensive plans and buying offensive forces."[17]

The 1914 Cult of the Offensive

Jack Snyder and Stephen Van Evera have found, in the origins of the First World War, the most extreme example of what can go wrong if the endemic military bias in favor of offensive doctrines is allowed to determine a state's strategy. Why did all the continental powers immediately launch offensives at the outburst of war? Snyder argues, "The offensive strategies of 1914 were largely domestic in origin, rooted in bureaucratic, sociopolitical, and psychological causes."[18] In France, the offensive nature of Plan XVII is seen as the

---

Die: Civil War Military Tactics and Southern Heritage (University, Ala.: University of Alabama Press, 1982).
15. Posen, *Sources of Military Doctrine*, pp. 47–48. It should be noted, however, that this advantage need not be a military bias, as the civilian leadership as well as the military would favor it.
16. Antoine-Henri Jomini, "L'art de la guerre," *Pallas: Eine Zeitschrift für Staats und Kreigs Kunst*, Vol. 1 (1808), pp. 32–40, as quoted in John I. Alger, *The Quest for Victory: The History of the Principles of War* (Westport, Conn.: Greenwood Press, 1982), p. 22.
17. Snyder, "Civil–Military Relations and the Cult of the Offensive," p. 119.
18. Ibid., p. 137.

result of the professional military's use of offensive doctrine as a defense of its institutional interests. After the Dreyfus affair, political leaders sought to "republicanize" the French military, hoping to create an army based largely on reservists and capable of conducting only defensive operations. The French military countered this threat to its institutional "essence," Snyder explains, by adopting the *offensive à outrance* doctrine which required the discipline and élan that only a professional standing army's long training and service together could provide.[19] The offensive strategy of the Schlieffen Plan, according to Snyder, was adopted in Germany because of the Prussian General Staff's bias in favor of decisive operations and because an offensive strategy promoted their power, prestige, and autonomy. In Russia, the ambitious war plans calling for offensive operations against both Germany and Austria–Hungary were adopted because the absence of a strong central political authority enabled each of the conflicting military factions, one favoring an offense against Germany and the other supporting an offense against Austria–Hungary, to pursue its own preferred campaign. Because a realistic assessment of the chances of success would have threatened their fundamental beliefs, Russian military leaders resorted to what Snyder calls "needful thinking," believing that both "necessary" offensives could succeed when objective assessment would have shown that they would fail.[20] The conclusion is clear:

> Strategic instability in 1914 was caused not by military technology, which favored the defender and provided no first-strike advantage, but by offensive war plans that defied technological constraints. The lesson here is that doctrines can be destabilizing even when weapons are not, since doctrine may be more responsive to the organizational needs of the military than to the implications of the prevailing weapon technology.[21]

While Snyder has focused on the causes of the "cult of the offensive," Van Evera's work concentrates on its consequences; his detailed analysis of the July crisis goes considerably further than earlier work on this subject in identifying precisely how the military plans of the European powers contrib-

---

19. Ibid., pp. 129–133; and Snyder, *Ideology of the Offensive,* chapters 2 and 3. For a different argument, stressing that offensive doctrine was adopted so that the morale of patriotic French recruits could compensate for superior German material strength, see Douglas Porch, *The March to the Marne: The French Army, 1871–1914* (Cambridge: Cambridge University Press, 1981), chapter 11.
20. Snyder, "Civil–Military Relations and the Cult of the Offensive," pp. 125–129, 133–137; and Snyder, *Ideology of the Offensive,* chapters 4–7.
21. Snyder, *Ideology of the Offensive,* pp. 10–11.

uted to strategic instability in 1914.[22] He finds the "cult of the offensive" to be the "mainspring" driving the numerous mechanisms that led the great powers to war and illustrates his argument by imagining what 1914 would have looked like had military and civilian leaders not had such strong beliefs about the efficacy of offenses. The expansionist aims of Germany, the perceived incentives for preventive war and preemptive strikes, and the dynamics of rapid escalation in the crisis could have all been avoided, Van Evera argues, had European leaders "recognized the actual power of the defense":

> German expansionists then would have met stronger arguments that empire was needless and impossible, and Germany could have more easily let the Russian military buildup run its course, knowing that German defenses could still withstand Russian attack. All European states would have been less tempted to mobilize first, and each could have tolerated more preparations by adversaries before mobilizing themselves, so the spiral of mobilization and counter-mobilization would have operated more slowly, if at all.[23]

Van Evera even concludes that the First World War might not have broken out at all if leaders had understood the strength of defenses: "If armies [had] mobilized, they might have rushed to defend their own trenches and fortifications, instead of crossing frontiers. . . ." Indeed, "In all likelihood, the Austro–Serbian conflict would have been a minor and soon-forgotten disturbance on the periphery of European politics."[24]

An Alternative Explanation: Strategic Interests and Alliance Commitments

Snyder's work has identified a number of ways in which the organizational interests of the military can affect strategic doctrine, and Van Evera's writings have persuasively demonstrated the alarming consequences that specific aspects of the offensive war plans of 1914 had on the political leadership's ability to control events during the July crisis. But their "cult of the offensive" analysis greatly exaggerates the degree to which the offensive doctrines in 1914 were caused by military-motivated biases or misperceptions of the of-

22. Van Evera, "The Cult of the Offensive and the Origins of the First World War." Also see Levy, "Organizational Routines and the Causes of War." For earlier interpretations, see Herman Kahn, *On Thermonuclear War* (Princeton: Princeton University Press, 1960), pp. 357–375; and Thomas Schelling's discussion of "the dynamics of mutual alarm" in Schelling, *Arms and Influence* (New Haven: Yale University Press, 1966), pp. 221–244.
23. Van Evera, "The Cult of the Offensive and the Origins of the First World War," p. 105.
24. Ibid. For a similar argument, that the European armies should have been "rushing to their own trenches rather than the enemy's territory" in 1914, see Jervis, "Cooperation under the Security Dilemma," p. 191.

fense/defense balance. Moreover, by focusing exclusively on the problems of the 1914 offensive military doctrines, they have overlooked the negative consequences that would have resulted if the great powers had adopted purely defensive military doctrines.

Three related problems exist with the "cult of the offensive" theory explanation. First, the theory exaggerates the probability that critical offensive military operations would fail. This theory is perhaps the strongest in explaining the French military's *offensive à outrance* doctrine, which even Field Marshall Joffre admitted was influenced by "le culte de l'offensive."[25] But it was *not* this *French* offensive doctrine that produced the dynamic of rapid escalation during the July crisis. For the French, a decision to mobilize was *not* a decision to attack Germany, and the Paris government specifically ordered its army not to move within 10 kilometers of the German border upon general mobilization.[26] Instead, the most critical offensive war plan in 1914 was that of Germany, for it was the German military's perceived need to mobilize quickly to implement the Schlieffen Plan's preemptive attack on Liège and attack in the West before Russian mobilization was complete that caused the crisis to move beyond control.

With 20-20 hindsight, however, it is too easy to argue that the German offensive plan for conquest was doomed to fail. Thus, when Snyder writes of the "*vain attempt* to knock France out of the war" and complains that the German General Staff "could not accept that a future war *would inevitably take the form of an inglorious, unproductive stalemate*," he assumes that the historical outcome, the defeat of the Schlieffen Plan, was the only one possible.[27] Similarly, when Van Evera writes, "Had statesmen understood that in reality the defense had the advantage, they also would have known that *the possession of the initiative could not be decisive*" or that "German expansionists then would have met stronger arguments that empire was *needless and impossible*," he assumes that the outcome of the war was a foregone conclusion.[28]

In fact, the Schlieffen Plan came very close to succeeding and the Germans almost did win the short war they had expected to fight. The French, who call the decisive battle outside Paris "the Miracle of the Marne," have a better

25. As quoted by Van Evera, "The Cult of the Offensive and the Origins of the First World War," p. 61, fn. 14.
26. See Luigi Albertini, *The Origins of the War of 1914* (London: Oxford University Press, 1957), Vol. 3, pp. 66–111.
27. Snyder, "Civil–Military Relations and the Cult of the Offensive," p. 128; and Snyder, *Ideology of the Offensive*, p. 17. Emphasis added.
28. Van Evera, "The Cult of the Offensive and the Origins of the First World War," pp. 75, 105. Emphasis added.

sense of the probability of German victory than do those who assume that the German attack was bound to fail. It would be beyond the scope of this article to reexamine the long-standing debate among military analysts and historians on whether Moltke's timidity and poor judgment ruined what could have been a major German victory in September 1914. A number of major participants and historians have maintained that the German offense would have succeeded if Moltke had moved more vigorously and had not weakened the strength of the attacking right wing by moving forces to defend Alsace–Lorraine; others have countered that, even if the Germans had won an overwhelming victory at the Marne, the French would not have capitulated and a stalemate would still have developed.[29] Defense-advocates have maintained that the Germans came close to an offensive victory only because the French launched their offensive Plan XVII.[30] Others have stressed, however, that knowledge that the French would attack in Alsace–Lorraine was the underlying premise of the Schlieffen Plan. The subtlety of the German offense was that, as Liddell Hart put it, "it would operate like a revolving door—the harder the French pushed on one side, the more sharply would the other swing around and strike their back."[31]

The key point for this critique of "cult of the offensive" theory is not how this counterfactual debate is resolved, but rather that it exists at all. For it demonstrates that the theory's assumption, that only gross misperceptions

---

29. Among participants, General von Kluck, who commanded the right wing, and Admiral Tirpitz both maintained that, had the German attack started a few days earlier, victory would have been much more likely. More recently, L.C.F. Turner has argued that the plan "offered a real prospect of forcing a decision in the West and avoiding the agonizing trench war deadlock" if it had been executed as Schlieffen had designed. Gordon A. Craig and Walter Goerlitz agree that the plan would have achieved "an overwhelming initial success" if carried out in its original form, but they maintain, as do Martin Van Creveld and Richard Ned Lebow, that the French and British could have made a German success on the Marne a Pyrrhic victory by continuing the war even if the Germans won the battle and Paris fell. It should be noted, however, that the Germans have attacked France three times since 1815: twice they took Paris and France eventually capitulated; once they failed and the French kept fighting. Von Kluck and Tirpitz quoted in Van Evera, "The Cult of the Offensive and the Origins of the First World War," p. 77, fn. 66. See also Turner, "The Significance of the Schlieffen Plan," in Paul M. Kennedy, ed., *The War Plans of the Great Powers 1880–1914* (Boston: Allen & Unwin, 1979), pp. 203–204; Gordon A. Craig, *The Politics of the Prussian Army, 1640–1945* (London: Oxford University Press, 1955), p. 280; Walter Goerlitz, *History of the German General Staff* (New York: Praeger 1953), p. 135; Martin Van Creveld, *Supplying War: Logistics from Wallenstein to Patton* (Cambridge: Cambridge University Press, 1977), p. 116; and Lebow, "The Soviet Offensive in Europe," pp. 62–65.
30. Snyder, *Ideology of the Offensive*, p. 9.
31. B.H. Liddell Hart, "Foreward" to Gerhard Ritter, *The Schlieffen Plan: Critique of a Myth* (London: Oswald Wolff, 1958), p. 6; see also Turner, "The Significance of the Schlieffen Plan," p. 204.

of the offensive/defensive balance can explain why offensive strategies were chosen in 1914, is questionable. In addition, it should be remembered that Schlieffen was acutely aware of the danger of frontal attacks against fortified positions, which was precisely why the German plan emphasized an enveloping attack on the French flanks.[32] Thus, although the professional military's excessive faith in "the offensive spirit" is likely to have played a role in the fruitless British and French frontal assaults along the Western Front during the war,[33] the "cult" explanation is far less persuasive in the case of the offensive Schlieffen Plan, which is what caused the dynamic of escalation in the July crisis.

The second major problem with the "cult of the offensive" explanation is that it ignores the fundamental issue of the military balances. After all, what do the terms used in this literature, "offense-dominance" and "defense-dominance," really mean? Robert Jervis's definition, that offense has the advantage when "it is easier to destroy the other's army and take its territory than it is to defend one's own" is problematic, because it has been generally recognized since Clausewitz that defense is almost always "easier" in land warfare because of advantages of cover and the capability to choose and prepare terrain and fortify positions.[34] This is why military analysts usually think in terms of the force ratios—the required superiority of the offensive forces (2:1, 3:1, etc.) in order to achieve victory—rather than in all-or-nothing terms such as "offense-dominance" or "defense-dominance."

By focusing on the effects of military technology on the "offense/defense balance," the "cult of the offensive" theory fails to consider adequately the quantity or quality of military forces opposed to one another in a particular territorial campaign. How "defense-dominant" was the world of 1914? Certainly the range and rate of fire of small-caliber magazine rifles, machine guns, and field artillery strongly favored the defensive. And yet, even if one assumes that the historical outcome of 1914–1918 was the likely one, it is

---

32. Ritter, *The Schlieffen Plan*, pp. 50–51.
33. Michael Howard argues, however, that "the worst losses were those due not to faulty doctrine but to inefficiency, inexperience, and the sheer organizational problems of combining fire and movement on the requisite scale." Howard, "Men Against Fire: The Doctrine of the Offensive in 1914," p. 526.
34. Jervis, "Cooperation under the Security Dilemma," p. 187. For an excellent review of literature on the offense/defense balance, see Levy, "The Offensive/Defensive Balance of Military Technology." Also see Jack Snyder, "Perceptions of the Security Dilemma in 1914," in Robert Jervis, Richard Ned Lebow, and Janice Gross Stein (with contributions from Patrick M. Morgan and Jack L. Snyder), *Psychology and Deterrence* (Baltimore: Johns Hopkins University Press, 1985), pp. 157–160.

doubtful that defenses were so dominant or advantageous as to make all states simultaneously secure if they had maintained defensive military doctrines.

The point is best made by imagining what would have happened if individual armies had rejected offensive doctrines and had, as Stephen Van Evera and Robert Jervis recommend, "rushed to their trenches" instead of others' territory. Serbia did, after all, adopt what was essentially a defensive strategy and was eventually conquered by the overwhelming forces of the Central Powers.[35] What if the French had no offensive plans but had "rushed to their trenches," staying completely on the defensive in 1914? The outcome of the war in the East suggests that the Central Powers could have defeated the Russians even more soundly if there had been no Western Front. If the Germans, after defeating the Russians, had turned and quickly attacked the French, would the technology of 1914 have proven the "dominance" of the defense? One can, of course, only speculate on such a question, but the near victory of the German offensive in 1918 against the French, a fully mobilized and deployed British army, and the arriving Americans suggests that a massive German offensive against the French alone (or the French and the small British Expeditionary Force) would have stood a strong chance of success.

This leads to the third problem. The "cult of the offensive" argument, by focusing primarily upon narrow issues of military planning, ignores the critical role of the states' political objectives in determining their military doctrines. Here I am not referring to the expansionist war aims of Germany and the other great powers, for, as Jack Snyder correctly notes, the desire to annex territory had its most important influence on decision-making *after* the war began, and the doctrinal decisions of the military officers who prepared the prewar offensive war plans in France, Germany, and Russia were not strongly influenced by such territorial ambitions.[36] Yet, the "cult of the offensive" argument overlooks a key point: offensive military doctrines are needed not only by states with expansionist war aims, but also by states that have a strong interest in protecting an exposed ally. Unless sufficient capability to protect an ally at the point of attack exists, "protector" powers require

---

35. For a brief review of the campaigns in Serbia, see Trevor N. Dupuy and Molly R. Mayo, *Campaigns in Southern Europe* (New York: Franklin Watts, 1967), pp. 9–20, 38–46.
36. Snyder, *Ideology of the Offensive*, pp. 19–20. Snyder does make an exception for Russia: the Russian offensive plans against Austria–Hungary were influenced by ambitions in the Balkans as well as military operational considerations. It should also be noted that Austria–Hungary, because of its punitive policy toward Serbia, required an offensive doctrine.

offensive strategies even if their goals are defensive in nature. This consideration is often recognized with respect to extended deterrence today; it also, however, lay at the heart of the offensive doctrines of 1914.

Thus, the Russians needed an offensive capability against Austria–Hungary, in order to be able to prevent the Austrians from attacking Serbia with overwhelming offensive superiority.[37] The French required offensive capabilities against Germany in order to support Russia. Germany needed an offense to protect Austria–Hungary if Russia launched an attack against Germany's ally. Particular conditions of the military balance in 1914—the slowness of the Russian mobilization and what General von Falkenhayn called "the almost unlimited power of the Russians to evade a final decision by arms"[38] due to its vast territory—resulted in the German plan to attempt to knock France out of the war first. This further complicated the issue, for the German offensive plans against France resulted in both a Russian need for an offensive against Germany and even Austrian plans to attack Russia as a "relief offensive" to give Moltke more time to defeat the French.

This need for offensive capabilities to provide support for allies can be seen as the root cause of the offensive war plans of the great powers. This argument does not mean that motivated biases, due to organizational or psychological factors, had no influence whatsoever on military doctrine prior to 1914. It does suggest that the states' strategic interests were dominant. For example, minor powers, such as Belgium and Serbia, may have had military cults of the offensive, but they did *not* have offensive military doctrines or war plans when war broke out in 1914.[39] In contrast, the great powers required offensive military doctrines because of their alliance com-

---

37. That the initial Austrian offensive against Serbia failed should be seen as a success of the Russian strategy, and not as proof that the Russians misperceived the offense/defense balance in the Balkans. The Austrian attack failed because Conrad was forced to send the Second Army north into Galicia instead of attacking Serbia in the south because of Russian intervention. See Norman Stone, "Moltke and Conrad: Relations between the Austro–Hungarian and German General Staffs, 1909–1914," in Kennedy, *The War Plans of the Great Powers, 1880–1914*, pp. 222–251.

38. General Erich von Falkenhayn, *The German General Staff and its Decisions, 1914–1916* (Freeport, N.Y.: Books for Libraries Press, 1971 reprint), p. 16.

39. As Van Evera notes, some Belgian officers proposed that Belgium attack Germany at the outbreak of war. Van Evera, "The Cult of the Offensive and the Origins of the First World War," p. 61; and Van Evera, "Why Cooperation Failed in 1914," p. 84. It is important to note, however, that the "temerity of such an operation" was immediately pointed out by the Belgian Chief of Staff, and Belgium pursued a defensive strategy despite such "cult of the offensive" influences. The best discussion is Albertini, *The Origins of the War of 1914*, Vol. 3, pp. 455–463, quoting Antonin Selliers de Moranville, *Contribution à l'histoire de la guerre mondiale 1914–19* (Paris, 1933), pp. 163–164.

mitments. This connection between alliance commitments and offensive doctrine was well understood in the years preceding the First World War. The Franco–Russian military convention, written in 1892, specified:

> If France is attacked by Germany or by Italy with Germany's support, Russia will bring all her available forces to bear against Germany. If Russia is attacked by Germany or by Austria with Germany's support, France will bring all her available forces to bear against Germany. . . . These troops will proceed to launch a vigorous and determined offensive, so that Germany will be forced to give battle in the East and West simultaneously.[40]

The spectre of Germany being able to defeat the French or the Russians in a piecemeal fashion haunted the allies before 1914: Joffre states that he replaced the more defensive-oriented Plan XVI with the offensive Plan XVII when he took office in part because the former failed to take into account "an eventuality which was altogether likely, namely, that the Germans might return to the old Von Moltke plan of an immediate offensive directed against the Russians." Foreign Minister Poincaré promised the Russians in 1912 that France would launch an immediate offensive against Germany if war broke out and demanded that the Russians give similar assurances in writing. Furthermore, the French and Russian General Staffs specifically rejected the possibility of war being "conducted defensively" and confirmed the need for immediate offenses in their joint meetings in 1911, 1912, and 1913.[41]

Similarly, the Russian initial offensive against the Germans in 1914 demonstrates "wishful thinking" due to faith in the offensive less than the critical need to stop Germany from amassing its forces against France alone. According to Joffre, in 1912 Grand Duke Nicholas "fully understood the necessity of the Russian Army taking the offensive rapidly, whatever the risks such an attitude might seem to involve; for it was essential to bring some relief to our front at any price. . . ."[42] Thus, in the summer of 1914 Russian Foreign Minister Sazonov and the military commanders of the East Prussian front believed that a hasty offensive against Germany might indeed fail, but that, as Sazonov reportedly argued, "we have no right to leave our Ally in

40. Georges Michon, *The Franco–Russian Alliance* (New York: Macmillan, 1929), p. 54.
41. *The Personal Memoirs of Joffre* (New York: Harper and Brothers, 1932), Vol. 1, p. 20; Jan Karl Tannenbaum, "French Estimates of Germany's Operational War Plans," in Ernest R. May, ed., *Knowing One's Enemies: Intelligence Assessment Before the Two World Wars* (Princeton: Princeton University Press, 1984), pp. 167–168; Michon, *The Franco–Russian Alliance*, p. 54; and L.C.F. Turner, "The Russian Mobilization in 1914," in Kennedy, *The War Plans of the Great Powers*, pp. 256–258.
42. *The Personal Memoirs of Joffre*, Vol. 1, p. 59.

danger, and it is our duty to attack at once, notwithstanding the indubitable risk of the operation as planned."[43] The Chief of Staff of the Russian field army also stressed this point, telling his troops to attack Germany "by virtue of the same inter-allied obligations" that were demonstrated by the French offensive.[44]

The logic of the strategic situation was similarly understood in Germany. Offenses were required to support Austria–Hungary, albeit an offensive against France first, to be followed by the combined attack on Russia. When confronted with Austrian complaints that German offense in the West would leave them unprotected, Schlieffen responded that "the fate of Austria will be decided not on the Bug but on the Seine," a line that Moltke repeated to Conrad in February 1913.[45] The need for offensive action to protect Germany's ally was also the central argument of Moltke's urgent memorandum to Bethmann-Hollweg on July 29, 1914, written after Moltke learned that Russia had instituted only *partial* mobilization against Austria–Hungary:

> If Germany is not to be false to her word and permit her ally to suffer annihilation at the hands of Russian superiority, she, too, must mobilize. And that would bring about the mobilization of the rest of Russia's military districts as a result.[46]

The German threat to Russia—that it would soon be forced to mobilize, which meant war, which meant the Schlieffen Plan's offensive, if Russia did not stop the *partial* mobilization against Austria–Hungary—underscores the importance of the alliance commitment in Berlin's calculations.[47] Moreover, it was the long-standing belief of the Russian military, which these threats reinforced in 1914, that the Germans would not passively tolerate Russian preparations for war against Austria–Hungary, which had led them to plan for and then in 1914 to insist upon, not a partial mobilization on the Austrian front, but rather a full mobilization against both Central Powers.[48] This

---

43. As quoted in Nicolas N. Golovine, *The Russian Army in the World War* (New Haven: Yale University Press, 1931), p. 213.
44. Norman Stone, *The Eastern Front, 1914–1917* (New York: Scribner, 1975), p. 48.
45. See Stone, "Moltke and Conrad: Relations between the Austro–Hungarian and German General Staffs, 1909–1914," pp. 224, 232.
46. The Grand General Staff to the Imperial Chancellor, July 29, 1914, *German Documents,* No. 349, p. 307.
47. See ibid., Nos. 342, 343, 401, and 490. It should be added, however, that both Bethmann-Hollweg and Jagow originally misled the Russians on this issue. See ibid., No. 219; and Albertini, *The Origins of the War of 1914,* Vol. 2, pp. 481–485 and Vol. 3, pp. 220–221.
48. As General Kokovtzov put it, arguing against *partial* mobilization in 1912: "no matter what

decision was critical, for once the full mobilization of the Russian army began, Bethmann-Hollweg called off the attempt to avert war by having Austro-Hungarian forces "Halt in Belgrade." Thus, the alliance system—or more properly, the strategic interests the great powers had in maintaining their alliance partners—led not only to the offensive doctrines of 1914 but even to one of the specific conditions that contributed to the dynamic of rapid mobilization and counter-mobilization that constrained last minute efforts to prevent the outbreak of war.

The failure of the "cult of the offensive" theory to examine the influence of alliances on military doctrines is significant. While the theory has done a service by highlighting a number of the risks and instabilities that can result from offensive doctrines, it has ignored the risks and instabilities that result from purely defensive military strategies. For states with a security interest in preserving an exposed ally, offensive forces and strategies are necessary. Defensive military doctrines may not produce the same degree of preventive war or preemptive war incentives, but they can undercut extended deterrence by denying a government sufficient capability to protect its allies. As France discovered in 1938 and 1939, when the lack of offensive capabilities against Germany enabled Hitler to conquer France's East European allies in a piecemeal manner, a purely defensive military doctrine can also prove strategically disastrous.

The Balance and British Intervention

If the European powers required offensive military doctrines to support their alliance commitments in 1914, does this suggest an added dimension of tragedy to the events of that summer? Did the offensive war plans, required to support alliance commitments, nevertheless inevitably produce the strategic instability that caused the July crisis to spiral out of control? The answer is, in my opinion, no. For the destabilizing consequences of the German war plans, which Van Evera skillfully analyzes, were *not* the result of the mere *offensive nature* of German military doctrine, but rather of *specific vulnerabilities*

we chose to call the projected measures, a mobilization remained a mobilization, to be countered by our adversaries with actual war." Quoted in Turner, "The Russian Mobilization in 1914," p. 255. For detailed examinations of the interaction of mobilization plans, see Van Evera, "The Cult of the Offensive and the Origins of the First World War," pp. 85–94; and Ulrich Trumpener, "War Premeditated? German Intelligence Operations in July 1914," *Central European History*, Vol. 9, No. 1 (March 1976).

in the military posture of Germany's adversaries. In other words, the German General Staff's perceived incentives to mobilize rapidly and start the war with a prompt and massive offensive campaign were the result of a number of weaknesses in the Entente and Belgium's military position.

Three specific factors were critical in the July crisis. The first was the inability of the Russian army to mobilize rapidly. Although Berlin required an offensive capability against Russia, to take pressure off Austria–Hungary if that state were attacked, it was the German General Staff's belief that it could knock France out of the war in the West, before the Russians were able to mount a full-strength attack in the East, that produced the Schlieffen Plan's emphasis on rapid mobilization and attacks through Luxembourg and Belgium. Without this particular aspect of the military balance (the weaker France exposed to military defeat while Russia was mobilizing), the German offensive might well have focused on Russia, with fewer incentives to move quickly and less perceived need to turn a war in the East immediately into a continental conflagration.[49]

The second factor was the lack of permanent Belgian defenses at the critical railway junction at Liège. The German General Staff understood that failure to take Liège with its tunnels and bridges intact would mean disaster for the Schlieffen Plan. As Moltke noted prior to the war, the advance through Belgium "will hardly be possible unless Liège is in our hands. . . . the possession of Liège is the *sine qua non* of our advance." Such a *coup de main* would only be possible, however, "if the attack is made at once, before the forts are fortified."[50] Without this preemptive opportunity, again, German incentives to mobilize and attack quickly would have been reduced.

The third critical factor contributing to instability in July 1914 was the uncertainty concerning British intentions to intervene in a continental war if Germany attacked France. Throughout the July crisis, the German government received sufficiently contradictory intelligence about London's intentions as to be highly uncertain about British intervention. Although Ambassador Lichnowsky did report Foreign Minister Grey's warning on July 29 that Britain could not remain uninvolved in a continental war, Prince Heinrich of Prussia had just reported on July 28 that King George V had explicitly told

49. Snyder and Van Evera both appear to accept this point. See Snyder, "Civil Military Relations and the Cult of the Offensive," p. 109; and Van Evera, "The Cult of the Offensive and the Origins of the First World War," p. 90.
50. Ritter, *The Schlieffen Plan*, p. 166. As quoted in Van Evera, "The Cult of the Offensive and the Origins of the First World War," p. 74.

him that the British government "shall try all we can to keep out of this and shall remain neutral." Moreover, on August 1, Lichnowsky reported that Grey had suggested that Britain would not only remain neutral, but would guarantee French neutrality if Germany fought in the East but refrained from attacking France. The resulting common assessment in Berlin was expressed well by the report of the Bavarian minister to Berlin: "England's attitude is mysterious."[51]

Sir Edward Grey's failure to present a clear and credible threat of British intervention early in the July crisis and the specific preemptive aspects of Germany's offensive war plans caused by the slow Russian mobilization and the Liège bottleneck are linked together as an immediate cause of the First World War. Bethmann-Hollweg began his belated effort to prevent the imminent war—by pressuring Austria-Hungary into stopping its advance into Serbia once Belgrade had been taken—only after Grey warned Berlin that Britain would not remain neutral in a continental conflict. The available evidence suggests that, while the Kaiser and the Chancellor were willing to precipitate a continental war, they wanted to avoid a world war with England fighting on the side of the allies. But Bethmann-Hollweg had precious little time to forge a difficult diplomatic solution at the last minute, for the military preparations of the Central Powers' adversaries were increasing rapidly. During the night of July 30, he called off the "Halt in Belgrade" effort upon being informed by the General Staff that the "military preparations of our neighbors especially in the east" prevented any further delays in German mobilization.[52] The General Staff had, by 11 p.m., not only received initial indications that Russian general mobilization had begun, but also was in-

---

51. On the contradictory warnings and Grey's "neutrality offer," see *German Documents*, Nos. 368, 374, 562, 574, 578, 607, and 613. Numerous historians have accepted at face value Grey's explanation that Lichnowsky's reports were due to the German Ambassador's "misunderstanding" Grey's offer. But this explanation is discredited by the fact that Grey raised the prospect of British neutrality in a German–Russian war in a telegram on August 1 to *British* Ambassador Bertie in Paris. See G.P. Gooch and Harold Temperley, eds., *British Documents on the Origins of the War* (hereinafter *British Documents*), Vol. 11, No. 419, 426, and 453. The best treatment of these incidents is Albertini, *The Origins of the War of 1914*, Vol. 2, pp. 429, 517, 687–688 and Vol. 3, pp. 380–386. Also see Harry F. Young, "The Misunderstanding of August 1, 1914," *The Journal of Modern History*, Vol. 48, No. 4 (December 1976), pp. 644–665. The final quotation from Minister at Berlin to President of Ministerial Council, July 31, 1914, *German Documents*, Supplement 4, No. 27, pp. 634–635.

52. *German Documents*, No. 451, p. 378. The best discussion remains Albertini, *The Origins of the War of 1914*, Vol. 3, pp. 1–65. It is important to note, however, that Bethmann-Hollweg had agreed, under pressure from Moltke, that a decision on German mobilization would be made no later than noon on July 31. Thus, even if the Czar had not ordered *general* mobilization on the night of July 30, the time available to reach a diplomatic "Halt in Belgrade" settlement would have been limited.

formed that the Belgians' preparations to defend Liège and destroy the critical bridges across the Meuse were under way.[53] Military necessity had taken over.

These events were not preordained, however, and should not lead to the conclusion that the outbreak of war was inevitable. A number of possible contingencies could have produced a very different result. *If* Grey had given a clear warning earlier, *if* the Czar had further delayed Russian mobilization against Austria and then Germany, and *if* the German offensive war plans had not been able to depend upon a preemptive *coup de main* against Liège and the decisive battle in France before Russian mobilization was completed in the East, *then* it is possible, just possible, that Bethmann-Hollweg would have had the time and the courage necessary to apply sufficient pressure on Vienna to accept the "Halt in Belgrade." And if this had occurred, 1914 might today appear as only another one of a series of Balkan crises that almost led to a world war.

Any analysis of 1914 that emphasizes the British failure to issue a credible warning to Germany must contend with Richard Ned Lebow's provocative thesis that the members of the Berlin government were so psychologically committed to believing in British neutrality that they utilized common defense mechanisms—such as denial, selective attention, and wishful thinking—to ignore the otherwise clear British intention to intervene. As Lebow has argued, the German miscalculation "ought to be recognized as a German problem for which there is no plausible external explanation":

> Given Britain's commitment to Belgium, her enduring interest in the balance of power on the continent, her prior support of France in two crises with Germany and the obvious political reasons that constrained her from speaking out, it should have been apparent to all but the most unsophisticated observer of British politics that no inferences about British intentions could be drawn from her reluctance to commit herself publicly to the defense of France.[54]

This view emphasizing the psychologically motivated bias of the German leadership is not, however, convincing. It overlooks the fact that the leadership in other nations, even those with the opposite interests and motivated biases, held beliefs similar to the Germans about the likelihood of British

---

53. Trumpener, "War Premeditated?," pp. 79–83.
54. Richard Ned Lebow, *Between Peace and War* (Baltimore: Johns Hopkins University Press, 1981), pp. 130–131.

intervention. Government officials in France,[55] Russia,[56] and Great Britain itself[57] were uncertain that London would join its allies against Germany. Indeed, Sir Edward Grey himself later recalled his own pessimistic assessment during the week preceeding the British decision to intervene:

> [If] war came, the interest of Britain required that we should not stand aside, while France fought alone in the West, but must support her. I knew it to be *very doubtful* whether the Cabinet, Parliament, and country would take this view on the outbreak of war, and through the whole of this week I had in view the *probable contingency that we should not decide at the critical moment to support France.*[58]

Finally, it is worth noting that the German military, as opposed to the Kaiser and Bethmann-Hollweg, did not bank on British neutrality. Moltke assumed, as did Schlieffen, that London would attempt to send the British Expeditionary Force (BEF) to the continent in support of France.[59] From the military perspective, the critical question was how to *delay* the British decision for as long as possible, in order to increase the likelihood that British military assistance would arrive too late to influence significantly the campaign in

55. Joffre judged British intervention as highly uncertain and did not, in Plan XVII, assume the British army's support. Samuel R. Williamson, "Joffre Shapes French Strategy, 1911–1913," in Kennedy, *The War Plans of the Great Powers, 1880–1914*, p. 146. Also see pp. 137–145. French cabinet member Alexander Millerand similarly noted, "In the event of war, the English soldiers ask only to fight. The machine is ready to go: will it be unleashed? Complete uncertainty. The cabinet is vulnerable in its domestic policies (Home Rule), very uncertain in foreign policy, not knowing what it wants to do." Quoted in Porch, *March to the Marne*, p. 228.
56. For example, Foreign Minister Sazonov told the Russian Council of Ministers on July 24, 1914 that risking a continental war was dangerous "since it is not known what attitude Great Britain would take in the matter." As quoted in D.C.B. Lieven, *Russia and the Origins of the First World War* (New York: St. Martin's Press, 1983), p. 142, citing Bark Ms. 7, pp. 7–13 (Columbia University, Bakhmetev Archive).
57. Lloyd George, Churchill, and Buchanan all expressed skepticism over the likelihood of British intervention. See John Grigg, *Lloyd George: From Peace to War 1912–1916* (Berkeley and Los Angeles: University of California Press, 1985), p. 140; Winston S. Churchill, *The World Crisis*, Vol. 1 (New York: Scribner, 1923), p. 211; and Maurice Paléologue, *An Ambassador's Memoirs*, trans. F.A. Holt (New York: Octagon Books, 1972), Vol. 1, p. 32. Also see Sir George Buchanan, *My Mission to Moscow* (Boston: Little, Brown, 1923), pp. 210–211; and *British Documents*, Vol. 11, No. 101. Buchanan had been informed in April by Sir Arthur Nicolson, Permanent Secretary of the Foreign Office, that the dispatch of a British expeditionary force in the event of war was "extremely remote." Nicolson to Sir George Buchanan, April 7, 1914, as quoted in Cameron Hazlehurst, *Politicians at War, July 1914 to May 1915* (London: Jonathan Cape, 1971), p. 88, fn 2.
58. Grey of Fallodon, *Twenty-Five Years* (New York: Frederick A. Stokes Co., 1925), p. 302.
59. Gerhard Ritter, *The Schlieffen Plan*, pp. 57, 61–63, 68–69; Albertini, *The Origins of the War of 1914*, Vol. 3, p. 239. For an argument that Bethmann-Hollweg shared this view, see Karl Dietrich Erdmann, "War Guilt 1914 Reconsidered: A Balance of New Research," in H.W. Koch, ed., *The Origins of the First World War: Great Power Rivalry and German War Aims*, 2nd ed. (London: Macmillan, 1984), pp. 363–364.

France.[60] This question of timing is important,[61] for it provided yet another incentive for the German General Staff to call for prompt mobilization and offensive action during the closing moments of the July crisis.

To summarize, what caused the perceived incentives in Berlin to mobilize quickly and rapidly launch an offensive campaign? What were the root causes in 1914 of what would today be called "crisis instability"? Certainly, if the German General Staff had designed a purely defensive strategy, there would have been less pressure to mobilize and strike quickly. Yet, even if the Germans had maintained an offensive doctrine to protect Austria–Hungary, as I believe was necessary, the incentives to mobilize and attack promptly might have been severely dampened if the French had not been exposed to a "knockout blow" while the slow Russian mobilization was taking place and if Liège had not been vulnerable to the *coup de main*. Finally, if the British early in the crisis had clearly and credibly placed their forces into the balance on the side of the Entente, it is at least possible that Bethmann-Hollweg and the Kaiser would have been able to restrain their Austrian ally, pulling back from the brink of war and seeking a diplomatic solution to the crisis. Churchill put the argument best: "An open alliance, if it could have peacefully been brought about on an earlier date, would have exercised a deterring effect on the German mind, or at least would have altered their military calculations."[62]

Across the Nuclear Divide: Offense Without Instability?

This article has emphasized the degree to which the political objectives of the European powers determined the offensive nature of their military doc-

60. This was the view that Admiral Tirpitz later expressed: "a delay of even a few days in the preparation of the English expeditionary force and its transport to France might have been of the greatest importance to us." Tirpitz, Document II, p. 13, as quoted in Albertini, *The Origins of the War of 1914*, Vol. 3, p. 242, fn. 2.
61. Lebow, unfortunately, does not deal with this issue of delaying British intervention and overlooks the importance of "timing" in his assessments. Indeed, as an example of wishful thinking, Lebow, citing an article by Konrad Jarausch, states that Bethmann-Hollweg assured the Kaiser on July 23 that "it was *impossible* that England would enter the fray." This is, however, a misquotation as Jarausch states that Bethmann-Hollweg told the Kaiser, "it is *improbable* that England will *immediately* enter the fray." The original German sentence reads: "Dass dies sofort geschieht, namentlich, dass England sich gleich zum Eingreifen entschliesst, is nicht anzunehmen." Lebow, *Between Peace and War*, p. 132 (emphasis added); Konrad Jarausch, "The Illusion of Limited War: Chancellor Bethmann Hollweg's Calculated Risk," *Central European History*, Vol. 2, No. 1 (March 1969), p. 62 (emphasis added); Bethmann-Hollweg an Wedel, July 23, 1914, *DD 125* in Imanuel Geiss, *Julikrise und Kriegsausbruch 1914* (Hannover: Verlang für Litteratur und Aeitgeschenhen GMBH, 1963), Vol. 1, No. 235, p. 305.
62. Churchill, *The World Crisis*, Vol. 1, p. 217.

trines prior to the First World War. Although the "cult of the offensive" theory is correct to note that the professional military glorified the offensive in 1914, it exaggerates the influence of such motivated military biases on the development of offensive doctrines. Continental powers that had territorial ambitions or a strong interest in protecting allies—Austria-Hungary, Germany, France, and Russia—had offensive military doctrines; smaller states that lacked such interests—Belgium and Serbia—had defensive doctrines in August 1914. The "cult" literature has also correctly emphasized the destabilizing consequences of the specific offensive war plans during the July crisis. But it has misplaced its emphasis by blaming the "offensive" nature of the prewar military doctrines, rather than the specific conditions that produced the "destabilizing" characteristics of Moltke's operational plan—the need for prompt mobilization, the immediate attack on Liège, and the plan to knock France out of the war before Russian mobilization was complete—for the lack of "crisis stability" seen in July 1914. These different interpretations of the origins of the First World War are illustrated, in simplified form, in Figure 1.

These models, however, lead not only to different explanations for 1914, but also to different lessons for U.S. strategy today. The "cult of the offensive" theory has led its proponents to argue against all offensive military doctrines today, which they view as destabilizing, and in favor of defensive military

Figure 1

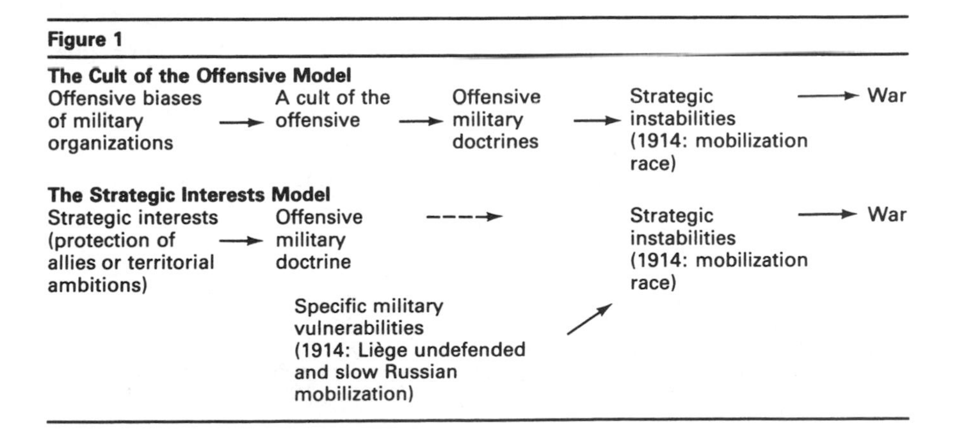

doctrines. At the nuclear level, as Van Evera writes, 1914 "warns us that we tamper with MAD at our peril":

> The 1914 case . . . warns both superpowers against the offensive ideas which many military planners in both countries favor. Offensive doctrines have long been dogma in the Soviet military establishment, and they are gaining adherents in the United States as well. This is seen in the declining popularity of the nuclear strategy of "assured destruction" and the growing fashionability of "counterforce" nuclear strategies, which are essentially offensive in nature.[63]

The debate over counterforce vs. countercity targeting is, of course, a familiar one, and there is no need to review all the arguments yet once again. It is important, nevertheless, to draw attention to the different "lessons" suggested by the contrasting views of 1914. Because of the assumption that offensive doctrines are caused by the biases of the professional military, the "cult of the offensive" literature calls for senior civilian authorities to review war plans and reject all counterforce strategies.[64] While civilian review of and central control over U.S. nuclear strategy is essential, this article suggests that there have been strong political and strategic reasons why *civilian* authorities in Washington have preferred counterforce targeting. In particular, it would suggest that the perceived need for counterforce options to enhance the credibility of NATO's first-use threat, which is required for extended deterrence, has been an important factor contributing to U.S. counterforce strategy. Strong evidence for this view can be found in the recently declassified 1961 Draft Memorandum for the President in which Secretary of Defense Robert McNamara specifically rejected "a 'minimum deterrence' posture . . . in which, after a Soviet attack, we would have a capability to destroy most of Soviet urban society, but . . . would not have a capability to counterattack against Soviet military forces."[65] "We should reject the 'minimum deterrence extreme,'" McNamara reported to President Kennedy, "for the following reasons:

---

63. Van Evera, "The Cult of the Offensive and the Origins of the First World War," pp. 106–107.
64. Ibid.; and Posen, *Sources of Military Doctrine*, pp. 241, 244.
65. Draft September 23, 1961, Appendix I to Memorandum for the President, Subject: Recommended Long Range Nuclear Delivery Forces 1963–1967, p. 4. OSD FOI (declassified with deletions).

a. Deterrence may fail, or war may break out for accidental or unintended reasons, and if it does, a capability to counter-attack against high-priority Soviet military targets can make a major contribution to the objectives of limiting damage and terminating the war on acceptable terms;
b. *By reducing to a minimum the possibility of a US nuclear attack in response to Soviet aggression against our Allies, a 'minimum deterrence' posture would weaken our ability to deter such Soviet attacks.*"[66]

Of course, as this document suggests, extended deterrence has not been the only reason for counterforce targeting. Civilian authorities have also approved capabilities and plans to attack Soviet military forces to enhance deterrence by denying Soviet war aims and to limit damage to the United States if deterrence fails. The "cult" literature suggests that all nuclear counterforce plans are highly destabilizing. The perspective on 1914 offered here—emphasizing the degree to which instability in the July crisis was caused by German incentives to mobilize and launch their offenses *quickly* before Liège was fortified and the Russian mobilization completed—leads neither to such a blanket condemnation of counterforce nor to an acceptance of assured destruction targetting. If future U.S. counterforce capabilities and plans are carefully designed to threaten Soviet military and leadership targets, without posing a credible threat of a disarming first strike, it would be possible to avoid throwing out the baby of crisis stability with the bathwater of MAD. This is not the place to examine in detail what such a U.S. second-strike counterforce posture would entail, but clearly U.S. strategic vulnerabilities would have to be reduced as well as U.S. offensive capabilities constrained. Specifically, efforts to reduce the vulnerability of the U.S. ICBM force and the fragility of U.S. command and control are a high priority to enhance deterrence and stability, by reducing both Soviet incentives to strike first and American pressures to launch promptly. In addition, a retaliatory counterforce posture—designed to threaten hardened Soviet leadership targets and disrupt and destroy Soviet follow-on attacks, launch failures, reserves, and reloads—requires fewer counterforce warheads than are needed for a credible first-strike capability and could place increased reliance on slowly arriving weapons (gravity bombs and cruise missiles) for many important missions.

At a more general level, Snyder argues that the disastrous consequences of 1914 suggest that a status quo power should "compete with one hand—its offensive hand—tied behind its back":

66. Ibid. Emphasis added.

In practical terms, this means that the status quo state must keep up its end of the power competition but do so by deploying defensive forces, forces that dissuade without the threat of escalation and create no first-strike advantage for either side. Moreover, it must avoid commitments and alliances that can be defended only by destabilizing means.[67]

Yet, for the foreseeable future, as long as the United States has a strong interest in maintaining extended deterrence—that is, as long as the NATO alliance continues to be in the United States' political, strategic, and moral interests—it will be necessary to have offensive military capabilities even if NATO has defensive political objectives. For unless NATO's conventional capability could be increased to offset traditional Warsaw Pact conventional superiority, extended deterrence will require "the threat of escalation." Again, as the French discovered when Hitler conquered their allies in 1938 and 1939, purely "defensive" doctrines can also be strategically destabilizing. This point is an important one, for it is often overlooked not only by the "cult of the offensive" theorists, but also by the most enthusiastic supporters of strategic defense today: if President Reagan's dream of making nuclear weapons impotent and obsolete ever becomes a reality for both superpowers, it will be essential that other means of protecting Western Europe from Soviet conventional superiority be found.

A review of the "lessons of the past" cannot illuminate all of our current defense problems nor provide specific answers to questions of how much counterforce is enough or whether a first-use policy or conventional retaliation policy is most advisable for NATO. Still, in the final analysis, if 1914 is a potent reminder of the strategic instabilities that offensive military doctrines can produce, the events of the late 1930s send an equally strong message about the dangers of purely defensive doctrines for states with extended deterrent interests. Thus, at the abstract level, the lesson is clear enough: the United States must confront both problems and avoid, not only a modern cult of the offensive, but also any potential cult of the defense.

---

67. Snyder, "Perceptions of the Security Dilemma in 1914," p. 179.

Correspondence

The Origins of Offense and the Consequences of Counterforce

Jack Snyder
Scott D. Sagan

To the Editors:

Scott Sagan, in "1914 Revisited: Allies, Offense, and Instability," in the Fall 1986 issue of *International Security*, presents two arguments that significantly advance the debate on the origins of offensive strategies and their consequences for strategic stability.[1] First, he argues that offense is needed to honor alliance commitments when sufficient forces cannot be moved to defend the threatened ally on its own territory. He suggests that this created strong incentives for Germany, France, and Russia to adopt offensive strategies in 1914, just as the requirements of extended deterrence may provide a legitimate rationale for nuclear first use and counterforce strategies today.[2] Offensive biases of military planners, unleashed and intensified by prevailing patterns of civil-military relations, were at most a secondary cause of the 1914 offensives, in his view. Second, he suggests that the destabilizing effects of offense can be mitigated by designing counteroffensive or limited offensive strategies, which the opponent can distinguish from plans for a totally disarming first strike.

I would agree that in some circumstances offense may be needed to aid allies and that counteroffensive strategies can sometimes be made distinguishable from first-strike plans, but I believe that it is worth reviewing the 1914 and contemporary cases to see whether they are of this type. In my

---

Jack Snyder is Assistant Professor of Political Science at Columbia University.
Scott D. Sagan is a Lecturer in the Government Department, Harvard University.

---

1. Scott D. Sagan, "1914 Revisited: Allies, Offense, and Instability," *International Security*, Vol. 11, No. 2 (Fall 1986). For the most part, Sagan is rebutting arguments presented in Jack Snyder, "Civil-Military Relations and the Cult of the Offensive, 1914 and 1984" and Stephen Van Evera, "The Cult of the Offensive and the Origins of the First World War," both in Steven E. Miller, ed., *Military Strategy and the Origins of the First World War: An* International Security *Reader* (Princeton: Princeton University Press, 1985); Snyder, *The Ideology of the Offensive: Military Decision Making and the Disasters of 1914* (Ithaca: Cornell University Press, 1984); and Van Evera, "Why Cooperation Failed in 1914," *World Politics*, Vol. 38, No. 1 (October 1985), pp. 80–117.
2. Sagan notes other reasons to adopt counterforce strategies, but this one is the most directly related to his arguments about 1914.

view, the Russians' need to come to the aid of France does partially explain their offensive against Germany,[3] but the offensives mounted by the other powers actually harmed the security interests of their allies. France would have best served itself and its ally by standing on the defensive.[4] Germany's optimal strategy was probably a counteroffensive, which under the prevailing circumstances could have been made distinguishable from a strategy of unlimited offensive aims. In the current NATO case, defensive measures should be adequate to protect allies, but among the limited offensive options that might be considered, limited hard-target kill capabilities add the least to extended deterrence and may be indistinguishable from first-strike preparations in Soviet eyes.

Whether offensives were needed to protect allies in 1914 depends, first, on whether the defender's tactical and operational advantages were large or small, and second, on whether incentives for offense associated with alliance commitments, if any, were substantial enough to outweigh its operational disadvantages.

Sagan warns against overstating the advantages of the defender, noting that the Schlieffen Plan almost succeeded, that minor adjustments could have made it work better, and that only the benefits of hindsight make it look like a reckless, self-defeating enterprise. I have argued, in contrast, that the Schlieffen Plan was not close to success. Despite the numerous operational and intelligence errors of the French command in August 1914, the French were able to deploy substantially superior forces against the Germans' crucial right flank at the battle of the Marne. This outcome was due to inherent advantages enjoyed by the defender and inherent disadvantages sapping the strength of the attacker. The Germans' offensive ensured that British, Belgian, and French rear-echelon reserve forces would be brought into play against them. The need to besiege fortresses and guard lines of communications subtracted from the size of the Germans' attacking force, while the long, forced-pace march and supply shortages debilitated its fighting power. Meanwhile, the French had the full use of their own rail system to shift forces rapidly to the decisive sector of the front.[5]

Hindsight was not necessary to understand these barriers to the success of the Schlieffen Plan. Schlieffen himself understood that it could easily fail

---

3. I develop this argument in *Ideology of the Offensive,* pp. 157–164.
4. Sagan accepts this judgment, except for the hypothetical case in which Germany would have attacked Russia first.
5. Snyder, *Ideology of the Offensive,* pp. 111–115.

for these reasons, but he chose to cover up the plan's flaws because he considered defensive alternatives to be unthinkable. In particular, the Liège operation, crucial to the plan's success, was not seen as a tempting walkover. The younger Moltke recognized that "the capture of a modern fortress by a *coup de main* would be something unprecedented in military history," but nonetheless felt it "must be attempted, for the possession of Liège is the *sine qua non* of our advance." Since he saw the Liège operation as essential if Germany were to have any chance for a decisive offensive in the West, Moltke counted on the gratuitously wishful (and, as it turned out, erroneous) hope that "once our troops have entered the town I believe that the forts will not bombard it but will probably capitulate."[6]

But did the requirements of alliance politics create incentives for offense that outweighed the operational advantages of the defender? Was this the motive behind the war plans of 1914? I believe that this is true only for the Russians, who had to relieve the pressure on France by attacking Germany's lightly defended eastern front. Even in this case, however, it is doubtful that the Russians' hasty, overcommitted attack, culminating in overextension and defeat at Tannenberg, was the most effective way to aid the French. The Russians' strategic thinking, colored by the cult of the offensive, overestimated the likelihood of a quick, decisive outcome on the western front. This led them to mount a premature offensive in the early days of the campaign, which hindered their ability to exert more effective pressure on Germany's rear in subsequent weeks.[7]

---

6. Ibid., p. 152. This kind of evidence convinces me that the Schlieffen Plan had foreseeably a very low chance of success and high costs of failure. However, my arguments about the cult of the offensive do not hinge primarily on the assertion that Schlieffen and other war planners picked the wrong strategies; rather, they hinge on the biased way in which the planners evaluated more defensive alternatives. I tried to show that the process of strategic assessment was systematically slanted through the use of double standards, logical inconsistencies, selective use of information, and rigged war-gaming in favor of offensive strategies that suited the military's organizational purposes. Despite this emphasis on military bias in the World War I cases, neither Van Evera nor I consider organizational biases of the military to be the sole or even the most important source of offensive, destabilizing strategic ideas in other cases. Van Evera's Ph.D. dissertation, "Causes of War" (University of California, Berkeley, 1984), includes a chapter on the strategic mythmaking propensity of the state apparatus as a whole, viewed as an organized interest group. My current research takes a similarly broader focus.

7. It is true that the hasty Russian attack frightened the Germans into shifting two corps from their rear echelon in Belgium to East Prussia, and that these forces were still in transit when the major battles were being fought on both fronts. Some have argued that this saved France and consequently that the losses at Tannenberg were worthwhile. Others have argued, in my view more persuasively, that the two corps could not have been transported to the Marne anyhow or that they would not have been decisive even if they could have been brought into battle.

In the German case, two separate questions must be distinguished. First, was the Schlieffen Plan necessitated by the Austrian alliance? Second, would some other kind of offensive have served the interests of the alliance better than a strictly defensive plan?

The answer to the first question is clearly no. The basic idea behind the Schlieffen Plan had nothing to do with the Austrian alliance. The stratagem of using Germany's interior position to defeat France and Russia piecemeal had occurred to the elder Moltke long before Germany and Austria became allied. If, as Moltke eventually concluded, a bid for a rapid victory over France was a dubious risk from a strictly German standpoint, it had even greater drawbacks as a strategy for the Austro–German alliance. The Austrians were understandably dubious about the idea of defending Vienna by sending 90 percent of the German army into France. They pressed the Germans to reexamine the feasibility of defending in the west and sending more forces east. In fact, the Schlieffen Plan was so divisive to the alliance that the Germans had to mask from the Austrians the extent to which the eastern front would be stripped. Detailed consultations between the two staffs lapsed for years because of this problem.[8]

The best strategy for the Austro–German alliance would have been a positional defense of the short frontier in the west, combined with either a counteroffensive or a positional defensive in the east. The Germans might have succeeded with a positional defense in the east, by deploying some of their own forces in Austrian Galicia to help fend off a Russian attack. The argument against this is that the long, crescent-shaped frontier would have made for a thin force-to-space ratio, making positional defense problematic. A better strategy, worked out by Schlieffen himself, would have deployed the main German force in Silesia, directly adjacent to the main Austrian forces in Galicia and forming a common front with them. If the Russians attacked the Austrians, the Germans in Silesia would have been positioned to smash the Russian flank with a counterattack. Schlieffen would have followed up this riposte with a limited advance into Poland, thus shortening the front and making a positional defense easier.

From the standpoint of strategic stability, the beauty of this plan is that it is so clearly distinguishable as limited and counteroffensive. A German deployment in Silesia would have threatened Russia only if Russia first attacked

8. Snyder, *Ideology of the Offensive*, p. 117; and Gerhard Ritter, *The Schlieffen Plan: Critique of a Myth* (London: Oswald Wolff, 1958), pp. 17–18, 30–32.

Austria. Ultimately, however, Schlieffen rejected limited, eastward-oriented schemes because they offered no prospect of a rapid, decisive victory, which he made into a dogmatic tenet of German strategy.[9]

The French did not need an offensive strategy to protect Russia from Germany, nor did they adopt their offensive plan with this motive in mind. The framers of Plan 17 never considered the possibility that Germany might defend in the west and concentrate its forces mainly against Russia. In their minds, the French had no extended deterrence problem. They had a self-defense problem that they decided to solve by offensive means.[10] Moreover, the French offensive vastly complicated Russia's strategic problems. With the French and the Germans both on the attack, a rapid decision on the western front—for better or worse—was made more likely. This increased the pressure on the slow-mobilizing Russians to mount their own offensive prematurely, before their supply trains and second-echelon forces could be deployed. The Russians' attempt to mitigate this problem by mobilizing first in July 1914 touched off the infamous mobilization race, which cut short the search for diplomatic solutions to the crisis.[11] Had France stood on the defensive, Russia would not have needed to race so hard to mobilize and attack. In short, in the French case as in the German one, the offensive strategy adopted in 1914 was harmful to the strategic interests of an ally.

Finally, how might the lessons of these historical cases illuminate the United States' extended deterrence problem in Europe? Certainly, the Russian case shows that, under some circumstances, offense may be needed to protect an exposed ally. But the United States' situation is not at all like that of Russia in 1914. The United States can choose to defend Europe by deploying defensive forces in Europe, whereas Russia obviously could not have defended France by sending its army to fight in France. Some analysts say that NATO's current capability for conventional defense is close to being adequate; others say it is not.[12] Either way, the lesson of 1914 is clear: since

9. Ritter, *The Schlieffen Plan*, pp. 22–33.
10. Snyder, *Ideology of the Offensive*, chapters 2 and 3. Sagan quotes Marshal Joffre's statement that France needed an offensive strategy in case Germany attacked Russia first. The French military did make statements to that effect, not only in Joffre's memoirs, but before the war as well. This served two purposes. First, it added yet another public justification for the offensive. Second, it made French pressure for a hasty Russian offensive look like a mutual commitment instead of the unilateral Russian concession that it really was. Actual war planning documents, however, reveal no consideration of the scenario that Joffre mentions.
11. Snyder, "Civil-Military Relations," pp. 113–115.
12. Assessments suggesting that current NATO forces may be adequate are found in Barry R. Posen, "Measuring the European Conventional Balance: Coping with Complexity in Threat

offense is such a dangerous tool for providing security, Western publics and governments should understand that all other remedies must be exhausted before turning to this last resort.

But assume, for the sake of argument, that positional conventional defenses must be supplemented with nuclear first-use options to deter Soviet conventional attack in Europe. What kind of options would provide deterrent credibility while also minimizing the destabilizing effects of offensive capabilities? One approach would be to look for nuclear options that would deny success to the Soviet land offensive without being offensive. The use of neutron warheads on German territory is one obvious example. If that were infeasible, limited offensive options might be considered. A force designed to stop the Red Army by destroying soft targets in Eastern Europe, but having no capability to destroy Soviet silos or command and control, could be easily distinguished from a first-strike force.

A second approach is Sagan's second-strike counterforce proposal, which would deploy forces capable of destroying some Soviet hard targets, including missile silos and leadership bunkers, but lacking the speed and numbers to execute a disarming first strike. This raises two questions. First, does this kind of hard-target kill capability succeed in addressing the extended deterrence problem in a meaningful way? In the days when the Strategic Air Command (SAC) packed something close to a disarming first-strike capability, counterforce did enhance extended deterrence, but nowadays the link is questionable.[13] Second, would this limited counterforce posture, whether adopted to support allies or for other reasons, be clearly distinguishable from an unlimited first-strike posture in Soviet eyes? To use the 1914 analogies, would it look like a Silesian counterattack or like a Schlieffen Plan?

In answering these questions, one's view of the notion of the "cult of the offensive" is important. Perhaps a clear-headed opponent would understand the fine distinction between a disarming first-strike capability and more limited capabilities for attacking missile silos and command and control centers in Moscow. But a myth-ridden Soviet opponent, partially imbued with the cult of the offensive and steeped in the history of U.S. counterforce strategy, might see those forces through lenses colored with preconceptions and deep suspicions. The strongest motive that the Soviets could have for

---

Assessment," *International Security*, Vol. 9, No. 3 (Winter 1984–85), pp. 47–88; and John J. Mearsheimer, "Why the Soviets Can't Win Quickly in Central Europe," *International Security*, Vol. 7, No. 1 (Summer 1982), pp. 3–39.

13. Robert Jervis, *The Illogic of American Nuclear Strategy* (Ithaca: Cornell University Press, 1984).

attacking NATO would be to excise a node of offensive military capability that might be used to disarm or decapitate them. The lesson of 1914 is that we should do everything possible to avoid feeding such fears.

—Jack Snyder
New York, N.Y.

The Author Replies:

The purpose of "1914 Revisited: Allies, Offense, and Instability" was to counter what I view as the "cult of the offensive" theory's inadequate analysis of the causes and the consequences of offensive military strategies in 1914 and in current U.S. nuclear doctrine. I acknowledged that Jack Snyder's work demonstrated how organizational interests, psychological distortions, and doctrinal oversimplification can produce offensive biases in the professional military and noted that Stephen Van Evera's analysis illuminated the destabilizing effects of the 1914 offensive war plans during the July crisis. My critique, however, was threefold. First, I argued that the "cult of the offensive" literature exaggerated the importance of such military biases as a cause of the 1914 offenses and underestimated the degree to which offensive strategies were determined by the governments' political ambitions and perceived needs to protect exposed allies.[1] Second, I argued that the "cult of the offensive" writers incorrectly blamed offense strategies per se for the significant "crisis instability" seen in July 1914 (rather than the very specific military vulnerabilities that produced such "preemptive" incentives) and ignored the grave dangers that would have developed if the European powers had adopted purely defensive strategies. Finally, I criticized Snyder and Van Evera for their blanket condemnation of offensive elements in U.S. strategy today, which I argued gave inadequate regard to prudent assessments of U.S. and allied security requirements.

It should be noted from the onset that a number of complex patterns of interaction can exist between two sources of military doctrine: the strategic interests of the government and the organizational biases of the professional

---

1. It is important to note that while Snyder's and Van Evera's analyses of the origins of offensives in 1914 largely overlook the role of alliance commitments as a cause of offensive doctrines, the theoretical literature on the subject of offense and strategic stability does discuss the impact of alliances. See, for example, Robert Jervis, "Cooperation Under the Security Dilemma," *World Politics*, Vol. 30, No. 2 (January 1978), pp. 167–214.

military. It is certainly possible that organizational biases can produce offensive doctrine when an unbiased assessment of the states' strategic interests would recommend a defensive doctrine. It is also possible, however, that prudent political calculations in favor of defensive strategies could overcome military biases in favor of offensives. In addition, an offensive doctrine, chosen for sound reasons at the strategic level, could be exaggerated at the operational level by professional military biases. Here the difficult task for the student of 1914 is to evaluate the various factors affecting the military strategies of the European powers in 1914. A related task for the analyst of U.S. nuclear strategy is to determine the degree to which U.S. security interests require an offensive strategy and, if the requirements are high, whether the potentially destabilizing consequences of such offenses can be minimized.

For some of the European governments in 1914, the strategic necessity of an offense was clear. For Russia, as Snyder acknowledges in his letter, "the requirements of alliance politics create[d] incentives for offense that outweighed the operational advantages of the defender" since the Russians "had to relieve the pressure on France by attacking Germany's lightly defended eastern front." I would also note that similar considerations led to Russia's offensive against Austria–Hungary to prevent Vienna from sending all its forces against Serbia. In addition, Austria–Hungary required an offense to punish the Serbians and, in the event of the Schlieffen Plan, a "relief offensive" against the Russians to help protect the exposed German frontier. For other governments in 1914, however, the advantages of defensive strategies were clear. The small powers, Belgium and Serbia, maintained defensive strategies in August 1914, despite "cult of the offensive" biases in their professional military. The British government also rejected biased military arguments in favor of offensive strategy against Germany. In 1911, the British Navy presented the Committee of Imperial Defense with a proposal to attack Germany in the event of war, with raids against the Baltic and Frisian coasts and a division strength attack on Heligoland. Prime Minister Asquith, however, recognized that the Navy proposal was "puerile," dismissed it as "wholly impractical," and maintained that the plan to use British troops to defend France was "the only alternative" if Britain went to war against Germany.[2]

---

2. Asquith to Haldane, August 31, 1911, as quoted in Samuel R. Williamson, Jr., *The Politics of Grand Strategy: Britain and France Prepare for War, 1904–1914* (Cambridge: Harvard University

The French and German decisions are more complex. Snyder is correct to emphasize the degree to which the French military's organizational and psychological biases led to the massive offenses into Alsace–Lorraine in August 1914. Anyone who reads French military writings of the period cannot ignore the unreasoned character of their war planning, and, as I stated in my article, "the cult of the offensive" theory is "the strongest in explaining the French military's *offensive à outrance* doctrine."[3] What France required, because of its alliance with Russia, was an offensive *option*: a capability to attack Germany in order to prevent the Germans from concentrating their entire army against Russia. In the event of a major German attack on France, however, I agree that the best French strategy would have been to remain on the defensive at least in the initial campaign.[4]

With respect to Germany, the disagreement between Snyder and me is *not* about whether biases in favor of offensive strategies were present in the General Staff. The disagreement is over whether such biases were the primary cause of Germany's offensive doctrine. With respect to the importance of alliance commitments, I stated in my article that "[a]lthough Berlin required an offensive capability *against Russia*, to take pressure off Austria–Hungary if that state were attacked, it was the German General Staff's belief that it could knock France out of the war in the West, before the Russians were able to mount a full-strength attack in the East, that produced the Schlieffen Plan's emphasis on rapid mobilization and attacks through Luxembourg and Belgium."[5] Snyder does not disagree with the first proposition but argues in his letter, with respect to the second point, that military "cult of the offensive" biases were the root cause of such General Staff beliefs.

Press, 1969), p. 193. General Nicholson had responded to the Navy proposal by asking "if the Admiralty would continue to press that view even if the General Staff expressed their considered opinion that military operations in which it was proposed to employ this division were madness." Ibid., p. 190.

3. Scott D. Sagan, "1914 Revisited: Allies, Offense, and Instability," *International Security*, Vol. 11, No. 2 (Fall 1986), p. 159.

4. One should not, however, dismiss the influence of alliance considerations on PLAN 17 altogether. Snyder's comment that war planning documents do not reveal consideration of the need for offenses in case Germany attacked Russia (footnote 10) is too narrow a focus. It disregards the issue of why the French political leadership—who were willing to "interfere" in operational issues as when they overruled Joffre's plan to violate Belgian neutrality—accepted and encouraged the *offensive à outrance*. For an analysis emphasizing the importance of alliance politics in determining French political support for Joffre's military plans, see Gerd Krumeich, *Armaments and Politics in France on the Eve of the First World War* (Warwickshire, U.K. and Dover, N.H.: Berg Publishers, 1984), p. 23 and *passim*.

5. Sagan, "1914 Revisited: Allies, Offense, and Instability," p. 167 (emphasis added). Also see p. 163.

Without the "cult of the offensive," the Germans would have rejected the Schlieffen Plan and concentrated on a limited offense against Russia.

This argument rests on shaky grounds. First, Snyder criticizes my view that the Germans accurately perceived the vulnerability of Liège, maintaining instead that Moltke relied upon the "gratuitously wishful (and, as it turned out, erroneous) hope" that the forts would capitulate once German troops entered Liège in a coup de main. If Snyder had quoted the full text of Moltke's memorandum, however, Moltke's views would appear far more cautious and much less "gratuitously wishful." For although the Chief of the General Staff did state that such a coup de main would be "unprecedented," he also added that "the heaviest artillery must be at hand, so that in case of failure we can take the fortress by storm."[6] Indeed, it was such contingency planning that did enable the Germans to take Liège in August 1914. In short, Moltke's views of the attack on Liège were not so biased as to provide adequate evidence to demonstrate, as Snyder puts it, that "the Schlieffen Plan had foreseeably a very low chance of success."

Secondly, and more importantly, Snyder's argument is based upon his view that the German General Staff's opposition to the eastern attack option against Russia was also due to a military bias, what he calls "the dogmatic tenet" that a rapid, decisive military victory was necessary in 1914. The belief that Germany had strong incentives to avoid a war of attrition is not, however, best seen as a biased dogma due to the organizational interests, desire for glory, or professional training of military officers. On the contrary, leading civilian authorities such as Chancellor Bethmann–Hollweg agreed with such

---

6. Moltke's full statement reflects more thoroughness and worst-case contingency planning, and less wishful thinking, than does the excerpt quoted by Snyder: "I think it possible to take it by a *coup de main*. Its salient forts are so unfavourably sited that they do not overlook the intervening country and cannot dominate it. I have had a reconnaissance made of all roads running through them into the centre of the town, *which has no ramparts*. An advance with several columns is possible without their being observed from the forts. Once our troops have entered the town I believe that the forts will not bombard it but will probably capitulate. Everything depends on meticulous preparation and surprise. The enterprise is only possible if the attack is made at once, before the areas between the forts are fortified. It must therefore be undertaken by standing troops immediately war is declared. The capture of a modern fortress by a *coup de main* would be something unprecedented in military history. But it can succeed and must be attempted, for the possession of Liège is the *sine qua non* of our advance. It is a bold venture whose accomplishment promises a great success. In any case the heaviest artillery must be at hand, so that in case of failure we can take the fortress by storm. I believe the absence of an inner rampart will deliver the fortress into our hands." Quoted in Gerhard Ritter, *The Schlieffen Plan: A Critique of a Myth* (London: Oswald Wolff, 1958), p. 166. Also see Jehuda L. Wallach, *The Dogma of the Battle of Annihilation: The Theories of Clausewitz and Schlieffen and Their Impact on the German Conduct of Two World Wars* (Westport, Conn.: Greenwood, 1986), pp. 93–96.

perspectives,[7] and the First World War proved such German fears to have not been unfounded. In short, the German military and civilian authorities alike accepted the Schlieffen Plan, despite its recognized dangers, over the attack in the east option because the eastern attack plan was likely to produce only a stalemate and a prolonged war of attrition. This does not mean that the Schlieffen Plan had no faults (for indeed it had many), only that General Staff "cult of the offensive" biases were not necessary for Germany to have chosen its offensive military strategy.

Finally, let me turn to the implications of these issues for the debate on U.S. nuclear strategy. In my article, I cautioned against a "cult of the defensive" and argued that Snyder and Van Evera's strong opposition to offensive elements in U.S. strategy failed to take into account U.S. alliance and security requirements and failed to distinguish adequately between more and less destabilizing offensive strategies. Snyder, I noted, had concluded that a status quo state must only deploy "defensive forces, forces that dissuade without the threat of escalation" and "must avoid commitments and alliances that can be defended only by destabilizing means."[8] I interpreted this to mean that the U.S. should abandon NATO if means are not found to cease the current reliance on the first use of nuclear weapons. The formulation in Snyder's letter, that "offense is such a dangerous tool" that "all other remedies must be exhausted before turning to this last resort," is a more moderate statement with which I would agree.

With respect to counterforce capabilities, however, Snyder's letter simply reinforces my position. I referred to three reasons why civilian authorities historically have approved of counterforce targeting: enhancing deterrence by denying Soviet war aims; limiting damage to the United States and its allies in a war; and adding credibility to the NATO threat of first use. Snyder's concern about the destabilizing consequences of offensive counterforce leads him, however, to propose only a force designed to "stop the Red Army by destroying soft targets in Eastern Europe, but having no capability to destroy Soviet silos or command and control." Such a nuclear posture would un-

7. After the war, Bethmann-Hollweg argued that "offense in the East and defense in the West would have implied that we expected at best a draw. With such a slogan no army and no nation could be led into a struggle for their existence." Quoted in Konrad Jarausch, *The Enigmatic Chancellor* (New Haven: Yale University Press, 1973), p. 195.

8. Jack Snyder, "Perceptions of the Security Dilemma in 1914," in Robert Jervis, Richard Ned Lebow, and Janice Gross Stein, *Psychology and Deterrence* (Baltimore: Johns Hopkins University Press, 1985), p. 179.

doubtedly avoid feeding Soviet fears of a NATO attack, but it would also severely weaken the Western deterrent.

Why is this the case? The Soviet Union has built, by DoD estimates, over 1,500 *hardened* alternative command posts and shelters for its leaders, key party and government personnel.[9] A U.S. nuclear posture that rejected counterforce capabilities altogether would therefore increase the likelihood that the Soviet leadership would survive and (most importantly for deterrence) that it would believe it would survive a U.S. retaliatory strike. Given the Soviet capability to launch its nuclear forces on warning, U.S. counterforce improvements cannot ensure that Soviet missiles would be destroyed and hence that major damage limitation would occur in the event of a U.S. counterforce attack. But a U.S. nuclear posture that eliminated our capability to disrupt and destroy Soviet follow-on attacks, launch failures, reserves, and reloads would also weaken deterrence by enabling the Soviet Union to carry out nuclear war plans with a higher probability of success, as it defines success.

In short, crisis stability is an important objective of U.S. nuclear strategy. It is not the only objective. Although even a carefully designed U.S. second-strike counterforce posture would inevitably appear somewhat ambiguous to the Soviets, the effort should nonetheless be made. A retaliatory force capable of destroying critical, hardened Soviet leadership and military targets, but incapable of executing a prompt disarming strike, would maximize our chances of maintaining both crisis stability and robust deterrence.

—Scott D. Sagan
Cambridge, Mass.

---

9. U.S. Department of Defense, *Soviet Military Power 1986* (Washington, D.C.: U.S. Government Printing Office, 1986), p. 53. Also see *Department of Defense Appropriations for 1984*, Hearings Before a Subcommittee of the Committee on Appropriations, House of Representatives, 98th Cong., 1st Sess., Part 8, p. 316.

Technology, Military Advantage, and World War I

A Case for Military Entrepreneurship

Jonathan Shimshoni

In formulating the military dimension of grand strategy, statesmen face a crucial decision: should they adopt an offensive or a defensive doctrine?[1] How most effectively can one approach this question? Should leaders first assess the technological state of the art, to see whether it favors offense or defense, and adapt their doctrines and grand strategies accordingly? Or should they determine independently their best grand strategy, and expect or direct that their military leaders will create the necessary supportive military advantage, be it offensive, defensive, or both?

The answer depends on where one believes military advantage comes from. The "bottom-up" approach posits that there is—at any given moment—a ubiquitous technological condition that determines, exogenous to the actors, whether offense or defense has the advantage. The "top-down" alternative recommends that advantages are manufactured and destroyed by the

I would like to thank Barry Posen, Steven Miller, Edward Rhodes, Stephen Van Evera, and John Mearsheimer for their comments on earlier drafts of this paper. I am also grateful to Samuel Huntington and Harvard University's Center for International Affairs for their assistance, and am most indebted to Barry Posen, Jack Ruina, and the entire staff at MIT's Center for International Studies for their financial and scholarly support. As ever, my mother Rose Shimshoni provided critical assistance.

Jonathan Shimshoni, the author of Israel and Conventional Deterrence: Border Warfare from 1953 to 1970 *(Cornell University Press, 1988), received his doctorate in public and international affairs from Princeton University. This article was written while he was a Visiting Scholar at MIT's Center for International Studies, and an Associate of Harvard's Center for International Affairs.*

1. *Grand strategy* is a state's overall "theory of how it can best 'cause' security for itself." Barry R. Posen, *The Sources of Military Doctrine: France, Britain and Germany Between the World Wars* (Ithaca: Cornell University Press, 1984), p. 13. Such a theory should be supported by a *military doctrine*, which is the national military organization's applied theory of victory. Military doctrine integrates and institutionalizes the principles of force organization and operation at all levels. In this article, I focus on the higher-level operational principles, such as offense or defense, and at a slightly more detailed level, on doctrines such as "elastic defense" or "blitzkrieg." The *operational level* refers to the operations of large units (corps and armies, for example) in pursuit

actors themselves, endogenously. If the first be true, as many authors have argued, then indeed a leader's job is to diagnose and *adapt;* if the second, as I argue in this essay, then he must analyze and *create.*

Jack Snyder writes of World War I that:

> Military technology should have made the European strategic balance in July 1914 a model of stability, but offensive military strategies defied those technological realities, trapping European statesmen in a war-causing spiral of insecurity and instability. As the Boer and Russo-Japanese Wars had foreshadowed and The Great War itself confirmed, prevailing weaponry and means of transport strongly favored the defender. . . . Why then were these self-defeating [offensive] war-causing strategies adopted?[2]

This passage reflects one prominent response to this essay's central questions. Snyder, not alone in the security field, is really arguing the following: (1) military technology in 1914 created an overall and system-wide defensive advantage, and therefore (2) a central failing of decision-makers at the time was the adoption of offensive military doctrines and strategies despite this diagnosis.[3]

This offense/defense balance approach is flawed. I argue that, while technology is important to warfare (and advantages surely exist), the first does

2. Jack Snyder, "Civil-Military Relations and the Cult of the Offensive, 1914 and 1984," in Steven E. Miller, ed., *Military Strategy and the Origins of the First World War: An* International Security *Reader* (Princeton: Princeton University Press, 1985), pp. 108–109.

3. There is much support for this approach, or slight variations on it, in the security literature. Most prominent are George H. Quester, *Offense and Defense in the International System* (New York: Wiley, 1977); Robert Jervis, "Cooperation Under the Security Dilemma," *World Politics,* Vol. 30, No. 2 (January 1978), pp. 187–214. The approach is directly tied to World War I by Snyder, "Civil-Military Relations"; and by Stephen Van Evera, "The Cult of the Offensive and the Origins of the First World War"; both in Miller, *Military Strategy and the Origins of the First World War,* pp. 108–146, pp. 58–107. The notion that grand strategy should rest on a perceived (though not only technologically determined) offense/defense balance is applied well by Stephen M. Walt, "The Case for Finite Containment: Analyzing U.S. Grand Strategy," *International Security,* Vol. 14, No. 1 (Summer 1989), pp. 5–49, esp. pp. 22–30. Much of the motivation for this literature has been the desire to prevent unintended escalation, or spiraling, to avoid World War III. See Bernard Brodie, *Strategy in the Missile Age* (Princeton: Princeton University Press, 1965), pp. 42–70; and the essays in Miller, *Military Strategy and the Origins of the First World War.* One direction taken by a number of intellectuals and political leaders has been to seek technological fixes by de-escalation or confidence-building measures, such as non-offensive defense in Europe. See, for example, Jack Snyder, "Limiting Offensive Conventional Forces: Soviet Proposals and Western Options," *International Security,* Vol. 12, No. 4 (Spring 1988), pp. 48–77; David Gates, "Area Defence Concepts: The West German Debate," *Survival,* Vol. 29, No. 4 (July/August 1987), pp. 301–317; and the entire September 1988 issue of *Bulletin of the Atomic Scientists.* For a critical survey of much of the technologically-motivated offense/defense balance literature, see Jack S. Levy, "The Offensive/Defensive Balance of Military Technology: A Theoretical and Historical Analysis," *International Studies Quarterly,* Vol. 28, No. 2 (June 1984), pp. 219–238.

not mechanically determine the second. Advantages, rather, are driven by the interaction of technology with doctrine and war plans. Most importantly, all three factors and the manner of their interaction are endogenous, motivated and manipulated by decision-makers. Thus, advantage is not an inherited product of nature; it is manufactured, or created. Furthermore, one should not expect a group of states to enjoy one particular advantage throughout the system. At a given moment some will enjoy an offensive, and others a defensive, advantage. Even a single state may enjoy simultaneously different advantages on different fronts, or on the same front at different moments, or as doctrine, war plans, or technological applications change.[4]

Therefore, instead of resting the choice of doctrine and grand strategy on an assessment of technology, decision makers should choose grand strategy as their point of departure. With the chosen grand strategy as their guide, policymakers should pursue an explicitly entrepreneurial course in which they manipulate, develop, and exploit available technology through doctrine and war plans in order to create the necessary advantages in light of the adversary's behavior. Such imaginative and competitive manipulation of resources I call "military entrepreneurship."

Offense/defense balance theorists have found natural pasture in World War I. As Snyder's passage suggests, the Great War does seem to be an overwhelming demonstration of the price to be paid when the technological dictates are not heeded. The technology of the time—railroads, machine guns, barbed wire—apparently created an immutable defensive advantage. Pursuit of offense by all the actors in 1914, despite the state of technology did, it appears, create a quagmire of futile attacks. Therefore, I reexamine the evidence of 1914–18 for the light it sheds on both the positive and prescriptive questions posed in this article.

In fact, the World War I experience actually undermines the offense/defense balance line of argument, and supports a rather strong indictment of

---

4. For other critical assessments of such theory see John J. Mearsheimer, *Conventional Deterrence* (Ithaca: Cornell University Press, 1983), pp. 24–28. See also Steven E. Miller, "Technology and War," *Bulletin of the Atomic Scientists,* Vol. 41, No. 1 (December 1985); Bernard Brodie, "Technological Change, Strategic Doctrine, and Political Outcomes," in Klaus Knorr, ed., *Historical Dimensions of National Security Problems* (Lawrence: University Press of Kansas, 1976); Colin Gray, "New Weapons and the Resort to Force," *International Journal,* Vol. 30, No. 2 (Spring 1975). For "improvements" or qualifications, see Stephen Van Evera, "Causes of War" (Ph.D. dissertation, University of California at Berkeley, 1984), p. 78; Jack Snyder, *The Ideology of the Offensive: Military Decision Making and the Disasters of 1914* (Ithaca: Cornell University Press, 1984), pp. 20–21.

the European policy makers—not for being misguided believers, but rather for being poor entrepreneurs: having decided that offense was required for reasons of grand strategy, they were neither inventive nor revolutionary in *creating* an offensive advantage. They believed that "the necessary was possible," instead of *making* it so. This was their sin.

My discussion unfolds as follows: First I examine the logical underpinnings of the offense/defense balance approach. Second, I suggest the alternative analysis of military advantage and the role of entrepreneurship in its creation. Third, I offer an analysis of World War I as a massive failure of military entrepreneurship. Finally, I offer some observations on why, despite its importance to warriors, military entrepreneurship is a rare phenomenon.

Technology and the Offense/Defense Balance: A Critique

At times explicitly, and in any event implicitly, the theory that the offense/defense balance is technologically motivated posits the following: In a given period there is a state of technology that creates a prevalent system-wide advantage for either offense or defense. By taking account of the available weaponry, an observer can deduce (*a priori*) whether an offensive or a defensive doctrine (and grand strategy) would be best.[5] Logically, this approach rests upon two premises.

First, regarding the nature of operations and their relationship to strategy, offense/defense balance theory presumes that offense and defense are separate and different phenomena, each imposing different requirements on technology. A further presumption is that, if the balance is to be a guide to doctrine and strategy, then whatever appears to promote tactical or operational offensive or defensive superiority must do so at the strategic level as well. In turn, this implies the existence of a useful universal yardstick with which to measure advantage, one that rests on the relative ease of offensive or defensive operations.

---

5. The idea that an effect is systemic is a main organizing theme of Quester, *Offense and Defense*, and is an important underpinning of Snyder "Civil-Military Relations"; and Van Evera, "The Cult of the Offensive." It is a prevalent notion, often incorporated "in passing"; for example, see Robert Gilpin, *War and Change in World Politics* (Cambridge: Cambridge University Press, 1981), pp. 59–63. On this general point see Levy, "Offensive/Defensive Balance," p. 227. Stephen Walt suggests that the prevailing balance be used as a guide in formulating strategy, in "The Case for Finite Containment," pp. 10, 22–30, as do Van Evera, "Cult of the Offensive," especially pp. 105–107, and Snyder, "Civil-Military Relations," pp. 108–112, 140–146.

The second premise concerns the characteristics, distribution, and effects of technology. Holding the gross distribution of military assets equal or constant, the qualitative nature of prevailing technology is believed to determine the advantage, one way or the other.[6] For this to work, it must be true that specific technological assets (or sets of assets) favor offense or defense, that this characteristic is inherent to the weapons and independent of the users, and that this effect is universal throughout the international system in question.

Both sets of beliefs or assumptions are wrong, as I argue in the next section.

OPERATIONS AND STRATEGY

Perhaps at the lowest tactical level there is sense in differentiating between offensive actions aimed to traverse, dislodge, and capture, and defensive actions whose purpose is to stand pat and prevent this from happening. We might think of such actions as modules of operation. However, at any level above this, operations are composed of accumulations of modules, and such discrimination becomes impossible. Military operations are composed of both offensive and defensive components—in parallel, series, or both.[7] As J.F.C. Fuller wrote: "the art of fighting depends upon the closest combination of the offensive and defensive, so closely as does the structure of a building depend upon bricks and mortar."[8] Often operations require sequential ordering of tactical offense and defense: in 1973 the Egyptian (strategically offensive) plan rested on an initial attack, capture of territory, and then defense against Israeli counterattacks. The Israeli (strategically defensive) plan required operational defense, mobilization, and then counterattack. Even more characteristic is the simultaneous execution of offensive and defensive actions. Defense of one's flanks while pursuing offensive operations is an old and trusted principle of warfare. The success of the Schlieffen Plan

6. Quester, *Offense and Defense*, p. 3. The attempts to make technologically-based distinctions at the World Disarmament Conference (Geneva, 1932) are of interest. See Marion W. Boggs, *Attempts to Define and Limit "Aggressive" Armament in Diplomacy and Strategy*, University of Missouri Studies XVI, No. 1 (1941); Levy "Offensive/Defensive Balance," pp. 225–226.
7. On the inseparability of offense and defense, see Ariel Levite, *Offense and Defense in Israeli Military Doctrine* (Boulder, Colo.: Westview Press, 1990), chap. 4. On Clausewitz's view that these two forms are necessarily integrated in warfare, see Raymond Aron, *Clausewitz: Philosopher of War*, trans. by Christine Booker and Norman Stone (Englewood Cliffs, N.J.: Prentice Hall, 1985), chap. 6. On Mao Tse Tung's view that offense and defense are complementary, see Samuel B. Griffith, "Introduction," to Sun Tzu, *The Art of War*, trans. by Samuel B. Griffith (Oxford: Oxford University Press, 1963), p. 53.
8. Major-General J.F.C. Fuller, *Armoured Warfare* (London: Eyr and Spottiswoode, 1943), p. 120.

(1914) depended on secure flanks for all the attacking armies, and the entire plan entailed an offensive on the right and defense on the left.

The terms "offense" and "defense" may conjure up specific rather stereotypical images of each, providing the illusion that one might successfully map given technologies into each form of war. One classic distinction has been the fire-power vs. mobility dichotomy. This view portrays defense as static, maximized by fire-power technologies, and offense as mobile and dynamic, made possible by maneuver technologies.[9] However, this oversimplifies matters; there really are no generic defense and offense, though there are some prototypical approaches or doctrines. Significantly, the variance among defensive (or offensive) doctrines is as great as the differences between the two classes of operations. Operational offense may be pursued by attrition, blitzkrieg, or a "limited aims strategy." It may be "baited," allowing the enemy to commit first and then exploiting his demonstrated weaknesses, or by initiated surprise. Operational defense may also take several basic forms: in-depth or forward; static, elastic, or mobile; with or without spoiling attacks; or attrition.[10] Each of these possibilities places different demands on technology; however, either defense and offense can be designed to stress fire-power or maneuver technologies, or both in various combinations.[11]

Although the aim here is to assess the systemic *strategic* advantage (offense or defense) on the basis of the *operational* advantage, the analysis so far suggests that the nature of operations severely muddles any attempt to label an operational advantage on the basis of technology, much less characterize the international system as having a universal advantage, in either direction. Thus, lacking a firm foothold at the operational level, we are unlikely to leap safely to the strategic level.

9. See Van Evera, "Causes of War," pp. 102–106; Levy, "Offensive/Defensive Balance," pp. 225, 226; Gunilla Hesolf, "New Technology Favors Defense," *Bulletin of the Atomic Scientists*, Vol. 44, No. 7 (September 1988), p. 42; Boggs, *Attempts to Define "Aggressive" Armament*, passim.

10. For discussions of various types of offensive and defensive operations, see B.H. Liddell Hart, *Thoughts on War* (London: Faber and Faber, 1943), chaps. 16, 17; Mearsheimer, *Conventional Deterrence*, pp. 30–58.

11. The tank has generally been viewed as the epitome of offensive technology, and artillery the mainstay of defense. This is, however, a much over-simplified classification. Tanks do provide maneuver, serving attackers well but also defenders, as the German defense in the east in World War II, and the Israeli defense of the Golan Heights in 1973, amply demonstrate. Artillery mainly provides concentrated fire, which can support friendly maneuver forces (in defense and offense), or protect the flanks of attacking or defending forces. Note Mearsheimer's argument that massive firepower is central to blitzkrieg: *Conventional Deterrence*, chap. 7. For varied opinions regarding the possible contribution and effects of tanks, see Mearsheimer, *Conventional Deterrence*, pp. 25, 26, 34; Snyder, *Ideology of the Offensive*, p. 29; Boggs, *Attempts to Define "Aggressive" Armament*, pp. 47–48.

Furthermore, introducing the strategic level delivers the *coup de grâce* to the idea of a universal yardstick of advantage, even at the operational level. There is little sense in assessing an operational advantage without reference to each state's war aims, as it is the latter that define operational requirements.[12]

Imagine, for example, that Hitler's 1941 objectives in Russia had been much more modest. Might this have guaranteed offensive success? What if Sadat had had a much more extensive purpose in 1973—say the destruction of the Israel Defense Forces (IDF) and forceful recapture of the Sinai. Would we now be asserting the advantage of the defense in that case? The point here is twofold. First, the actual operational advantage (or prospects for success) may be heavily influenced by the extent or definition of grand-strategic war aims. Second, in order to assess an operational advantage one must first ask: Advantage in doing *what?*

TECHNOLOGY

One could argue that the impact of technology is so powerful that it has a leveling effect, washing out the variance suggested in the discussion of operations and strategy. And perhaps in the realm of nuclear weapons and confrontation this is so: it may be a MAD world (with a defensive advantage) no matter what mere earthlings do.[13] But certainly in the world of conventional weapons, the distribution of technology is critical; so is how it is used.

Most importantly, technology is either merely ideas (or capabilities), or, if applied, piles of machinery and equipment in warehouses and on parking lots. Only when technology is developed, manufactured, and put to use does it make a difference. *People* do these things, and they do them differently. Therefore, the impact of technology is by nature inequitable.

An important characteristic of technological application is that small differences matter a lot. Offense/defense balance theorists often correctly iden-

---

12. It is in this sense that Jervis's definition of the offensive/defensive balance will not do. In "Cooperation Under the Security Dilemma," p. 187, he writes that, "when we say the offense has the advantage, we simply mean that it is easier to destroy the other's army and take its territory than it is to defend one's own." But we fight to win specific goals, not simply to *fight*. Therefore the question must be, "is it easier to win?" That question requires a definition of "winning" that is nation-specific and situation-specific.

13. "MAD," mutual assured destruction, refers to the condition in which nuclear powers have invulnerable capability to retaliate; thus all nuclear attacks are deterred. For analyses of MAD arguing that it is defense-dominant, see Shai Feldman, *Israeli Nuclear Deterrence: A Strategy for the 1980s* (New York: Columbia University Press, 1982), chap. 1; Van Evera, "Causes of War," chap. 13.

tify a prominent and ubiquitous technology in a state system—such as "machine guns and barbed wire," "tanks," "artillery," "surface-to-air missiles," or "precision guided munitions." But the effect of the technological variance possible within each such category is large enough to justify much finer differentiation. A NATO artillery force composed of multiple-launch rocket systems (MLRS) firing cluster bomb munitions, Seek-Search-and-Destroy Armor (SADARM)[14] and scatterable mines, and self-propelled guns firing laser-guided munitions, is not comparable to a Soviet artillery force employing regular rocket launchers, some self-propelled guns, and mostly towed howitzers, all firing high explosive (HE) shells. Furthermore, it is not single weapons with which we make war, but rather systems of interdependent hardware. The probability that two states would create identical systems and sets of systems, and then purchase them in similar proportions, is virtually nil.

Even if this purely technological variance could be controlled, and assets were somehow distributed evenly in quantity and quality, technological *impact* would remain—almost by definition—heterogeneous in nature. The operational effect of weapons depends greatly on their interaction with a whole host of factors, some controlled by decision-makers, and others not. The most outstanding factors not subject to leaders' authority are political culture and geography. The former determines the ability to concentrate energy for war at the (macro) national level, and the ability to operate weapons and systems at the personal or small unit (micro) level. An early example is the energy for imperial expansion generated by Mohammed's ability to make of so many tribes a people united by ideology and purpose. They possessed no special technological advantage, but their political culture made them a formidable war-making force.[15] In modern times, the Israelis have had a considerable advantage in confronting their Arab neighbors because of social-cultural differences. Israeli society, steeped in western liberalism, educated, industrialized, and urban, has an advantage in deploying, using, and maintaining sophisticated mechanical and electronic tools of war, and in operating

14. For analysis of the special contributions of these technologies, see U.S. Congress, Office of Technology Assessment, *Technologies for NATO's Follow-On Forces Attack Concept—Special Report,* OTA-ISC-312 (Washington, D.C.: U.S. Government Printing Office, July 1986).
15. William H. McNeill, *The Pursuit of Power: Technology, Armed Force and Society Since A.D. 1000* (Chicago: University of Chicago Press, 1982), p. 21. For other European examples, see Gilpin, *War and Change,* p. 63; Michael Howard, *War in European History* (Oxford: Oxford University Press, 1976), pp. 55–58.

with the small-unit independence and overall coordination necessary for modern warfare, both offensive and defensive.[16]

The geography and topography of states in a system, similarly beyond decision-makers' control, heavily influences the demands on, and effects of, technological assets.[17] Armor could produce positive results for the Germans when, early in World War II, they confronted the shallow and finite fronts of Northern France and the Polish theater. But in the nearly infinite expanses of Russia these same tanks ultimately "got lost" and could produce no decisive results.[18] In the same war, the Germans discovered that aircraft well-suited for the short-range missions of short duration required in the Battle of France were later unsuitable for the Battle of Britain across the Channel.[19]

In contrast, doctrine and war plans are directly subject to leaders' control. These have considerable impact on the utility and efficacy of technological assets. Examples abound of protagonists who succeeded in maximizing their "bang for the buck" by purely doctrinal means, and achieved superiority without resorting to technological manipulation. The German blitzkrieg of World War II essentially applied existing armored assets in a new and concentrated manner instead of the accepted infantry-centered approach of the time. Thus, similar armored fleets of the Allies and Germans had very different operational effect. The counterpart in classical times was the Greek phalanx, which concentrated existing infantry into a form previously unaccepted, and was unstoppable.[20] At sea, in both World Wars, the organization

---

16. See Dan Horowitz, "Flexible Responsiveness and Military Strategy: The Case of the Israeli Army," *Policy Sciences*, Vol. 1, No. 2 (Summer 1970). On the role of relative administrative skill in producing military power, see Klaus Knorr, *The Power of Nations: The Political Economy of International Relations* (New York: Basic Books, 1975), pp. 63–67. On the inability of certain societies to apply the same doctrines as other societies, see Howard, *War in European History*, pp. 78–79.
17. On the importance of geography, see Posen, *The Sources of Military Doctrine*, pp. 39, 50–51, 65–67; Sun Tzu, *The Art of War*, pp. 64, 118, 127–128; and Mearsheimer, *Conventional Deterrence*, pp. 43–44.
18. On the problems for blitzkrieg in the east, see Larry H. Addington, *The Blitzkrieg Era and the German General Staff, 1865–1941* (New Brunswick: Rutgers University Press, 1971), pp. 192–193, 209, 216; Martin Van Creveld, *Supplying War: Logistics from Wallenstein to Patton* (Cambridge: Cambridge University Press, 1977), chap. 5. Note that Michael Howard argues that in World War I, cavalry became obsolete quickly in the dense Western Theatre, yet remained effective in the expansive East. *War in European History*, p. 104.
19. Posen, *Sources of Military Doctrine*, p. 97.
20. Often the critical point of leverage is purely organizational. Examples include: Le Tellier's establishment (in the seventeenth century) of a large civil bureaucracy to administer the logistics of the French army; invention of the "division" by the French in the eighteenth century; invention of the Great General Staff in ninteenth century Prussia. See Howard, *War in European*

of Allied ships into convoys gave these defenseless machines a defensive advantage. In World War II, existing air power was wedded to existing naval technology and used in a manner that revolutionized the power and projection of naval forces.

The specific war plan that leaders choose to execute brings technology and doctrine together in a manner that greatly influences military advantage. For example, destruction of Arab air forces on the ground on June 5, 1967, was not a technological feat, nor was the strategic surprise effected by the German attack through the Ardennes on May 10, 1940. Both are examples of how specific plans can amplify the effect of technology so that although it is equitably distributed, its impact may be completely skewed.

IN SUMMARY: TECHNOLOGY DOES NOT DETERMINE ADVANTAGE

What this analysis suggests is that there is no good way to diagnose an offensive or defensive advantage by observing the prevailing technology, much less to use such a measurement to recommend an overall doctrine or strategy. Indeed, the analyst who would attempt to measure an advantage cannot himself really escape reference to specific developments, production, and doctrinal application of weapons—explicitly or not—and should therefore expect to be just as surprised after the fact as adversaries often are. We remember, uncomfortably perhaps, that American technological analysis did not predict the outcome of the Vietnam War accurately. With similar discomfort Allied analysts of the 1930s must have looked back at their pre-war predictions that the technology of the time (armor included) would create a defensive advantage in the war to come.[21]

All of this does not mean that there are no offensive or defensive advantages, merely that a state system is unlikely to enjoy a single characteristic one, be it defensive or offensive. Furthermore, a single state in the system may simultaneously enjoy an offensive advantage on one front and not on another; it may have a defensive advantage at one level of strategic goal yet not on another; it may have an offensive advantage given one operational plan yet not another; or with one doctrine and not another.

History, pp. 64–65, 100–101; McNeill, *Pursuit of Power*, pp. 162–163; and Dallas J. Irvine, "The Origin of Capital Staff," *Journal of Modern History*, Vol. 10, No. 2 (June 1938), pp. 161–173.

21. See John J. Mearsheimer, *Liddell Hart and the Weight of History* (Ithaca: Cornell University Press, 1988), chap. 6; Brian Bond and Martin Alexander, "Liddell Hart and De Gaulle: The Doctrines of Limited Liability and Mobile Defense," in Peter Paret, ed., *Makers of Modern Strategy: From Machiavelli to the Nuclear Age* (Princeton: Princeton University Press, 1986), p. 612.

I also do not argue that technology is not a central pillar of warfare, for surely we fight with "things." But it is a bad foundation on which to build a theory of military advantage, for it is but one of a number of factors that, by interacting, create advantages or disadvantages. Advantages, then, are not exogenously determined for political-military leaders, but rather created by them, as they manipulate the factors that they can: technology, doctrine, and war plans.

Advantage and Entrepreneurship

Various threads of the critique of the offense/defense balance paradigm pull together to create an alternative approach to military advantage. From this critique, what have we learned about military advantage? First, it must serve a grand strategic goal, or purpose. Second, it must be or may be created or engineered. Third, in producing it, leaders must take account of factors outside their control, adapt to these, and concentrate on manipulating three basic elements within their control: doctrine, war plans, and technology. Fourth, "advantage" is relative to an opponent and fleeting, and therefore creating it is a never-ending competitive enterprise, as I discuss in the concluding paragraphs of this article. This characteristic is perhaps the most significant. I argue that there is room for true entrepreneurship in this competitive process, and that the military advantage normally goes to the more entrepreneurial state.

How does a military entrepreneur think about, and go about creating, advantage? As Barry Posen suggests, the point of departure should be at the top, starting from balance of power considerations. States do (or should) determine their military solutions in response to their strategic environment and in pursuit of grand strategic goals. Potential enemies, alliances, and the distribution of power constitute the strategic environment; also important are geography and topography, and socio-cultural constraints on (or advantages in) the use and application of force. Always remembering that "advantage" is a relative notion, a central piece of the analysis must be devoted to one's opponents—their environment, their capabilities, their military doctrine. Armed with goals and this analysis, military leaders must find a way to execute the strategy. This entails the construction or creation of an integrated system of technology, doctrine, and war plans.[22] As they turn to

22. A clear discussion of the need to bring current strategic doctrine and organization into line

implementation, military leaders may discover that the goals as defined are not militarily achievable. In this case there is room for another iteration, perhaps requiring the redefinition of grand strategic goals, or some other manipulation at the strategic level. However, even after such reiteration, the problem of finding a solution is in the military leaders' lap.

The framework just described encapsulates the *modus operandi* of a competitive state in search of advantage. However, this process is not, in itself, necessarily entrepreneurial, for states, like corporations, may choose to engage either in regular competition or in the entrepreneurial variety. What is the difference? "The reasonable man adapts himself to the world; the unreasonable one persists in trying to adapt the world to himself," wrote George Bernard Shaw. "Therefore, all progress depends on the unreasonable man." This "lack of reason" is entrepreneurship. Joseph Schumpeter argues that economic progress results, not from the small adjustments that competitive firms routinely make in response to changes in their markets, but by the great innovative leaps made by individuals. Reasonable firms seek out and adjust to existing equilibria, trying to do their best around them. But truly entrepreneurial firms engage in a process of "creative destruction," which promotes them (and capitalism generally), as they seek economic advantage via "the new consumers' goods, the new methods of production or transportation, the new markets, the new forms of industrial organization."[23] Significantly, because firms live in a competitive environment, this process is endless, infinitely iterative. To stay ahead, a firm must be permanently entrepreneurial, for an advantage created today should be the prime target of other competitive entrepreneurial firms tomorrow,[24] and will decay if not nourished constantly.

---

with emerging technology—in particular with precision guided munitions (PGMs)—can be found in Steven Canby, *The Alliance and Europe,* Part IV: *Military Doctrine and Technology,* Adelphi Paper No. 109 (London: International Institute of Strategic Studies [IISS], 1974/75); see also Richard Burt, *New Weapons Technologies: Debate and Directions,* Adelphi Paper No. 126 (London: IISS, 1976).

23. Joseph A. Schumpeter, *Capitalism, Socialism and Democracy,* 3rd ed. (New York: Harper and Row, 1950), chap. seven. On the idea of drastic versus marginal change (in this case as a result of technological innovation), see Nathan Rosenberg, "The Impact of Technological Innovation: A Historical View," in Ralph Landau and Nathan Rosenberg, eds., *The Positive Sum Strategy: Harnessing Technology for Economic Growth* (Washington, D.C.: National Academy Press, 1986). On the entrepreneur as someone who engages in "purposeful innovation" away from the status quo and the present equilibrium, see Peter F. Drucker, *Innovation and Entrepreneurship: Practice and Principles* (New York: Harper and Row, 1985), esp. chap. 1.

24. Drucker, *Innovation and Entrepreneurship,* chap. 12. On the problems of staying ahead, see Michael E. Porter, *Competitive Advantage: Creating and Sustaining Superior Performance* (New York: Free Press, 1985), esp. chap. 5.

I recommend that national leaders approach military advantage within this "entrepreneurship" paradigm.[25] A military entrepreneur must constantly ask two questions that entail difficult leaps of imagination: Given the environmental analysis discussed earlier, (1) what would war look like if fought today, and (2) how can I "engineer" the next war away from (1) so as to maximize my relative advantages and bypass those of my competitors? The second leap is the essence of military entrepreneurship. In the broad sense it represents a constant search for surprise, and captures the dynamic nature of military competition as potential opponents seek to overcome each other's advantages and create their own.

Within this framework, technology (doctrine) and applications (war plans) interact in a cyclical manner, in the quest for integration. On one hand, currently available technology should influence the principles and techniques chosen for the application of force. On the other hand, the way one wishes to fight should create demand for particular technological development and innovation. In the short run effective military leaders will innovate in applications; in the long run they will innovate in research, developments, and the acquisition of technological assets.

The entrepreneurs' desire to create advantage by departure from the norm dictates a particular approach to, or attitude towards, the evidence upon which to base their "leaps." As in most areas of international relations, much of the data for analysis are historical events. While leaders may normally be tempted (mistakenly) to see in previous military encounters (and especially their outcomes) indications of where technological and overall advantage lie—an arrow pointing in the direction of desirable adaptation—the military entrepreneur would use the historical record more analytically. For him battles of the past will not be repeated—he will not expect them to, nor will he let them be; he should use their record warily, to suggest problems, opportunities, potential avenues to solutions, and routes to advantage—an arrow towards consistently rejuvenated theories of victory.[26]

---

25. A trap critical to avoid is the belief that entrepreneurial behavior is somehow reserved for the small business just starting up. Entrepreneurship is a state of mind, or an approach, and has been crucial to the long-term success of such "small" businesses as Johnson and Johnson or IBM. See Drucker, *Innovation and Entrepreneurship*, chap. 13. See Tom Peters, *Thriving on Chaos: Handbook for a Management Revolution* (New York: Harper and Row, 1988), Section III, on how to make and keep large organizations innovative.

26. For discussions of the utility and correct (and incorrect) uses of historical evidence in general and for strategic analysis, see David Hackett Fischer, *Historians' Fallacies: Towards a Logic of Historical Thought* (New York: Harper and Row, 1970), p. 258; see chap. 9 on the use and misuse

In the Battle of France, entrepreneur met "reasonable" competitor and won.[27] For grand-strategic reasons the Germans required a quick offensive decision in France. Facing general technological parity (in quality and quantity) with his opponents in the West, Hitler created an offensive advantage by integrating all three pillars: technology, doctrine, and war planning. Technology was modified to some extent—notably the production of close support aircraft instead of heavy bombers—yet for the most part, existing weapons were used. They were integrated into a revolutionary blitzkrieg doctrine, and all brought to maximum effect through a war plan (Operation Yellow) that maximized strategic surprise. In looking back at the historical record of World War I, the Germans identified severe problems for offensive operations; as effective entrepreneurs, they concentrated on creating a new war, not in adapting to the old one.

The German approach worked exceptionally well because it confronted an opponent conceptually locked in the old equilibrium. The French, like the British, had learned in World War I that tanks and aircraft were important, so they bought lots of them between the wars. However, they imagined a future war as a replay of the previous one, and so failed to produce an innovative and integrated system with which to confront the Germans.[28]

Probably the best recent example of military entrepreneurship is the Egyptian preparation for the 1973 War. The overall grand-strategic goal, never disputed in Egypt,[29] was offensive in nature and called for the reacquisition of the Sinai Peninsula. However, military analysis and the experience of 1967–70 indicated that the Israelis would enjoy a defensive advantage regardless of possible Egyptian efforts and innovations, should the Egyptians attempt to reconquer the Sinai. The military balance of power dictated a

---

of analogy. See also Mearsheimer, *Liddell Hart and the Weight of History,* p. 219. For discussions of Clausewitz's views on theory and history, and his use of analogy. See Aron, *Clausewitz,* chap. 8; and Michael Howard, *Clausewitz* (Oxford: Oxford University Press, 1983), pp. 30–31.

27. On the differential preparation for the Battle of France and descriptions of the results, see Mearsheimer, *Conventional Deterrence,* chaps. 3, 4: Posen, *Sources of Military Doctrine,* chaps. 3, 4, 6; and Addington, *Blitzkrieg Era,* chap. 5.

28. The Battle of Britain witnessed similar asymmetry, but with the roles reversed: the British prepared in a skillful entrepreneurial style, while the Germans did not. See Posen, *Sources of Military Doctrine,* pp. 94–102 and chap. 5; and Liddell Hart, *History of World War Two* (New York: Putnam, 1971), chap. 8.

29. On the process of Egyptian planning, see Lieutenant General Saad el-Shazly, *The Crossing of the Suez* (San Francisco: American Mideast Research, 1980), chaps. 1–5; and Hassan el-Badri, Taha el-Magdoub, and Mohammed Dia el-Din Zohdy, *The Ramadan War, 1973* (Dunn Loring, Va.: T.N. Dupuy, 1978), Part I.

number of adjustments of grand strategy: (1) grand strategic offense would be pursued through the political instability created by limited military offense, followed by defense; (2) an ally (Syria) would be mobilized to improve the overall balance of power; and (3) the Egyptian economy and population would be mobilized to an unprecedented extent in order to establish the required balance.

Now with a military mission, Egyptian military entrepreneurship turned to the task of creating the necessary advantages. The Egyptians drew on their unhappy historical experience with the Israelis just as an entrepreneur should: not as proof of Israeli superiority, but rather as indicative of where Egypt should seek opportunities to create advantages. Thus, undisputed Israeli superiority in the execution of high technology and mobile armored warfare drove the Egyptians to pursue a doctrine of low-technology, infantry-intensive, and static operations, incorporating massive use of anti-tank guided missiles (ATGMs) and infantry instead of armor. Israeli overall superiority in the air led the Egyptians to pursue local instead of general air superiority, through the use of surface-to-air missiles (SAMs) instead of aircraft. For the specific missions required by their plan and doctrine—such as the Canal crossing and breaching of the sand ramparts—the Egyptians equipped their forces with rubber boats, ladders, and water cannons. On the ground, a very large regular Egyptian army stood opposite the IDF, whose main force was reserves. Together with the Canal barrier, this made it possible for Egypt to complete its preparation while avoiding direct friction, enabling the Egyptians to achieve both strategic and operational surprise. These aspects of the war plan were crucial to Egyptian success.

East of the Canal the Israelis performed a traditional, non-entrepreneurial analysis. Based on known technologies, traditional doctrines, and what seemed to be reasonable war plans, it was clear to them that an operational and hence strategic defensive advantage existed.[30] They were thinking in the offense/defense balance paradigm. They were wrong.

In sum, then, the Egyptians integrated technology, war plans, and doctrine in a system that created an advantage by purposefully departing from the historical equilibrium. The Israelis failed to prepare and plan in a similar manner. In the final showdown, the traditional "true believers" could not stop the dynamic entrepreneurs.

30. For Israeli failures in preparation, see Chaim Herzog, *The War of Atonement: October 1973* (Boston: Little, Brown, 1975), chaps. 1–4, 18.

World War I and Military Entrepreneurship

For offense/defense balance theorists, World War I is keystone evidence, and at first glance does seem to lend almost incontestable support for their paradigm. Nevertheless, I think it provides even better support for a theory of military entrepreneurship; it is a best case through which to test entrepreneurship precisely because it seems to be so supportive of the former approach. Offense/defense balance theorists argue thus: The Great War provides excellent illustrations (a) of strategic choice that attempts to "buck" the prevailing (defensive) advantage, and (b) of the extreme price exacted as a result of such poor choice. War resulted from a system-wide choice of offense, and was waged in a futile and costly manner, because this choice was wrong. In early twentieth-century Europe, "true believers" chose offense despite clear historical evidence that the defense was dominant. They failed in their analysis; the result was the cream of an entire generation slain and buried, a continent laid waste, for nought. Therefore, concludes the argument, they should have chosen defensive strategies and doctrines. I agree—up to a point.

I concur with the depiction of European leadership and agree that in 1914—at least to some extent—there may have been a defensive advantage. But I differ sharply with the offense/defense theorists' view that this advantage dictated the entire course of the war. European leaders did prepare poorly for war, took sloppy account of technology and historical evidence, and armed themselves with cults and myths instead of analytically derived integrated and innovative systems for victory. In short, they were true believers and not entrepreneurs, and if there was a defensive advantage it was for this reason: they did not *create* offensive capabilities. Examination of the actual conduct of the war suggests that operational offense was possible, and was even achieved in such unlikely arenas as the Western Front when leaders put their minds to it in an imaginative manner. The viability of offense, demonstrated in the war, together with the gloomy picture of how war was prepared for, points to an alternative conclusion: if offensive strategy and operations were required for grand-strategic reasons, European leaders should have gone about producing an offensive advantage in an entrepreneurial manner.

In this section, first I examine the war itself, to demonstrate whether, in fact, the defensive advantage was strong, system-wide, and immutable in the 1914–18 period; whether effective offensive action was possible; whether any offensive advantage was created during the war where the defense had been dominant; and whether there was room for entrepreneurship.

Having shown that entrepreneurship could have made a difference, I then examine the pre-war period to see how the problems exposed as the war began followed from stagnant, non-competitive, and non-innovative preparation.

CONDUCT OF THE WAR

Focusing in from the grand strategic level may have a sobering effect on the confidence with which one so readily labels the First World War as an instance of defense-dominance. It would be fair to ask: just who exactly *won*, or achieved their war aims, through the defense? It would hardly do to score the Allies with success, when they had to spend four years in national attrition or total war, having surrendered Belgium and a good proportion of France's most productive territory. In fact no country gained its strategic purpose through defense, and this assessment is true regardless of what one makes of the relative operational advantage of offense and defense. If strategic goals are the yardstick of advantage, then one cannot ignore the fact that the two critical German fronts were pacified by *offensive* operations. The Germans effectively knocked Russia out of the war through a series of offensives in 1915–16. In the west the Allies brought the war to a close through the offensive of the "last hundred days," in summer 1918.

The common identification of a *strategic* defensive advantage in 1914, based on an *operational* defensive advantage, stems from the extensive reach of grand strategic goals. Imagine our assessment of German offensives, had the Schlieffen Plan called for *limited* offensive victories followed by negotiated settlement, much as the elder Moltke had envisaged.[31] The limited (and possibly near total) victory of the Schlieffen offensive in France, and the effective neutralization of Russia in 1915–16 by German attack, were decisive enough that a more moderate grand strategy might have had us today believing in the offensive advantage in World War I.

For years there was a defensive advantage on the Western Front; one could hardly argue with this observation. But if we put off—for just a moment—discussion of this sad phenomenon, even a cursory look through World War I reveals plenty of operational offense that worked, and worked well. The

31. Addington, *Blitzkrieg Era*, pp. 10–12.

two most critical and decisive strategic defensive successes in the opening days of the war were achieved by offensive operations. These were the Battle of the Marne in the west and the Battle of Tannenberg in the east. At Tannenberg the German Eighth Army defeated two numerically superior attacking Russian armies through offensive maneuvers at all levels.[32] At the Marne, the defensive "miracle" was effected by an unexpected allied attack on the flank of the German First Army and through a critical seam between the German First and Second Armies.[33] The eastern theatres witnessed a number of other successful operational offenses: the German offensive in Poland (1915), Russia's Brusilov Offensive into Galicia and the Carpathians (1916); and the Austro-German conquest of Romania (1916).[34]

Even on the Western Front, especially as the war progressed into its second half, much of the defensive success rested on offensive activity. From 1916, and especially after the Battle of the Somme, German defensive doctrine changed from stereotypically static to elastic defense-in-depth. It rested on the principle of surrendering territory purposefully in order to set the attacker up for counterattack.[35] When traditional British offense met the new German elastic defense at Arras (April 1917), the result was an overwhelming German success. Similar results obtained at the infamous Nivelle offensive of April–May 1917, failure of which led to the temporary disintegration of the French army.[36] Later, in 1918, the Allies learned how to use motorized and mechanized units to counterattack the flanks of successful penetrations of the

32. James E. Edmonds, *A Short History of World War I* (Oxford: Oxford University Press, 1951), pp. 50–52; B.H. Liddell Hart, *The Real War 1914–1918* (Boston: Little, Brown, 1930), pp. 103–114; Barbara W. Tuchman, *The Guns of August* (New York: Bantam, 1976), chap. 16; and Norman Stone, *The Eastern Front 1914–1917* (New York: Scribner's, 1975), chap. 3.
33. Edmonds, *A Short History of World War I*, pp. 41–49; Liddell Hart, *The Real War 1914–1918*, pp. 82–102; and Tuchman, *Guns of August*, chaps. 20–22.
34. For descriptions of these operations, see Edmonds, *A Short History of World War I*, pp. 31–33, 50–52, 64–68, 99–103, 172–176, 197–201; Geoffrey Jukes, *Carpathian Disaster: Death of an Army* (New York: Ballantine, 1971), chap. 6; and Liddell Hart, *The Real War 1914–1918*, pp. 103–114, 131–134, 224–226, 261–266.
35. Timothy T. Lupfer, *Dynamics of Doctrine: The Changes in German Tactical Doctrine During the First World War*, Leavenworth Papers No. 4 (Fort Leavenworth, Kansas: U.S. Army Command and General Staff College, July 8, 1981), pp. 1–36; Tim Travers, *The Killing Ground: The British Army, the Western Front, and the Emergence of Modern Warfare, 1900–1918* (London: Allen and Unwin, 1987), pp. 257–259; and Liddell Hart, *The Real War 1914–1918*, pp. 197–198, 242, 321–329, 342–343, 390. For descriptions of the Somme, see ibid, pp. 227–248; Travers, *Killing Ground*, pp. 127–199.
36. Liddell Hart, *The Real War 1914–1918*, pp. 321–329; Edmonds, *A Short History of World War I*, chaps. 19, 20.

German spring offensive, thereby neutralizing several German offensive successes.[37]

However, the Western Front did witness two terrible years of immobility and useless carnage, and we shall, therefore, turn our attention to that theatre, where the defensive advantage was so pronounced. The undisputable demonstrations of the unbreachable defense climaxed with the German offensive at Verdun and the Allied (British) offensive at the Somme (1916). After these defensive "victories," both sides undertook most seriously to create offensive advantages. Attempts by the Germans and especially by the British to do so demonstrate the extent to which offense was not a total impossibility in World War I.

In 1917 the German Army High Command (OHL) began the purposeful development of an offensive capability for the great strategic offensive planned for 1918. German preparation has been well documented elsewhere,[38] so I will simply recapitulate the central principles. The Germans consciously chose not to pursue a technological solution, but rather to innovate in doctrine and war plans. Previously applied doctrines of envelopment and broad frontal attrition attack were obviously, by then, not promising, and so the OHL developed "infiltration tactics," intended to break through the Allied defensive crust and enable strategic exploitation. The driving principles of this doctrine were: (1) create a narrow gap by incision instead of causing a general collapse by attrition or envelopment; (2) purchase surprise by forgoing preparatory bombardments; (3) pursue enemy weak points instead of strong ones; (4) exploit success instead of reinforcing failure; (5) bypass resistance where possible. Implementation of this doctrinal innovation required organizational changes. Infiltration was effected by small independent units: "detachments" and "groups" within special "storm battalions." The storm battalions spearheaded operations, and were heavily armed with existing light machine guns and flamethrowers. The role and hence methods of fire support were changed in order to bolster the new doctrine. As a general departure from the past, planning and doctrine focused on ensuring successful infantry operations; even artillery doctrine was

37. Kenneth Macksey, *Tank Warfare: A History of Tanks in Battle* (London: Rupert Hart-Davis, 1971), pp. 55–58; Lupfer, *Dynamics of Doctrine,* p. 53.
38. The following discussion is based on Lupfer, *Dynamics of Doctrine,* pp. 37–54; and Liddell Hart, *The Real War 1914–1918,* pp. 387–410.

forced to change to provide better support for infantry maneuver. Light artillery was made integral ordnance at the battalion level, in order to improve its independence and coordination. German artillery was organized and trained to provide effective neutralization fire instead of the traditional expensive and ineffective attempts to destroy, to fire for effect without prior registration (to improve surprise), to fire at enemy assets that directly infringed on friendly infantry, and to move forward rapidly so as to assist in the exploitation phase and in defense against counterattacks.

Initial German success in the 1918 offensive was much enhanced by practiced deception which resulted in complete surprise of the Allies in time and place. Such operational surprise against a defender who was also not appropriately prepared in doctrine or organization enabled the Germans to make unprecedented penetrations of forty miles at Amiens and fifteen at Ypres. Ultimately, however, the German offensive petered out. At the grand-strategic level, the Germans were exhausted and could not replace their losses. At the strategic level the Germans failed to pursue their major operational successes. At the operational level they had not worked out a satisfactory doctrinal or technological solution to the problem of exploitation of the penetration, and their enemies learned to use mechanized and motorized forces to attack the German flanks and to array themselves in depth.

Changes in British doctrine between 1916 and 1918 also had a remarkable effect on the battlefield, often unnoticed by offense/defense theorists.

The Somme epitomized traditional offense by attrition and when that came up against the as-yet static German defense, the resultant failure was attended by such a holocaust of death and destruction that the British were compelled to seek an alternative doctrine.[39] The British responded to the challenge, step by step. First they undertook to frustrate the evolving German elastic defense-in-depth by playing on its weaknesses. In the fall of 1917 the British mounted a number of limited-aims attacks. Since the new German doctrine purposely left front lines weak, the shallow British attacks were successful.[40] Then, beginning at Cambrai (November 1917), the British began

39. At the Somme about 600,000 Allied troops and 500,000 Germans lost their lives. British doctrine and planning for the Somme are described by Travers, *Killing Ground*, chaps. 6 and 7; and Shelford Bidwell and Dominick Graham, *Fire Power: British Army Weapons and Theories of War, 1904–1945* (London: Allen and Unwin, 1982), pp. 80–83, 112–114. For a vivid description of doctrine and of the actual fighting, see John Keegan, *The Face of Battle: A Study of Agincourt, Waterloo and the Somme* (Harmondsworth, U.K.: Penguin, 1978), chap. 4. See also Lupfer, *Dynamics of Doctrine*, pp. 4–8; Macksey, *Tank Warfare*, pp. 33–38; and works cited in n. 34, above.
40. Lupfer, *Dynamics of Doctrine*, p. 35; Edmonds, *A Short History of World War I*, pp. 250–252.

to employ a revolutionary departure in offensive doctrine and organization, so that in 1918, Allied and especially British offensives bore no resemblance to the earlier debacles. In contrast to the Germans' approach, an important component of developing British offensive capability was technological: the tank.

The evolving British doctrine was predicated on surprise and deep penetration.[41] Surprise was achieved by careful force concentration and by forgoing the traditional lengthy preparatory bombardments. Penetration was effected by the deployment of armor in large concentrations and in full coordination with infantry forces. Also, as with German "infiltration tactics," the infantry was organized to fight in platoons, not lines, and instructed to seek out and exploit the enemy's weak points. Machine guns were given to the attacking infantry units, and artillery-infantry coordination was much improved. The artillery also concentrated on battery suppression, important for sustained attacks. Air support was expanded to provide pre-battle and real-time intelligence, and more efficient spotting for artillery fire direction. In rudimentary form at Cambrai these developments proved effective, and were most impressive in the offensives of the Last Hundred Days, August–November 1918.

The Germans tried to defend with anti-tank ditches and anti-tank guns; the British countered these with thicker armor and by carrying rolls of bracken on the tanks to drop into the ditches as causeways. The British also overcame improved German front-line defenses with "small" tactical innovations, such as night attack, opting for extremely narrow frontal penetration, and repeatedly surprising in time and place of attack.

PREPARING FOR WAR

The fact that operational offensive advantage was constructed and innovated successfully on the Western Front, but only as the war progressed and after years of defensive stalemate, tells us that: (1) there was room for offensive advantage in 1914; and (2) therefore, the protagonists must have prepared

---

41. Actually, the British introduced the tank at the Somme. A small number of tanks were employed in that series of battles, though without an identifiable doctrine, and were dispersed among the infantry in small clusters with no standard techniques for coordination or concerted action. Naturally, they had no operational effect at the Somme. Evolving British concepts and doctrine are described by Bidwell and Graham, *Fire Power,* pp. 125–144. See Macksey, *Tank Warfare,* chaps. 3–5, for evolving British concepts, descriptions of encounters, and German responses in 1917–1918. See also, Edmonds, *A Short History of World War I,* chap. 28; and Liddell Hart, *The Real War 1914–1918,* pp. 118, 249–260, 344–356, 432.

poorly for the war. There must have been a dearth of entrepreneurship in the manner in which the major combatants prepared for conflict.

For grand strategic reasons the major European actors opted for offensive military doctrines in the early years of the century.[42] However, they failed to proceed, as entrepreneurs would have, to create viable offensive capacities and advantages. Indeed, with the possible and only partial exception of Germany, the decision makers of 1914 were not truly competitive; they were non-innovative, failed to consider the implications of new fire technologies, and misused, even abused, historical evidence in devising their approaches to the coming war.

Ironically, the most damning descriptions of this early decision making as non-entrepreneurial are made by the offense/defense balance theorists. Jack Snyder, Stephen Van Evera, and others characterize the prevailing doctrines as inappropriate and one-sided. These writers fault the French and Belgians for "*offensive à outrance*," describe the French Plan Seventeen as mindless, ridicule the British for relying on a "psychological battlefield" theory of victory, scold the Russians for strategic overextension, and criticize the Schlieffen Plan as strategically overextended and a catalyst of early British intervention. Those responsible are faulted for failing to analyze historical data and the implications of emerging technology, and for acting on cults and mystiques.[43] I agree. But these criticisms, therefore, indicate *not* that there was an inherent defensive advantage, but rather that there was much room for creative and analytical preparation for any kind of war, offensive or defensive.

The French approached 1914 in an almost purely non-competitive mode, for at any meaningful level they essentially ignored their expected enemy. Plan Seventeen was an offensive operational plan that was not conceived or designed to overcome a particular set of supposed German contingencies. It was really more of a plan of deployment than one of operations; it did not emerge from a detailed study of the problem at hand and did not, therefore, provide a detailed solution.[44] This is perhaps best revealed by the religiously

42. For an argument that the offensive plans of 1914 reflected and correctly served prevailing balance-of-power considerations, see Scott D. Sagan, "1914 Revisited: Allies, Offense, and Instability," *International Security*, Vol. 11, No. 2 (Fall 1986), pp. 151–175.
43. For these "faults" see Snyder, *Ideology of the Offensive;* Van Evera, "The Cult of the Offensive"; Travers, *Killing Ground*, chaps. 2, 3; and Michael Howard, "Men Against Fire: Expectations of War 1914," in Miller, *Military Strategy;* Van Creveld, *Supplying War*, chap. 4.
44. For critiques of Plan Seventeen and French preparation for war, see S.R. Williamson, "Joffre Reshapes French Strategy, 1911–1913," in Paul M. Kennedy, ed., *The War Plans of the Great*

and singularly *offensive* nature of the plan and of French doctrine. One cannot take war and one's enemy seriously and at the same time believe that offense can exist without defense, or vice versa. In fact, the French ignored concrete evidence about the likely German plans—notably the "wheel" through Belgium and the massive use of the reserves—and did little to counter these. The technological dimensions of the approaching war were also largely ignored. French plans and doctrine ignored the fact that the Germans had relatively large numbers of machine guns and heavy artillery, as well as the potential of such assets for integration into the French offensive doctrine and organization. And, although France had the largest civilian motor-car industry at the time, the potential benefits of motorization for a large offensive campaign were neglected.

France entered the war without having performed either of the two mental leaps required in entrepreneurial thought. First, the French failed to foresee the probable face of a future war. Perhaps this fact is best reflected in their 1914 dress: the traditional *pantalons rouges*. Second, they made no attempt to "engineer" the future battlefield to their advantage on any of the three dimensions: technologically, doctrinally, or strategically. In sum, in 1914 the French did not have a creative theory of victory, but rather a blind faith in one, which led them, in the event, to have a cult not merely of the offensive, but of the "reckless offensive."[45]

The British, too, failed to foresee completely the face of the coming war, although they do seem to have understood that the new fire technologies would be difficult to overcome. But instead of attempting to devise or create advantage through the manipulation of technology and doctrine, they resorted to a psychological battlefield notion, in which victory was to be the fruit of superior morale and will-power.[46] Failure to integrate the machine gun and understand its operational and tactical implications is exemplary: although well acquainted and experienced with these weapons, the British failed until well into the war to adapt both doctrine and organization in order to maximize the effect of machine guns or to acquire them in sufficient quantities, and the British were slow in reacting to their use by the Germans. In confronting the German submarines, again the British failed to foresee

Powers, 1880–1914 (Boston: Allen and Unwin, 1985); Snyder, *Ideology of the Offensive,* chap. 3; Tuchman, *Guns of August*, pp. 208–210, chap. 13, pp. 266, 293, 295; Howard, "Men against Fire"; and Theodore Ropp, rev'd. ed., *War in the Modern World* (New York: Collier, 1962), pp. 227–229.
45. Ropp, *War in the Modern World*, p. 263.
46. Travers, *Killing Ground*, chaps. 2,3.

their potential application and strategic menace, and not till well into the war were effective countermeasures developed.

German preparation for the war was the most entrepreneurial of the three. The Germans came closest to completing the required leaps of the imagination, and they were, in fact, the only combatants to achieve offensive success of any kind in the early phases of the war. Given a grand-strategically derived offensive mission, the German army devised the Schlieffen Plan: an attempt to create a viable theory of victory given the expected difficulties on the battlefield and the prevailing technologies. At the highest level, in order to create the required physical balance, the Germans innovated by using reserves in large numbers in combat missions. Also at that level, and for the same reason, the Germans planned to create simultaneously an offensive advantage in the West and a defensive one in the East. At the operational level in the West, the plan's envelopment of the French left wing was meant to obviate the need for frontal attack; meanwhile, by tying down the French right, the plan intended to overcome the danger of lateral rail movement in defense. At the operational and tactical levels, the Germans integrated heavy artillery and machine guns into their offensive doctrine to an extent unmatched by their opponents.

However, the Germans did not push their entrepreneurial thought to its logical conclusions, and here they failed. They foresaw three critical problems: the tactical one of frontal attack; the potential for French lateral movement by rail; and the logistical strains of the German right wing. But their solution came up short.[47] An attacker's envelopment aimed to neutralize his enemy's frontal defense must be of such size and speed that it can unbalance the defender. The Schlieffen Plan was too big an envelopment given its slow speed. Thus, as an operational solution it could not overcome tactical problems: the French could move operationally to meet the operational "trick" in time to *re*-present the Germans with the tactical problems of frontal attack. In the event, the Germans did not stick to their plan of "sucking in" the French right wing, and could not withstand the temptation of attacking on

47. Criticism of the Schlieffen Plan as an unsuccessful operational solution to these problems, which only created more problems, may be found in: L.C.F. Turner, "The Significance of the Schlieffen Plan," in Kennedy, *The War Plans of the Great Powers*, pp. 204, 212; Gerhard Ritter, *The Schlieffen Plan: Critique of a Myth*, trans. Andrew and Eva Wilson (New York; Praeger, 1958), esp. "Foreword" by B.H. Liddell Hart; Van Creveld, *Supplying War*, chap. 4; Addington, *Blitzkrieg Era*, pp. 15–22; Van Evera, "Causes of War," pp. 301–302; and Gunther R. Rothenberg, "Moltke, Schlieffen, and the Doctrine of Strategic Envelopment," in Paret, *Makers of Modern Strategy*, pp. 316–325.

their own left. Their unsuccessful attacks left French forces intact, available, and close enough to be transported to the Marne in time for the "miracle" that defended Paris. I have found no evidence that the Germans actively attempted to obstruct such rail transport.

The Germans also failed to address practically the logistic fragility of their right-wing envelopment. It is here that their thinking was most non-entrepreneurial, reverting to faith that "the necessary is possible";[48] in a sense wishing away, in Anglo-French style, instead of solving a problem they well understood. It is here, also, that the German failure to mobilize new technologies is most egregious: the First Army had only some 162 trucks for its transport missions. This is simply poor preparation to execute in 1914 a plan that had its origins in 1897!

One prevalent complaint about the actors of 1914 is that they ignored or misused the lessons of history—of the American Civil War, the Boer War, and the Russo-Japanese War. These charges are apposite but not, as one might suspect, always as intended by those who make them. The military leaders are often faulted for not diagnosing these wars as symptomatic of an overall defensive advantage. But this is not their central failing. Rather, it is that they did not treat these wars analytically. Scant was the detailed study aimed to uncover particular problems or ideas and opportunities to engineer a future advantage. For the most part their study of recent conflict was motivated—by political, social, or organizational imperatives—to find proof of, and support for, their preferred doctrines. Proponents of defense found in history proof that defense was best; proponents of offense found in it evidence for a general offensive advantage.[49] Both sides were wrong.

Of the historical lessons, the ones most missed by the leaders of 1914 were those regarding the impact of evolving technologies. It is really quite remarkable just how poorly these decision makers understood the implications of new (and old) weapons. The Germans have a mixed record on this count, somewhat better than that of their adversaries. To their credit, they did understand the potential power of machine guns and heavy artillery for offense and defense. In 1914 the German battalion had six machine guns,

---

48. Snyder, *Ideology of the Offensive*, chap. 5.
49. On the misuse of historical evidence before 1914, see Snyder, *Ideology of the Offensive*, pp. 77–81; Howard, "Men against Fire"; Brodie, *Strategy in the Missile Age*, pp. 42–52; T.H.E. Travers, "Technology, Tactics, and Morale: Jean de Bloch, the Boer War, and British Military Theory, 1900–1919," *Journal of Modern History*, Vol. 51, No. 2 (June 1979), pp. 264–286; and Brodie, "Technological Change," pp. 283–292.

whereas its British counterpart had two. (But even the German six are negligible when the potential benefits are considered; as of 1914 neither side—least of all the British—had really figured out how to utilize this technology within infantry doctrine.) The Germans armed their first-line divisions with twice as many field guns as did the French and Russians, all of higher caliber. At corps level the Germans made available to their divisions heavy artillery (150 mm), while the French were armed with the famous 75s. The Russians and the British also went to war with very few pieces of heavy field artillery. Analysis of the obstacles to offensive operations led the Germans, quite appropriately, to develop and build a small number of heavy siege guns for assaults on major fortresses. Finally, German artillery relied substantially on indirect fire, while the French (and the British to an extent) relied on direct fire from close range. Indeed, in the early stages of the war, the Germans capitalized well on the advantages provided by these differences on both fronts.[50] The Germans also foresaw the potential strategic use of submarines, and being alone in this insight nearly knocked the British out of the war before the Allies found the appropriate counter-measures.[51]

However, all the major actors failed to foresee or apply the opportunities for massive capitalization of the intended land offensives. The knowledge and basic technology to produce tanks or armored cars was surely available to all by 1912. But the British only introduced them in 1916, and only after the original development was pushed, oddly enough, by the Admiralty, while the Germans refused to introduce tanks until the bitter end. Similarly, all failed to foresee the potential of motorization for logistics and troop transport in the coming war, despite the well-developed European motorcar industries at the time.[52]

---

50. For details of artillery development, armament and operations during the war, see Bidwell and Graham, *Fire Power,* Books I and II, and pp. 296–297; Lt.-General Nikolai N. Golovin, *The Russian Army in the World War* (New Haven: Yale University Press, 1931), pp. 33–34, 133–142; and Tuchman, *The Guns of August,* pp. 190–195, 218–220, 234–235, 262–263.
51. Brodie, "Technological Change," pp. 280–282. On the other hand, the Germans invested in a large surface fleet whose only effect was to cause a strategic "spiral" with England, and which may have catalyzed Britain's entrance into the war, all at the expense of capitalization of the land army. On the strategic implications of the German naval build-up, see Paul M. Kennedy, "Tirpitz, England and the Second Navy Law of 1900: A Strategical Critique," *Militärgeschichtliche Mitteilungen,* Vol. 2 (1970).
52. In the decade before the war, a number of British, Russian, and German officers and strategists lobbied for motorization to solve the problems of mobility they foresaw, especially for logistics. The French at the time were the most advanced and capable in automobile production and could have motorized their forces in 1914 had they decided to do so. See Travers, "Technology, Tactics, and Morale," p. 271; Snyder, *Ideology of the Offensive,* pp. 191–192; Van

Whether the explanations lie at the level of organizational-bureaucratic conservative behavior (for example, the difficulty of assimilating the machine gun into doctrine and tactics), organizational politics (for example, the exalted role of horse cavalry),[53] or domestic politics (for example, the German insistence on a large surface fleet, or the French army's insistence on an offensive doctrine),[54] there was nothing exogeneously predetermined about the impact of these technologies. Rather, their influence resulted from biased historical analysis and a reluctance or inability to figure out how best to counter the central and emerging technologies of the day, and how most effectively to exploit their promise. Military leaders failed as entrepreneurs; they stumbled instead of making the necessary leaps of the imagination. The first leap—"what would a war look like today?"—was made by some,[55] if incompletely; but the second—"how, therefore, do I *engineer* my way to offensive victory?"—was performed poorly, and in most cases not at all.

Conclusion: Cults and Revolution

An entrepreneurial military organization must be in a state of permanent revolution; change must be its constant condition. Successful innovation demands leaps of the imagination, the adoption and integration of untried doctrines, organizational changes, and technological application, all in response to each other and to guesses about similar developments by one's opponents. Entrepreneurship involves the intellectual honesty, imagination, and courage to exploit historical failure and success as indications of how to innovate, and to treat history analytically and not in a biased manner. Most importantly, military leaders must force themselves to remain impartial towards offense or defense, and must be prepared to create superior capability in either or both, as grand strategy demands.

These are steep demands. Like business entrepreneurship, the military variety does not come naturally to either individuals or organizations, for it

---

Creveld, *Supplying War*, pp. 110–119; and Paul Kennedy, *The Rise and Fall of the Great Powers: Economic Change and Military Conflict from 1500 to 2000* (New York: Random House, 1987), p. 221.

53. Edward L. Katzenbach, Jr., "The Horse Cavalry in the Twentieth Century," in Robert J. Art and Kenneth N. Waltz, eds., *The Use of Force: Military Power and International Politics*, 3d ed. (Lanham, Md.: University Press of America, 1988).

54. McNeill, *The Pursuit of Power*, pp. 303–305; Snyder, "Civil-Military Relations," pp. 129–133; Alfred Vagts, *A History of Militarism, Civilian and Military*, rev. ed. (New York: Free Press, 1959), chap. 10.

55. Most notably, of course, by Jean de Bloch. See Howard, "Men Against Fire."

requires "unnatural" behavior. Large bureaucratic organizations are normally conservative; they reduce uncertainty by adhering to standard operating procedures, sticking to the tried and true, and learning in a process that is linear, sequential, and cybernetic.[56] Even more than other organizations, militaries are likely to display these traits. The simultaneous orientation towards promotion and discipline make the socialization to the "particular way of doing business"[57] natural to army officers. Changing the way of doing business is not likely in a hierarchical organization whose leaders reached the top by virtue of success in the past. The way of the past is the way in which they are experts, and the source of their authority.

This conservatism is further encouraged by the fact that—especially in peace time—armies are "input organizations"; in the business world a chief executive officer is tested by the profits he produces, or "outputs," while a military leader is tested by how well he trains, by the morale and discipline of his troops, and by the quality of maintenance of his weapons. These are all "inputs," and the comparison of commanders for promotion requires that these all be performed against given and accepted standards or procedures.

Further inertia and obstruction of innovation have their roots in the role of armies in the modern world. These organizations have been important to the economic, social, and political activity of states and their regimes, and therefore have been deeply influenced by national politics in which they participate.[58] The central role of modern armies, of course, is to guarantee the longevity of the state against external threats; risk aversion is extreme in an organization whose profession is about the life and death of the nation. Armies are not perceived, by themselves and others, as just "another business" that might succeed or fail with impunity. This may have a number of related consequences. Armies are unlikely to part with whatever worked in the past. This attitude is further encouraged by the objective difficulty of performing decisive peacetime tests of new doctrine and technology.[59] A similar difficulty in assessing potential adversaries' developments turns military organizations inward, rendering their behavior less competitive and

56. John D. Steinbruner, *The Cybernetic Theory of Decision: New Dimensions of Political Analysis* (Princeton: Princeton University Press, 1974). On the barriers to innovation in public service organizations, see Drucker, *Innovation and Entrepreneurship,* chap. 14.
57. Posen, *Sources of Military Doctrine,* p. 44.
58. See note 50 above and accompanying text, on the politically-motivated reluctance in Germany to equip the ground forces with capital (heavy equipment, trucks, tanks), and the German insistence on building a large surface fleet.
59. Katzenbach, "The Horse Cavalry in the Twentieth Century," pp. 162–163.

innovative. The desire to reduce risk and uncertainty also amplifies the desire to impose standard procedures and scenarios and to "control" events. This leads soldiers to prefer and adopt either an offensive or defensive doctrine; usually the former because it appears to allow them to fight and to terminate wars according to plan.[60]

It should not be surprising, therefore, that realization of an entrepreneurial approach has often required immense external pressure, whether from an undeniable previous failure, or from above by a strong, authoritarian, and imaginative personality like Napoleon or Hitler. At times the change in paradigm required of the organization may be too great or sudden for the prevailing cadre of military leaders. This phenomenon has induced some armies (like the IDF) to insist on short careers for its officers, and has forced other leaders to purge the military elite, as Sadat did in preparation for the 1973 War.[61] But a military organization does not have the option of retaining the prevailing paradigm until the anomalies attending it create an evident crisis *and* a better alternative is presented and tested, as in the scientific revolutions described by Thomas Kuhn.[62] A military organization can wait for neither. It may be too late.

The antithesis of such permanent revolutionary thought is the cult—a theological belief that a particular approach, paradigm, or doctrine is correct. Such orthodoxy may be comforting, but it neglects competition, and true believers may fall easy prey to hungry entrepreneurs. Indeed falling into any cult—of the *offensive* in 1914, the *defensive* in 1939, or the *status quo* in the 1990s—may be the greatest gift one could bestow on his enemies.

---

60. Snyder, *Ideology of the Offensive*, chap. 1; Posen, *Sources of Military Doctrine*, pp. 41–50; Major Robert L. Maginnis, "Harnessing Creativity," *Military Review*, Vol. 66, No. 3 (March 1986).
61. Muhammad Hasanayn Haykal, *The Road to Ramadan* (New York: Quadrangle/New York Times Books, 1975), pp. 180–182.
62. Thomas S. Kuhn, *The Structure of Scientific Revolutions*, 2nd ed. (Chicago: University of Chicago Press, 1970), chaps. 5–8.

Part III: Offense-Defense Theory: The Contemporary Debate

Offense, Defense, and the Causes of War

Stephen Van Evera

Is war more likely when conquest is easy? Could peace be strengthened by making conquest more difficult? What are the causes of offense dominance?[1] How can these causes be controlled? These are the questions this article addresses.

I argue that war is far more likely when conquest is easy, and that shifts in the offense-defense balance have a large effect on the risk of war. Ten war-causing effects (summarized in Figure 1) arise when the offense dominates. (1) Empires are easier to conquer. This invites opportunistic expansion even by temperate powers (explanation A). (2) Self-defense is more difficult; hence states are less secure. This drives them to pursue defensive expansion (explanation B). (3) Their greater insecurity also drives states to resist others' expansion more fiercely. Power gains by others raise larger threats to national security; hence expansionism prompts a more violent response (explanation C). (4) First-strike advantages are larger, raising dangers of preemptive war (explanation D). (5) Windows of opportunity and vulnerability are larger, raising dangers of preventive war (explanation E). (6) States more often adopt fait accompli diplomatic tactics, and such tactics more often trigger war (explanation F). (7) States negotiate less readily and cooperatively; hence negotiations fail more often, and disputes fester unresolved (explanation G). (8) States enshroud foreign and defense policy in tighter secrecy, raising the risk of

Stephen Van Evera teaches international relations in the Political Science Department at the Massachusetts Institute of Technology.

Thanks to Robert Art, Charles Glaser, and an anonymous reviewer for their comments on this article. It is distilled from *Causes of War, Volume 1: The Structure of Power and the Roots of War* (Ithaca, N.Y.: Cornell University Press, forthcoming 1999).

1. In this article "offense dominant" means that conquest is fairly easy; "defense dominant" means that conquest is very difficult. It is almost never easier to conquer than to defend, so I use "offense dominant" broadly, to denote that offense is easier than usual, although perhaps not actually easier than defense. I use "offense-defense balance" to denote the relative ease of aggression and defense against aggression. As noted below, this balance is shaped by both military and diplomatic/political factors. Two measures of the overall offense-defense balance work well: (1) the probability that a determined aggressor could conquer and subjugate a target state with comparable resources; or (2) the resource advantage that an aggressor requires to gain a given chance of conquering a target state. I use "offense" to refer to strategic offensive action—the taking and holding of territory—as opposed to tactical offensive action, which involves the attack but not the seizure and holding of territory.

miscalculation and diplomatic blunder (explanation H). (9) Arms racing is faster and harder to control, raising the risk of preventive wars and wars of false optimism (explanation I). (10) Offense dominance is self-feeding. As conquest grows easier, states adopt policies (e.g., more offensive military doctrines) that make conquest still easier. This magnifies effects 1–9 (explanation J).

The perception of offense dominance raises these same ten dangers, even without the reality. If states think the offense is strong, they will act as if it were. Thus offense-defense theory has two parallel variants, real and perceptual. These variants are considered together here.

How does this theory perform in tests? Three single case-study tests are performed below. They corroborate offense-defense theory[2] and indicate that it has large theoretical importance: that is, shifts in the offense-defense balance—real or perceived—have a large effect on the risk of war. The actual offense-defense balance has marked effects; the effects of the perceived offense-defense balance are even larger.

What causes offense and defense dominance? Military technology and doctrine, geography, national social structure, and diplomatic arrangements (specifically, defensive alliances and balancing behavior by offshore powers) all matter. The net offense-defense balance is an aggregate of these military, geographic, social, and diplomatic factors.

How can offense dominance be controlled? Defensive military doctrines and defensive alliance-making offer good solutions, although there is some tension between them: offensive forces can be needed to defend allies. Offense dominance is more often imagined than real, however. Thus the more urgent question is: How can illusions of offense dominance be controlled? Answers are elusive because the roots of these illusions are obscure.

On balance, how does offense-defense theory measure up? It has the attributes of good theory. In addition to having theoretical importance, offense-defense theory has wide explanatory range and prescriptive richness. It explains an array of important war causes (opportunistic expansionism, defensive expansionism, fierce resistance to others' expansion, first-strike advantage,

---

2. I use "offense-defense theory" to label the hypothesis that war is more likely when conquest is easy, plus explanatory hypotheses that define how this causation operates. The classic work on the topic is Robert Jervis, "Cooperation under the Security Dilemma," *World Politics,* Vol. 30, No. 2 (January 1978), pp. 167–214 at 169. An overview is Sean M. Lynn-Jones, "Offense-Defense Theory and Its Critics," *Security Studies,* Vol. 4, No. 4 (Summer 1995), pp. 660–691. The theory I frame here subsumes and elaborates on Jervis's theory.

windows of opportunity and vulnerability, faits accomplis, negotiation failure, secrecy, arms races, and offense dominance itself) that were once thought to be independent. In so doing, offense-defense theory explains the dangers that these war causes produce and the wars they cause. This simplifies the problem of power and war: a number of disparate dangers are fed by a single taproot. Moreover, both the reality and the perception of easy conquest can be shaped by human action; hence offense-defense theory offers prescriptions for controlling the dangers it frames.

The next section outlines offense-defense theory's ten explanations for war. The following section identifies causes of offense and defense dominance. The fourth section frames predictions that can be inferred from offense-defense theory, and offers three case studies as tests of the theory: Europe since 1789, ancient China during the Spring and Autumn and Warring States periods, and the United States since 1789. The final section assesses the general quality of offense-defense theory.

Hypotheses on the Effects of Offense Dominance

A host of dangers arise when conquest is easy. Some are obvious and some more subtle, some are direct and some indirect. Together they make war very likely when the offense dominates.

A: OPPORTUNISTIC EXPANSIONISM

When conquest is hard, states are dissuaded from aggression by the fear that victory will prove costly or unattainable. When conquest is easy, aggression is more alluring: it costs less to attempt and succeeds more often.[3] Aggressors can also move with less fear of reprisal because they win their wars more decisively, leaving their victims less able to retaliate later. Thus even aggressor states are deterred from attacking if the defense is strong, and even quite benign powers are tempted to attack if the offense is strong.

B AND C: DEFENSIVE EXPANSIONISM AND FIERCE RESISTANCE TO EXPANSION

When conquest is hard, states are blessed with secure borders; hence they are less aggressive and more willing to accept the status quo. They have less need

3. Suggesting this hypothesis are Ivan S. Bloch, *The Future of War,* trans. R.C. Long, pref. W.T. Stead (New York: Doubleday and McClure, 1899), pp. xxx–xxxi, lxxix; also George H. Quester, *Offense and Defense in the International System* (New York: John Wiley and Sons, 1977), p. 9. A corroborating test is John J. Mearsheimer, *Conventional Deterrence* (Ithaca, N.Y.: Cornell University Press, 1983).

for wider borders because their current frontiers are already defensible. They have less urge to intervene in other states' internal affairs because hostile governments can do them less harm.

Conversely, when conquest is easy, states are more expansionist because their current borders are less defensible.[4] They covet others' geographic strong points, strategic depth, and sources of critical raw materials. They worry more when hostile regimes arise nearby because such neighbors are harder to defend against. These motives drive states to become aggressors and foreign intervenors.[5] States also resist others' expansion more fiercely when conquest is easy. Adversaries can parlay smaller gains into larger conquests; hence stronger steps to prevent gains by others are more appropriate. This attitude makes disputes more intractable.

The basic problem is that resources are more cumulative when conquest is easy. The ability to conquer others and to defend oneself is more elastic to one's control over strategic areas and resources. As a result, gains are more additive—states can parlay small conquests into larger ones—and losses are less reversible. Hence small losses can spell one's demise, and small gains can open the way to hegemonic dominance. States therefore compete harder to control any assets that confer power, seeking wider spheres for themselves while fiercely resisting others' efforts to expand.

This problem is compounded by its malignant effect on states' expectations about one another's conduct. When conquest is hard, states are blessed with neighbors made benign by their own security and by the high cost of attacking others. Hence states have less reason to expect attack. This leaves states even more secure and better able to pursue pacific policies. Conversely, when the offense dominates, states are cursed with neighbors made aggressive by both temptation and fear. These neighbors see easy gains from aggression and danger in standing pat. Plagued with such aggressive neighbors, all states face

---

4. As Robert Jervis notes, "when the offense has the advantage over the defense, attacking is the best route to protecting what you have...and it will be hard for any state to maintain its size and influence without trying to increase them." Jervis, "Cooperation under the Security Dilemma," p. 211; see also pp. 168–169, 173, 187–199.

5. It also seems possible that states should be more careful to avoid war when conquest is easy, because war then brings greater risk of total defeat. If so, offense dominance should cause more caution than belligerence among states, and should lower the risk of war. Advancing this argument is James Fearon, "The Offense-Defense Balance and War since 1648," paper prepared for the annual meeting of the International Studies Association, Chicago, February 1995, pp. 18–24. Fearon's argument seems deductively sound, but history offers very few examples of policymakers who argued that offense dominance was a reason for caution. This is one of many cases where deduction and the historical record point in opposite directions.

greater risk of attack. This drives them to compete still harder to control resources and create conditions that provide security.

Thus states become aggressors because their neighbors are aggressors. This can proceed reciprocally until no state accepts the status quo.

D: MOVING FIRST IS MORE REWARDING

When conquest is easy, the incentive to strike first is larger because a successful surprise attack provides larger rewards and averts greater dangers. Smaller shifts in ratios of forces between states create greater shifts in their relative capacity to conquer and defend territory. (A reversal in the force ratio between two states from 2 to 1 to 1 to 2 means little if attackers need a 3 to 1 advantage to conquer; it means everything if an attacker needs only a 1.5 to 1 advantage.) Hence a surprise strike that shifts the force ratio in the attacker's favor pays it a greater reward. This expands the danger of preemptive war and makes crises more explosive. States grow more trigger-happy, launching first strikes to exploit the advantage of the initiative, and to deny it to an opponent.[6]

Conversely, if the defense dominates, the first-move dividend is small because little can be done with any material advantage gained by moving first. Most aggressors can be checked even if they gain the initiative, and defenders can succeed even if they lose the initiative. Hence preemptive war has less attraction.

E: WINDOWS ARE LARGER AND MORE DANGEROUS

When conquest is easy, arguments for preventive war carry more weight.[7] Smaller shifts in force ratios have larger effects on relative capacity to conquer or defend territory; hence smaller prospective shifts in force ratios cause greater hope and alarm. Also, stemming decline by using force is more feasible because rising states can be overrun with greater ease. This bolsters arguments for shutting "windows of vulnerability" by war. As a result, all international change is more dangerous. Events that tip the balance of resources in any direction trigger thoughts of war among states that face relative decline.

Conversely, if the defense dominates, arguments for preventive war lose force because declining states can more successfully defend against aggressors even after their decline, making preventive war unnecessary. States are also

---

6. The classic discussion of these dangers is Thomas C. Schelling, *Arms and Influence* (New Haven, Conn.: Yale University Press, 1966), pp. 221–259.

7. For a discussion of the dangers of preventive war, see Jack S. Levy, "Declining Power and the Preventive Motivation for War," *World Politics*, Vol. 40, No. 1 (October 1987), pp. 82–107.

deterred from preventive war by the likelihood that their attack will fail, defeated by their enemy's strong defenses.

F: FAITS ACCOMPLIS ARE MORE COMMON AND MORE DANGEROUS

When conquest is easy, states adopt more dangerous diplomatic tactics—specifically, fait accompli tactics—and these tactics are more likely to cause war.

A fait accompli is a halfway step to war. It promises greater chance of political victory than quiet consultation, but it also raises greater risk of violence.[8] The acting side moves without warning, facing others with an accomplished fact. It cannot retreat without losing face, a dilemma that it exploits to compel the others to concede. But if the others stand firm, a collision is hard to avoid. Faits accomplis also pose a second danger: because they are planned in secret, the planning circle is small, raising the risk that flawed policies will escape scrutiny because critics cannot quarrel with mistaken premises.

Faits accomplis are more common when the offense dominates because the rewards they promise are more valuable. When security is scarce, winning disputes grows more important than avoiding war. Leaders care more how spoils are divided than about avoiding violence, because failure to gain their share can spell their doom. This leads to gain-maximizing, war-risking diplomatic strategies—above all, to fait accompli tactics.

Faits accomplis are more dangerous when the offense dominates because a successful fait accompli has a greater effect on the distribution of international power. A sudden resource gain now gives an opponent more capacity to threaten its neighbors' safety. Hence faits accomplis are more alarming and evoke a stronger response from others. States faced with a fait accompli will shoot more quickly because their interests are more badly damaged by it.

G: STATES NEGOTIATE LESS AND REACH FEWER AGREEMENTS

When conquest is easy, states have less faith in agreements because others break them more often; states bargain harder and concede more grudgingly, causing more deadlocks; compliance with agreements is harder to verify; and

---

8. On fait accompli strategies, see Alexander L. George, "Strategies for Crisis Management," in Alexander L. George, *Avoiding War: Problems of Crisis Management* (Boulder, Colo.: Westview, 1991), pp. 377–394 at 382–383, also pp. 549–550, 553–554. Other discussions of faits accomplis include R.B. Mowat, *Diplomacy and Peace* (London: Williams and Norgate, 1935), chap. 10 (on "sudden diplomacy"); Richard Ned Lebow, *Between Peace and War: The Nature of International Crisis* (Baltimore, Md.: Johns Hopkins University Press, 1981), pp. 57–97 (on "brinkmanship"); and Thomas C. Schelling, *Strategy of Conflict* (New York: Oxford University Press, 1963), pp. 22–28 (on games of "chicken").

states insist on better verification and compliance. As a result, states negotiate less often and settle fewer disputes; hence more issues remain unsettled and misperceptions survive that dialogue might dispel.

States break agreements more quickly when the offense dominates because cheating pays larger rewards. Bad faith and betrayal become the norm. The secure can afford the luxury of dealing in good faith, but the insecure must worry more about short-term survival. This drives them toward back-alley behavior, including deceits and sudden betrayals of all kinds—diplomatic faits accomplis, military surprise attacks, and breaking of other solemn agreements. Hence compliance with agreements is less expected.

When states do negotiate, they bargain harder and concede less when the offense dominates. Agreements must be more finely balanced to gain both sides' agreement, because a relative gain by either side poses greater risks to the other's safety.

Verification of compliance with agreements is both more necessary and more difficult when the offense dominates. States insist on better verification of the other's compliance because smaller violations can have larger security implications; for example, an opponent might convert a small advantage gained by cheating on an arms control agreement into a larger offensive threat. At the same time, verification of compliance is harder because states are more secretive when security is scarce (see explanation G). As a result, the range of issues that can be negotiated is narrowed to the few where near-certain verification is possible despite tight state secrecy.

As a net result, states let more disputes fester when the offense dominates.

H: STATES ARE MORE SECRETIVE

Governments cloak their foreign and defense policies in greater secrecy when conquest is easy. An information advantage confers more rewards, and a disadvantage raises more dangers: lost secrets could risk a state's existence. Thus states compete for information advantage by concealing their foreign policy strategies and military plans and forces.

Secrecy in turn is a hydra-headed cause of war. It can lead opponents to underestimate one another's capabilities and blunder into a war of false optimism.[9] It can ease surprise attack by concealing preparations from the victim. It opens windows of opportunity and vulnerability by delaying states' reac-

---

9. On wars of false optimism, see Geoffrey Blainey, *The Causes of War*, 3d ed. (New York: Free Press, 1988), pp. 35–56.

Figure 1. Offense-Defense Theory
Prime hypothesis: War is more likely when conquest is easy.

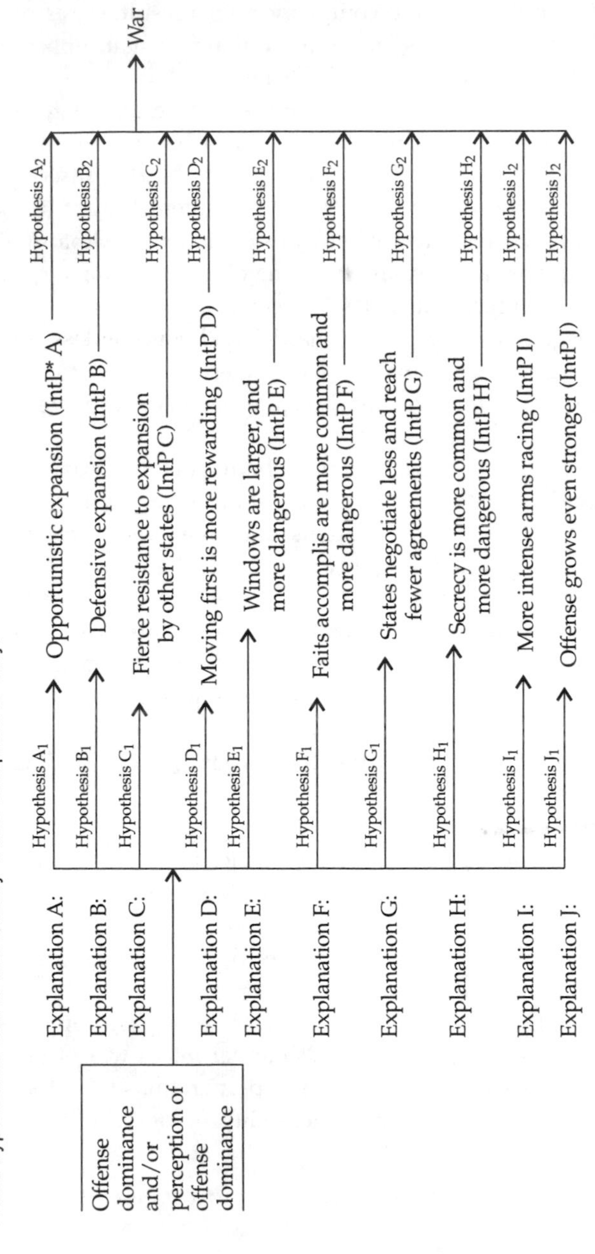

* IntP = Intervening phenomenon.

tions to others' military buildups, raising the risk of preventive war. It fosters policy blunders by narrowing the circle of experts consulted on policy, increasing the risk that flawed policies will survive unexamined. It prevents arms control agreements by making compliance harder to verify.

I: STATES ARMS RACE HARDER AND FASTER

Offense dominance intensifies arms racing, whereas defense dominance slows it down.[10] Arms racing in turn raises other dangers. It opens windows of opportunity and vulnerability as one side or the other races into the lead. It also fosters false optimism by causing rapid military change that confuses policymakers' estimates of relative power. Thus offense dominance is a remote cause of the dangers that arms racing produces.

States have seven incentives to build larger forces when the offense is strong.

- Resources are more cumulative (see explanations B and C). Wartime gains and losses matter more: gains provide a greater increase in security, and losses are less reversible. Therefore the forces that provide these gains and protect against these losses are also worth more.
- Self-defense is more difficult because others' forces have more inherent offensive capability. Hence states require more forces to offset others' deployments.
- States are more expectant of war. Their neighbors are more aggressive (see explanation B), so they must be better prepared for attack or invasion.
- The early phase of war is more decisive. Lacking time to mobilize their economies and societies in the event of war, states maintain larger standing forces.[11] The possibility of quick victory puts a premium on forces-in-being.[12]
- States transfer military resources from defense to offense because offense is more effective (see explanation J). Others then counterbuild because their neighbors' capabilities are more dangerous and so require a larger response. States also infer aggressive intent from their neighbors' offensive buildups, leading them to fear attack and to build up in anticipation.
- States hold military secrets more tightly when the offense dominates (see explanation H). This causes rational overarming, as states gauge their defense efforts to worst-case estimates of enemy strength, on grounds that

10. See Jervis, "Cooperation under the Security Dilemma," pp. 172, 188–190.
11. See ibid., pp. 172, 189.
12. General Joseph Joffre argued for a larger French standing force in 1913, because "the affair will already have been settled" by the time reservists were mobilized in three to four weeks. David G. Herrmann, *The Arming of Europe and the Making of the First World War* (Princeton, N.J.: Princeton University Press, 1996), p. 193.

underspending is disastrous whereas overspending is merely wasteful. It also allows national militaries to monopolize defense information more tightly. Given that militaries are prone to inflate threats, states will overspend groundlessly when militaries have an information monopoly that lets them alone assess the threat. Thus "action-reaction" becomes "action-overreaction-overreaction."

- States reach fewer arms control agreements when the offense dominates, because agreements of all kinds are fewer (see explanation G). Hence states are less able to limit arms competition through agreement.

If the defense dominates, things are reversed. States build smaller offensive forces because offense is less effective, and because other states have less aggressive aims. States are safe without wider empires; hence offensive forces that could provide empires lose utility. The national military therefore grows defense-heavy. This causes other states to feel safer, which in turn makes them less aggressive, further lowering all states' insecurity—hence their need for empire and for offense—up to a point.

States also reduce defensive forces when the defense dominates because defense is easier and attack seems more remote. Moreover, as their neighbors buy less offense, they need even less defense because their defense faces less challenge.

In short, states buy smaller forces in general, and less offense in particular, when the defense dominates. This leads to still smaller forces and still less offense. If information were perfect, arms racing would slow to a crawl if the defense strongly dominated.

J. CONQUEST GROWS STILL EASIER

Offense dominance is self-reinforcing[13] for three main reasons. First, states buy relatively more offensive forces when the offense dominates. They prefer the more successful type of force, so they buy defensive forces when the defense is strong and offensive forces when the offense is strong.[14] This reinforces the initial dominance of the defense or the offense.

---

13. Making this argument is Jervis, "Cooperation under the Security Dilemma," pp. 188, 199, 201.
14. Thus Clausewitz explained: "If attack were the stronger form [of war], there would be no case for using the defensive, since its purpose is only passive. No one would want to do anything but attack. Defense would be pointless." Carl von Clausewitz, *On War*, ed. and trans. Michael Howard and Peter Paret, intro. by Paret, Howard, and Bernard Brodie, commentary by Brodie (Princeton, N.J.: Princeton University Press, 1976), p. 359.

Second, alliances assume a more offensive character[15] when the offense dominates because aggressors can more easily drag their allies into their wars of aggression.[16] Insecure states can less afford to see allies destroyed, so they must support even bellicose allies who bring war on themselves. Knowing this, the allies feel freer to get into wars. As a net result, even de jure defensive alliances operate as defensive-and-offensive alliances. Alliances also assume a more offensive character if the allies adopt purely offensive military doctrines. This hamstrings states that would demand that their allies confine themselves to defensive preparations in a crisis, given that all preparations are offensive.

Third, status quo states are less able to protect their allies from conquest when the offense dominates because attackers can overrun defenders before help can arrive.

Thus offense dominance raises the danger of greater offense dominance. Once entered, an offense-dominant world is hard to escape.

Military offense dominance has one self-limiting effect: it leads status quo powers to cooperate more closely against aggressors.[17] They jump to aid an aggressor's victims because each knows that its neighbor's demise could lead more directly to its own undoing. Conversely, when states think that the defense dominates, they do less to save others from aggression because each expects it can defend itself alone even if others are overrun. As a result, aggressors can more often attack their victims seriatim, which is far easier than defeating a unified coalition. This countervailing effect, however, is more than offset by the several ways that offense dominance feeds itself.

These are the dangers raised by offense dominance. As noted above, these same ten dangers arise when the offense is weak but governments think it dominates. They then act as if it dominates, with comparable effects.

Are offensive capabilities always dangerous? The one-sided possession of offensive capabilities by status quo powers that face aggressors can lower rather than raise the risk of war under some conditions. Most important, status quo powers often need offensive capabilities to defend other states against

15. A defensive alliance is conditioned on defensive behavior by the ally; the alliance operates if the ally is attacked but not if it attacks. A defensive-and-offensive alliance operates in the event of war regardless of which side started it. The distinction began with Thucydides, who used "empimachy" to denote defensive alliance, "symmachy" for defensive-and-offensive alliances. G.E.M. de Ste. Croix, *The Origins of the Peloponnesian War* (Ithaca, N.Y.: Cornell University Press, 1972), pp. 60, 72–73, 106–108, 184, 298–302, 328.
16. Developing this point are Thomas J. Christensen and Jack Snyder, "Chain Gangs and Passed Bucks: Predicting Alliance Patterns in Multipolarity," *International Organization*, Vol. 44, No. 2 (Spring 1990), pp. 137–168.
17. Making this argument is ibid.

aggressors (e.g., as France required some offensive capability to defend Czechoslovakia and Poland from Germany in 1938–39). Offensive capabilities in the hands of status quo powers also may provide more deterrence than provocation if the aggressor state knows that it provoked the status quo power's hostility, if the aggressor knows that the status quo power has no bedrock aggressive intentions, and if the aggressor cannot remove the status quo power's offensive threat by force. These conditions are not unknown but they are rare. Hence offensive capabilities usually create more dangers than they dampen.

Causes of Offense and Defense Dominance

The feasibility of conquest is shaped by military factors, geographic factors, domestic social and political factors, and the nature of diplomacy. Discussions of the offense-defense balance often focus on military technology, but technology is only one part of the picture.[18]

MILITARY FACTORS

Military technology, doctrine, and force posture and deployments all affect the military offense-defense balance.[19] Military technology can favor the aggressor or the defender. In past centuries, strong fortification techniques bolstered the defense, and strong methods of siege warfare strengthened the offense. Technologies that favored mass infantry warfare (e.g., cheap iron, allowing mass production of infantry weapons) strengthened the offense because large mass armies could bypass fortifications more easily, and because mass armies fostered more egalitarian polities that could raise loyal popular armies that would not melt away when sent on imperial expeditions. Technologies that favored chariot or cavalry warfare (e.g., the stirrup) strengthened the defense, because cavalry warfare required smaller forces[20] that were more easily stopped by

18. For a discussion of the causes of offense and defense dominance, see Jervis, "Cooperation under the Security Dilemma," pp. 176, 194–199.

19. Several measures of the military offense-defense balance could be adopted, such as: (1) the probability that an offensive force can overcome a defensive force of equal cost; (2) the relative cost that attackers and defenders must pay for forces that offset incremental improvements by the other; or (3) the loss ratio when an offensive force attacks a defensive force of equal cost. All three measures (and more are possible) capture the concept of relative military difficulty of conquest and defense. For a list of possible measures, see Charles L. Glaser and Chaim Kaufmann, "What Is Offense-Defense Balance and How Can We Measure It?," *International Security,* Vol. 22, No. 4 (Spring 1998), pp. 44–82.

20. Cavalry warfare was capital intensive; hence it was usually waged by small forces of tax-supported specialists-knights in shining (and expensive) armor on expensive horses. Infantry warfare is more manpower intensive, and is usually waged by larger, less capitalized armies.

fortifications, and fostered hierarchic societies that could not raise armies that would remain loyal if sent on quests for empire.[21] In modern times, technology that gave defenders more lethal firepower (e.g., the machine gun) or greater mobility (e.g., the railroad) strengthened the defense. When these technologies were neutralized by still newer technologies (motorized armor), the offense grew stronger.

Thus when fortresses and cavalry dominated in the late Middle Ages, the defense held the advantage. Cannons then made fortifications vulnerable and restored the strength of the offense. In the seventeenth and eighteenth centuries new fortification techniques strengthened the defense. The mercenary armies of the age also remained tightly tied to logistical tails that kept them close to home: one historian writes that an eighteenth-century army "was like a diver in the sea, its movements strictly limited and tied by the long, slender communicating tube which gave it life."[22] Then revolutionary France's mass armies strengthened the offense because they had greater mobility. Their size let them sweep past border forts without leaving the bulk of their manpower behind for siege duty, and their more loyal troops could be trusted to forage without deserting, so they needed less logistical support. After the conservative restoration in France, Europe abandoned the mass army because it required, and fostered, popular government. This restored the power of the defense, which then waned somewhat as Europe democratized and large mass armies reappeared in the mid-nineteenth century.[23]

The combined effects of lethal small arms (accurate fast-firing rifles and machine guns), barbed wire, entrenchments, and railroads gave the defense an enormous advantage during World War I. The first three—lethal small arms, barbed wire, and trenches—gave defenders a large advantage at any point of attack. The fourth—railroads—let defenders reinforce points of attack faster than invaders could, because invaders could not use the defenders' railroads (given that railroad gauges differed across states, and defenders destroyed rail lines as they retreated) while the defenders had full use of their own lines. During 1919–45 the power of the offense was restored by motorized armor and an offensive doctrine—blitzkrieg—for its employment; this overrode machine

21. On the effects of the stirrup on warfare and society in the Middle Ages, see Lynn White, Jr., *Medieval Technology and Social Change* (New York: Oxford University Press, 1964), pp. 1–38. On the general effect of military technology on social stratification, see Stanislav Andreski, *Military Organization and Society* (Berkeley: University of California Press, 1971), pp. 20–74.
22. Harold Temperley, quoted in Blainey, *Causes of War*, p. 188.
23. Large armies aid the offense only up to a point, however. Once armies grow so big that they can cover an entire frontier (as on the western front in World War I), their size aids the defense because offensive outflanking maneuvers against them become impossible.

guns, trenches, and barbed wire. Then after 1945 thermonuclear weapons restored the power of the defense—this time giving it an overwhelming advantage.[24]

Technology and doctrine combined to define these tides of offense and defense. Sometimes technology overrode doctrine, as in 1914–18 and in 1945–91 (when the superpowers' militaries embraced offensive doctrines but could not find offensive counters to the nuclear revolution). Sometimes doctrine shaped technology, as in 1939–45, when blitzkrieg doctrine fashioned armor technology into an offensive instrument.

States shape the military offense-defense balance by their military posture and force deployments. Thus Stalin eased attack for both himself and Hitler during 1939–41 by moving most of the Red Army out of strong defensive positions on the Stalin Line and forward into newly seized territories in Poland, Bessarabia, Finland, and the Baltic states.[25] This left Soviet forces better positioned to attack Germany and far easier for Germany to attack, as the early success of Hitler's 1941 invasion revealed. The U.S. eased offense for both itself and Japan in 1941 when it deployed its fleet forward to Pearl Harbor and bombers forward to the Philippines.[26] Egypt eased Israel's assault by its chaotic forward deployment of troops into poorly prepared Sinai positions in the crisis before the 1967 war.[27]

States also can change the offense-defense balance through their wartime military operations. Aggressive operations can corrode key enemy defenses, and reckless operations can expose one's own defenses. Thus the dangers of offense dominance can be conjured up by unthinking wartime policymakers. For example, General Douglas MacArthur's reckless rush to the Yalu River in 1950 created an offensive threat to China's core territory and, by exposing badly deployed U.S. forces to attack, eased a Chinese offensive.[28]

---

24. Jack Levy provides synoptic history of the military offense-defense balance in "The Offensive/Defensive Balance of Military Technology: A Theoretical and Historical Analysis," *International Studies Quarterly*, Vol. 28, No. 2 (June 1984), pp. 219–238 at 230–234. Other discussions include Quester, *Offense and Defense in the International System;* and Andreski, *Military Organization and Society*, pp. 75–78. A detailed history is needed.
25. Peter Calvocoressi and Guy Wint, *Total War: The Story of World War II* (New York: Pantheon Books, 1972), p. 168.
26. Jonathan G. Utley, *Going to War with Japan, 1937–1941* (Knoxville: University of Tennessee Press, 1985), pp. 84, 163.
27. Donald Neff, *Warriors for Jerusalem: The Six Days That Changed the Middle East* (New York: Simon and Schuster, 1984), pp. 141, 168.
28. Likewise, during the Cold War some worried that NATO might inadvertently threaten the Soviet Union's strategic nuclear deterrent in its effort to defend NATO's Atlantic sea-lanes during

GEOGRAPHY

Conquest is harder when geography insulates states from invasion or strangulation. Hence conquest is hindered when national borders coincide with oceans, lakes, mountains, wide rivers, dense jungles, trackless deserts, or other natural barriers that impede offensive movement or give defenders natural strong points. Human-made obstacles along borders, such as urban sprawl, can also serve as barriers to armored invasion. Conquest is hindered if foes are separated by wide buffer regions (third states or demilitarized zones) that neither side can enter in peacetime. Conquest is hindered when national territories are mountainous or heavily forested, and when populations live mainly in rural settings, easing guerrilla resistance to invaders. Conquest is hindered when states are large and their critical war resources or industries lie far in their interior, where they cannot be quickly overrun. Conquest is hindered when states are invulnerable to economic strangulation. Hence conquest is hindered when states are self-sufficient in supplies of water, energy, food, and critical raw materials, or when their trade routes cannot be severed by land or sea blockade.

The geography of Western Europe, with its mountain ranges and ocean moats, is less favorable to conquest than the exposed plains of Eastern Europe or the open terrain of the Middle East. Israel's geography is especially unfortunate: physically small, its frontiers have few obstacles and much of its industry and population lie on exposed frontiers. Israeli territory is not conducive to guerrilla resistance, and its economy is import dependent. Germany's borders are better but still relatively poor: its eastern frontier is open; its economy is import dependent; and its trade routes are vulnerable. Britain, France, and Italy have formidable frontier barriers that make them relatively defensible. The United States' vast size, ocean-moat frontiers, and independent economy bless it with very defensible geography.

SOCIAL AND POLITICAL ORDER

Popular regimes are generally better at both conquest and self-defense than are unpopular regimes, but these effects do not cancel out. On net, conquest is probably harder among popular than unpopular regimes today, but in past centuries the reverse was likely true.

---

an East-West conventional war. Barry R. Posen, *Inadvertent Escalation: Conventional War and Nuclear Risks* (Ithaca, N.Y.: Cornell University Press, 1991), pp. 129–158. On a related danger, see ibid., pp. 28–67.

Popular governments can better raise larger, more loyal armies that can bypass others' border forts and can operate far from home with less logistical support. This gives popular regimes greater offensive power. Popular regimes can better organize their citizens for guerrilla resistance, making them harder to conquer. Citizen-defense guerrilla strategies are viable for Switzerland or China, but not for Guatemala or ancient Sparta, because these unpopular governments cannot arm their people without risking revolution. The citizens of unpopular oligarchies may actively assist advancing invaders. This gives attackers more penetrating power and makes early losses less reversible. Thus Sparta feared an invading army might grow if it entered Spartan territory, because Spartan slaves and dissident tribes would desert to the enemy.[29]

Unpopular regimes are more vulnerable to subversion or revolution inspired from abroad. Subversion is a form of offense, and it affects international relations in the same way as do offensive military capabilities. Frail regimes are more frightened of unfriendly neighbors, making them more determined to impose congenial regimes on neighboring states. The French revolutionary regime and the oligarchic Austrian regime worried that the other side might subvert them in 1792, causing both sides to become more aggressive.[30] After the Russian Revolution similar fears fueled Soviet-Western conflict, as each side feared subversion by the other.

On balance, is conquest easier in a world of popular or unpopular regimes? Popularity of regimes probably aided offense before roughly 1800 and has aided defense since then. The reversal stems from the appearance of cheap, mass-produced weapons useful for guerrilla war—assault rifles and machine guns, light mortars, and mines. The weapons of early times (sword and shield, pike and harquebus, heavy slow-firing muskets, etc.) were poorly adapted for guerrilla resistance. Guerrilla warfare has burgeoned since 1800 partly because the mass production of cheap small arms has tipped the balance toward guerrillas, allowing the hit-and-run harassment that characterizes guerrilla operations. The defensive power of popular regimes has risen in step with this increase in guerrilla warfare.

---

29. De Ste. Croix, *Origins of the Peloponnesian War,* pp. 89–94. Likewise, Hannibal hoped to defeat Rome by recruiting dissident tribes as he penetrated the Italian peninsula. See R.M. Errington, *Dawn of Empire: Rome's Rise to World Power* (Ithaca, N.Y.: Cornell University Press), pp. 62–64.
30. Stephen M. Walt, *Revolution and War* (Ithaca, N.Y.: Cornell University Press, 1996), pp. 123–124; and T.C.W. Blanning, *The Origins of the French Revolutionary Wars* (London: Longman, 1986), pp. 76, 85–86, 99–101, 111.

DIPLOMATIC FACTORS

Three types of diplomatic arrangements strengthen the defense: collective security systems, defensive alliances, and balancing behavior by neutral states. All three impede conquest by adding allies to the defending side.

States in a collective security system (e.g., the League of Nations) promise mutual aid against aggression by any system member. Such aggressors will face large defending coalitions if the system operates.[31]

States in a defensive alliance promise mutual aid against outside aggressors, leaving such aggressors outnumbered by resisting opponents. Thus during 1879–87 Bismarck wove a network of defensive alliances that discouraged aggression and helped preserve peace throughout central and eastern Europe.

Collective security systems and defensive alliances differ only in the kind of aggressor they target (system members versus outside aggressors). Both kinds of aggressors could be targeted at once, and a hybrid system that did this would offer defenders the most protection.

Neutral states act as balancers when they join the weaker of two competing coalitions to restore balance between them. Aggression is self-limiting when neutrals balance because aggressors generate more opposition as they expand. Britain and the United States traditionally played balancers to Europe, providing a counterweight to potential continental hegemons.

Balancing behavior is more selective than defensive alliance. Balancers balance to avert regional hegemony; hence pure balancers oppose expansion only by potential regional hegemons. Smaller states are left free to aggress. But balancing does contain hegemons and leaves their potential victims more secure. Conversely, if states bandwagon—join the stronger coalition against the weaker one—conquest is easier because aggressors win more allies as they seize more resources.[32]

Diplomatic arrangements have had a large influence on the offense-defense balance in modern Europe, and shifts in diplomatic arrangements have pro-

---

31. An introduction to collective security is Inis L. Claude, Jr., *Swords into Plowshares: The Problems and Progress of International Organizations*, 4th ed. (New York: Random House, 1971), pp. 411–433. A recent advocacy of collective security is Charles A. Kupchan and Clifford A. Kupchan, "Concerts, Collective Security, and the Future of Europe," *International Security*, Vol. 16, No. 1 (Summer 1991), pp. 114–163.

32. On balancing, bandwagoning, and other theories of alliances, see Stephen M. Walt, *The Origins of Alliances* (Ithaca, N.Y.: Cornell University Press, 1987). Historians have often suggested that a "breakdown in the balance of power" caused war. They usually mean (and should recast their claim to say) that states failed to engage in balancing behavior, which made aggression easier, causing war. War occurs not when the balance of power breaks down, but when balancers fail to balance, leaving aggressors unchecked, as in the late 1930s.

duced large shifts in the overall offense-defense balance. Collective security was never effective, but defensive alliances came and went, erecting barriers to conquest when they appeared. Balancing behavior rose and fell as the power and activism of the two traditional offshore balancers, Britain and the United States, waxed and waned. When the United States and/or Britain were strong and willing to intervene against aspiring continental hegemons, conquest on the continent was difficult. To succeed, a hegemon had to defeat both its continental victims and the offshore power. But when Britain and the United States were weak or isolationist, continental powers could expand against less resistance, leaving all states less secure.

Tests of Offense-Defense Theory

What predictions can be inferred from offense-defense theory? How much history does offense-defense theory explain?

PREDICTIONS AND TESTS

Offense-defense theory's predictions can be grouped in two broad types, *prime predictions* and *explanatory* predictions. The theory's prime predictions derive from its prime hypothesis ("War is more likely when conquest is easy"; or, for the theory's perceptual variant, "War is more likely when states think conquest is easy"). Tests of these predictions shed light on whether offense dominance (or perceptions of offense dominance) causes war.

Offense-defense theory's explanatory predictions derive from the hypotheses that comprise its ten explanations. Tests of these predictions shed light on both *whether* and *how* offense dominance (or perceptions of offense dominance) causes war.

PRIME PREDICTIONS. Three prime predictions of offense-defense theory are tested here.

1. War will be more common in periods when conquest is easy or is believed easy, less common when conquest is difficult or is believed difficult.
2. States that have or believe they have large offensive opportunities or defensive vulnerabilities will initiate and fight more wars than other states.
3. A state will initiate and fight more wars in periods when it has, or thinks that it has, larger offensive opportunities and defensive capabilities.

These predictions are tested below in three case studies: Europe since 1789 (treated as a single regional case study), ancient China during the Spring and

Autumn and Warring States eras, and the United States since 1789. I selected these cases because the offense-defense balance (or perceptions of it) varies sharply across time in all three, creating a good setting for "multiple within-case comparisons" tests that contrast different periods in the same case; because the United States is very secure relative to other countries, creating a good setting for a "comparison to typical values" tests that contrasts U.S. conduct with the conduct of average states;[33] and because two of these cases are well recorded (Europe since 1789 and the United States since 1789).

The case of Europe since 1789 allows tests of prime predictions 1 and 2.[34] We can make crude indices of the offense-defense balances (actual and perceived) for Europe over the past two centuries, and match them with the incidence of war (see Table 1). Offense-defense theory predicts more war when conquest is easy or is believed easy. We can also estimate the offensive opportunities and defensive vulnerabilities of individual powers—for example, since 1789 Prussia/Germany has been more vulnerable and has had more offensive opportunity than Spain, Italy, Britain, or the United States—and can match these estimates with states' rates of war involvement and war initiation. Offense-defense theory predicts that states with more defensive vulnerability and offensive opportunity will be more warlike.

The ancient China case allows a test of prime prediction 1. The offense-defense balance shifted markedly toward the offense as China's Spring and Autumn and Warring States periods evolved. Offense-defense theory predicts a parallel rise in the incidence of warfare during these periods.

The U.S. case allows testing of prime predictions 2 and 3. The United States is less vulnerable to foreign military threats than are other states; hence offense-defense theory predicts that it should start fewer wars and be involved in fewer wars than other states. Americans have also felt more vulnerable to foreign military threats in some eras than in others. The U.S. propensity for war involvement and war initiation should co-vary with this sense of vulnerability.

EXPLANATORY PREDICTIONS. Offense-defense theory posits that offense dominance leads to war through the war-causing action of its ten intervening phenomena A–J: opportunistic expansionism, defensive expansionism, fierce

---

33. I say more about the logic of within-case comparisons and comparison to typical values tests in Van Evera, *Guide to Methods for Students of Political Science* (Ithaca, N.Y.: Cornell University Press, 1997), pp. 58–63. On case selection criteria, see pp. 77–88.

34. In principle, prime prediction 3 could also be tested with this case. This, however, would require tracing and describing trends in each state's sense of vulnerability over time—a large task that would fill many pages.

Table 1. The Offense-Defense Balance among Great Powers, 1700s–Present.

Era	Military realities favored	Military realities were thought to favor	Diplomatic realities favored	Diplomatic realities were thought to favor	In aggregate military and diplomatic realities favored	In aggregate military and diplomatic realities were thought to favor	Amount of warfare among great powers
Pre–1792	Defs.	Defs.	Med.	Med.	Med.	Med.	Medium
1792–1815	Aggrs.	Aggrs.	Med.	Aggrs.	Aggrs.	Aggrs.***	High
1816–56	Defs.	Defs.	Defs.	Defs.	Defs.	Defs.	Low
1856–71	Med.	Med.	Aggrs.	Aggrs.	Aggrs.	Aggrs.	Medium
1871–90	Defs.	Med.	Defs.	Defs.	Defs.	Defs.***	Low
1890–1918	Defs.	Aggrs.	Aggrs.	Aggrs.	Defs.	Aggrs.	High
1919–45	Aggrs.	Mixed*	Aggrs.	Aggrs.**	Aggrs.	Aggrs.****	High
1945–1990s	Defs.	Med.	Defs.	Defs.	Defs.	Defs.***	Low

Aggrs.: The factor favors aggressors.
Defs.: The factor favors defenders.
Med.: A medium value: things are somewhere in between, cut both ways.
Mixed: Some national elites saw defense dominance, some saw offense dominance.

The perceptions entries are an average of the perceptions of the great power elites. In some cases, the perceptions of these elites varied sharply across states, for example, perceptions of military realities in the 1930s.

* Things varied across states. The German elite recognized the military power of the offensive in the late 1930s; the elites of other great powers thought the defense was dominant.

** Things varied across states. The German elite (above all Hitler) exaggerated the considerable actual diplomatic weakness of the defense; the elites of other great powers recognized this weakness but did not overstate it. These beliefs average to a perception of substantial diplomatic offense dominance.

*** Elites exaggerated the strength of the offense during 1792–1815, 1871–90, and 1945–1990s, but not by enough to give the realities and perceptions of the offense-defense balance different scores.

**** When we aggregate perceptions of the offense-defense balance, the errors of Germany and the other powers cancel each other out. Germany's exaggeration of the diplomatic power of the offense offsets other powers' exaggeration of the military power of the defense, leaving an aggregate perception fairly close to the offense-dominant reality.

resistance to others' expansion, first-strike advantages, windows of opportunity and vulnerability, faits accomplis and belligerent reactions to them, reluctance to solve conflicts through negotiation, policies of secrecy, intense arms racing, and policies that ease conquest, such as offensive force postures and offensive alliances. If offense-defense theory is valid, these intervening phenomena should correlate with the real and perceived offense-defense balance. Two explanatory predictions can be inferred.

1. Phenomena A–J will be more abundant in eras of real or perceived offense dominance: the ten phenomena should increase as offense strengthens and diminish as offense weakens.
2. States that have or believe they have large offensive opportunities or defensive vulnerabilities will more strongly embrace policies that embody phenomena A–J.[35]

Two of the case studies presented here shed light on these explanatory predictions. The case of Europe allows a partial test of both. We can code only two of offense-defense theory's ten intervening phenomena (IntPs A and B, opportunistic and defensive expansionism) for the whole period. We have fragmentary data for values on the other eight intervening variables. Hence the case lets us test explanations A and B fairly completely and offers scattered evidence on explanations C–J. To test explanations A and B, we ask if expansionism correlates over time with periods of real or perceived offense dominance, and if states that were (or believed they were) less secure and more able to aggress were more expansionist.

The case of the United States since 1789 allows a more complete, if rather weak, test of explanatory prediction 2.

TEST 1: EUROPE 1789–1990S

A composite measure of the offense-defense balance in Europe since 1789 can be fashioned by blending the histories of Europe's military and diplomatic

---

35. Explanatory predictions 1 and 2 are inferred from the "left side" of offense-defense theory, that is, from hypotheses A_1–J_1, which frame the claim that offense dominance causes intervening phenomena A–J (see Figure 1). Predictions could also be inferred from hypotheses A_2–J_2, which comprise the "right side" of the theory, and frame the claim that intervening phenomena A–J cause war. For example, we could infer that (6) warfare will be more common in eras and regions where phenomena A–J are more prevalent, and (7) states that embrace policies that embody phenomena A–J will be involved in more wars and will initiate more wars than other states. I leave "right side" hypotheses untested here because the effects of phenomena A–J are less debated than their causes. Most agree that they cause trouble.

offense-defense balances, as outlined above.[36] In sum, the offense-defense balance went through six phases comprising three up-down oscillations after 1789. Conquest was never easy in an absolute sense during these two centuries. Conquest was, however, markedly easier during 1792–1815, 1856–71, and 1930s–1945 than it was during 1815–56, 1871–1920s, and 1945–1990s.

Elite perceptions of the offense-defense balance track these oscillations quite closely, but not exactly. Elites chronically exaggerated the power of the offense, but did so far more in some periods than in others. Most important, they greatly exaggerated the power of the offense during 1890–1918: elites then wrongly thought conquest was very easy when in fact it was very hard. Thus the pattern of reality and perception run roughly parallel, with the major exception of 1890–1918.

Tides of war and peace correlate loosely with the offense-defense balance during this period, and tightly with the perceived offense-defense balance. Expansionism and war were more common when conquest was easy than when it was difficult, and were far more common when conquest was believed easy than when it was believed difficult. Moreover, states that believed they faced large offensive opportunities and defensive vulnerabilities (especially Prussia/Germany) were the largest troublemakers. They were more expansionist, they were involved in more wars, and they started more wars than other states.

1792–1815. During 1792–1815 the offense was fairly strong militarily, as a result of France's adoption of the popular mass army (enabled by the popularity of the French revolutionary government).[37] Moreover, European elites widely exaggerated one another's vulnerability to conquest: at the outset of the War of 1792 all three belligerents (France, Austria, and Prussia) thought their

36. My composite index represents my own "author's estimates" based on sources provided throughout this article. I measured the actual and perceived Europe-wide offense-defense balances by asking: (1) Did military technology, force posture, and doctrine favor the offense or the defense? Did elites and publics believe these factors favored the offense or the defense? (2) Did geography and the domestic social and political order of states favor the offense or the defense? Did elites and publics believe they favored the offense or defense? (3) How numerous and powerful were balancer states, and how strongly did they balance? Did elites believe that other states would balance or bandwagon? (4) Did defensive alliances form, and did they operate effectively? Did elites believe that they operated effectively? I gave these factors the same rough relative weight they receive in standard historical accounts.

37. A discussion of the military offense-defense balance in this era is Quester, *Offense and Defense in the International System*, pp. 66–72.

opponents were on the verge of collapse and could be quickly crushed.[38] Defense-enhancing diplomacy was sluggish: Britain, Europe's traditional balancer, stood by indifferently during the crisis that produced the War of 1792, issuing a formal declaration of neutrality.[39] Moreover, French leaders underestimated the power of defense-enhancing diplomacy because they widely believed that other states would bandwagon with threats instead of balancing against them.[40] In short, military factors helped the offense, and this help was further exaggerated; political factors did little to help bolster defenders, and this help was underestimated.

1815–56. After 1815 both arms and diplomacy favored defenders, as outlined above. Mass armies disappeared,[41] British economic power grew, and Britain remained active on the continent as a balancer. Continental powers expected Britain to balance and believed British strength could not be overridden.

This defense-dominant arrangement lasted until midcentury. It began weakening before the Crimean War (1853–56). When war in Crimea broke out, military factors still favored defenders, but elites underestimated the power of the defense: Britain and France launched their 1854 Crimean offensive in false expectation of quick and easy victory.[42] In general, diplomatic factors favored the defense (Britain still balanced actively), but during the prewar crisis in

---

38. Blanning, *Origins of the French Revolutionary Wars*, p. 116. Austrian and Prussian leaders were assured that revolutionary France could be quickly smashed. Ibid., p. 114. One Prussian leader advised his officers: "Do not buy too many horses, the comedy will not last long. The army of lawyers will be annihilated in Belgium and we shall be home by autumn." Ibid., p. 116. Meanwhile, French revolutionaries wrongly expected a pro-French revolutionary uprising of the oppressed peoples of feudal Europe. Ibid., p. 136; R.R. Palmer, *World of the French Revolution* (New York: Harper and Row, 1971), p. 95; and George Rudé, *Revolutionary Europe, 1783–1815* (Glasgow: Fontana/Collins, 1964), p. 209.

39. Blanning, *Origins of the French Revolutionary Wars*, pp. 131–135.

40. As Steven Ross notes, French expansionists thought they could intimidate Europe into coexisting with an expanded French empire in the 1790s: "By inflicting rapid and decisive defeats upon one or more members of the coalition, the [French] directors hoped to rupture allied unity and force individual members to seek a separate peace." Steven T. Ross, *European Diplomatic History, 1789–1815* (Garden City, N.Y.: Anchor Doubleday, 1969), p. 186.

Later Napoleon thought he could compel Britain to make peace by establishing French continental dominion, proclaiming after the Peace of Amiens, "With Europe in its present state, England cannot reasonably make war on us unaided." Geoffrey Bruun, *Europe and the French Imperium, 1799–1814* (New York: Harper and Row, 1938), p. 118. See also Blanning, *Origins of the French Revolutionary Wars*, p. 109.

41. On the post-1815 restoration of pre-Napoleonic warfare, see Quester, *Offense and Defense in the International System*, pp. 73–74; and Michael Howard, *War in European History* (London: Oxford University Press, 1976), pp. 94–95.

42. Richard Smoke, *War: Controlling Escalation* (Cambridge, Mass.: Harvard University Press, 1977), p. 191.

1853, diplomacy favored the offense because Britain and France blundered by giving Turkey unconditional backing that amounted to an offensive alliance. This encouraged the Turkish aggressions that sparked the war.[43]

1856–71. After the Crimean War the offense-defense balance shifted further toward the offense. Changes in the military realm cut both ways. Mass armies were appearing (bolstering the offense), but small arms were growing more lethal and railroads were expanding (bolstering the defense). In the diplomatic realm, however, the power of defenders fell dramatically because defense-enhancing diplomacy largely broke down. Most important, Britain entered an isolationist phase that lasted into the 1870s, and Russia lost interest in maintaining the balance among the western powers.[44] As a result, diplomatic obstacles to continental conquest largely disappeared, giving continental aggressors a fairly open field. This diplomatic change gave France and Sardinia, and then Prussia, a yawning offensive opportunity, which they exploited by launching a series of wars of opportunistic expansion—in 1859, 1864, 1866, and 1870. But defense-enhancing diplomacy had not disappeared completely, and it helped keep these wars short and limited.

In 1859 British and Russian neutrality gave France and Sardinia a free hand, which they used to seize Lombardy from Austria.[45] In 1864 British, Russian, and French neutrality gave Prussia and Austria a free hand, which they used to seize Schleswig-Holstein from Denmark.[46] In 1866 British, French, and Russian neutrality gave Prussia carte blanche against Austria, which Prussia used to smash Austria and consolidate its control of North Germany.[47] Even after war broke out, major fighting proceeded for weeks before any outside

---

43. Ibid., pp. 167, 179–181, 185; Richard Smoke, "The Crimean War," in George, *Avoiding War*, pp. 36–61 at 48–49, 52. The motives of the powers also illustrate offense-defense dynamics. The main belligerents (Britain, France, Russia, and Turkey) were impelled in part by security concerns that would have been allayed had they believed the defense more dominant. Smoke, *War*, pp. 149, 155, 158–159, 162, 190.

44. The harsh Crimean War settlement Britain imposed on Russia turned it into a non–status quo power. Overthrowing that settlement became Russia's chief aim in European diplomacy, superseding its interest in preserving order to the west. M.S. Anderson, *The Eastern Question, 1774–1923* (London: Macmillan, 1966), pp. 144–146.

45. A.J.P. Taylor, *The Struggle for Mastery in Europe 1848–1918* (London: Oxford University Press, 1971), pp. 108, 110.

46. Ibid., pp. 146–154. Britain would have backed Denmark had it found a continental ally but none was available. Ibid., pp. 146–148.

47. Smoke, *War*, pp. 85–92. Britain remained in a semi-isolationist mood in 1866, and Napoleon III thought France would profit from the long, mutually debilitating Austro-Prussian war he expected. Like the Soviets in 1939, Napoleon underestimated the danger of a quick, lopsided victory by either side. Ibid., pp. 87–90.

state even threatened intervention.[48] As A.J.P. Taylor notes, Bismarck's 1866 diplomatic opportunity—a wide-open field for unopposed expansion—was "unique in recent history."[49]

In 1870 Bismarck ensured the neutrality of the other European powers by shifting responsibility for the war to France and convincing Europe that the war stemmed from French expansionism.[50] As a result, Prussia again had a free hand to pursue its expansionist aims. It used this to smash France, seize Alsace-Lorraine, and consolidate control over South Germany.[51]

1871–90. For some twenty years after the Franco-Prussian War, the defense dominated because of Bismarck's new diplomacy and Britain's renewed activism. In the military area the cult of the offensive had not yet taken hold. In diplomacy Bismarck wove a web of defensive alliances that deterred aggressors and calmed status quo powers after 1879.[52] British power waned slightly, but this was offset by the recovery of Britain's will to play the balancer. The "war-in-sight" crisis of 1875 illustrates the change: Britain and Russia together deterred a renewed German attack on France by warning that they would not allow a repeat of 1870–71.[53]

1890–1919. After 1890 military realities increasingly favored the defense, but elites mistakenly believed the opposite. Diplomatic realities swung toward the offense, and elites believed they favored the offense even more than they did.

---

48. Ibid., p. 86.
49. Taylor, *Struggle for Mastery,* p. 156. Moreover, Bismarck stopped the 1866 war partly because he feared French or Russian intervention if Prussia fought on too long or conquered too much. Smoke, *War,* pp. 101–102. Thus lack of defense-enhancing diplomacy helped cause the war while Prussian fear of such diplomacy shortened and limited the war.
50. William Carr, *The Origins of the Wars of German Unification* (London: Longman, 1991), p. 202; and Michael Howard, *The Franco-Prussian War: The German Invasion of France, 1870–1871* (New York: Granada, 1961), p. 57. Austria also stayed neutral because Hungarian Magyar influence was growing inside the Dual Monarchy, and the Magyars felt that the more Austria was pushed out of Germany, the stronger the position of the Magyars within it would be. R.R. Palmer and Joel Colton, *A History of the Modern World,* 4th ed. (New York: Alfred A. Knopf, 1971), p. 574.
51. On Prussia's free hand, see Smoke, *War,* pp. 133–136; Norman Rich, *The Age of Nationalism and Reform, 1850–1890,* 2d ed. (New York: W.W. Norton, 1977), p. 140; and W.E. Mosse, *European Powers and the German Question* (New York: Octagon, 1969), pp. 291, 295.
52. Bismarck formed defensive alliances with Austria, Italy, and Romania, and a more limited defensive accord with Russia—specifically, a reciprocal agreement not to join a war against the other unless the other attacked France (in the German case) or Austria (in the Russian case). Synopses include Paul M. Kennedy, *The Rise and Fall of the Great Powers: Economic Change and Military Conflict from 1500 to 2000* (New York: Random House, 1987), pp. 249–250; and Robert E. Osgood and Robert W. Tucker, *Force, Order, and Justice* (Baltimore, Md.: Johns Hopkins University Press, 1967), pp. 80–81. For a longer account, see Taylor, *Struggle for Mastery,* pp. 258–280, 316–319.
53. Imanuel Geiss, *German Foreign Policy, 1871–1914* (Boston: Routledge and Kegan Paul, 1976), p. 28.

European militaries were seized by a "cult of the offensive." All the European powers adopted offensive military doctrines, culminating with France's adoption of the highly offensive Plan XVII in 1913 and with Russia's adoption of the highly offensive Plan 20 in 1914. More important, militaries persuaded civilian leaders and publics that the offense dominated and conquest was easy. As a result, elites and publics widely believed the next war would be quickly won by a decisive offensive.

Bismarck's defensive alliances withered or evolved into defensive-and-offensive alliances after he left office in 1890, largely because the cult of the offensive made defensive alliances hard to maintain. Pacts conditioned on defensive conduct became hard to frame because states defended by attacking, and status quo powers shrank from enforcing defensive conduct on allies they felt less able to lose. For example, Britain and France felt unable to enforce defensive conduct on a Russian ally that defended by attacking and that they could not afford to see defeated. Elites also thought that aggressors could overrun their victims before allies could intervene to save them, making defensive alliances less effective. Thus Britain seemed less able to save France before Germany overran it, leading Germany to discount British power. Lastly, German leaders subscribed to a bandwagon theory of diplomacy, which led them to underestimate others' resistance to German expansion. Overall, the years before 1914 were the all-time high point of perceived offense dominance.

Nine of the ten intervening phenomena predicted by offense-defense theory (all except phenomenon G, nonnegotiation) flourished in this world of assumed offense dominance. Opportunistic and defensive expansionist ideas multiplied and spread, especially in Germany. Russia and France mobilized their armies preemptively in the 1914 July crisis. That crisis arose from a fait accompli that Germany and Austria instigated in part to shut a looming window of vulnerability. This window in turn had emerged from a land arms race that erupted during 1912–14. The powers enshrouded their military and political plans in secrecy—a secrecy that fostered crisis-management blunders during July 1914. These blunders in turn evoked rapid, violent reactions that helped drive the crisis out of control. Belief in the offense fueled offensive military doctrines throughout the continent and impeded efforts to restrain allies. Together these dangers formed a prime cause of the war: they bore the 1914 July crisis and helped make it uncontrollable.

1919–45. The interwar years were a mixed bag, but overall the offense gained the upper hand by 1939, and the German elite believed the offense even stronger than in fact it was.

Military doctrine and technology gave the defense the advantage until the late 1930s, when German blitzkrieg doctrine combined armor and infantry in an effective offensive combination. This offensive innovation was unrecognized outside Germany and doubted by many in Germany, but the man who counted most, Adolf Hitler, firmly believed in it. This reflected his faith in the offense as a general principle, imbibed from international social Darwinist propaganda in his youth.[54]

More important, the workings of interwar diplomacy opened a yawning political opportunity for Nazi expansion. Britain fell into a deep isolationism that left it less willing to commit this declining power to curb continental aggressors.[55] The United States also withdrew into isolation, removing the counterweight that checked Germany in 1918.[56] The breakup of Austria-Hungary in that year created a new diplomatic constellation that further eased German expansion. Austria-Hungary would have balanced against German

---

54. Hitler often echoed international social Darwinist slogans on the short, precarious lives of states, for example, "Politics is in truth the execution of a nation's struggle for existence," and "Germany will either be a world power or there will be no Germany." Quoted in P.M.H. Bell, *The Origins of the Second World War in Europe* (London: Longman, 1986), p. 81; and in Anthony P. Adamthwaite, *The Making of the Second World War* (London: George Allen and Unwin, 1977), p. 119.

Hitler's faith in the offensive differed from that of the pre-1914 cultists of the offensive in three ways. First, he saw offensive capabilities arising from a long search for offensive methods, not from permanent properties of war. In his mind offense could be created, but also had to be; Germany would discover offensive answers only after a long effort. In contrast, the pre-1914 cultists thought offense inherently easier than defense; deep thought need not be given to how to make it superior, because it already was. Second, Hitler's offensive optimism was based on racism and social prejudice, as well as on assessment of military factors. Specifically, his contempt for Slavs and Jews led him to expect that the Soviets would quickly collapse under German attack. Third, Hitler's concerns for German security focused on fear of conquest by economic strangulation, not conquest by French or Soviet blitzkrieg. He thought German security was precarious, but for reasons rooted more in the political economy of war than in the nature of doctrine or weaponry. These differences aside, the logical implications of Hitler's offensive cult were the same as those of the pre-1914 cult. He exaggerated both German insecurity and the feasibility of imperial solutions to redress it.

55. Prime Minister Neville Chamberlain of Great Britain said in 1937 that he "did not believe we could, or ought . . . to enter a Continental war with the intention of fighting on the same lines as in the last," meaning that Britain would deploy no large ground force on the continent. Bell, *Origins of the Second World War in Europe,* p. 177. Britain had only two divisions available to send to the continent during the 1938 Munich crisis, and the four-division force it actually sent in 1939 was smaller and less well trained than its small expeditionary force of 1914. These four divisions were a drop in the bucket relative to the 84 French and 103 German divisions then deployed. Ibid., p. 175.

56. The United States also proclaimed this isolationism in four neutrality laws passed during 1935–39, giving Hitler a clear if misleading signal of American indifference to his aggression. On these laws a synopsis is Thomas A. Bailey, *A Diplomatic History of the American People,* 9th ed. (Englewood Cliffs, N.J.: Prentice-Hall, 1974), pp. 701–702, 715.

expansion, but its smaller successor states tended to bandwagon.[57] This let Hitler extend German influence into southeast Europe by intimidation and subversion.

The Soviet Union and the Western powers failed to cooperate against Hitler.[58] Ideological hostility divided them. Britain also feared that a defensive alliance against Hitler would arouse German fears of allied encirclement, spurring German aggressiveness. This chilled British enthusiasm for an Anglo-French-Soviet alliance.[59]

Hitler exaggerated the already-large advantage that diplomacy gave the offense because he thought bandwagoning prevailed over balancing in international affairs. This false faith colored all his political forecasts and led him to vastly underestimate others states' resistance to his aggressions. Before the war he failed to foresee that Britain and France would balance German power by coming to Poland's rescue.[60] Once the war began he believed Germany could intimidate Britain into seeking alliance with Germany after Germany crushed France—or, he later held, after Germany smashed the Soviet Union.[61] He thought the United States could be cowed into staying neutral by the 1940 German-Japanese alliance (the alliance had the opposite effect, spurring U.S. intervention).[62] In short, Hitler's false theories of diplomacy made three of his most dangerous opponents shrink to insignificance in his mind.

These realities and beliefs left Hitler to face temptations like those facing Bismarck in 1866 and 1870. Hitler thought he could conquer his victims seriatim. He also thought his conquests would arouse little countervailing opposition from distant neutral powers.[63] As a result, he believed he faced a yawning opportunity for aggression.

---

57. Explaining why weaker states are more prone to bandwagon than are stronger states is Walt, *Origins of Alliances*, pp. 29–30.
58. Bell, *Origins of the Second World War in Europe*, pp. 172, 224, 260; and Adamthwaite, *Making of the Second World War*, pp. 60, 69. This failure greatly eased Hitler's aggressions, because geography made Britain's 1939 guarantees to Poland and Romania unenforceable without a Soviet alliance. Ibid., pp. 86, 91.
59. Raymond J. Sontag, *A Broken World, 1919–1939* (New York: Harper and Row, 1971), p. 361.
60. On August 22, 1939, Hitler assured his generals that "the West will not intervene" to defend Poland. Jeremy Noakes and Geoffrey Pridham, eds., *Nazism, 1919–1945: A History in Documents and Eyewitness Accounts*, 2 vols. (New York: Schocken Books, 1988), vol. 2, p. 741.
61. See Jack Snyder, *Myths of Empire: Domestic Politics and International Ambition* (Ithaca, N.Y.: Cornell University Press, 1991), p. 94.
62. Noakes and Pridham, *Nazism*, vol. 2, p. 797. Some German leaders also hoped that Germany could win decisively in Europe before the United States could bring its power to bear. Thus in September 1940 Hitler's naval commander in chief voiced the hope that Britain could be beaten "before the United States is able to intervene effectively." Ibid., p. 794.
63. The fine-grained pattern of events during 1938–40—who attacked whom and when—also fits the predictions of offense-defense theory (specifically, prime prediction 3). The Western allies stood

Unlike 1914, the late 1930s were not a pure case of perceived offense dominance. Instead, the 1930s saw status quo powers' perceptions of defense dominance create real offensive opportunities for an aggressor state. Hitler thought the offense strong and even exaggerated its strength, but other powers (the Soviet Union, Britain, and France) underestimated its strength. Their perceptions of defense dominance relaxed their urge to jump the gun at early signs of threat (as Russia did in 1914); this made things safer. But this perception also relaxed their will to balance Germany, because they found German expansion less frightening. This weakened the coalition against Hitler, leaving him wider running room.[64]

1945–1990s. After 1945 two changes swung the offense-defense balance back toward the defense. First, the end of American isolationism transformed European political affairs. The United States replaced Britain as continental balancer, bringing far more power to bear in Europe than Britain ever had. As a result, Europe in the years after 1945 was unusually defense dominant from a diplomatic standpoint.

Second, the nuclear revolution gave defenders a large military advantage—so large that conquest among great powers became virtually impossible. Conquest now required a nuclear first-strike capability (the capacity to launch a nuclear strike that leaves the defender unable to inflict unacceptable damage in retaliation). Defenders could secure themselves merely by maintaining a second-strike capability (the capacity to inflict unacceptable damage on the attacker's society after absorbing an all-out strike). The characteristics of nuclear weapons—their vast power, small size, light weight, and low cost—ensured that a first-strike capability would be very hard to attain, while a second-strike capability could be sustained at little cost. As a result, the great powers became essentially unconquerable, and even lesser powers could now stand against far stronger enemies. Overall, the nuclear revolution gave defenders an even more lopsided advantage than the machine gun–barbed wire–entrenchments–railroad complex that emerged before 1914.

---

without attacking Germany in 1938 and again in 1939–40 because they doubted they could win a decisive victory. Germany stood without attacking westward in the fall of 1939 for the same reason, and finally attacked in May 1940 after German military leaders developed a plausible plan for decisive attack. Mearsheimer, *Conventional Deterrence*, pp. 67–133.

64. Would the risk of war have fallen had all powers believed the offense was dominant in the late 1930s? This seems unlikely. The status quo powers would have balanced harder against Hitler, offering him more discouragement, but they also would have been jumpier, making early crises more dangerous. One of these crises—Hitler's remilitarization of the Rhineland, the Spanish civil war, or the German seizure of Austria or Czechoslovakia—probably would have served as the "Sarajevo" for World War II, with the Allies moving first as Russia did in 1914.

American and Soviet policymakers grasped this cosmic military revolution only slowly, however. At first many feared nuclear weapons would be a boon to aggressors. When this fear proved false, the vast advantage they gave defenders was only dimly recognized, partly because scholars strangely failed to explain it. Thus the nuclear revolution changed realities far more than they did perceptions. As a result, state behavior changed only slowly, and both superpowers competed far harder—in both Central Europe and the third world—than objective conditions warranted. The Cold War was far more peaceful than the preceding forty years, but could have been still more peaceful had Soviet and U.S. elites understood that their security problems had vastly diminished and were now quite small.

In sum, the events of 1789–1990s clearly corroborate offense-defense theory predictions—specifically, prime predictions 1 and 2, as well as both explanatory predictions. These conclusions rest on rather sketchy data—especially regarding the explanatory predictions—but that data confirm offense-defense theory so clearly that other data would have to be very different to reverse the result.

- The incidence of war correlates loosely with the offense-defense balance and very tightly with perceptions of the offense-defense balance (for a summary see Table 1).
- Europe's less-secure and more offensively capable continental powers were perennial troublemakers, while more secure and less offensively capable offshore powers were perennial defenders of the status quo. Prussia/Germany was cursed with the least defensible borders and faced the most offensive temptations. It started the largest number of major wars (1864, 1866, 1914, 1939, and shared responsibility for 1870 with France). France and Russia, with more defensible borders and fewer temptations, started fewer major wars.[65] Britain and the United States, blessed with even more insulating borders, joined a number of European wars but started none.[66] Spain, Sweden, and Switzerland, also insulated from other powers by mountains or oceans, fought very little.

Thus the timing of war and the identities of the belligerents tightly fit prime predictions 1 and 2.

---

65. France can be assigned prime responsibility for 1792 and 1859, and shared responsibility for Crimea and 1870. Russia deserves prime responsibility for the Cold War and shared responsibility for Crimea and the 1904–05 Russo-Japanese War.
66. Britain does share responsibility for the Crimean War with Russia, France, and Turkey.

- Sketchy evidence suggests that opportunistic and defensive expansionism were more prominent during the periods of perceived offense dominance (1792–1815, 1859–71, 1890–1914, 1930s–1945) than at other times. The years 1792–1815 saw a strong surge of French expansionism, nearly matched at the outset by parallel Prussian expansionism.[67] The mid-nineteenth century saw large opportunistic expansionism in Prussia and some French expansionism. The years 1890–1914 saw vast expansionist ambitions develop in Wilhelmine Germany,[68] matched by fierce resistance to this German expansionism in Russia and France, and by lesser French and Russian expansionism. Large German expansionism then reappeared under the Nazis in the 1930s. During other periods European expansionism was more muted: European powers had smaller ambitions and acted on them less often. This supports explanatory prediction 1.
- Opportunistic and defensive expansionism were prominent among those states that saw the clearest defensive vulnerability and offensive opportunity (especially Prussia/Germany, also revolutionary France), while being more muted among states with more secure borders and fewer offensive opportunities (Britain, the United States, the Scandinavian states, and Spain). This corroborates explanatory prediction 2.

How strong is this test? The strength of a passed test depends on the uniqueness of the predictions tested. Do other theories predict the outcome observed, or is the prediction unique to the tested theory? The predictions tested here seem quite unique. There is no obvious competing explanation for the periodic upsurges and downsurges in European expansionism and warfare outlined above. Offense-defense theory has the field to itself. Particular domestic explanations have been offered to explain the aggressiveness of specific states—for example, some argue that Wilhelmine Germany was aggressive because it was a late industrializer, that revolutionary France was aggressive because its regime came to power through mass revolution, and so forth[69]—but no competing theory claims to explain the general cross-time and cross-state pattern of war involvement that we observe. Hence this test seems strong.

What importance does this evidence assign to offense-defense theory? That is, how potent is offense dominance as a cause of war? In Europe since 1789, the nature of international relations has gyrated sharply with shifts in the

67. On Prussia's expansionism, see Blanning, *Origins of the French Revolutionary Wars,* pp. 72–82; on French expansionism, see ibid., passim.
68. A summary of Wilhelmine German aims and policies is Geiss, *German Foreign Policy.*
69. On Germany as late industrializer, see Snyder, *Myths of Empire,* pp. 66–111; and on France as a revolutionary state, see Walt, *Revolution and War,* pp. 46–128.

perceived offense-defense balance. War is far more common when elites believe that the offense dominates, and states are far more belligerent when they perceive large defensive vulnerabilities and offensive opportunities for themselves. This indicates that perceptions of the offense-defense balance have a large impact on international relations. Offense-defense theory is important as well as valid.

How much history does this evidence suggest that offense-defense theory can explain? Explanatory power is partly a function of the prevalence of the theory's cause: abundant causes explain more history than scarce causes. In Europe since 1789 the offense has seldom been really strong, but it was believed strong quite often—often enough to cause considerable trouble.

TEST 2: ANCIENT CHINA

The ancient Chinese multistate system witnessed a long-term shift from defense dominance to offense dominance across the years 722–221 BCE.[70] Offense-defense theory predicts that warfare should have increased as this transformation unfolded (see prime prediction 1). This prediction is fulfilled: diplomacy grew markedly more savage and international relations grew markedly more violent as the power of the offense increased.

Before roughly 550 BCE the defense held the upper hand among China's many feudal states. Four related changes then strengthened the offense: feudalism declined,[71] mass infantry replaced chariots as the critical military force, conscription was introduced, and armies grew tremendously in size.[72] The two largest Chinese states deployed enormous armies of more than a million men, and some smaller states had armies numbering in the hundreds of thousands.[73] As armies grew, border forts had less stopping power against infantry because invaders could sweep past, leaving a smaller portion of their force behind to besiege the forts. Forts also lost stopping power as improved siege-engines appeared—battering rams, catapults, and rolling towers—that further eased the conquest of fortified positions.[74] The decline of feudalism eased offensive operations by reducing social stratification, which increased troop loyalty to

70. Concurring is Andreski, *Military Organization and Society*, p. 76.
71. Noting the decline of feudalism are Samuel B. Griffiths, "Introduction," in Sun Tzu, *The Art of War* (London: Oxford University Press, 1971), p. 33; and Dun J. Li, *The Ageless Chinese: A History*, 3d ed. (New York: Charles Scribner's Sons, 1978), p. 64.
72. On the growth of armies, the introduction of conscription, and the rise of infantry, see Li, *Ageless Chinese*, p. 56; Griffiths, "Introduction," pp. 28, 33; and Wolfram Eberhard, *A History of China* (Berkeley: University of California Press, 1977), p. 49.
73. Li, *Ageless Chinese*, p. 56.
74. Andreski, *Military Organization and Society*, p. 76.

regimes; this meant troops could be trusted to conduct long-distance offensive operations without deserting.

The outcomes of battles and wars reveal the shift toward the offense that these technical and social changes produced. The number of independent Chinese states declined from two hundred in the eighth century BCE to seven in the late fifth century, to one in the late third century—a clear measure of the growing power of the offense.[75] Before 550 BCE defenders were often victorious. Thus the states of Tsin and Ch'i fought three great battles, in 632, 598, and 567 BCE, each won by the defender. Dun J. Li concludes, "If the three battles indicate anything, they meant that neither side was able to challenge successfully the other's leadership in its own sphere of influence."[76] In contrast, the state of Ch'in conquered all of China in a rapid campaign lasting only nine years at the end of the Warring States period (230–221 BCE).[77]

This increase in the power of the offense coincides with a stark deterioration in international relations. During the Spring and Autumn period (722–453 BCE) interstate relations were fairly peaceful, and wars were limited by a code of conduct. The code confined warfare to certain seasons of the year and forbade killing enemy wounded. It was considered wrong to stoop to deceit, to take unfair advantage of adversaries, to "ambush armies," or to "massacre cities."[78] The subsequent Warring States period (453–221 BCE) was perhaps the bloodiest era in Chinese history. Warfare raged almost constantly,[79] becoming a "fundamental occupation" of states.[80] Restraints on warfare were abandoned. Casualties ran into hundreds of thousands, and prisoners of war were massacred en masse.[81] Diplomatic conduct deteriorated; one historian writes that "diplomacy was based on bribery, fraud, and deceit."[82]

In short, the shift toward offense dominance in China during 722–221 BCE correlates tightly with a dramatic breakdown of China's international order.

TEST 3: UNITED STATES 1789–1990S

Since 1815 the United States has been by far the most secure of the world's great powers, blessed with two vast ocean moats, no nearby great powers, and

75. Li, *Ageless Chinese*, pp. 50, 59.
76. Ibid., p. 52.
77. Ibid., p. 59.
78. Griffiths, "Introduction," p. 30.
79. Ibid., p. 21.
80. Ibid., p. 24, quoting Shang Yang, Prime Minister of Ch'in, who conceived war and agriculture to be the two fundamental occupations.
81. Li, *Ageless Chinese*, pp. 56, 58–59.
82. Griffiths, "Introduction," p. 24.

(after 1890) the world's largest economy. In the nineteenth century the United States also had substantial offensive opportunities, embodied in chances for continental and then Pacific expansion against weak defenders. However, America's security endowments were quite extraordinary, while its offensive opportunities were more ordinary. Offense-defense theory predicts that such a state will exhibit perhaps average offensive opportunism but markedly less defensive belligerence than other states. Hence, on net, it will start fewer wars and be involved in fewer wars than others (see prime prediction 2).

This forecast is confirmed, although not dramatically, by the pattern of past U.S. foreign policy. The United States has fought other great powers only three times in its two hundred–year history—in 1812, 1917, and 1941—a low count for a great power.[83] The 1812 war stemmed mainly from U.S. belligerence, but the wars of 1917 and 1941 resulted mainly from others' belligerence. The United States did start some of its lesser wars (1846 and 1898), but it joined other wars more reactively (Korea and Vietnam).

Offense-defense theory also predicts that while the United States will pursue some opportunistic expansionism (intervening phenomenon A), it will embrace few policies that embody offense-defense theory's other intervening phenomena (B–J) (explanatory prediction 2). Where the record allows judgments, this forecast is borne out. Regarding expansionism, the United States has confined itself largely to opportunistic imperialism against frail opponents. Defensive expansionism has been muted, and overall, expansionist ideas have held less sway in the United States than in other powers. This is reflected in the relatively small size of the U.S. empire. The modern American empire has been limited to a few formal colonies seized from Spain in the 1890s and an informal empire in the Caribbean/Central American area, with only intermittent control exerted more widely—a zone far smaller than the vast empires of the European powers.

The U.S. impulse to engage in preemptive and preventive war has been small. In sharp contrast to Germany and Japan, the United States has launched a stealthy first strike on another major power just once (in 1812) and has jumped through only one window of opportunity (in 1812). Surprise first strikes and window-jumping were considered on other occasions (e.g., preventive war was discussed during 1949–54, and surprise attack on Cuba was considered during the Cuban missile crisis), but seldom seriously.

83. Britain, France, Russia, and Prussia/Germany fought other great powers an average of five times over the same two hundred years, by my count. None fought as few as three times.

American diplomacy has been strikingly free of fait accompli tactics. American foreign and security policy has generally been less secretive than those of the European continental powers, especially during the late Cold War, when the United States published military data that most powers would highly classify as state secrets. The U.S. arms raced with the Soviet Union energetically during the Cold War, but earlier maintained very small standing military forces—far smaller than those of other great powers. Overall, intervening phenomena B–J of offense-defense theory are strikingly absent in the U.S. case.

In sum, the United States has not been a shrinking violet, but it has been less bellicose than the average great power. Compare, for example, U.S. conduct with the far greater imperial aggressions of Athens, Rome, Carthage, Spain, Prussia/Germany, Japan, Russia, and France.

Offense-defense theory further predicts that levels of American bellicosity should vary inversely with shifts over time in America's sense of security and directly with the scope of perceived external threats (see prime prediction 3)—as in fact they have.

During 1789–1815 the United States saw large foreign threats on its borders and large opportunities to dispel them with force. It responded with a bellicose foreign policy that produced the 1812 war with Britain.

During 1815–1914 the United States was protected from the threat of a Eurasian continental hegemon by Britain's active continental balancing, and protected from extracontinental European expansion into the Western hemisphere by the British fleet, which was the de facto enforcer of the Monroe Doctrine. The United States responded by withdrawing from European affairs and maintaining very small standing military forces, although it did pursue continental expansion before 1898 and limited overseas imperial expansion after 1898.

During 1914–91 Britain could no longer maintain the European balance. This deprived the United States of its shield against continental European aggressors. Then followed the great era of American activism—fitful at first (1917–47), then steady and persistent (1947–91). This era ended when the Soviet threat suddenly vanished during 1989–91. After 1991 the United States maintained its security alliances, but reduced its troops stationed overseas and sharply reduced its defense effort.

WHAT THESE TESTS INDICATE

Offense-defense theory passed the tests these three cases pose. Are these tests positive proof for the theory or mere straws in the wind?

We learn more from strong tests than from weak ones. The strength of a passed test is a function of the uniqueness of the predictions that the test corroborated. The more numerous and plausible are contending explanations for the patterns that the test theory predicted and the test revealed, the weaker the test.

The three case study tests reported here range from fairly weak to quite strong. They each lack Herculean power but in combination they pose a strong test. The test posed by the ancient China case is weak because our knowledge of ancient Chinese society and politics is fairly thin. This leaves us unable to rule out competing explanations for the rise of warfare in the Warring States period that point to causes other than the rise of offense. The test posed by the U.S. case is a little stronger but still rather weak overall. Alternative explanations for the rise and fall of American global activism are hard to come up with, leaving the offense-defense theory's explanation without strong competitors, so this element of the test posed by the U.S. case is fairly strong. Plausible contending explanations for other aspects of the U.S. case can be found, however. For example, some would argue that America's more pacific conduct is better explained by its democratic domestic structure than by its surfeit of security. Others would contend that the United States has fewer-than-average conflicts of interest with other powers because it shares no borders with them, and it fights fewer wars for this reason. Hence this element of the test posed by the U.S. case is weak: U.S. lower-than-average bellicosity is only a straw in the wind.

As noted above, the case of Europe since 1789 offers a fairly strong test. Some competing explanations for Germany's greater bellicosity are offered—as noted above, the lateness of German industrialization is sometimes suggested as an alternative cause, as is German culture. However, there is no obvious plausible competing explanation for the main pattern we observe in the case—the rise of warfare during 1792–1815, 1856–71, and 1914–45, and the greater periods of peace in between. The fit of this pattern with prime prediction 1 of offense-defense theory lends it strong corroboration.

WHAT PRESCRIPTIONS FOLLOW?

If offense dominance is dangerous, policies that control it should be pursued. Governments should adopt defensive military force postures and seek arms control agreements to limit offensive forces. Governments should also maintain defensive alliances. American security guarantees in Europe and Asia have made conquest much harder since 1949 and have played a major role in preserving peace. A U.S. withdrawal from either region would raise the risk of conflict.

Conclusion: Offense-Defense Theory in Perspective

Offense-defense theory has the attributes of a good theory. First, it has three elements that give a theory claim to large explanatory power. (1) Large importance, that is, its posited cause has large effects. Variance in the perceived offense-defense balance causes large variance in the incidence of warfare. Variance in the actual offense-defense balance has less impact because policymakers often misperceive it, but it has a potent effect when policymakers perceive it accurately. (2) Wide explanatory range. The theory explains results across many domains of behavior—in military policy, foreign policy, and crisis diplomacy.[84] It governs many intervening phenomena (e.g., expansionism, first-move advantage, windows, secrecy, negotiation failures, crisis management blunders, arms races, tight alliances) that have been seen as important war causes in their own right. Thus offense-defense theory achieves simplicity, binding a number of war causes under a single rubric. Many causes are reduced to one cause with many effects. (3) Wide real-world applicability. Real offense dominance is rare in modern times, but the perception of offense dominance is fairly widespread. Therefore, if perceived offense dominance causes war it causes lots of war, and offense-defense theory explains much of international history.

Second, offense-defense theory has large prescriptive utility, because the offense-defense balance is affected by national foreign and military policy; hence it is subject to political will. Perceptions of the offense-defense balance are even more malleable, being subject to correction through argument. Both are far more manipulable than the polarity of the international system, the strength of international institutions, the state of human nature, or other war causes that have drawn close attention.

Third, offense-defense theory is quite satisfying, although it leaves important questions unanswered. In uncovering the roots of its ten intervening phenom-

84. Moreover, offense-defense theory might be usefully adapted for application beyond the domain of war, for example, to explain international economic competition (or cooperation), or even intra-academic competition. Suggesting its application to economics is Jitsuo Tsuchiyama, who writes of the "prosperity dilemma"—a cousin of the security dilemma in which measures taken by one state to increase its economic well-being decrease another's economic well-being. See Jitsuo Tsuchiyama, "The U.S.-Japan Alliance after the Cold War: End of the Alliance?" unpublished manuscript, Olin Institute, Harvard University, 1994, p. 27.

ena, offense-defense theory offers a more satisfying (and simpler) explanation than do interpretations pointing directly to these phenomena. However, it also raises another mystery: Why is the strength of the offense so often exaggerated?

History suggests that offense dominance is at the same time dangerous, quite rare, and widely overstated. It further suggests that this exaggeration of insecurity, and the bellicose conduct it fosters, are prime causes of national insecurity and war. States are seldom as insecure as they think they are. Moreover, if they are insecure, this insecurity often grows from their own efforts to escape imagined insecurity.

The rarity of real insecurity is suggested by the low death rate of modern great powers. In ancient times great powers often disappeared, but in modern times (since 1789) no great powers have permanently lost sovereignty, and only twice (France in 1870–71 and in 1940) has any been even temporarily overrun by an unprovoked aggressor.[85] Both times France soon regained its sovereignty through the intervention of outside powers—illustrating the powerful defensive influence of great-power balancing behavior.

The prevalence of exaggerations of insecurity is revealed by the great wartime endurance of many states that enter wars for security reasons, and by the aftermath of the world's great security wars, which often reveal that the belligerents' security fears were illusory. Athens fought Sparta largely for security reasons, but held out for a full nine years (413–404 BCE) after suffering the crushing loss of its Sicilian expedition—an achievement that shows the falsehood of its original fears. Austria-Hungary held out for a full four years under allied battering during 1914–18, a display of toughness at odds with its own prewar self-image of imminent collapse. With twenty-twenty hindsight we can now see that modern Germany would have been secure had it only behaved itself. Wilhelmine Germany was Europe's dominant state, with Europe's largest and fastest-growing economy. It faced no plausible threats to its sovereignty except those it created by its own belligerence. Later, interwar Germany and Japan could have secured themselves simply by moderating their conduct. This would have assured them of allies, hence of the raw materials supplies they sought to seize by force. America's aggressive and often costly Cold War interventions in the third world now seem hypervigilant in light of the defensive benefits of the nuclear revolution, America's geographic

85. France helped trigger the 1870 war; hence one could argue for removing France in 1870 from the list of unprovoking victims of conquest, leaving only France in 1940.

invulnerability, and the strength of third world nationalism, which precluded the Soviet third world imperialism that U.S. interventions sought to prevent.

Paradoxically, a chief source of insecurity in Europe since medieval times has been this false belief that security was scarce. This belief was a self-fulfilling prophecy, fostering bellicose policies that left all states less secure. Modern great powers have been overrun by unprovoked aggressors only twice, but they have been overrun by provoked aggressors six times—usually by aggressors provoked by the victim's fantasy-driven defensive bellicosity. Wilhelmine and Nazi Germany, Imperial Japan, Napoleonic France, and Austria-Hungary were all destroyed by dangers that they created by their efforts to escape from exaggerated or imaginary threats to their safety.[86]

If so, the prime threat to the security of modern great powers is . . . themselves. Their greatest menace lies in their own tendency to exaggerate the dangers they face, and to respond with counterproductive belligerence. The causes of this syndrome pose a large question for students of international relations.

---

86. Mussolini also provoked his own destruction, but his belligerence was not security driven.

What Is the Offense-Defense Balance and Can We Measure It?

Charles L. Glaser and Chaim Kaufmann

Offense-defense theory (or security dilemma theory) is a quite optimistic theory of international politics, since it argues that when defense has the advantage over offense major war can be avoided. In addition, the likelihood of arms races and war can sometimes be further reduced by carefully designed arms control. Over the past two decades the theory has come to play an increasingly important role in both international relations scholarship and the analysis of foreign policy.[1] Scholars have employed the theory to address a wide array of theoretical and policy issues, including alliance behavior, comparative grand strategy, military doctrine, military competition and cooperation, nuclear strategy and policy, and conventional arms control.[2] Offense-defense logic has also been used to

Charles L. Glaser is Associate Professor in the Irving B. Harris Graduate School of Public Policy Studies, University of Chicago. Chaim Kaufmann is Assistant Professor of International Relations at Lehigh University.

We would like to thank Stephen Biddle, James Fearon, Keir Lieber, Sean Lynn-Jones, John Mearsheimer, Bruce Moon, Karl Mueller, Robert Pape, Dan Reiter, Norrin Ripsman, Ivan Toft, and participants in the Program in International Political Economy and Security seminar at the University of Chicago for comments on early drafts.

1. Robert Jervis, "Cooperation under the Security Dilemma," *World Politics*, Vol. 30, No. 2 (January 1978), pp. 167–214, is usually credited with launching this body of work. George Quester, *Offense and Defense in the International System* (New York: Wiley, 1977), makes many of the same arguments.
2. On balancing versus bandwagoning, see Stephen M. Walt, *The Origins of Alliances* (Ithaca, N.Y.: Cornell University Press, 1987); on the tightness of alliances, see Thomas J. Christensen and Jack Snyder, "Chain Gangs and Passed Bucks: Predicting Alliance Patterns in Multipolarity," *International Organization*, Vol. 44, No. 1 (Spring 1990), pp. 137–168; and Thomas J. Christensen, "Perceptions and Allies in Europe, 1865–1940," *International Organization*, Vol. 51, No. 1 (Winter 1997), pp. 65–97. On alliance choices of small powers, see Karl Mueller, "Alignment Balancing and Stability in Eastern Europe," *Security Studies*, Vol. 5, No. 1 (Autumn 1995), pp. 38–76. On comparative grand strategy, see Stephen Van Evera, *Causes of War, Volume 1: The Structure of Power and the Roots of War* (Ithaca, N.Y.: Cornell University Press, forthcoming), pp. 191–245 (of the September 1994 draft). On military doctrine, see Barry R. Posen, *The Sources of Military Doctrine* (Ithaca, N.Y.: Cornell University Press, 1984). On competition and cooperation see George W. Downs, David M. Rocke, and Randolph L. Siverson, "Arms Races and Cooperation," in Kenneth A. Oye, ed., *Cooperation under Anarchy* (Princeton, N.J.: Princeton University Press, 1986), pp. 118–146; Charles L. Glaser, "Political Consequences of Military Strategy: Expanding and Refining the Spiral and Deterrence Models," *World Politics*, Vol. 44, No. 4 (July 1992), pp. 497–538; and Robert Powell, "Guns, Butter, and Anarchy," *American Political Science Review*, Vol. 87, No. 1 (March 1993), pp. 115–132. On nuclear strategy, see Shai Feldman, *Israeli Nuclear Deterrence: A Strategy for the 1980s* (New York: Columbia University Press, 1982); and Charles L. Glaser, *Analyzing Strategic Nuclear Policy* (Princeton, N.J.: Princeton University Press, 1990). Although not cast in offense-

explain the causes of World War I, the causes and possible solutions of ethnic and civil wars, and the foreign policies of revolutionary states; to criticize U.S. grand strategy; and to predict the future of political relations in post–Cold War Europe as well as the size and number of independent states in the international system.[3]

Despite the theory's status as a growth industry, critics continue to question its utility. First, they argue—correctly—that the foundations of the theory are underdeveloped, holding, most important, that we lack an agreed definition of the theory's key independent variable, the offense-defense balance, which results in inconsistent application and testing of the theory.[4] Second, and more important, they contend that the theory contains inherent flaws, the most serious of which is that the offense-defense balance cannot be measured because the outcomes of wars are so uncertain.[5]

---

defense terms, offense-defense differentiation plays a central role in qualitative arms control; see Thomas C. Schelling and Morton H. Halperin, *Strategy and Arms Control* (New York: Twentieth Century Fund, 1961). The 1972 ABM treaty and certain provisions of the START treaties were justified on offense-defense logic. On conventional arms control see Jack Snyder, "Limiting Offensive Conventional Forces: Soviet Proposals and Western Options," *International Security*, Vol. 12, No. 4 (Spring 1988), pp. 48–77; Ivan Oelrich, *Conventional Arms Control: The Limits and Their Verification* (Cambridge, Mass.: Center for Science and International Affairs, Harvard University, 1990); and Anders Boserup and Robert Neild, eds., *The Foundations of Defensive Defence* (New York: St. Martin's, 1990).

3. On World War I, see Jack L. Snyder, *The Ideology of the Offensive: Military Decision Making and the Disasters of 1914* (Ithaca, N.Y.: Cornell University Press, 1984); and Stephen Van Evera, "The Cult of the Offensive and the Origins of the First World War," *International Security*, Vol. 9, No. 1 (Summer 1984), pp. 58–107. On ethnic conflict see Barry R. Posen, "The Security Dilemma in Ethnic Conflict," in Michael E. Brown, ed., *Ethnic Conflict and International Security* (Princeton, N.J.: Princeton University Press, 1993), pp. 103–124; and Chaim Kaufmann, "Possible and Impossible Solutions to Ethnic Civil Wars," *International Security*, Vol. 20, No. 4 (Spring 1996), pp. 136–175. On revolutionary states, see Stephen M. Walt, *Revolution and War* (Ithaca, N.Y.: Cornell University Press, 1996). On U.S. grand strategy, see Stephen M. Walt, "The Case for Finite Containment: Analyzing U.S. Grand Strategy," *International Security*, Vol. 14, No. 1 (Summer 1989), pp. 5–49; and Barry R. Posen, *Inadvertent Escalation: Conventional War and Nuclear Risks* (Ithaca, N.Y.: Cornell University Press, 1991). On the future of Europe, see Stephen Van Evera, "Primed for Peace: Europe after the Cold War," *International Security*, Vol. 15, No. 3 (Winter 1990/91), pp. 7–57. On the number of states, see Stanislav Andreski, *Military Organization and Society* (London: Routledge and Keegan Paul, 1968), pp. 75–76; and Robert Gilpin, *War and Change in World Politics* (Cambridge, U.K.: Cambridge University Press, 1981), p. 61.

4. Jack S. Levy, "The Offensive/Defensive Balance of Military Technology: A Theoretical and Historical Analysis," *International Studies Quarterly*, Vol. 38, No. 2 (June 1984), pp. 219–238.

5. Other criticisms include that ingenuity, not structural constraints, determine the balance; that state behavior is determined by perceptions, not the "objective" offense-defense balance; and that offense and defense cannot be distinguished. Many of these criticisms are addressed by Sean M. Lynn-Jones, "Offense-Defense Theory and Its Critics," *Security Studies*, Vol. 4, No. 4 (Summer 1995), pp. 660–691.

If the critics are right, a growing and influential body of international relations theory literature must be heavily discounted. Moreover, important policy implications would flow from the critics' insights. For example, the potential of arms control to maintain peace would be significantly lower and the probability of future major-power wars in Europe and Asia could be far greater than offense-defense theorists suggest.[6]

This article responds to the critics by providing needed development of the theory's foundations and by showing that the claim that the offense-defense balance cannot be measured is simply incorrect. First, we argue that the offense-defense balance should be defined as the ratio of the cost of the forces that the attacker requires to take territory to the cost of the defender's forces. This definition of the balance is especially useful because the offense-defense balance then provides an essential link between a state's power and its military capability, that is, its ability to perform military missions.

Next, we explain six key assumptions and specifications that are required to operationalize any definition of the offense-defense balance, whether stated in terms of cost ratios or not. These are required to ensure that the balance is well defined and therefore measurable. Perhaps most important, we explain that the offense-defense balance should be assessed assuming optimality—that is, countries choose the best possible strategies and force postures for attack and defense. Offense-defense theory requires this assumption because it focuses on the effects of the constraints and opportunities presented by the international environment. As a result, states' decisions do not influence the offense-defense balance. Although poor choices about military doctrine or force posture will influence a state's military capabilities, this shortfall reflects the state's lack of military skill, not a change in the offense-defense balance.

Third, we argue that the basic logic of offense-defense theory requires what we term a "broad approach" to measuring the offense-defense balance. Some analysts favor a narrow approach in which military technology and geography are the only factors that influence the balance. However, once we define the offense-defense balance as the cost-ratio of offense to defense, all factors that could significantly shift this ratio—including such variables as the size of forces, the cumulatively of resources, and nationalism—should be included.

---

6. For example, Van Evera's optimistic assessment of the future of Europe in his "Primed for Peace" relies on the judgment that the offense-defense balance greatly favors defense.

We do, however, explain why two possible candidates—the nature of international alliance behavior and first-move advantages—should be excluded.

Fourth, we explore whether states can measure the offense-defense balance and find strong grounds for concluding that they can. The analytic tasks required to measure the offense-defense balance are essentially the same as those required to perform military net assessments. Most analysts believe that, within reasonable bounds, net assessment has been a feasible task and, therefore, should also believe that the offense-defense balance can be measured. We offer anecdotal evidence that supports our optimism, but in-depth empirical studies of how well states can conduct net assessments are required to resolve this question.

This article proceeds as follows. The first section provides a brief overview of offense-defense theory. The second section explains why we prefer the ratio definition and the assumptions required for operationalizing any definition of the offense-defense balance. The third section explores factors that influence the offense-defense balance, and the fourth presents our assessment of the challenge involved in measuring it. The final section recommends directions for further research.

What Is Offense-Defense Theory?

Before proceeding we briefly summarize the propositions offered by offense-defense theory. The purpose of this article is not primarily to extend or modify this already large body of propositions. Instead, our contributions should provide confidence that these propositions are meaningful—that they are built on concepts that are well defined and that can be productively tested and applied to real-world security problems.

As originally described by Robert Jervis, the two key variables in the theory are (1) the offense-defense balance—whether it is "easier" to take territory or to defend it, and (2) offense-defense distinguishability—whether the forces that support offensive missions are different from those that support defensive missions. The basic predictions concerning the offense-defense balance are that as the advantage of offense increases, the security dilemma becomes more severe, arms races become more intense, and war becomes more likely.[7] When offense has the advantage, it is impossible for states of equal size to enjoy high

7. Jervis, "Cooperation under the Security Dilemma," pp. 187–194, 199–206.

levels of security simultaneously; arms races will be intense because when one country adds forces its adversary will have to make a larger addition to restore its ability to defend.[8] Offense advantage makes war more likely for a variety of reasons: war will be quick and decisive and therefore profitable, so greedy states will find war more attractive; states will be more insecure, making expansion more valuable, so security-seeking states will find war more attractive; and the advantage of striking first grows with offense advantage, which increases the probability of crises escalating via preemptive attacks and accidents.[9]

Stephen Van Evera has added hypotheses on how offense advantage fuels preventive war and encourages styles of diplomacy that increase the probability of war.[10] Offense advantage makes shifts in power more significant, which increases incentives for preventive war. When offense has the advantage states negotiate less and use fait accompli tactics more, and states become more secretive, which increases the probability of war by fueling miscalculations of both military capabilities and interests.[11]

We envision offense-defense theory as a partial theory of military capabilities, that is, of a state's ability to perform the military missions that are required to successfully attack, deter, and defend. A more complete theory would include two additional variables: (1) power, measured in terms of relative resources; and (2) what we term "military skill," that is, a country's ability to effectively employ military technology, including designing military strategy and assessing adversaries' forces and strategy.[12] Offense-defense theory does

8. For related analysis, see Malcolm W. Hoag, "On Stability in Deterrent Races," *World Politics*, Vol. 13, No. 4 (July 1961), pp. 505–527.

9. On the relationship between first strike-advantages, preemption, and accidents, see Schelling and Halperin, *Strategy and Arms Control*, pp. 14–16.

10. Van Evera, *Causes of War*, pp. 191–245.

11. The deductive strength of this body of offense-defense hypotheses has gone largely unchallenged, but recent work has questioned the relationship between offense advantage and the frequency of war. Because the risks of war could be greater for the attacker (not only the defender) when offense has the advantage, potential attackers should face countervailing pressures that make them more cautious, especially when considering large wars. James D. Fearon, "Rationalist Explanations for War," *International Organization*, Vol. 49, No. 3 (Summer 1995), pp. 402–403.

12. A variety of literatures bear on military skill, including work on states' evaluative capabilities, such as Stephen Van Evera, "Why States Believe Foolish Ideas: Non-Self-Evaluation by Government and Society" (Paper presented at the annual meeting of the American Political Science Association, Washington, D.C., 1988); organization theory, for example, Snyder, *Ideology of the Offensive;* Posen, *Sources of Military Doctrine;* and culture, including Trevor N. Dupuy, *A Genius for War: The German Army and General Staff, 1807–1945* (Englewood Cliffs, N.J.: Prentice-Hall, 1977); Elizabeth Kier, *Imagining War: French and British Military Doctrines between the Wars* (Princeton, N.J.: Princeton University Press, 1997); and Stephen P. Rosen, "Military Effectiveness: Why Society Matters," *International Security*, Vol. 19, No. 4 (Spring 1995), pp. 5–31.

not claim that the offense-defense balance is in general a more important determinant of military capabilities than is power or skill. Rather, each of the three variables has the potential to overwhelm the others in certain circumstances.

Offense-defense variables play a central role in recent work on structural realism—for example, defensive and contingent realism.[13] Unlike Kenneth Waltz's version of structural realism,[14] which focuses on power or the distribution of resources, these versions of realism focus on states' abilities to perform necessary military missions. Consequently, these theories need to introduce a variable that reflects a state's ability to convert power into military capabilities. This variable is the offense-defense balance.

These alternative versions of structural realism produce a number of explanations and predictions that diverge significantly from standard power-based structural realist analyses. For example, balance-of-power theory is indeterminate about the tightness of alliances, whereas defensive realism predicts that alliances will be tight when offense has the advantage, but loose when defense has the advantage. In addition, neorealist balance-of-power theory is pessimistic about the prospects for peace in a multipolar world, but contingent realism is optimistic if technological or other conditions strongly favor defense over offense, as can be the case in a world of strategic nuclear weapons.

Defining the Offense-Defense Balance

To measure the offense-defense balance, we must first define it. Adequately developing a definition is more complex than may be apparent, requiring that we spell out a number of assumptions and specifications that are rarely, if ever, recognized. This exercise is essential, because a variable that is not well defined cannot be measured reliably. In this section we make the case for defining the offense-defense balance as the cost ratio of attacker forces to defender forces. We then explore the additional requirements that are a necessary part of any definition of the balance.

13. See Charles L. Glaser, "Realists as Optimists: Cooperation as Self-Help," *International Security,* Vol. 19, No. 3 (Winter 1994/95), pp. 50–90; Sean M. Lynn-Jones, "Rivalry and Rapprochement: Accommodation between Adversaries in International Politics" (Ph.D. dissertation, Harvard University, in progress); Jack Snyder, *Myths of Empire* (Ithaca, N.Y.: Cornell University Press, 1991), pp. 11–12; and Van Evera, *Causes of War.* For similar efforts that combine perceptions of the offense-defense balance and power, see Christensen and Snyder, "Chain Gangs and Passed Bucks," and Christensen, "Perceptions and Allies in Europe, 1865–1940."
14. Kenneth N. Waltz, *Theory of International Politics* (Reading, Mass.: Addison-Wesley, 1979).

THE COST (INVESTMENT) RATIO OF OFFENSE TO DEFENSE

The offense-defense literature includes many different definitions of the offense-defense balance. Among these are that offense has the advantage (1) when "it is easier to destroy the other's army and take its territory than it is to defend one's own";[15] (2) when the defender has to outspend the attacker to offset an investment in offensive forces;[16] (3) when the costs of capturing territory are less than the value of the territory itself;[17] (4) when there is an incentive to strike first rather than to absorb the other's first strike;[18] (5) when a large portion of states' territory is likely to change hands as a result of war;[19] and (6) when weapons possess certain characteristics, for example, long range and especially mobility.[20]

We prefer to define the offense-defense balance as the ratio of the cost of the forces the attacker requires to take territory to the cost of the forces the defender has deployed.[21] That is, if the defender invests X in military assets, how large an investment Y must the attacker make to acquire the forces necessary for taking territory?[22] The offense-defense balance is the ratio Y/X.[23] Larger ratios indicate a balance more in favor of defense.

---

15. Jervis, "Cooperation under the Security Dilemma," p. 178.
16. Ibid.; and Gilpin, *War and Change in World Politics,* p. 62; see also Quester, *Offense and Defense in the International System,* for a definition cast in terms of force ratios instead of investment.
17. Gilpin, *War and Change in World Politics,* p. 63.
18. Jervis, "Cooperation under the Security Dilemma," p. 178. Fearon, "The Offense-Defense Balance and War since 1648," paper prepared for the Annual Meeting of the International Studies Association, Chicago, 1995, defines the balance in terms of the expected costs and benefits of attacking versus defending.
19. Stephen W. Van Evera, "Causes of War" (Ph.D. dissertation, University of California, Berkeley, 1984), p. 78.
20. Levy, "The Offense/Defense Balance of Military Technology," pp. 222–230, reviews works that use this definition as well as the ones above.
21. We do not claim that this is the only workable definition of the offense-defense balance. Instead, we explain the advantages of the cost-ratio definition and develop some of its important, less apparent features.
22. Although in principle "investment" is a straightforward concept, in practice it raises some complicated issues. For instance, what countries pay for resources may not reflect the value they place on them. An obvious example is when two countries pay different amounts to field soldiers of comparable quality, as can happen when one country has a draft system and the other has a volunteer army. However, although these systems allow countries to pay different amounts for soldiers, we may prefer to say that adding an equal number of equal-quality soldiers represents an equal investment.
23. The offense-defense balance should be calculated in terms of the countries' peacetime force requirements: by what ratio must the attacker outspend the defender before the start of the war, taking into account likely wartime mobilization by both sides? This choice is best for two reasons: first, the main questions addressed by the security dilemma focus on judgments made about peacetime national security policy choices, such as: how secure can my state be? and, how intensely

We arrive at our choice by focusing on the theoretical role that the balance needs to play. As we reviewed above, the offense-defense balance is one of two key variables, along with power, that determines states' abilities to perform military missions.[24] When we analyze whether a state can protect itself against potential adversaries, we need to know not only the relative resources (wealth, population, etc.) of the state and its adversaries, but also how effectively these resources can be used to produce offensive and defensive military capabilities.

Given this ratio definition, the attacker's power (i.e., the ratio of the attacker's resources to the defender's resources) divided by the offense-defense balance indicates the attacker's prospects for successful offense. All else being equal, the larger this quotient, the greater the attacker's prospects for success.

Defined as the cost ratio, the offense-defense balance plays the proper role, in combination with power (and skill), in determining a state's potential military capability and therefore its ability to maintain its security, as well as attain other nonsecurity goals. For example, the balance can sometimes overcome disparities in states' resources. When defense has a large advantage, even a state that is much smaller than its adversaries may still be able to afford effective defense. Conversely, power imbalances can sometimes overwhelm the offense-defense balance. Even if defense has a large advantage, a much wealthier attacker might still be able to outspend a defender by a sufficient margin to gain an effective offensive capability.

SPECIFICATIONS AND ASSUMPTIONS

Any definition of the offense-defense balance, whether or not based on cost ratios, requires a number of additional specifications and assumptions without which the balance is not well defined and cannot be measured. This section explains the requirements.[25]

COST OF FIGHTING. The attacker-to-defender cost ratio is not well defined until we specify the costs of fighting that the attacker would incur. This is because the cost of forces required to take territory varies with the costs of

---

do I need to arms race? Second, if defenders have an advantage in wartime extraction, for example, because of nationalism, incorporating this wartime spending would have the perverse effect of making the balance appear to shift toward offense when conquest was actually becoming more difficult.

24. As we have suggested, a more complete theory would include a third key variable—military skill. Structural theories do not focus on this variable, assuming that states have high and roughly equal levels of skill.

25. To be precise, the first specification below is required for some but not all definitions of the balance; the others are required for all definitions.

fighting the attacker would incur: all else being equal, more capable—and expensive—forces can usually take territory at lower costs of fighting than can less capable forces.[26]

Consequently, we need to set a cost of fighting at which the offense-defense balance will be defined. A useful way is to employ the conservative defense planning standards that are traditionally employed by defenders, which assume that other states may have very ambitious expansionary objectives and may be deterred only by the prospect of suffering extremely high losses in a war for such objectives.[27] With the cost of fighting set this way, the offense-defense balance will be the minimum investment ratio at which the attacker can not only take territory but can do so at an acceptable cost of fighting.

Operationalizing "extremely high" requires making subjective, contentious choices, but standards that are frequently used in analyzing security policy provide reasonable guidelines.[28] In the realm of modern conventional war, a typical standard is that virtually all states will value territory less than the costs of fighting a war of attrition, but that a determined expansionist might be willing to accept the costs of a successful blitzkrieg.[29] Importing this assumption into offense-defense theory means that we calculate the offense-defense balance by comparing the cost of forces the attacker requires to launch a successful blitzkrieg to the cost of the defender's forces.

In the nuclear realm, we focus on comparisons of the attacker's value for territory to the costs that the attacker would incur as a result of nuclear retaliation against its society. During the Cold War, an assured destruction capability was the standard most commonly used in gauging the adequacy of

---

26. The same observation applies to the probability of taking territory: forces that are expected to succeed in taking territory with a higher probability will be more expensive.
27. Even the theories that we think of as most purely structural must make assumptions about actors' values. Waltz, *Theory of International Politics*, pp. 105–106, for example, assumes that states value security far more than expansion. Randall Schweller, "Neorealism's Status Quo Bias: What Security Dilemma?" *Security Studies*, Vol. 5, No. 3 (Spring 1996), pp. 90–121, argues that changing this assumption would lead to quite different predictions.
28. Because the level at which the cost of taking territory is set can dramatically influence the offense-defense balance, it is essential that users of offense-defense theory be explicit about this choice.
29. John J. Mearsheimer, *Conventional Deterrence* (Ithaca, N.Y.: Cornell University Press, 1983), applies this standard to large-scale offensive campaigns that aim to take a large part or all of the defender's territory; willingness to accept costs for limited gains would presumably be lower. Consensus on this is not complete. For example, some participants in the Cold War debate over the conventional balance in Europe argued that Soviet willingness to pay costs for expansion might be higher than Mearsheimer's standard. See Samuel P. Huntington, "Conventional Deterrence and Conventional Retaliation in Europe," *International Security*, Vol. 8, No. 3 (Winter 1983/84), pp. 32–56; and Eliot A. Cohen, "Toward Better Net Assessment: Rethinking the European Conventional Balance," *International Security*, Vol. 13, No. 1 (Summer 1988), pp. 50–89.

U.S. forces.[30] Given this standard, taking territory at an acceptable cost of fighting translates into the ability to eliminate the defender's assured destruction capability. The offense-defense balance would be the ratio of the cost of forces required to undermine the defender's assured destruction capability to the cost of the defender's forces. If we chose a lower level of retaliatory damage as our standard, the attacker's forces would have to be more effective, shifting the offense-defense balance further toward defense advantage.[31]

ATTACKER'S TERRITORIAL GOAL. The offense-defense balance depends on how much territory the attacker is trying to take. More ambitious offensive missions, those that are designed to take more territory, tend to be more difficult than less ambitious ones. Facing a given defensive force, the offensive force required for a more ambitious mission will have to be larger, more technologically advanced, or both, than would be required for a less ambitious mission. Consequently, the cost ratio of offense to defense increases with the ambition of the offensive mission, which shifts the offense-defense balance toward defense advantage. Therefore states will often face different offense-defense balances for different territorial goals, which may influence their behavior.[32]

Two examples illustrate the mission dependence of the offense-defense balance. The first compares the conventional forces required for gaining limited amounts of territory and for unlimited-aims offensives.[33] Limited aims are usually easier to achieve. Among the reasons are that the greater the distance the attacking forces must advance, the longer their supply lines and the shorter the defender's; and that the larger the area of enemy territory the attacker must occupy, the greater the cost of occupation. Thus the balance for a limited aims offensive often is more favorable to offense than would be the balance for an unlimited goal.[34]

30. Many would consider this a highly conservative standard, although some argued that the United States needed to be able to destroy Soviet forces and leadership, as well as Soviet society. Reviewing this debate is Charles L. Glaser, "Why Do Strategists Disagree about the Requirements of Deterrence?" in Lynn Eden and Steven E. Miller, eds., *Nuclear Arguments* (Ithaca, N.Y.: Cornell University Press, 1989).

31. The preceding discussion makes clear why, although the balance is defined in terms of the relative cost of forces, the cost of fighting a war nevertheless influences the balance. If the costs of fighting increase, the balance shifts toward defense because the attacker must pay more to acquire forces that enable it to take territory at the specified cost of fighting.

32. This implies that the offense-defense theory can be useful in predicting not only the likelihood of war, but also the scale of likely wars.

33. On limited aims strategies, see Mearsheimer, *Conventional Deterrence*, pp. 53–56.

34. For a model of some of the issues involved, see Stephen D. Biddle, "The Determinants of Offensiveness and Defensiveness in Conventional Land Warfare" (Ph.D. dissertation, Harvard University, 1992). Limited offensives are not always easier than unlimited ones. Early conquests

The second example compares the nuclear forces required for an attacker to successfully challenge a defending state's minor interests to those required to successfully challenge its vital interests. When only minor interests are at stake, the balance of interests, and therefore the balance of resolve, may not favor the defender. Thus an attacker's nuclear threat could be effective even if the defender's nuclear retaliation would inflict equal or greater damage on the attacker, so the attacker may not need nuclear forces more capable or more expensive than the defender's. In contrast, a nuclear attacker challenging a defender's homeland would almost certainly face an unfavorable balance of resolve, and therefore would require a nuclear counterforce capability that could ensure that it would suffer far less damage in a nuclear war than would the defender. Attaining such a capability would certainly require the attacker to spend much more than the defender.

WARS, NOT BATTLES. We argue that of the three levels of war recognized by most analysts—strategic, operational, and tactical—the offense-defense balance should be defined at the strategic level.[35] Offense-defense theory addresses states' decisions about whether to go to war based on their judgments about whether war is likely to be successful. Consequently, when we say "take territory" we really mean "take and hold territory against counterattacks," because seizing territory only to lose it thereafter would not seem worthwhile to most attackers.[36] Thus the offense-defense balance should be defined in terms of final war outcomes, not the results of intermediate battles or campaigns.[37]

could promote rather than retard additional gains if the attacker can extract militarily useful benefits that exceed the costs of occupation (see the discussion of cumulative resources below). Also, if the tactical and operational balances favor the offense at least as strongly as the strategic balance, then it may be easier to achieve total victory than to hold early gains against later counterattacks. For instance, given the powerful punch but small size of the Germans' mechanized forces in 1940, it may have been more practical for them to conquer all of France than to try to hold just Belgium and northern France.

35. Strategy is generally understood as decisions concerning the achievement of ultimate war goals, such as mobilization of forces and their allocation to different campaigns or theaters of conflict. Operations concerns the movement of forces within a campaign or theater in order to ensure that battles are fought on favorable terms. Tactics concerns actions taken within a particular battle in order to win that battle. Edward N. Luttwak, *Strategy: The Logic of War and Peace* (Cambridge, Mass.: Harvard University Press, 1987), pp. 69–70. On the difference between strategic and tactical offense-defense balances, see Ted Hopf, "Polarity, the Offense-Defense Balance, and War," *American Political Science Review,* Vol. 85, No. 2 (June 1991), pp. 475–493.

36. Quincy Wright, *A Study of War* (Chicago: University of Chicago Press, 1965), p. 808; and Biddle, "Determinants of Offensiveness and Defensiveness," p. 59.

37. Any answer to the question of how long must the attacker hold its territorial gains to count as offensive success must be somewhat arbitrary. In keeping with the traditional primarily military

This is not to say that tactical and operational analyses are irrelevant to the balance. To gain territory an attacker must conduct offensive tactical battles and offensive operations, and to hold territory once gained may have to engage in tactical and operational defense.[38]

OPTIMALITY. As a structural theory, which attempts to predict states' behavior by focusing on the constraints and opportunities presented by their external environment, offense-defense theory must assume that states act optimally.[39] That is, within reasonable limits of analysis, states make the best possible decisions for attack or defense, taking into account their own and their opponents' options for strategy and force posture. In other words, all relevant countries are assumed to have a high level of military skill.[40] Thus the offense-defense balance is the cost ratio of the attacker's best possible offense to the defender's best possible defense.

Because the offense-defense balance is defined assuming optimality, military doctrines and force deployments cannot influence the balance. Instead, when states act optimally, doctrine and deployments merely reflect the balance; they are outputs of the optimization process, given the constraints imposed by the offense-defense balance and the distribution of resources. Suboptimal choices will influence a state's deployed capabilities but not the offense-defense balance.[41]

---

focus of offense-defense theory, we prefer to consider an offensive a success if the territorial gain is maintained until the end of continuous military efforts to take it back, that is, until the end of the current war. One could make a case for a longer time horizon—for example, until the defender loses interest in retaking the territory—on the grounds that the attacker must expend effort even in peacetime to hold the territory as long as a revanchist campaign is imaginable. For instance, our definition would count Napoleon's victories over Prussia and Austria as offensive successes, although one could argue that he was never secure in the fruits of those campaigns. The problem with such a long time horizon is that it would be extremely difficult to measure, and in some cases might never expire.

38. The need to integrate across levels of warfare raises issues that are addressed on pages 73–74.

39. The meaning of "structure" used here is softer than the one favored by some structural realists, which excludes all unit-level attributes. Rather, we use "structure" to mean that states judge adversaries' goals and therefore the threat they pose based on the information provided by their military and foreign policies, not properties of their domestic systems. We do not, however, preclude states from considering properties of other states or their own state—such as the degree of technological sophistication and nationalism—in assessing military capabilities.

40. Jonathan Shimshoni, "Technology, Military Advantage, and World War I: A Case for Military Entrepreneurship," *International Security*, Vol. 15, No. 3 (Winter 1990/91), pp. 187–215, argues that the offense-defense balance has little practical meaning, because states can design strategy to shift it. Shimshoni's criticism is flawed, however, because he relaxes the assumption of optimality. All of his examples of military entrepreneurship hinge on states having significant advantages in military skill over their opponents.

41. On this point, we disagree with Van Evera, *Causes of War*, pp. 261–273; and Jervis, "Cooperation under the Security Dilemma," p. 212.

This also means that when states do engage in suboptimal behavior, our ability to determine the offense-defense balance by observing military policies and war outcomes is greatly reduced. Because states have chosen the wrong forces and/or doctrines, comparing the states' capabilities for offense and defense sheds little light on the offense-defense balance. For example, early in World War II the Germans had already deployed forces and doctrines that took advantage of improved motor vehicles, portable radios, and other interwar innovations that made the blitzkrieg possible, but the Allies had not. Consequently, to evaluate the impact of these innovations on the offense-defense balance, we should focus not on the evidence from 1939–40, but instead on evidence from 1943–45 when the Allies had also realized the uses of these technological advances and deployed appropriate forces and doctrines on a broadly even footing with the Germans.[42]

Actual state behavior is not always optimal, but analysis performed assuming optimality remains useful. First, if state behavior is usually responsive to structural constraints, even if not optimal, a structural theory may predict well.[43] Second, the optimality assumption is useful in formulating policy because states should often assume that their opponents will act optimally. Finally, we need to assume optimality in order to assess whether states have acted suboptimally and to appreciate the implications of policy errors. For example, if under optimality defense has a large advantage, then states will have to adopt very bad policies before offensives will be successful. Deploying nuclear weapons in vulnerable basing modes is clearly suboptimal, but given the enormous destructive potential of nuclear weapons, even highly vulnerable basing may leave adequate capabilities for deterrence.

Although the offense-defense balance is most often applied to situations in which both countries have access to the same means for converting resources into military capabilities, in practice states do not always have access to the same geography, technology, or forms of political organization, especially if they have unequal resources.[44] Thus states face asymmetric opportunities and, even with optimal behavior by both sides, the different constraints may result

42. On the degree to which German success in 1940 depended on suboptimal Allied doctrines, see Len Deighton, *Blitzkrieg* (New York: Ballantine, 1979).
43. Posen, *Sources of Military Doctrine*, pp. 228–236, argues that the greater the degree of threat, the closer states' military doctrines approach optimal choice.
44. We argue that such asymmetries should not simply be treated as a difference in military skill, because they are sometimes (not always) the result of differences in scientific or gross economic resources or rigidities in societal structure that even optimal policy cannot rectify quickly.

in asymmetries in states' force postures. Examples include the inability of revolutionary France's absolutist enemies to recruit equally committed citizen-soldiers, and the inability of Japan or Germany to deploy nuclear weapons during World War II.

Under such asymmetries, the offense-defense balance will depend on which country is the attacker, because asymmetric development of an innovation will usually enable the advantaged country to do better both on offense and on defense, even if symmetric deployment of that innovation would strongly favor defense (or offense). For example, although the nuclear revolution strongly favors defense, in the 1940s the United States might have been able to use its nuclear monopoly to conquer the Soviet Union. Consequently, under such asymmetries the offense-defense balance shifts toward offense when the advantaged state is the attacker and toward defense when it is the defender.[45]

The offense-defense balance that would exist if both countries had deployed the innovation, that is, the balance for the symmetric case, does provide information that is useful for understanding the asymmetric case. Because technology lags usually erode, the symmetric case helps us understand the future offense-defense balance that the innovation will eventually create. States' behavior under asymmetry will be influenced by their expectations about this future balance.[46]

DYADIC, NOT SYSTEMIC BALANCES. The offense-defense balance is well defined only for specific dyads of states, not for the entire international system. As we discuss below, the offense-defense balance depends on a number of diverse factors—including geography, cumulativity of resources, and nationalism—some of which are often not shared across dyads. Consequently, the offense-defense balance will frequently vary across dyads.

Whether the international system or a region within it can be usefully characterized in terms of a single value of the offense-defense balance depends on the nature and extent of variation in the factors that influence the relevant dyadic balances. Offense-defense theory will make predictions that hold across the system when factors that do not vary across the system dominate those

45. Such asymmetries can yield situations where, even with equal resources, one state of a dyad may be able to defeat the other on either offense or defense. Predictions about such cases will be more uncertain than other equal-resource cases, because they require greater information about states' goals; greedy states might attack when security seekers would not.

46. For example, the United States was less likely to use its nuclear monopoly to prevent Soviet development of nuclear weapons and an assured destruction capability because nuclear superiority was going to decay into a world of mutual assured destruction capabilities, (MAD), in which defense has a large advantage. See Glaser, *Analyzing Strategic Nuclear Policy*, chap. 5.

factors that do vary. The most obvious example is nuclear weapons: this technology so heavily favors defense that when all the major powers have nuclear weapons variation in other factors becomes relatively unimportant.

THE COMPOUND OFFENSE-DEFENSE BALANCE. As noted in our discussion of asymmetry, within each dyad of states there are really two offense-defense balances: one with state A as the attacker and state B as the defender, and one with B attacking A. These two "directional balances," as we call them, will sometimes be identical—the symmetric case—but will be unequal when one state has advantages over the other in one or more of the factors that influence the balance (e.g. technology or geography). For example, before 1914 the Vosges Mountains on the French side of the Franco-German border would have hindered any German offensive against France on that front but not vice versa.[47]

In such cases the directional balances alone can answer some of the questions posed by offense-defense theory (e.g., whether a certain greedy state is likely to succeed in conquering a particular target), but will not be sufficient to answer the two most important questions posed by the security dilemma: whether a security-seeking state should choose a defensive strategy for protecting its territory, and whether there are possible force postures that would allow both states to be secure.

To answer these questions, we introduce the "compound balance." A status quo state's preference for offense or defense depends on the cost of defending its territory with a defensive strategy compared to the cost of doing so with an offensive strategy. Each of the two directional balances, however, can supply only half of the answer. Within a dyad, the balance with the state as defender tells us how much it must invest to succeed on the defense, but nothing about the cost of an offensive approach. The other directional balance—with the same state as the attacker—tells us the cost of an offensive strategy but not that of a defensive one. The state's strategy preference is determined by the arithmetic product of the two balances—the compound balance. If the product is greater than 1, the state can defend at lower cost than it can attack and therefore should prefer a defensive strategy; if less than 1, it can attack at lower cost than it can defend and should prefer an offensive strategy.[48]

47. Since war outcomes are determined by power and skill as well as the offense-defense balance, this does not necessarily mean that France would have been more likely to win a war.

48. Given that the function of the offense-defense balance is to compare the efficacy of offensive strategies to defensive ones, situations where both sides choose offensive strategies create anomalies for offense-defense theory. This can happen in two ways: when the compound balance is less

For instance, suppose that the directional balance with the state on defense is 1 to 2 (favoring offense) and the one with it on offense is 6 to 1 (favoring defense). The compound balance is $1/2 \times 6 = 3$. Because this value is greater than 1, the state should prefer defense, even though the directional balance with the state on defense favors offense. Given that both directional balances are unfavorable, this state cannot be secure unless it outspends its opponent, which it may not be able to do. No matter how adequate or inadequate the state's resources, however, it will be better off choosing defense, because offensive strategies would cost three times as much to reach the same level of capability.

This discussion clarifies three important points. First, the most common usage of the concept of the offense-defense balance—the directional balance—does not determine whether security-seeking states should prefer offense or defense. This depends not on either directional balance but on whether their product is greater or less than 1. Second, security-seeking states' preferences for offense or defense are often much stronger than would be suggested by looking at either directional balance, even in the case of symmetry. Given that the relative cost of offense and defense is provided by the product of the two directional balances, whenever both are greater than 1 (or both are less than 1) the compound balance will reveal a more extreme offense-defense cost ratio than either directional balance alone. For example, consider a dyad for which the offense-defense balance is 3 to 1 in each direction. Assume that one state spends X on offense. To have an adequate defense, the second state needs to spend only $X/3$. If the second state chooses to fight on the offense, however, then the 3 to 1 ratio works against it, and it must spend $3X$ to succeed. The compound balance is $3 \times 3 = 9$; successful offense would cost 9 times as much as successful defense. Third, whether or not the directional balances in a dyad are symmetric, both countries always face the same compound balance. There-

than 1, both sides could choose offense on efficiency grounds. The theory then predicts the states' strategy choices, but since the theory does not define an "offense-offense balance," it may tell us little about likely war outcomes. Second, when the compound balance is not much greater than 1 and one state chooses offense for nonsecurity reasons (i.e., because it is greedy), then the theory may not be able to predict the second state's strategy choice. Normally, if the compound balance is greater than 1 the second state would choose defense on efficiency grounds. We should expect, however, that an offensive strategy by the second state will do better against the first state's offense (since it is not optimized for defense) than it would against an opposing defense (which, by definition, is), but because we have no way of estimating how much better, we cannot predict the second state's best choice or whether it can achieve security. We do not have a solution for this, and are unaware of discussions that have addressed this issue.

fore, if both are pure security seekers, they will always have the same preference for offense versus defense.[49]

Factors That Influence the Offense-Defense Balance

In this section we make three points. First, we argue that the basic logic of offense-defense theory requires a "broad" approach to operationalizing the offense-defense balance, in contrast to what we call the "narrow" approach that includes only technology and geography. Second, we enumerate the major causal factors that we believe should be included in operationalizations of the offense-defense balance—technology, geography, force size, nationalism, and cumulativity of resources. Third, we explain why two factors that are sometimes included in a broad approach—alliance behavior and first-move advantages—should be excluded.

BROAD VERSUS NARROW APPROACHES TO OPERATIONALIZING THE BALANCE

Offense-defense analysts are divided over how widely to cast their nets in identifying the causal factors that determine the balance. Advocates of the narrow approach argue that measures of the balance should incorporate only technology and geography variables,[50] whereas proponents of the broad approach include many additional, often diverse factors.[51]

Critics of the broad approach argue that it makes the offense-defense balance impossible to calculate and the theory impossible to use.[52] This objection has merit because increasing the number of factors does heighten the difficulty of measuring the balance. Moreover, given that offense-defense theory focuses on military capabilities, it may seem natural to limit analysis to the most obviously military inputs.

---

49. At first glance, this may seem surprising, but there is an intuitive explanation. Whenever we say that one side can do better on defense than on offense, we must also be saying that the second side would do worse on offense (against the first side's defense) than on defense (against the first side's offense); thus both will prefer defense to offense.

50. Jervis, "Cooperation under the Security Dilemma," pp. 194–199. Quester, *Offense and Defense in the International System,* focuses mainly on technology. Both Lynn-Jones, "Offense-Defense Theory and Its Critics," p. 668; and Levy, "The Offensive/Defensive Balance of Military Technology," pp. 225–227, 229, argue explicitly for limiting the concept to technology.

51. See, for example, Hopf, "Polarity, the Offense-Defense Balance, and War"; and Van Evera, *Causes of War,* pp. 261–273. Stephen Walt, *Revolution and War,* pp. 37–43, suggests that the permeability of societies to ideas as a cause of offense dominance.

52. Additional arguments include claims that only technology is constant across the international system, and that the narrow definition is already the accepted one in the scholarly community. See Lynn-Jones, "Offense-Defense Theory and Its Critics," p. 668.

Nevertheless, the purpose of the theory, which is to explain states' decisions based on expectations of how structural constraints will mold military outcomes, requires the broad approach because these expectations are often influenced by factors in addition to technology. An attempt, for example, to measure the strategic offense-defense balance in Europe in 1939 that omitted the possible impact of cumulative resources would not have been helpful in estimating the likely outcome of a major war or in predicting the behavior of Germany or of other states.[53]

Equally important, the broad approach is required because appropriate operationalizations must follow from the definition of the balance. Once we have defined the offense-defense balance as the cost ratio, in principle our operationalization should include all material factors that can have a sizable impact on this ratio. In practice, however, we may have to exclude certain factors that make measurement especially difficult or create problems of logical consistency. In addition, parsimony versus power trade-offs are always present, so for some research questions it may be useful to omit certain factors even though they are tractable. For instance, a measure of incentives for a strategic nuclear arms race in the year 2000 could probably ignore geography and cumulativity of resources, but a measure of incentives for an arms race in Europe in 1900 could not.

CRITICAL FACTORS

Changes in any one of the following factors can have a significant effect on the offense-defense balance, making offensive projects feasible that would be otherwise be infeasible and vice versa: technology, geography, force size, nationalism, and the cumulativity of resources.

TECHNOLOGY. The factor most frequently cited as influencing the offense-defense balance is technology. The offense-defense impact of a specific weapons or technology innovation cannot be assessed simply by considering its performance properties in isolation; rather, we must assess its impact on states' abilities to perform offensive and defensive missions. The most critical question in this process is how the innovation differentially affects advancing forces and nonadvancing forces.[54] Innovations that are usable only or primarily by

53. Analysts who reject the broad approach appreciate that additional factors besides technology and geography do influence the cost ratio of offensive to defensive strategies. Lynn-Jones, "Offense-Defense Theory and Its Critics," pp. 668–670. Thus this is not a deep disagreement about how the world works, but rather over how best to build theories.

54. A related, although not identical, distinction is the effect of a technology on moving versus nonmoving forces. The impact of both distinctions on the feasibility of offensive and defensive

nonadvancing forces will tend to favor defense, while innovations that are equally usable by forces that are advancing into enemy-controlled territory will favor the offense.[55] Six major areas of technology are relevant: mobility, firepower, protection, logistics, communication, and detection.

The most widely agreed proposition is that improvements in mobility favor offense.[56] Only offense inherently requires mobility; a force that cannot move cannot attack, and a defender that can hold its positions need not move. The critical issue, however, is the relationship between the two stages of a successful offensive: the attacker must first achieve a breakthrough by defeating or destroying a section of the defender's front; the attacker must then exploit this breakthrough to advance into the defender's rear.[57] Breakthrough is logically and temporally prior to exploitation, and substantively more important because the issue of exploitation arises only if and when breakthrough succeeds.

Improvements in operational mobility (the ability to move, supply, and concentrate forces for battle), such as the introduction of motor trucks, mobile bridging equipment, and long-range combat aircraft, improve the attacker's ability to outflank the defender or concentrate to assault the defender's weakest points.[58] Increases in tactical mobility (the ability to move under fire and survive), such as the introduction of tanks, reduce the attacker's losses while

---

missions, however, is in most instances so similar that we often characterize the offense-defense implications of a given technology based on either distinction.

55. At least compared with technologies which are only usable by nonadvancing forces; in practice technologies that are actually *more* usable by advancing forces are rare because almost all tasks are easier to carry out while stationary than while moving.

56. Basil H. Liddell Hart, "Aggression and the Problem of Weapons," *English Review* 55 (1932), p. 73; Marion W. Boggs, "Attempts to Limit 'Aggressive' Armament in Diplomacy and Strategy" (Ph.D. dissertation, University of Missouri, 1941), p. 85; Quester, *Offense and Defense in the International System*, pp. 2–3; Oelrich, *Conventional Arms Control*, pp. 14–25; Anders Boserup, "Mutual Defensive Superiority and the Problem of Mobility along an Extended Front," in Boserup and Nield, eds., *Foundations of Defensive Defense*, pp. 63–78; and Catherine M. Kelleher, "Indicators of Defensive Intent in Conventional Force Structures and Operations in Europe," in Lawrence Freedman, ed., *Military Power in Europe* (New York: St. Martin's, 1990), pp. 165–168.

57. This rather abstract discussion does not distinguish between attritional and blitzkrieg-style offensives, although actually both proceed in essentially the same two stages. The difference is the attacker's confidence that a breakthrough can be produced using a narrow versus a wide-front assault, and can be produced quickly and easily. On the two types of offensives, see Mearsheimer, *Conventional Deterrence*, pp. 33–43. On the role of breakthrough in attrition offensives, see C.R.M.F. Cruttwell's treatment of the 1917 Cambrai assault in *A History of the Great War, 1914–1918*, 2d ed. (Oxford, U.K.: Clarendon Press, 1936), pp. 467–477.

58. Another way to put this is that mobility multiplies the attacker's advantage of the initiative. There is always a time lag between the initiation of an offensive action and the beginning of effective response by the defender because of the time needed to (1) detect the action, (2) assess the threat, (3) decide on a response, and (4) disseminate instructions to begin implementing the response. Increased mobility means that the attacker can accomplish more with an initiative lag of any given duration. For a formal model illustrating this, see Robert Nield, "The Relationship of Mobility to Defensive Stability," *Defense Analysis*, Vol. 8, No. 2 (August 1992), pp. 199–201. Mear-

assaulting defending positions.[59] Thus increases in mobility make breakthroughs more likely and therefore generally favor offense.

The implications of mobility for exploitation and counterexploitation operations are less clear, because in this phase both sides may be maneuvering to concentrate or attack enemy forces simultaneously. Within the exploitation stage, it is not clear whether the attacker or defender is favored by increases in mobility. The likelihood of reaching this stage, however, depends on the offense-defense balance in the breakthrough stage, including mobility. Thus, whenever achieving a successful breakthrough is difficult or uncertain, mobility improvements will favor offense. Only when attack is so easy that a successful breakthrough is virtually assured does the impact of mobility become indeterminate.

Nearly all historical advances in military mobility—chariots, horse cavalry, tanks, motor trucks, aircraft, mobile bridging equipment—are generally considered to have favored the offense, while major countermobility innovations—moats, barbed wire, tank traps, land mines—have favored defense. The effect of mobility improvements, however, does depend on whether they are equally usable by attackers and defenders, which in turn depends on their dependence on infrastructure. Railroads, which depend on elaborate networks of infrastructure which can easily be destroyed by retreating defenders but which cannot be extended quickly, are much more useful to forces operating in friendly controlled territory than they are to advancing forces. Thus they favor defense in comparison with motor trucks or helicopters, which require less infrastructure and so can more easily operate at or near the spearhead of an advance.

---

sheimer, *Conventional Deterrence*, p. 26, disagrees with this logic, arguing that the attacker can use the advantage of the initiative to position its forces at leisure, while once the offensive begins the defender must redeploy with great speed to meet the threat; thus increases in operational mobility actually favor defenders. This argument, however, has two weaknesses. First, it requires a huge initiative lag, sufficient to allow attacker forces to reposition virtually anywhere in the theater, before the defender detects anything. Second, it does not touch the differential value of tactical mobility to attackers.

59. Strategic mobility (the ability to transport and support large forces far from one's centers of mobilization) may not have much effect in the offense-defense balance in wars between small or medium-sized powers, but is essential to any long-distance offensive. See George Modelski and William R. Thompson, *Seapower in Global Politics, 1494–1983* (Seattle: University of Washington Press, 1987), on the concept of the power/distance gradient; improvements in strategic mobility can be understood as a "flattening" of this gradient. In the special case where two sides share a traversable land border but one relies more heavily than the other on reinforcement from overseas territories and allies—for example, France in 1939–40—better strategic mobility will simply favor the power of that side more than it favors either defense or offense.

Improvements in firepower are generally considered to favor defense on essentially the reverse of the logic that applies to mobility. In battle, attackers are usually more vulnerable to fire than are defenders because they must advance, often in plain sight of defenders, making them easy to detect and to hit, whereas defenders are often well dug-in and camouflaged. Firepower innovations usually considered to have favored defense include machine guns in World War I, infantry antitank weapons during World War II, and antitank guided missiles (ATGMs) and surface-to-air missiles today. At the operational level, the need to concentrate forces to achieve local superiority means attackers are often more vulnerable than defenders to area-effect weapons such as artillery and tactical nuclear weapons.[60] As with mobility, there can be exceptions when specific firepower innovations are differentially useful against defenders, such as heavy siege artillery before World War II, whose main use was against fixed fortifications, or today's antiradiation missiles, whose function is to attack air defenses.[61]

The effects of innovations in protection, logistics, communication, and detection are more varied, depending on how specific innovations interact with force behavior; those whose full benefit can be realized only by nonadvancing forces or only against advancing ones will favor defense, whereas those with benefits that are equally available to both advancing and nonadvancing forces will favor offense (at least compared with technologies of unequal usefulness). The earliest tanks strengthened the offense mainly because they provided troops with protection that, unlike trenches and bomb shelters, they could take with them as they advanced. Military communications based on land-line telephones favor defense compared with systems based on portable radios. Early radar, which could detect incoming enemy aircraft but not stationary ground targets, favored defense compared with modern downward-looking systems such as airborne warning and control systems and joint surveillance target attack radar systems, which can detect low-altitude aircraft and land vehicles deep in enemy territory.

GEOGRAPHY. The implications of geography are perhaps the least controversial of all the factors that affect the offense-defense balance.[62] Generally speaking, barriers to movement, cover, and distance all favor defenders more than does the absence of these conditions.

60. Another reason why tactical nuclear weapons and the most powerful conventional munitions favor defense is that they destroy transportation infrastructure, thus reducing mobility.
61. On heavy artillery, see Liddell Hart, "Aggression and the Problem of Weapons," p. 73.
62. A standard discussion is Jervis, "Cooperation under the Security Dilemma," pp. 194–196.

First, terrain that slows or channelizes movement, or that strains logistics, strengthens the defense more than terrain that does not. This includes forests and swamps with few roads, mountains with few passes, and rivers with few bridges, or simply any region with sparse infrastructure. Such barriers channelize advances into the few roads, bridges, or passes that are available, thus reducing the defender's intelligence difficulties as well as shortening the length of front requiring serious defense. Sparse infrastructure also limits logistic throughput, decreasing the amount of force that the attacker can deploy forward of the barrier even after crossing it. NATO plans for defending West Germany focused on the North German plain and the Fulda Gap in the south because these were the only two places along the inner German border where major mechanized offensives seemed feasible.

Second, terrain that provides cover in which defenders can hide—such as forests, mountains, and cities—strengthens defense. Cover reduces the speed at which an attacker can advance because the attacker must reconnoiter every possible location that could hide an ambush. The denser the cover—the more hiding places per unit of area—the more the attackers' speed is reduced.[63] Attackers often prefer to avoid areas of dense cover if at all possible, in which case the cover has a channelizing effect rather than a delaying effect.[64]

Finally, distance favors defense. If the attacker must travel a considerable distance just to reach the defender's territory, the amount of force it can project is reduced by the costs of transporting and supplying the projected force, as well as the costs of defending long lines of communication. The Atlantic and Pacific Oceans have always made it difficult to invade the United States, and would make it difficult for the United States to attack opponents overseas unless it had allies in the region. If the intervening distance is not water but land occupied by neutral or hostile states, the costs of overcoming their resistance may further reduce the attacker's ability to project power against the ultimate defender.[65]

---

63. Very dense cover, especially cities, is often extremely costly to capture because there is no way for advancing units to avoid exposing themselves to fire from defended positions, often at very short range. The German attackers at Stalingrad in 1942 and the Russians at Grozny in 1994–95 both suffered heavily despite possessing much more force than the defenders.

64. Heavy cover, such as stone or concrete buildings, or mountainous terrain in which caves or tunnels can be dug, can further delay an attacker by requiring very heavy firepower to destroy defenders even after they are located, although precision guided missiles are making this less important.

65. Some have argued that a neutral Eastern Europe would provide the United States and Russia better protection against each other than would expansion of NATO. See Mueller, "Alignment Balancing and Stability in Eastern Europe," pp. 67–76.

More than other factors, the impact of geography is often asymmetric. When Syria held the Golan Heights, they reduced Israel's offensive options against Syria but not vice versa. Before World War II, Britain was secured by a water barrier against German attack, while Germany's open western frontier still left it vulnerable to Britain and its continental allies. In such cases, the effect of geography is to shift one directional balance of the dyad toward defense, while the other is unaffected or even shifted toward offense.

FORCE SIZE. The offense-defense balance can depend on the size of the forces deployed. At least two examples of this phenomenon are readily available. First, many analysts argue that the prospects for success in conventional offensives depend not only on force-to-force ratios but also on force-to-space ratios, that is, the size of forces in relation to the length of the front. The basic reason is that the defender's ability to compel the attacker to make expensive frontal assaults depends on being able to man all viable axes of advance thickly enough that the attacker cannot achieve a quick breakthrough by overwhelming a weakly held area. Thus the offense-defense balance generally shifts in favor of the defense as force size increases.[66]

Second, nuclear offense, which is usually evaluated in terms of the ability to reduce an opponent's retaliatory capability below a specified "assured destruction" level, becomes more difficult as force size increases. As the size of the defender's force increases, maintaining a given probability of success requires either a larger ratio of attacker forces to defender forces or more sophisticated attacker forces, or both. Thus nuclear offense-defense balances shift in favor of the defense as force size increases.

NATIONALISM. Nationalism can influence the offense-defense balance in two ways. First, to the extent that people are imbued with nationalist consciousness, they may become willing to fight harder for territory that they understand to be part of their national homeland and less willing to fight for other territories.[67] As a result, provided that prewar borders more or less match nationalist claims, an attacker will have to invest more to defeat a given

66. Liddell Hart, "The Ratio of Troops to Space," *Military Review*, Vol. 40 (April 1960), pp. 3–14; Mearsheimer, "Why the Soviets Can't Win Quickly in Central Europe," *International Security*, Vol. 7, No. 1 (Summer 1982), pp. 3–39 at 26–30; and Biddle, "Offensiveness and Defensiveness," pp. 164–169. Biddle suggests that at extremely low force-to-space ratios, still lower densities can strengthen defense, down to the point at which offense is impossible because there are no forces at all.

67. Jervis, "Cooperation under the Security Dilemma," p. 204. Examples include the many Confederate soldiers who deserted the Army of Northern Virginia at the Potomac in 1862 because "they felt that they were fighting to defend Virginia's soil, not to invade the North" and Hitler's unwillingness to risk imposing full war mobilization on Germany until the failure of Barbarossa opened the possibility that Germany's own homeland security might be threatened. Bruce Catton,

defender than would otherwise be the case.[68] Second, nationalism can provide the defender with a wartime extraction advantage, enabling the defender to forgo some peacetime military investment, thereby further shifting the balance. One implication of nationalism is that, at least in much of Europe, South and East Asia, and the Americas, values of the offense-defense balance should generally be more favorable to defense in today's "age of nationalism" than they once were.

CUMULATIVITY OF RESOURCES. Attackers can sometimes pay part of the cost of achieving expansionist objectives with resources extracted during the war from parts of the target territory itself. Such "cumulative resources" reduce the requirement for military investment before the war, thus shifting the balance in favor of offense.[69] Examples include Germany in 1939 and Japan in 1941, both of which realized that their domestic resources were insufficient to achieve their planned conquests and expected to "bootstrap" their campaigns using captured resources.[70] Relevant resources may include both natural resources extracted directly from occupied territory, as well as productive effort which can be obtained from the occupied population.[71]

The extent to which resources from captured territories are cumulative depends not only on the resources extracted by the attacker, but also on the costs to the attacker of extracting them, including costs of controlling the occupied population and of repairing sabotage. The degree of cumulativity is therefore affected by the fierceness of popular resistance and by the organization and technological level of the local economy.[72] If resistance is fierce, it may

---

Mr. Lincoln's Army (Garden City, N.Y.: Doubleday, 1951), p. 252; and Alan S. Milward, *War, Economy, and Society 1939–1945* (Berkeley: University of California Press, 1977), pp. 76–80.

68. Where there are significant irredenta, however, nationalists may fight just as hard on offense as on defense. More important, where multinational empires rule disgruntled subject people, nationalism may actually favor offense by raising the probability that invaders will not be resisted but welcomed as liberators. Examples include subject towns of both the Romans and Carthaginians which opened their gates to the other side in the First and Second Punic Wars, as well as the contrast between the collaboration of some Soviet subject peoples from 1941 to 1944 and the fierce resistance of ethnic Russians. Polybius, *The Histories* Vol. 4, trans. W.R. Taton (Cambridge, Mass.: Harvard University Press, 1925), pp. 445–447, 455.

69. The decisiveness of cumulative resources to the outcome of a particular war depends on how quickly they can be converted to military power and on the length of the war.

70. Milward, *War, Economy, and Society*, pp. 18–36; and Michael A. Barnhart, *Japan Prepares for Total War: The Search for Economic Security, 1919–1941* (Ithaca, N.Y.: Cornell University Press, 1987).

71. Cumulative resources need not be directly military in nature if they substitute for domestic resources which the attacker can then shift to military production. For instance, German strategic plans in both world wars counted on acquiring grain from Eastern Europe to release German workers for war industries, and Germany's war effort in 1918 was severely hampered by failure to gain control of agricultural areas in the Ukraine in time to collect the 1917 harvest.

72. The importance of popular resistance means that cumulativity depends in part on the intensity of nationalism. On the economic side, Van Evera, *Causes of War*, pp. 180–183, argues that the

be impossible for the attacker to achieve a net gain in resources. In addition, the extent of cumulativity depends on the resources lost by the dispossessed defender. If the attacker can seize or destroy resources the defender planned to employ, then the attacker can invest less, which shifts the balance toward offense. The overall degree of cumulativity equals the attacker's net gain in resources plus the defender's loss of resources.

The term "cumulativity of resources" is also often used in a second sense, referring to the postwar effects on the international balance of power of resource changes resulting from conquest, rather than to intrawar effects. In fact this use of the term is more common than ours.[73] This kind of cumulativity, however, does not affect the requirements for the conquest itself, and therefore does not directly affect the offense-defense balance. Although it does not reduce the prewar investment ratio required to take territory, postwar cumulativity may still affect the probability of war by increasing prospective attackers' assessments of the value of taking territory. Thus reasonable assumptions about states' values for territory in an environment where cumulativity is high will be different than when it is low.

EXCLUDED VARIABLES

Although both the nature of international alliance behavior and first-move advantages can have large effects on war outcomes, we exclude both from our operationalization of the offense-defense balance, although for different reasons. We exclude alliance behavior on theoretical grounds; although alliances affect the probability of offensive success, they do so in ways that lie outside the offense-defense balance. First-move advantages, by contrast, should be excluded because including them would introduce logical incoherence in the definition of the balance that we see no way to resolve.

NATURE OF INTERNATIONAL ALLIANCE BEHAVIOR. Some offense-defense theorists suggest that the balance is affected by the nature of international alliance behavior—that is, whether balancing or bandwagoning is the norm. Bandwagoning is said to favor offense, balancing defense.[74] While balancing cer-

---

transition to information-age economies may reduce cumulativity because these economies cannot function without a level of free communication which an occupier cannot safely permit. Peter Liberman, *Does Conquest Pay? The Exploitation of Advanced Industrial Societies* (Princeton, N.J.: Princeton University Press, 1996), argues oppositely that increasing specialization makes it easier for attackers to use control of the food supply to compel urban workers to produce.

73. See Liberman, *Does Conquest Pay?;* and Van Evera, *Causes of War,* pp. 161–183.

74. Van Evera, *Causes of War;* and Hopf, "Polarity, the Offense Defense Balance, and War."

tainly does increase an attacker's problems, we conclude that alliance behavior is best excluded from the offense-defense balance for three reasons.

First, and most fundamental, including alliance behavior in the offense-defense balance is inconsistent with the central goal of offense-defense theory.[75] The goal of offense-defense theory (and of structural theories generally) is to explain the behavior of states based on the material constraints imposed by the international system. Although other states' alliance decisions could influence an attacker's prospects, including this effect in the offense-defense balance would mean that the balance would then be translating one type of state behavior into another type of behavior, not structural constraints into behavior. Because alliance choices can depend on nonstructural factors, states' intentions among them, including alliance behavior would create an offense-defense balance that lies outside the structural boundary of the theory.

Second, alliance behavior may often be determined in part by the offense-defense balance itself, so that including it as a factor influencing the balance would create an intractable circularity in which the balance depends on alliance behavior, which in turn depends on the balance. A theory of alliance behavior that could separate endogenous and exogenous effects would be complex, and has not been developed.[76]

Third, a complete theory of alliance behavior requires dyadic offense-defense balances as an input variable, because states decide their alliance choices based in part on the offense-defense balance between likely frontline adversaries.[77] Consequently, we must start with an offense-defense balance that excludes alliance behavior.

Our argument has significant implications for the current literature. The most extensive test of offense-defense arguments has been performed by Van Evera, who attempts to predict outcomes based on both a "military offense-defense balance" (which corresponds to the offense-defense balance as we define it) and a "diplomatic offense-defense balance" (whether states tend to balance or bandwagon).[78] Van Evera finds strong support for the main propo-

75. Fearon, "The Offense-Defense Balance and War Since 1648," pp. 11–12, makes a similar argument.
76. A theory of alliance behavior is required because the behavior that matters for measuring the balance is the behavior that states can reasonably expect in advance of war (and actually as far back as its decisions to procure forces and design its strategy), which may not be the same as what actually happens.
77. Christensen and Snyder, "Chain Gangs and Passed Bucks"; and Christensen, "Perceptions and Allies in Europe, 1865–1040."
78. Van Evera, *Causes of War*, pp. 261–273.

sitions of offense-defense theory. He does not, however, explore how well the military offense-defense balance does on its own, nor does he offer a theory that explains how much of alliance behavior is explained by the offense-defense balance and how much by other factors. As a result, we cannot determine from his analysis how much variation in the incidence of war is explained by an offense-defense balance that is defined only in terms of structural constraints.

In addition, a state's contribution to Van Evera's combined offense-defense balance (i.e., the balance including the effects of alliance behavior) is a function not only of its willingness to balance but also of its capability, thereby including power in his definition of the offense-defense balance. Bundling in power, however, would mean that the balance then represents not the relative ease of offense compared to defense, but instead measures an attacker's prospects for winning a particular war. For example, when Van Evera says that the relative decline of British power in the late nineteenth century shifted the balance in favor of offense, he does not mean that European states in general became easier to conquer, but rather that given Britain's likely alliance choices, British decline improved Germany's chances of winning in an offensive war against France or Russia.[79] Thus Van Evera is describing an effect of a shift in relative power, not of a shift in the offense-defense balance as we define it.

A better way to incorporate alliance behavior into a structural theory of military capabilities is to understand it entirely in terms of its impact on the balance of power between opposing sides. Given a theory of alliances, which could depend in part on dyadic offense-defense balances, states could then predict which alliances were likely, thus identifying the likely sides in a conflict. We could then define a "coalition offense-defense balance" as the ratio of investment required by the attacking side to take territory from the defending side in a conflict.[80] Given this formulation, adding allies to a side increases that side's power, but might or might not yield a coalition offense-defense balance different from the frontline dyadic balance. The coalition balance might be different if, for instance, the war were likely to be fought on multiple fronts with differing geography, or if the nationalism of some members of one or both alliances was more fully engaged in the dispute than that of others.[81]

79. Ibid., p. 270.
80. We are indebted to Karl Mueller for this insight.
81. For example, the formation of NATO not only increased the power resources available to resist a possible Soviet invasion of West Germany, but may also have shifted the offense-defense balance toward defense because of the likelihood that some NATO members might have been unwilling to participate in offensive military operations aimed at unifying East and West Germany.

FIRST-MOVE ADVANTAGES. A first-move advantage—sometimes called a "first-strike advantage"—exists when the sides' military prospects in a potential conflict differ depending on which side acts first. We prefer the term "first-move advantage" because the important action need not involve actual combat; examples of first moves include raising standing forces to a higher level of alert, calling up reserves, moving forces to positions from which they can defend (or attack) more effectively, and actually firing on opposing forces or territory.[82]

Many formulations of offense-defense theory have treated first-move advantages as an important factor that influences the offense-defense balance. The argument is that the larger the first-move advantage, the greater the dominance of the offense.[83] We believe, however, that treating first-move advantages as part of the offense-defense balance would create confusion. First, whenever there is a first-move advantage, a directional balance that attempts to incorporate its impact will not be well defined.[84] By definition, if there is a first-move advantage, the investment ratio that the attacker requires for success is different depending on which side moves first. The attacker—the side attempting to take territory from the other—will not always be the side that moves first, however. Whenever a first-move advantage exists, the defender will have an incentive to move first, even if the defender has no ambition, or even any possibility, of taking any territory from the attacker. First-move advantages are always mutual; both sides have an incentive to move first, if only to avoid the consequences of letting the other side move first.[85] The result is that a first-move advantage will not always help the attacker, because its existence makes uncertain which side will move first. Thus incorporating a first-move advantage into the offense-defense balance would yield two balances and no obvious way to choose between them.

Second, including first-move advantages in the offense-defense balance would mean that the balance would vary with the level of alert at which states

---

82. Van Evera, ibid., pp. 52–54, uses the term the same way we do.
83. Jervis, "Cooperation under the Security Dilemma," pp. 188–189; and Van Evera, *Causes of War,* unnumbered page after p. 52 and pp. 73–74. Important works that predate offense-defense theory include Albert Wohlstetter, "The Delicate Balance of Terror," *Foreign Affairs,* Vol. 37, No. 2 (January 1959), pp. 211–234; and, Schelling, *Arms and Influence,* pp. 221–259. Dan Reiter, "Exploding the Powder Keg Myth: Pre-emptive Wars Almost Never Happen," *International Security,* Vol. 20, No. 2 (Fall 1995), pp. 5–34, argues that preemptive wars have been rare.
84. Much of the existing offense-defense literature has ignored this problem, apparently assuming that attackers always go first.
85. For example, if one side can destroy the other's unalerted air force on the ground, the other has an incentive to alert its planes simply to avoid this, even if it has no prospect of destroying the first side's air force.

place their forces, with the result that the balance would vary with political conditions and would change during crises. Consequently, even if we knew which side was going to move first, the balance would not be well defined.

For these reasons, we define the directional balance by assuming that first-move advantages are zero. To assess the balance given this assumption, we envision wars in which offensives are launched after the country that moves second has had the opportunity to react to the other's first move, thereby offsetting the advantage of the first move. For example, when first-move advantages reflect the value of mobilizing first, we assess the balance assuming that both countries fully mobilize before the attacker launches its offensive. We choose this assumption not because it is empirically reasonable, but because it is useful for separating the effects of the offense-defense balance and of first-move advantages.

Measuring the Offense-Defense Balance

A common criticism of offense-defense theory is that the offense-defense balance cannot be measured, and therefore, the theory cannot predict or explain anything. According to this argument, war is too complex and too poorly understood to predict outcomes with any degree of confidence. For example, Colin Gray argues that "policymakers can never know reliably in advance the costs of the military decision they seek."[86] If we cannot estimate whether an attacker will succeed, then we cannot measure the offense-defense balance.[87]

We argue that critics have overstated the difficulty of measuring the balance for four reasons. First, they have exaggerated the difficulty of integrating across levels of warfare. Second, the analytical tasks required to measure the offense-defense balance are the same as those required for military net assessments, a task for which capable analytic tools exist. Third, the precision with which the theory requires the balance to be measured is often lower than critics may realize: ballpark estimates of the balance are sometimes sufficient to provide

86. Colin Gray, *Weapons Don't Make War: Policy, Strategy, and Military Technology* (Lawrence: University Press of Kansas, 1993), p. 23. For a skeptical view of the feasibility of reliable net assessment, see Cohen, "Toward Better Net Assessment."

87. To be more precise, the accuracy of our estimates of the offense-defense balance depends on the completeness of our model of the determinants of war outcomes. The better our understanding of war, the narrower the bands of uncertainty around our estimates of the balance. This problem is a central one in military history, with a large and active literature; a few political scientists are also beginning to take an interest, for example, Biddle, "Determinants of Offensiveness and Defensiveness."

the theory with substantial predictive and explanatory capabilities. Fourth, critics are mistaken when they argue that the indistinguishability of offense and defense makes measurement of the balance impossible.

INTEGRATING ACROSS LEVELS OF WARFARE

Since the offense-defense balance is defined at the level of whole wars, but many of the factors that influence it have their direct impact only at the tactical or operational levels, measuring the offense-defense balance requires integrating effects across the tactical, operational, and strategic levels. This integration problem breaks down into two parts: direction and magnitude.

Assessing the direction of effect is normally straightforward: a change that shifts the balance in a given direction at one level will usually also shift it in the same direction at all higher levels. Since any strategic offensive necessarily requires offensive operations, and offensive operations require offensive tactical battles, a change that makes tactical offense harder will usually also make operational offense harder, which in turn makes strategic offense more difficult.[88]

Some analysts contest this point, arguing that, because defenders need counteroffensive capabilities for retaking lost territory, in practice the strategic defender may actually engage in just as much or more offensive action at the operational and tactical levels as the strategic attacker. Therefore it is unclear whether a tactical or operational offense-defense balance that favors, say, defense will have the same effect—or indeed any predictable effect—on the strategic offense-defense balance.[89]

This objection is overstated because offense is logically and temporally prior to counteroffense, and substantively more important to the outcomes of strategic offensives; counteroffense only arises if and to the extent that initial offensives succeed. If the attacker's initial offensive operations fail, the likelihood that a hypothetical counteroffensive might also fail is irrelevant because the defender will not need to undertake it. Conversely, arguments that offense-strengthening innovations might make the difference in the defender's ability to recover lost ground must confront the possibility that they might also

88. Exceptions are possible where two or more different factors interact. For instance, an improvement in naval transport might normally favor offensive operations, but could favor defense at the strategic level if the major beneficiary was an offshore balancer that was too weak to attack any continental power but strong enough to assist certain of them in defense against others.
89. Shimshoni, "Technology, Military Advantage, and World War I," p. 192; and Gray, *Weapons Don't Make War*, pp. 36–37.

multiply the attacker's initial success to the point that the defender would be unable to regain the initiative. Thus the more the operational and tactical offense-defense balances favor defense, the less likely initial offensives are to succeed, the less counteroffensive capabilities matter, and the more the overall strategic balance favors defense.

Compared to the direction of effect, assessing the magnitude of the effect of lower-level changes on the strategic offense-defense balance is more complex. Even where the effect of certain factors on tactical- or operational-level outcomes is relatively well understood—such as the impact of ATGMs on tank battles or of trucks in place of railroads on troops' rate of advance—the magnitude of the effect on the strategic offense-defense balance will depend on how states' optimal strategies are affected by complex combinations of operational and tactical constraints and opportunities. Evaluating these effects requires detailed strategic net assessments that incorporate the key factors at all levels of war.

We can, however, often take advantage of a shortcut. Assessments of large-scale conflicts can often be simplified by focusing on one campaign or theater of operations whose outcome will be critical for the outcome of the entire conflict. Most net assessments of a Warsaw Pact offensive against NATO's Central Front, for example, did not assess the entire campaign but instead only its first phase, the breakthrough battle. They did this partly for tractability, but also because the breakthrough battle would be the most important phase in determining the final outcome. While Pact success in the breakthrough battle would not guarantee victory, failure would make it very unlikely. For the same reason, analyses of strategic nuclear wars focus on the initial rounds of nuclear counterforce exchanges.

THE BALANCE IS MEASURED WITH NET ASSESSMENT TECHNIQUES

Measuring the offense-defense balance is no harder (or easier) than performing net assessments—analyses of the ability of a country's forces to perform military missions against the forces of an opponent. Measuring the offense-defense balance requires working through essentially the same steps as performing a net assessment: for given military missions, the analyst develops a model of how the forces will interact in combat, and explores the predictions of the model under different scenarios.

Although estimating the offense-defense balance requires the same analytical tools as standard net assessments, they are employed slightly differently. Net assessment was developed, and is most commonly used, to analyze the adequacy of particular force postures. The question is usually put in the form: "Are our existing forces sufficient to defeat this contingency? If not, would this alternative force be sufficient?" In contrast, to estimate the offense-defense balance we posit forces for the defender, then manipulate the size of the attacker's forces, asking "Exactly how much force would the attacker need to succeed?"[90] This difference, however, does not pose any additional analytic problems. If reliable net assessment is feasible, then reliable estimates of the offense-defense balance are also feasible.[91]

To illustrate that net assessments address many of the factors that influence the balance, we consider a sophisticated example from the unclassified literature. Barry Posen's net assessment of a possible Warsaw Pact conventional offensive in the 1980s required assessments of contemporary weapons, transportation, communications, and detection technologies in order to make judgments about likely mobilization and concentration rates, as well as about casualty exchange ratios.[92] Judgments about the number of plausible Warsaw Pact attack sectors required consideration of the effects of geography and force-to-space ratios. Posen's decision to analyze a breakthrough attempt rather than an attrition campaign or a limited-aims offensive reflected assumptions about the Warsaw Pact's likely military objectives, as well as its tolerance for losses in an effort to conquer Western Europe. These are the same sorts of assumptions about the attacker's territorial goals and the costs of fighting to take territory that are needed for measuring the offense-defense balance.

For example, Posen's assessment that a Warsaw Pact offensive with a 1.2 to 1 advantage in forces would be unlikely to succeed is equivalent to a finding that the offense-defense balance for that scenario was greater than 1.2. By

---

90. To be more precise, after answering this question, we then calculate the cost of these forces relative to the cost of the defender's forces.

91. Military assessments are sometimes stated in ways which are especially easily translatable into offense-defense terms. For example, Basil Liddell Hart's finding that in World War II Allied ground offensives rarely succeeded unless the attackers had a superiority of at least 5 to 1, together with domination of the air, amounts to an a statement that the offense-defense balance, at least at the tactical level, for these offensives was greater than 5. Liddell Hart, *Deterrent or Defense* (London: Stevens, 1960), p. 179. The recent discussion of the "3 to1 rule" favored by the U.S. military and others is similar. *FM 100-5: Operations* (Washington, D.C.: U.S. Department of the Army, 1976); Major H.F. Stoeckli, *Soviet Operational Planning: Superiority Ratios vs. Casualty Ratios* (Sandhurst, U.K.: Royal Military Academy, 1985); John J. Mearsheimer, "Assessing the Conventional Balance: The 3:1 Rule and Its Critics," *International Security*, Vol. 13, No. 4 (Spring 1989), pp. 54–89; and Joshua Epstein, "The 3:1 Rule, the Adaptive Dynamic Model, and the Future of Security Studies," *International Security*, Vol. 13, No. 4 (Spring 1989), pp. 90–127.

92. Barry Posen, "Measuring the European Conventional Balance: Coping with Complexity in Threat Assessment," *International Security*, Vol. 9, No. 3 (Winter 1984/85), pp. 47–88.

rerunning the model and increasing Warsaw Pact forces until they do prevail, we could estimate the balance for that scenario. Given the essential identity of the two modes of analysis, those who criticize offense-defense theory on the ground that the balance cannot be measured must argue that reliable net assessment is infeasible.[93]

Although it is beyond the scope of this article to settle the question of whether net assessment is feasible, there are important reasons for thinking that net assessment techniques can be used to effectively evaluate military capabilities. One reason for optimism is simply that many analysts have concluded that it is a useful tool. Military organizations have been investing substantial resources in net assessment for centuries, joined in recent decades by civilian experts inside and outside of government. Sophisticated analytic tools have been developed that incorporate information about terrain and the performance of weapons systems, extrapolations from previous battles and wars, and educated guesses about difficult-to-quantify factors such as human performance. During the Cold War, net assessments performed by civilian analysts established the foundation for extensive debate over NATO's prospects for defeating a possible Soviet offensive in Central Europe.[94]

More important, the historical record suggests that reasonably reliable net assessment has been feasible in the past. Although there has been extensive criticism of the quality of certain past net assessments, much of this criticism finds that the flaws were politically and/or bureaucratically motivated, not due to any inherent infeasibility of the task. Certainly the most famous of all net assessment failures are the overestimations before 1914 by the French, German, Russian, and other European militaries of the prospects for successful offensives against each other, generally known as the "cult of the offensive."[95] While all European militaries recognized that increased firepower and higher

---

93. Thus Mearsheimer's insistence that the balance cannot be measured is belied by his net assessment work.

94. Examples include Mearsheimer, "Why the Soviets Can't Win Quickly in Europe"; Posen, "Measuring the European Conventional Balance"; Joshua Epstein, *Measuring Military Power: The Soviet Air Threat to Europe* (Princeton, N.J.: Princeton University Press, 1984); and Cohen, "Toward Better Net Assessment," as well as official studies such as James Blaker and Andrew Hamilton, *Assessing the NATO/Warsaw Pact Military Balance* (Washington, D.C.: Congressional Budget Office, December 1977).

95. Van Evera, "The Cult of the Offensive and the Origins of the First World War." On the generally better, but often still flawed, net assessments before World War II, see Williamson Murray and Allan Millett, eds., *Calculations: Net Assessments and the Coming of World War II* (New York: Free Press, 1992).

force densities would make frontal assaults drastically more expensive, certain militaries, because of combinations of bureaucratic incentives, class interests (especially in Britain), and domestic political threats (especially in France), chose to believe that "morale" would somehow overcome bullets. In fact, however, this escape from the structural constraint was not only infeasible, but its infeasibility was knowable in advance. The evidence from the American Civil War, Franco-Prussian War, Russo-Turkish War, Boer War, and Russo-Japanese War was already in. Unbiased observers, both civilians as well as some junior officers, correctly predicted that frontal assaults would be impossible, as did the German military.[96]

The German army, recognizing that frontal assaults would be infeasible, but motivated by bureaucratic and political needs to find an offensive solution to a two-front war against France and Russia, chose to pin its hopes on a wide flanking maneuver through Belgium (the Schlieffen Plan). This, however, was logistically impossible, given the distances that the advancing Germans and their supplies would have to cover by foot and horse-drawn wagon, while the French and British defenders could react by rail. This too was knowable, and in fact known to the General Staff. Chief of Staff Helmuth von Moltke, who was maligned for weakening Schlieffen's original commitment to an overwhelmingly strong "Right Wing," was later only recognizing logistic reality.[97] Thus the errors of 1914 were avoidable; accurate net assessment, and therefore accurate estimation of the offense-defense balance, was feasible.

Finally, many debates over particular net assessments are not about the validity of the analytic tools but rather about scenarios—whether the scenario being modeled is actually a likely contingency—or data, such as the sides'

96. Michael Howard, "Men against Fire: Expectations of War in 1914," *International Security*, Vol. 9, No. 1 (Summer 1984), pp. 41–57. Snyder, *Ideology of the Offensive*, and Van Evera, *Causes of War*, pp. 316–394, argue that prewar estimates were wrong because of bias, not inherent complexity. Scott Sagan, "1914 Revisited: Allies, Offense, and Instability," *International Security*, Vol. 11, No. 2 (Fall 1986), pp. 151–176, argues that European powers' prewar estimates were reasonably accurate. On the British, see Edward L. Katzenbach, Jr., "The Horse Cavalry in the Twentieth Century," *Public Policy* (1958), pp. 120–149. John Ellis, *The Social History of the Machine Gun* (New York: Arno, 1975), argues that European racism contributed to the underestimation of firepower. On the evidence from late-nineteenth- and early-twentieth-century wars and the reports of military observers, see William McElwee, *The Art of War: Waterloo to Mons* (Bloomington: Indiana University Press, 1974), pp. 147–255. The most famous, accurate, contemporary civilian analysis is Ivan Bloch, *The Future of War in Its Technical, Economic, and Political Aspects: Is War Now Impossible?* trans. R. C. Long (New York: Doubleday and McClure, 1899).

97. Martin Van Creveld, *Supplying War: Logistics from Wallerstein to Patton* (Cambridge, U.K.: Cambridge University Press, 1977), pp. 109–141.

orders of battle. Debates about these sorts of questions are relevant to whether a given net assessment is useful for a particular policy purpose, but not to the validity of net assessment as a technique, or to whether it can be used to estimate the offense-defense balance.[98]

BALLPARK ESTIMATES OF THE BALANCE MAY BE SUFFICIENT

Critics have overstated the difficulty of measuring the offense-defense balance by overestimating the degree of measurement precision required by the theory. Most of the predictions of offense-defense theory are qualitative, not quantitative. For example, the theory expects more intense arms racing, tighter alliances, and a higher likelihood of war when the offense-defense balance is close to 1 (i.e., relatively favorable to offense) than when it is much greater than 1 (i.e., greatly favors defense); this prediction does not depend on knowing the exact numerical value of the balance.

Because the offense-defense balance is stated in terms of attacker to defender cost ratios, the boundaries of the meaningful ranges depend on the resource ratios between states. If the offense-defense balance for a particular attacker versus a certain defender is much greater than the ratio of the state's resources, then the defender should be able to achieve a high degree of security; if the balance is close to or less than the resource ratio, then the defender will be unable to achieve a high degree of security. Thus, if we can estimate whether the offense-defense balance is close to or far from 1, and especially how it compares to resource ratios, this is often sufficient for explanation, prediction, and policy use.

Some important net assessments have strongly suggested that the offense-defense balance often lies within a range that makes predictions possible. The easiest case concerns nuclear weapons. Repeated detailed examinations of possible U.S.-Soviet nuclear exchanges showed that neither country could get close to having a damage-limitation capability, implying that the defense had

98. Cohen, "Toward Better Net Assessment," argued that even the best net assessments of the Central Front did not do justice to the complexity of the problem. Most of his specific criticisms of Mearsheimer, Posen, and Epstein, however, concern scenario issues, such as inclusion of certain Soviet units, possible Warsaw Pact advances through Austria or Denmark, the likelihood of airdrops in the NATO rear, or the possibility that NATO might not mobilize in time. A few of Cohen's criticisms—for example, that these authors overestimated force-to-space constraints on Warsaw Pact ground offensives as well as the effectiveness of NATO compared to Warsaw Pact airpower—are relevant to the quality of our understanding of modern land warfare, although the weight of opinion on these questions seems to be against Cohen. See John J. Mearsheimer, Barry R. Posen, and Eliot A. Cohen, "Correspondence: Reassessing Net Assessment," *International Security*, Vol. 13, No. 4 (Spring 1989), pp. 128–179.

a large advantage.[99] It seems likely that unbiased net assessment before World War I would have shown that the advantage of defense was so large that no European major power could attack another with much chance of success. Nicholas Spykman's estimate in 1942 that even if Germany and Japan succeeded in subjugating and integrating the whole economic potential of Eurasia, the United States would still be able to defend North America against military invasion amounted to a judgment that the offense-defense balance was more favorable to defense than the resource ratio between the United States and Eurasia.[100]

Another way in which critics exaggerate the precision required of net assessments is by overlooking the value of relative net assessment, which is easier to perform than absolute net assessment. Even in circumstances where the balance cannot be measured even within the broad ranges just discussed, relative measures can often be made. That is, it is often possible to say whether the offense-defense balance for one dyad (or one time period) favors defense more or less than the balance for another. This is useful because many of the predictions and policy implications of offense-defense theory are cast in relative terms—for example, shifting the balance toward defense reduces incentives to arms race and to attack. Thus, even when we cannot say quantitatively how much the balance has shifted from one situation to another, knowing in which direction it has shifted tells us something about the behavior we expect in the second situation compared to the first.

MEASUREMENT DOES NOT REQUIRE THAT OFFENSE AND DEFENSE BE DISTINGUISHABLE

Critics have argued that the indistinguishability of offense and defense prevents measurement of the offense-defense balance. For example, John Mearsheimer argues that determining the offense-defense balance is problematic because "it is very difficult to distinguish between offensive and defensive weapons."[101] This criticism is mistaken, however, because it conflates distin-

99. See, for example, Congressional Budget Office, *Counterforce Issues for the U.S. Strategic Offensive Forces* (Washington, D.C.: U.S. Government Printing Office, 1978); and Michael M. May, George F. Bing, and John D. Steinbruner, "Strategic Arsenals after START: The Implications of Deep Cuts," *International Security*, Vol. 13, No. 1 (Summer 1988), pp. 90–133.
100. Nicholas Spykman, *America's Strategy in World Politics: The United States and the Balance of Power* (New York: Harcourt Brace Jovanovich, 1942), esp. 389–457.
101. John J. Mearsheimer, "The False Promise of International Institutions," *International Security*, Vol. 19, No. 3 (Winter 1994/95), p. 23; see also Shimshoni, "Technology, Military Advantage, and World War I," pp. 190–191.

guishability with the offense-defense balance—two analytically separable concepts. Whether or not particular weapons are distinguishable has no effect on our ability to calculate the offense-defense balance.[102]

To assess the offense-defense balance, we assume that the attacker and the defender act optimally, deploying the weapons that best enable them to achieve their respective missions. The attacker's and the defender's optimal force structures may or may not include some of the same types of weapons. Either way, given these forces, measuring the balance then requires performing the kind of net assessment described above. The ability to performing this net assessment is not impeded by having some of the same types of weapons on both sides.[103]

To appreciate this, imagine the extreme hypothetical case in which only one type of weapon exists. All states must rely on this weapon for both offense and defense, which in effect leaves offense and defense indistinguishable. Assessing the offense-defense balance might still be easy, however. Consider two worlds: in the first, the only available weapons are fast but rather thin-skinned tanks; in the second, only artillery pieces so heavy that they cannot be moved are available. In both worlds offense and defense are indistinguishable. It is clear, however, even from a causal inspection that defense has a much greater advantage in the second world than in the first; net assessment is in no way impeded by the indistinguishability of the weapons.

The Offense-Defense Balance and the Status of Offense-Defense Theory

We believe that this article should lay to rest the concern that the underdevelopment of the concept of the offense-defense balance could undermine offense-defense theory. In addition, we show that the barriers to measuring the offense-defense balance are less severe than critics suggest. The key point is that the skills required for net assessment are essentially the same as those required to measure the balance, and net assessment is often feasible.

---

102. In a forthcoming article we argue that offensive and defensive weapons and force postures generally are distinguishable.

103. Lynn-Jones, "Offense-Defense Theory and Its Critics," pp. 674–676, makes essentially the same argument. Biddle, "Determinants of Offensiveness and Defensiveness," pp. 17–19 and 341–343, concludes that defense has the advantage, although he argues that distinguishing offensive and defensive weapons is hard.

We do, however, anticipate that our approach may generate one new criticism. The understanding of the offense-defense balance that we have developed in this article is complex, so much so that some readers may conclude that it is too complex to operationalize, or that if operationalized it will not generate enough additional explanatory power to warrant the effort. While appreciating the theory's complexity, we do not see this as a crippling problem. First, most international relations theories are complex, if one looks carefully. Offense-defense theory is not significantly more complex than balance-of-power theory, once one takes operationalization of power seriously. In balance-of-power theory, power means above all the ability to win wars. Operationalizing power thus requires many of the same steps as measuring the offense-defense balance. To measure either, we must know how resources of various kinds are converted into military forces and, most important, how different kinds of military capability are related to war outcomes.[104] If it were true that the offense-defense balance cannot be measured because we do not understand the determinants of war outcomes, then all structural theories of war and peace would be in trouble.

Our suggestions for research reflect our conclusion that the feasibility of measuring the offense-defense balance depends directly on the feasibility of net assessment. First, we need research that improves our understanding of the factors, including cumulativity and nationalism, that influence states' abilities to convert power into offensive and defensive capabilities.[105] For some of this we will need to reach beyond political science disciplinary boundaries to military scientists and historians, economic historians, and students of the sociology of nationalism.

Second, we need additional research on the feasibility and accuracy of past net assessments. We have argued, based on a preliminary examination, that reasonably effective net assessments have often been feasible. Given the importance of this issue, especially given critics' claim that the offense-defense theory is of little value because the balance cannot be measured, thorough studies of feasibility would be very valuable.[106]

104. On the difficulty of operationalizing power and the relevant literature, see William C. Wohlforth, *The Elusive Balance: Power and Perceptions during the Cold War* (Ithaca, N.Y.: Cornell University Press, 1993); see also Aaron Friedberg, *The Weary Titan: Britain and the Experience of Relative Decline, 1895–1905* (Princeton, N.J.: Princeton University Press, 1988).
105. For example, Liberman's study of the net resources that Germany was able to extract during World War II contributes to a research agenda on cumulativity; see Liberman, *Does Conquest Pay?*
106. For example, Biddle's work on the determinants of the relative success and failure of different stages of the German Spring Offensive of 1918 bears on the relative explanatory power of

In closing, this article has placed offense-defense theory on a firmer foundation by thoroughly exploring the cost-ratio definition of the offense-defense balance and the assumptions that are required to operationalize any definition of the balance, and by emphasizing the close relationship between the feasibility of net assessment and the feasibility of measuring the offense-defense balance. The result should be increased confidence in the potential of the growing literature that employs offense-defense theory to tackle important historical, theoretical, and policy questions.

---

technology, terrain, numbers, and skill; Biddle, "Determinants of Offensiveness and Defensiveness," pp. 241–311. Sean Lynn-Jones in a personal communication has suggested that "experimental net assessment" using the sorts of techniques developed for the U.S. Army's National Training Center (NTC) might be a useful measurement approach for a variety of tactical- and operational-level questions. On the methods used at NTC, see Daniel Bolger, *Dragons at War* (Novato, Calif.: Presidio, 1986).

Correspondence

Taking Offense at Offense-Defense Theory

James W. Davis, Jr.
Bernard I. Finel
Stacie E. Goddard

Stephen Van Evera
Charles L. Glaser and
Chaim Kaufmann

To the Editors (James W. Davis, Jr. writes):

In his article "Offense, Defense, and the Causes of War,"[1] Stephen Van Evera claims that "offense-defense theory" is "important," has "wide explanatory range. . . . wide real-world applicability. . . . large prescriptive utility. . . . [and] is quite satisfying" (p. 41).

Van Evera's conclusions are, however, unwarranted. First, his reformulation of influential arguments made prominent by Robert Jervis stretches the meaning of key concepts such that interesting avenues of empirical inquiry are closed off rather than opened. Second, the hypotheses—or "prime predictions"—Van Evera derives from the theory are themselves products of faulty deductive logic. Furthermore, they are non-testable, presumably nonscientific in Van Evera's understanding of the term.[2] Van Evera's results are thus of little use to the social scientist who is interested in understanding the myriad causes of war and conditions facilitative of peace.

In his classic article, "Cooperation under the Security Dilemma," Jervis argued that the security dilemma is more virulent and the international system less stable when offense enjoys an advantage over defense. By contrast, when defense is more potent, status quo powers find it easier to adopt compatible security policies, and the pernicious effects of international anarchy are greatly diminished.[3] Although the operation-

James W. Davis, Jr., is Assistant Professor of International Politics at Ludwig-Maximilians-Universität in Munich, Germany. He is also a NATO Research Fellow.

Bernard I. Finel is Associate Director of the National Security Studies Program and Visiting Assistant Professor of National Security Studies and International Affairs at Georgetown University's Edmund A. Walsh School of Foreign Service. *He thanks Dieter Dettke, Robert Haffa, Tim Hoyt, Jeffrey Lord, Kristin Lord, James Ludes, Gary Schaub, and Brent Sterling for their comments.*

Stacie E. Goddard is a doctoral candidate in the Department of Political Science at Columbia University.

Stephen Van Evera teaches International Relations in the Political Science Department at the Massachusetts Institute of Technology. He is a member of the MIT Security Studies Program.

Charles L. Glaser is Professor and Deputy Dean in the Irving B. Harris Graduate School of Public Policy Studies at the University of Chicago. Chaim Kaufmann is Associate Professor of International Relations at Lehigh University.

1. Stephen Van Evera, "Offense, Defense, and the Causes of War," *International Security*, Vol. 22, No. 4 (Spring 1998), pp. 5–43. Subsequent citations to this article appear in parentheses in the text.
2. See Stephen Van Evera, *Guide to Methods for Students of Political Science* (Ithaca, N.Y.: Cornell University Press, 1997).
3. See Robert Jervis, "Cooperation under the Security Dilemma," *World Politics*, Vol. 30, No. 2 (January 1978), pp. 167–214.

alization of the offense-defense balance has been the subject of considerable debate,[4] the concept as originally employed by Jervis referred to the modalities of battlefield conquest: military tactics, strategy, technology, and the state's geography. The argument's appeal derives from its elegance and parsimony, as well as its explanatory range. Through variations in a rather simple—basically material—relationship, we appear to gain leverage over a wide range of behavioral outcomes.

In his reformulation of the offense-defense balance, however, Van Evera adds "diplomatic factors" to the military and geographic factors identified by Jervis. In doing so, Van Evera subsumes under the offense-defense balance much of what we thought the balance helped explain. When "collective security systems, defensive alliances, and balancing behavior by neutral states" (pp. 21–22) are all constitutive of the offense-defense balance, we are no longer in a position to ask which military and geographic factors promote balancing, bandwagoning, or efforts at collective security; how they do so; or how the balance between offense and defense interacts with these diplomatic variables to produce such outcomes as war, peace, or overall system stability. Van Evera's redefinition of the offense-defense balance is a step backward, a regressive reformulation of a heretofore useful concept.

A second problem emerges because Van Evera fails to keep the material or "objective" offense-defense balance analytically distinct from the balance as it is perceived by the actors. That actors might not apprehend the true or objective state of the offense-defense balance was already recognized by Jervis.[5] The manifest difficulties that discrepancies between the objective and perceptual balance raise for attempts to use the concept in actual empirical investigations were, however, only later appreciated.[6]

The individual theorist may come down on one or the other side of the objective/perceptual divide, or she may choose to test which of the two variants accounts for outcomes in a given case. Logically ruled out, however, is the combination of both in a given hypothesis. Yet this is precisely what Van Evera attempts to do: "War will be more common in periods when conquest is easy or is believed easy, less common when conquest is difficult or is believed difficult" (p. 22). As formulated, the hypothesis is imprecise, internally incoherent, and as a result cannot be tested in any meaningful fashion.

---

4. See, for example, Jack S. Levy, "The Offense/Defense Balance of Military Technology: A Theoretical and Historical Analysis," *International Studies Quarterly*, Vol. 28, No. 2 (June 1984), pp. 219–238; Scott D. Sagan, "1914 Revisited: Allies, Offense, and Instability," *International Security*, Vol. 11, No. 2 (Fall 1986), pp. 151–175, esp. p. 161; Sean M. Lynn-Jones, "Offense-Defense Theory and Its Critics," *Security Studies*, Vol. 4, No. 4 (Summer 1995), pp. 660–691; and Charles L. Glaser and Chaim Kaufmann, "What Is the Offense-Defense Balance and Can We Measure It?" *International Security*, Vol. 22, No. 4 (Spring 1998), pp. 44–82.

5. Jervis, "Cooperation under the Security Dilemma," pp. 190–194.

6. See, for example, the rather ad hoc justification that Thomas Christensen and Jack Snyder offer for adopting the perceptual balance in their amended Waltzian model of the balance of power in Christensen and Snyder, "Chain Gangs and Passed Bucks: Predicting Alliance Patterns in Multipolarity," *International Organization*, Vol. 44, No. 2 (Spring 1990), p. 145. See also Snyder, "Perceptions of the Security Dilemma in 1914," in Robert Jervis, Richard Ned Lebow, and Janice Stein, eds., *Psychology and Deterrence* (Baltimore, Md.: Johns Hopkins University Press, 1985), pp. 153–179; and Richard Ned Lebow, *Between Peace and War* (Baltimore, Md.: Johns Hopkins University Press, 1981), chap. 7.

If two dimensions are at work—one objective and one perceptual—then logically we have four possible combinations. That is, conquest can be (1) easy and believed to be easy; (2) easy but believed to be difficult; (3) difficult but believed to be easy; or (4) difficult and believed to be difficult.

At least two cases contemplated by the permutation of the two variants of the offense-defense balance cannot be included in the same hypothesis, because they stand in logical opposition to each other. Thus we cannot have a hypothesis that simultaneously predicts war to be relatively frequent because people mistakenly believe offense to be dominant and rare because the defense is in fact dominant. Similarly, we cannot have a situation where war is predicted to be rare because the defense is believed to be dominant, but where in fact offense is dominant and the hypothesis simultaneously predicts wars to be more frequent. And if perceptions always track the "objective" offense-defense balance, then parsimony would dictate we leave perceptions out of our theory and thus reject the two classes of cases emerging from the prime prediction that are not ruled out by logic (i.e., offense is dominant and believed to be dominant, and defense is dominant and believed to be dominant). Moreover, such cases would be uninformative if we are interested in finding out how perceptions matter.

Third, Van Evera overstates the extent to which his theory stands up to empirical tests. He argues that "the strength of a passed test depends on the uniqueness of the predictions tested. Do other theories predict the outcome observed, or is the prediction unique to the tested theory? The predictions here seem quite unique. There is no obvious competing explanation for the periodic upsurges and downsurges in European expansionism and warfare outlined above. Offense-defense theory has the field to itself" (p. 35).

Every hypothesis is, however, tested against a competing explanation, even if merely a hypothetical counterfactual.[7] But given that every outcome is in some way consistent with Van Evera's hypothesis, one cannot even formulate a hypothetical counterfactual.[8] Moreover, it is generally accepted that one is justified in ascribing some plausibility to a theory's explanatory claims only after it has been tested against a competing theory.[9] Theories and hypotheses are "fortified" or "strengthened" to the degree to which they pass tests that are suggested in light of competing explanations.[10] Van Evera is thus

7. See James D. Fearon, "Counterfactuals and Hypothesis Testing in Political Science," *World Politics*, Vol. 43, No. 2 (January 1991), pp. 169–195.

8. This has the effect of closing off a traditional escape route for structural theorists (i.e., the argument that their theory explains only tendencies and not particular outcomes, because tendencies are demonstrable only to the extent to which we can clearly identify outliers).

9. For an argument with roots in Popper, see Paul Feyerabend, "Problems of Empiricism," in Robert Colodny, ed., *Beyond the Edge of Certainty* (Englewood Cliffs, N.J.: Prentice-Hall, 1965); Feyerabend, "Reply to Criticism," in Robert S. Cohen, ed., *Boston Studies in the Philosophy of Science*, Vol. 2 (New York: Humanities Press, 1965), pp. 223–261, esp. p. 227; and Imre Lakatos, "Methodology of Scientific Research Programmes," in Lakatos and Alan Musgrave, eds., *Criticism and the Growth of Knowledge* (Cambridge: Cambridge University Press, 1970), p. 190. Even Thomas Kuhn stressed his acceptance of this criterion, although under certain limited conditions. See Kuhn, "Logic of Discovery or Psychology of Research?" in ibid., pp. 1–23.

10. For discussions of hard or crucial tests in the social sciences, see Harry Eckstein, "Case Study and Theory in Political Science," in Fred Greenstein and Nelson Polsby, eds., *Handbook of Political*

promoting an unorthodox understanding of hard tests when he writes: "Alternative explanations for the rise and fall of American global activism are hard to come up with, leaving the offense-defense theory's explanation without strong competitors, so this element of the test posed by the U.S. case is fairly strong" (p. 40).

Of course, serious alternative explanations for variations in war propensity abound. For example, based on a study of the European states system from 1640 to 1990, Andreas Osiander concluded that stability is a function of the coherence of the principal (normative) assumptions upon which an international system is founded.[11] And although he does not dismiss the effects of "size, structure, power, and geographic position of the various European states," Paul Schroeder argues that the chief difference between the relatively war-prone late eighteenth century and the more peaceful Concert of Europe a generation later was the lack of consensus among the great powers on legitimate principles of conduct and an equitable balance of power prior to the Napoleonic Wars.[12]

Given the existence of competitors, Van Evera's discovery that he has the field to himself suggests that he is either lost, or is playing something more akin to solitaire than to science.

—*James W. Davis, Jr.*
Munich, Germany

To the Editors (Bernard I. Finel writes):

Several recent articles have provided textured considerations of offense-defense theory and the impact of the offense-defense balance on state behavior.[1] These works have tightened the conceptual logic and added much-needed refinements to the argument. Four major problems with offense-defense theory remain, however. First, offense-defense theory ignores interaction effects in warfare. Second, it makes ill-considered assumptions about the links between control of territory, conquest, and victory in war. Third, the theory is still neither well conceptualized nor operationalized. Finally, the approach lacks parsimony.

Science, Vol. 7, *Strategies of Inquiry* (Reading, Mass.: Addison-Wesley, 1975), esp. pp. 118–120; and Arend Lijphart, "Comparative Politics and the Comparative Method," *American Political Science Review*, Vol. 65, No. 3 (September 1971), esp. pp. 692–693.

11. Andreas Osiander, *The States System of Europe, 1640–1990: Peacemaking and the Conditions of International Stability* (Oxford: Clarendon Press, 1994).

12. Paul W. Schroeder, *The Transformation of European Politics, 1763–1848* (Oxford: Clarendon Press, 1994), quotation at p. 10. For a similar argument, see Henry A. Kissinger, *A World Restored: Metternich, Castlereagh, and the Problems of Peace, 1812–1822* (New York: Grosset and Dunlap, 1964).

1. Most notably, Stephen Van Evera, "Offense, Defense, and the Causes of War," *International Security*, Vol. 22, No. 4 (Spring 1998), pp. 5–43; Charles L. Glaser and Chaim Kaufmann, "What Is the Offense-Defense Balance and Can We Measure It?" *International Security*, Vol. 22, No. 4 (Spring 1998), pp. 44–82; and Sean M. Lynn-Jones, "Offense-Defense Theory and Its Critics," *Security Studies*, Vol. 4, No. 4 (Summer 1995), pp. 660–691.

The Importance of Interaction Effects

The offense-defense balance is not a structural variable.[2] Rather it can be influenced by immediate decisions about deployments and employment strategies. This fact creates difficulties for Charles Glaser and Chaim Kaufmann as they try to use the tools of net assessment to operationalize and potentially quantify the offense-defense balance.[3] Their article is vague about the meaning of net assessment. Are Glaser and Kaufmann referring to net assessment as used by Eliot Cohen, Andrew Krepinevich, and Andrew Marshall—that is, broad-based, subjective analyses of nonmilitary as well as military factors?[4] Or are they referring to campaign analysis—that is, the use of mathematical models to predict the results of highly specified force-on-force engagements?[5] Campaign analysis would certainly fit their goal, but the problem is that campaign analysis usually relies on Forward Edge of the Battle Area (FEBA) models. FEBA models are useful in explaining the results of attrition warfare, but not necessarily of dynamic, maneuver-based warfare.

To predict the outcome of dynamic, maneuver-based warfare, it is possible to use complex war games. These war games rarely create reproducible results, and they are extremely sensitive to modification in the initial rules. However, war games usually demonstrate that different strategies, and more important, the interaction effects of different strategies, make a big difference. If we take the war-game approach seriously, then we must conclude that the offense-defense balance is not a structural variable, but an outgrowth of strategic interaction.[6]

The effectiveness of battlesystems depends on employment strategies, doctrine, and training and tactics. Changes at these three distinct levels of analysis are potentially independently capable of altering the course of a battle.[7] For example, the Schlieffen Plan determined the initial course of World War I in the West at the strategic level. It determined how the Germans would mobilize their forces, their concentration points, and their operational goals. In the end, the Schlieffen Plan's flaws—overextension and an uncovered right flank—doomed the German attack. Doctrine refers to the conceptual

2. Cf. Glaser and Kaufmann, "What Is the Offense-Defense Balance?" p. 55.
3. Ibid., p. 76.
4. Andrew W. Marshall, *Problems of Estimating Military Power* (Santa Monica, Calif.: RAND, 1996); Eliot A. Cohen, "Net Assessment: An American Approach," unpublished paper presented as JCSS (Jaffee Center for Strategic Studies) Memo No. 29 (April 1990); and Cohen, "Toward Better Net Assessment," *International Security*, Vol. 13, No. 1 (Summer 1988), pp. 50–89.
5. Robert P. Haffa, Jr., *Rational Methods, Prudent Choices: Planning U.S. Forces* (Washington, D.C.: National Defense University Press, 1988); Joshua M. Epstein, *The Calculus of Conventional War: Dynamic Analysis without Lanchester Equations* (Washington, D.C.: Brookings Institution, 1985); Epstein, *Conventional Force Reductions: A Dynamic Assessment* (Washington, D.C.: Brookings Institution, 1990); and Alain C. Enthoven and Wayne K. Smith, *How Much Is Enough? Shaping the Defense Program, 1961–1969* (New York: Harper and Row, 1971).
6. As an empirical example, the German attack on France in 1940 did not succeed quickly because of offensive dominance. Rather the German advance through the Ardennes to the English Channel coast was particularly effective because the Franco-British forces were pivoting into Belgium at the time. Martin van Creveld et al., *Air Power and Maneuver Warfare* (Maxwell Air Force Base, Ala.: Air University Press, 1994), p. 41.
7. John J. Mearsheimer, *Conventional Deterrence* (Ithaca, N.Y.: Cornell University Press, 1983).

basis for a tactical battlesystem. The blitzkrieg doctrine was a complex melding of armor, airpower, and disruptive penetrating advances. The development of the blitzkrieg concept allowed for the exploitation of the emerging technologies of the pre–World War II period, thus leading to a discontinuous increase in military effectiveness. Training and tactics refers to how forces actually fight. Are subordinate commanders trained to take the initiative or wait for orders? Do units engage or bypass enemy strong points? Do forces launch preparatory artillery barrages, or do they seek to maintain the element of surprise? The adoption of infiltration tactics, for example, jumpstarted the German offensive of March 1918 on the western front. Significantly, all three levels can vary independently of the current forces in being. Not only are there almost always several plausible tactical battlesystems and usage doctrines at any level of technology, but these battlesystems and doctrines generate a system of strategic interaction.

Warfare is fundamentally a "rock, paper, scissors" game. Choices are only dominant vis-à-vis other states' choices. Historically, light missile infantry dominated heavy infantry, while heavy infantry armed with pikes or spears was invulnerable to heavy cavalry. But heavy shock cavalry always dominated light infantry, which lacked the ability to resist charges. If the opponent was fielding a heavy cavalry force, the best defensive countermeasure was a heavy infantry force. But if the opponent was armed with heavy infantry, the best countermeasure was light infantry.[8] The optimal choice depends on the opponent's decisions.

Combined arms warfare is the response to this fact, but the effectiveness of a specific balance of forces in a combined arms system is also subject to strategic interaction. Furthermore, the effectiveness of any weapons system and any combined arms system depends on how the forces are being used. Are the forces being used as raiders or as holders of territory? Are they being used offensively or defensively?[9] The dynamics here are harder to illustrate, but consider this simple example: a strategic plan geared toward defeating an enemy army will work only if the opponent is willing to stand and fight. If the opponent chooses to use a Fabian strategy of avoiding conflict instead, the plan may come to naught.

Territory, Conquest, and Victory: Unpacking the Assumptions

This last point about the success of war plans raises a second problem with both the Van Evera and the Glaser and Kaufmann articles. Both articles assume that when seizing territory is easier, there is a greater propensity to use force.[10] The logic is flawed. John Mearsheimer has argued convincingly that the expected rapidity of victory is the crucial determinant in decisions to use force.[11] The problem is that seizing territory is

8. Archer Jones, *The Art of War in the Western World* (New York: Oxford University Press, 1987), p. 494.
9. For an extended examination of warfare in the West that focuses on different strategies and interaction effects, see ibid., *passim*.
10. Glaser and Kaufmann, "What Is the Offense-Defense Balance?" p. 47.
11. Mearsheimer, *Conventional Deterrence*, p. 64.

not synonymous with victory. Indeed, seizing and holding territory is neither necessary nor sufficient to win a war.

Wars are won under two conditions. First, it is possible to win a war by effectively eradicating the ability of the other side to resist. With the exception of World War II in Europe, however, no war in modern history has ended as a result of the absolute destruction and occupation of a country's territory. The second way to win a war is by either inflicting higher costs than the other side can accept or threatening credibly to do so. In many cases, this cost tolerance is not an objective measure, but rather a set of social constructions.[12]

It is possible to inflict these costs without occupying territory. Indeed, for much of recorded history, the norm was for the losing side to concede the issue of the war following defeat in a major battle, even when this did not, in any significant and lasting way, undermine the losing side's ability to wage war. In addition, historically, many wars have been won using raiding strategies in which control of territory is not sought. In ancient Greece, raiding was the dominant strategy. Alexander's defeat of Persia was not the result of his ability to control territory. Even in the American Civil War, the devastating impact of William Tecumseh Sherman's march to the sea had nothing to do with his ability to control territory.[13]

To the extent that it is not necessary to control territory to win a war, offense-defense theory begins to break down. If defeating enemy armies and inflicting costs are major priorities, a fundamentally defense-dominant world, given a low enough force-to-space ratio, can lead to very rapid victories. In 1866 Prussia defeated Austria by winning a relatively indecisive victory at Königgrätz—the Austrians were able to withdraw in good order and link up with reinforcements from Italy[14]—thereby causing a political crisis in Vienna. In 1870 the Prussians won a set of decisive victories against France by using turning maneuvers to wage an offensive strategy using the tactical defensive. These two very rapid and low-cost victories occurred in an era of extreme defense dominance in terms of tactical military factors. The explanation is that success in war and the ability to seize and hold territory are not coterminous.

The Implication of Complexity: The Problem of Post Hoc Justifications

Although one might argue that the offense-defense balance is worth examining on its own terms, offense-defense theory is often invoked as a concise way to expand the richness of systemic models of international relations, conflict, and even foreign policy.[15] Thus much of the value of offense-defense theory derives from its contribution to building rich, powerful, and parsimonious explanatory and predictive theories.

12. John Keegan, *A History of Warfare* (New York: Alfred A. Knopf, 1993), pp. 23–60.
13. Jones, *The Art of War in the Western World,* p. 417.
14. Ibid., p. 397.
15. See, for instance, Thomas J. Christensen and Jack Snyder, "Chain Gangs and Passed Bucks: Predicting Alliance Patterns in Multipolarity," *International Organization,* Vol. 44, No. 2 (Spring 1990), pp. 137–168.

In his article, Van Evera cites military factors, geography, social and political order, and diplomatic factors as causes of offense and defense dominance.[16] Although he attempts to aggregate these into a single offense-defense measure, these variables are vague, too disparate to aggregate, and extremely dependent on subjective assessments. They produce a wide variety of conflicting theoretical predictions, most of which can be resolved only by empirical analysis.

It is difficult to avoid analytical bias in this process. For instance, Van Evera argues that "popularity of regimes probably aided offense before roughly 1800 and has aided defense since then. The reversal stems from the appearance of cheap, mass-produced weapons useful for guerrilla warfare—assault rifles and machine guns, light mortars, and mines. The weapons of early times (sword and shield, pike and harquebus, heavy slow-firing muskets, etc.) were poorly adapted for guerrilla resistance."[17] There were, however, many guerrilla campaigns before 1800.[18] Longbows and crossbows were adequate guerrilla weapons.[19] In this case, it is difficult to see how the coding can be done a priori. If Van Evera used the fact that guerrilla campaigns were more common and successful after 1800 as a basis for the judgment, then he may have conflated causes and outcomes. In any case, the argument is underspecified because the basic coding criteria are not explicit.

This lack of explicit criteria creates ambiguities in Van Evera's article. Van Evera suggests that military factors favoring mass infantry enhance the offense, but he also stresses the limitations on offensive action imposed by the logistical demands of large forces.[20] Cavalry forces, because they are expensive and hence limited in number, are argued to favor the defense despite their greater mobility, but tactically analogous armored forces (although with even larger logistical requirements than cavalry) are said to favor the offense. Mass infantry in the Napoleonic Era favors the offense. Mass infantry during World War I favors the defense. According to Van Evera, this distinction is the result of "lethal small arms, barbed wire, and trenches."[21] However, Borodino and Waterloo—Napoleon's two major setbacks and the two most prominent battles where he fought a steady foe in a frontal assault—demonstrate the defensive power of mass infantry even in the early 1800s. Another example of this coding problem can be found by comparing the Van Evera and Glaser and Kaufmann articles. Whereas Glaser and Kaufmann argue that "the most widely agreed proposition is that improvements in mobility favor offense,"[22] Van Evera argues that chariots, cavalry, and railroads—all

16. Van Evera, "Offense, Defense, and the Causes of War," pp. 16–22.
17. Ibid., p. 20.
18. Keegan, *A History of Warfare,* pp. 5–11; Jones, *The Art of War in the Western World*, pp. 55–65; and Donald Kagan, "Athenian Strategy in the Peloponnesian War," in Williamson Murray, MacGregor Knox, and Alvin Bernstein, eds., *The Making of Strategy: Rulers, States, and War* (Cambridge: Cambridge University Press, 1994), p. 44.
19. Bernard Brodie and Fawn Brodie, *From Crossbow to H-Bomb* (Bloomington: Indiana University Press, 1973), pp. 35–39. Crossbows are extremely easy to use, and longbows were widely available hunting weapons in areas where hunting supplemented local food production.
20. Van Evera, "Offense, Defense, and the Causes of War," pp. 16, 17.
21. Ibid., p. 17.
22. Glaser and Kaufmann, "What Is the Offense-Defense Balance?" p. 62.

systems that enhance mobility—help the defense.[23] The other variables cited by Van Evera and Glaser and Kaufmann—the impact of geography, social and political order, and diplomatic factors—are even less susceptible to clear coding.

A larger problem is Van Evera's belief that these variables can be aggregated. This is troubling from a methodological perspective. Van Evera does not present any conceptual explanation for how he actually measures the offense-defense balance in each area. Instead, he presents a laundry list of things to look for. Nor does he explain the relative weights he uses in aggregating his offense-defense variables. This leads to such confusing passages as, "Sometimes technology overrode doctrine, as in 1914–18 and in 1945–91 (when the superpowers' militaries embraced offensive doctrines but could not find offensive counters to the nuclear revolution). Sometimes doctrine shaped technology, as in 1939–45, when blitzkrieg doctrine fashioned armor technology into an offensive instrument."[24] Without a set of contingent generalizations about the conditions that define the "sometimes," the theory assumes what it ought to demonstrate.

Offense-defense theory represents what Giovanni Sartori called "concept misformation."[25] Sartori argued, "The lower the discriminating power of a conceptual container, the more the facts are misgathered, i.e., the greater the misinformation."[26] By defining the balance as being a function of a vast, unrelated grab bag of conditions and variables, Van Evera and Glaser and Kaufmann have created a situation where the empirical referents become merely a menu of items to choose from to justify a preexisting assessment of what the offense-defense balance is at a given point in time. This problem is exacerbated because most observers begin their research with significant knowledge and preconceptions about what the offense-defense balance was during the particular period they study. Familiarity with the cases almost certainly leads to bias in interpretations. Glaser and Kaufmann compound this problem with their methodological approach. They assume that if you throw the right variables together and do some net assessment, the offense-defense balance will emerge. This sort of naked empiricism does not advance the cause of theory building.

A Parsimonious Addition?

It might be possible for a historically knowledgeable and methodologically sophisticated scholar to develop a comprehensive model of the causes of the offense-defense balance. The extraordinarily complex resulting model would not, however, be parsimonious.

A fully specified model of the sources of offense and defense dominance would take into account the interaction of different possible battlesystems and different usage options. This process would create a broad typological theory. Then by examining the cost of the competing system, we might be able to derive a crude offense-defense

23. Van Evera, "Offense, Defense, and the Causes of War," pp. 16–17.
24. Ibid., p. 18.
25. Giovanni Sartori, "Concept Misformation in Comparative Politics," *American Political Science Review*, Vol. 64, No. 4 (December 1970), pp. 1033–1053.
26. Ibid., p. 1039.

balance for a given typological space. This result would then need to be validated by some sort of empirical analysis across cases. Given that the process of operationalizing the balance is unwieldy, involving complicated theorizing and budgetary and campaign analysis, it is reasonable to question whether the offense-defense balance adds sufficient richness and explanatory power to justify the very significant loss in parsimony.

Conclusion

Offense-defense theory is methodologically flawed and conceptually muddled. Although Van Evera and Glaser and Kaufmann push offense-defense theory forward, it is time to ask whether offense-defense theory in fact moves the field forward, or whether it represents instead the security studies' version of the emperor's new clothes.

Defenders of offense-defense theory will likely make several responses. First, they will correctly argue that I fail to address the perceptual variant of offense-defense theory. However, if the offense-defense balance is not an objective or structural condition, but instead resides purely in the realm of perceptions, then it ought to be integrated into a cognitive-processes framework rather than held apart as a special sort of (mis)perception.

Second, defenders of offense-defense theory will claim that I overstate the difficulty of operationalizing the balance or that I overemphasize the complex interaction effects of tactical battlesystems, doctrines, and usage decisions. If that is the case, however, then I would simply ask them to demonstrate the operationalization process in a systematic, reproducible manner. Until that time, I will remain a skeptic.

Third, offense-defense theorists will argue that I overstate the problem with relying on seizing territory as a measure of offensive success. They may claim that substituting "victory" for "conquest" or "seizing territory" is a simple change that does not harm the theoretical construct. I would argue, however, that once one breaks the link between ease of seizing territory and victory, one is left with the fact that none of the empirical indicators operates in a consistent fashion. This is not a semantic distinction. Rather, all the "causes" of the offense-defense balance are derived from the ease of seizing territory. Once that link is broken, the entire logic of the argument is questionable.

Finally, and linked to the previous point, offense-defense theorists will claim that even if my arguments about interaction and mobile warfare are valid, the theory is still useful because it explains the conditions under which breakthroughs occur. These breakthroughs, offense-defense theorists might argue, are a prerequisite for any kind of military victory, and they require the sort of frontal, attrition assaults the relative costs of which offense-defense theory claims to measure. The problem with this argument is threefold. First, although the costs of breakthrough may be relatively higher or lower, breaking through on a narrow front may not raise the costs significantly for the campaign as a whole. Second, this sort of breakthrough implies a situation of high force-to-space ratios and a continuous front. Empirically, this is not a common condition. Third, breakthroughs do not necessarily require attrition and frontal assaults. Fronts can be broken by infiltration tactics and can usually be outflanked by a sufficiently imaginative foe.

So what can be done? Can offense-defense theory be saved? The short answer is, no. Offense-defense theory has too many critical and fatal flaws. If we cannot "uninvent" offense-defense theory, then we must be very cautious about how we use it. There is an unfortunate tendency in the field to believe that offense-defense theory is a cheap and easy way to add predictive power to an explanatory model. In reality, the issues raised by offense-defense theory are extremely complex and difficult to parse effectively. The theory creates more conceptual holes than it fills, and should come with a strong warning label attached.

—Bernard I. Finel
Washington, D.C.

To the Editors (Stacie E. Goddard writes):

Stephen Van Evera's and Charles Glaser and Chaim Kaufmann's recent contributions to *International Security* are welcome expansions of offense-defense theory.[1] Both articles recognize that although hypotheses presented in this literature are intuitively and empirically plausible, offense-defense theory has suffered from a lack of methodological rigor: definitions are often tautological; the variables of offense and defense dominance are continuously conflated with other factors significant to international relations theory; and at times hypotheses seem to be nothing more than "folk theorems" derived from the popular case of World War I.[2] Unfortunately, while both articles attempt to address and overcome these critiques, neither satisfactorily resolves the methodological problems mentioned above. Most notably, these authors do not distinguish the offense-defense balance from factors such as the balance of power and military skill. This in turn leaves them vulnerable to tautological propositions, overdetermination, and difficulties with empirical measurement and testing.

---

1. See Stephen Van Evera, "Offense, Defense, and the Causes of War," *International Security*, Vol. 22, No. 4 (Spring 1998), pp. 5–43; and Charles L. Glaser and Chaim Kaufmann, "What Is the Offense-Defense Balance and Can We Measure It?" *International Security*, Vol. 22, No. 4 (Spring 1998), pp. 44–82. Offense-defense theory holds that under conditions of offense dominance, war is more likely between states. The seminal work on offense-defense theory is Robert Jervis, "Cooperation under the Security Dilemma," *World Politics*, Vol. 30, No. 2 (January 1978), pp. 167–214. Other works include Stephen Van Evera, *Causes of War*, Vol. 1: *The Structure of Power and the Roots of War* (Ithaca, N.Y.: Cornell University Press, forthcoming); Jack Snyder, *The Ideology of the Offensive: Military Decision Making and the Disasters of 1914* (Ithaca, N.Y.: Cornell University Press, 1984); Stephen M. Walt, *Revolution and War* (Ithaca, N.Y.: Cornell University Press, 1996); Quincy Wright, *A Study of War*, 2d rev. ed. (Chicago: University of Chicago Press, 1965), pp. 792–808; George H. Quester, *Offense and Defense in the International System* (New York: Wiley, 1977); and Sean M. Lynn-Jones, "Offense-Defense Theory and Its Critics," *Security Studies*, Vol. 4, No. 4 (Summer 1995), pp. 660–691.
2. Important criticisms of the program can be found in Jack S. Levy, "The Offensive/Defensive Balance of Military Technology: A Theoretical and Historical Analysis," *International Studies Quarterly*, Vol. 28, No. 2 (June 1984), pp. 219–238; John J. Mearsheimer, *Conventional Deterrence* (Ithaca, N.Y.: Cornell University Press, 1983); Jonathan Shimshoni, "Technology, Military Advantage, and World War I: A Case for Military Entrepreneurship," *International Security*, Vol. 15, No. 3 (Winter 1990/91), pp. 187–215; and Colin S. Gray, *Weapons Don't Make War: Policy, Strategy, and Military Technology* (Lawrence: University Press of Kansas, 1993).

I begin by examining Van Evera's article, arguing that his conceptualization of offense dominance as the "ease of conquest" confuses the offense-defense balance with the probability of success in war. This conceptual problem has serious ramifications for the explanatory hypotheses and empirical evidence he brings to bear in his theory. I then turn to Glaser and Kaufmann, noting that while the authors are cognizant of previous methodological critiques, the theorists cannot avoid conflation with other variables in their framework. Specifically, their broad definition of the offense-defense balance, combined with their assumption of "optimal doctrine," is insufficient to distinguish offense dominance from either power or doctrine. Following these critiques, I conclude by offering definitions and suggestions for empirical testing that might help avoid these methodological problems in future research.

The Ease of Conquest and the Probability of War

In "Offense, Defense, and the Causes of War," Van Evera argues that "war is far more likely when conquest is easy, and that shifts in the offense-defense balance have a large effect on the risk of war."[3] He offers ten explanatory hypotheses, including the temptation to strike first and increased incentives for expansionism, that further link his definition with outbreaks of war. After testing these explanatory propositions against three periods in history (Europe since 1789, ancient China during the Spring and Autumn and Warring States periods, and the United States since 1789), Van Evera concludes that "offense-defense theory has the attributes of a good theory," explaining large amounts of international history with a single variable.[4]

Although many of the theoretical and empirical propositions are compelling, serious methodological flaws detract from Van Evera's argument. First, his definition of offense dominance—"conquest is easy"—conflates offense dominance with a host of other variables, most notably with the balance of power. Simply put, Van Evera has defined offense dominance in terms of war outcomes, confusing the offense-defense balance with the probability that an attacking state will prevail in the event of war.[5] To say that an attacking state has a high probability of defeating its opponent says nothing, however, about the relative efficacy of offensive and defensive operations per se. One can easily imagine a scenario in which defensive strategies have an advantage, yet the attacking state prevails because of superior logistical support, deeper economic resources, or an overwhelming advantage in the number of forces. For example, conquest certainly appeared "easy" for the North at the close of the American Civil War. This was not a function of the strength of offensive versus defensive strategies, but of the greater social, economic, and logistical support of the population.[6] In short, in order to

3. Van Evera, "Offense, Defense, and the Causes of War," p. 5.
4. Ibid., p. 41.
5. Also making this critique are Glaser and Kaufmann, "What Is the Offense-Defense Balance?" p. 70; however, I argue that they too are vulnerable to these criticisms.
6. See Michael Howard, "The Forgotten Dimensions of Strategy," *The Causes of Wars and Other Essays* (Cambridge, Mass.: Harvard University Press, 1983), pp. 101–115; and Theodore Ropp, *War in the Modern World* (New York: Collier Books, 1971), pp. 175–194.

distinguish the offense-defense balance from the balance of power, Van Evera needs to cast his definition in relative terms (i.e., the value of attacking compared to the value of defending), rather than focus on the absolute value of attacking for a state.

These conceptual problems are compounded in Van Evera's explanatory hypotheses. For instance, he asserts that "when conquest is easy, aggression is more alluring: it costs less to attempt and succeeds more often."[7] Therefore "resources are more cumulative when conquest is easy. . . . As a result, gains are more additive."[8] It may be the case that conquests are self-reinforcing; however, this hypothesis cannot logically be derived from an assessment of offensive and defensive strategies. Although the offense-defense balance can tell us the relative costs of attacking versus defending, it measures neither the absolute probability of success nor the absolute value of conquest.

Furthermore, Van Evera's causal explanations come close to tautologies: it often seems he is arguing that when conquest is easy or perceived to be easy, states will attempt to conquer. Indeed, myriad variables—including state behavior—are subsumed in his conceptualization. According to Van Evera, "Military technology and doctrine, geography, national social structure, and diplomatic arrangements (specifically, defensive alliances and balancing behavior by offshore powers) all matter" in determining the offense-defense balance.[9] One is left wondering what factors could be excluded from this definition to show the causal autonomy of the offense-defense balance. Moreover, after subsuming all of these behavioral variables into his conceptualization, he uses the offense-defense balance to explain behavior among states. For instance, after the Crimean War "the power of defenders fell dramatically because defense-enhancing diplomacy largely broke down."[10] In the end, Van Evera is using the behavior of states (eschewing defensive alliances in favor of offensive diplomacy) to explain the behavior of states (offensive diplomacy until 1871).[11]

Finally, these methodological flaws are strikingly evident in his empirical accounts. Most important, Van Evera makes no attempt to avoid conflating the offense-defense balance with the balance of power or military forces. For example, he argues that "during 1792–1815 the offense was fairly strong militarily, as a result of France's adoption of the mass army (enabled by the popularity of the French government)."[12] Although it is certainly plausible that Napoleonic doctrine and tactics related to mass armies favored the offense, the size of the army itself is indeterminate of the offense-defense balance. Arguably, Napoleon's mass army would have been more effective defending French soil, rather than searching out offensive campaigns across Europe.[13]

---

7. Van Evera, "Offense, Defense, and the Causes of War," p. 7.
8. Ibid., p. 8.
9. Ibid., p. 6.
10. Ibid., p. 28.
11. Van Evera avoided this tautology in "The Cult of the Offensive and the Origins of the First World War," *International Security*, Vol. 9, No. 1 (Summer 1984), pp. 58–107. In this article, alliances are a clearly distinguished dependent variable. The independent variable is perception of offense dominance, and thus Van Evera does not incorporate states' behavior into his explanation.
12. Van Evera, "Offense, Defense, and the Causes of War," p. 26.
13. For an example of this argument, see Jean Jaures, *L'armée nouvelle* (Paris: Editions Sociales, 1977).

That France could deploy a mass army tells us a lot about the power of the centralized state, but very little about the offense-defense balance.

In sum, while Van Evera argues that offense-defense theory offers both unique and plausible predictions, methodological flaws with his approach undermine this claim. By subsuming power, military skill, diplomacy, doctrine, social structure, and domestic political structure into his conceptualization of the offense-defense balance, Van Evera cannot argue that the offense-defense balance has more explanatory power than any of these variables taken separately. Needless to say, a much narrower definition and systematic hypotheses are necessary before these claims to progress can be empirically tested.

Power, Skill, and Strategy: The Offense-Defense Balance and the Optimality Assumption

In "What Is the Offense-Defense Balance and Can We Measure It?" Glaser and Kaufmann are particularly concerned with previous methodological critiques of offense-defense theory. They note that critics have questioned the utility of the theory on the grounds that "the foundations of the theory are underdeveloped" and that "the theory contains inherent flaws, the most serious of which is that the offense-defense balance cannot be measured because the outcomes of war are so uncertain."[14] In response to these criticisms, Glaser and Kaufmann state that the offense-defense balance should be defined relatively: "the ratio of the cost of the forces that the attacker requires to take territory to the cost of the defender's forces."[15] They argue that this definition, combined with six key assumptions, allows offense-defense theorists to avoid conflation with other variables. Moreover, the theorists contend that the offense-defense balance can be measured by using the analytical toolbox of military net assessment, thus allowing empirical tests of the theory.

Clearly, distinguishing the offense-defense balance from other factors is a crucial task for Glaser and Kaufmann. They state that they "envision offense-defense theory as a partial theory of military capabilities. . . . A more complete theory would include two additional variables: (1) power, measured in terms of relative resources; and (2) what we term 'military skill,' that is, a country's ability to effectively employ military technology."[16] Indeed, Glaser and Kaufmann do not claim that the offense-defense balance is the only or even primary determinant of military outcomes, but rather "each of these three variables has the potential to overwhelm the others in certain circumstances."[17]

Although Glaser and Kaufmann realize that distinguishing the offense-defense balance from power and skill is important, they fail to do so adequately in their methodological framework. First, their adoption of a broad definition of the balance subsumes competing factors, such as force size and even nationalism. Although these are critical factors in determining the outcome of a war, they are more likely reflective of the power

14. Glaser and Kaufmann, "What Is the Offense-Defense Balance?" p. 45.
15. Ibid., p. 46.
16. Ibid., pp. 48–49.
17. Ibid., p. 49.

of a centralized state, and the ability of the state to effectively mobilize resources and forces, than they are of the relative efficacy of offensive and defensive strategies. As I argue above, it is plausible that an attacking state with popular support and a mass army will defeat a country lacking these attributes. This, however, does not tell us the relative value of offensive and defensive strategies for either of the actors involved.

Glaser and Kaufmann face similar methodological problems when trying to distinguish the offense-defense balance from military skill. They argue that "the offense-defense balance should be assessed assuming optimality—that is, countries choose the best possible strategies and force postures for attack and defense. Offense-defense theory requires this assumption because it focuses on the effects of the constraints and opportunities presented by the international environment."[18] The problem with this assumption is that it invariably leads to a *post hoc, ergo propter hoc* fallacy—we can only ascertain the optimal strategy after observing which strategies succeeded during a war. By measuring the balance in terms of successful strategies, we can neither determine the balance ex ante nor satisfactorily separate the balance from power and skill.

For instance, consider the offense-defense balance during World War I. Most would concur that the balance heavily favored the defense, both before and during the war, although this balance was misperceived by statesmen and military leaders alike.[19] Militarily, this defensive advantage is epitomized by the 1916 Battle on the Somme, an attack launched by the British on July 1, 1916, and lasting through November of that same year. Over this five-month span, and at a cost of approximately 500,000 casualties (the largest number ever of British casualties in battle), the British were able to move the front only seven miles. Two years later, however, on March 21, 1918, the Germans achieved a massive breakthrough on the same terrain, using the same type of weapons available to the British in 1916. After breaking through the juncture of French and British troops, the Germans employed a creeping barrage and infiltration tactics to gain more ground on the first day of attack (approximately 140 square miles) than the British had in 140 days. The German strategy in World War I would therefore be considered by Glaser and Kaufmann to be the optimal doctrine. In fact, using these criteria would mean that World War I was "objectively" offense dominant.[20]

---

18. Ibid., p. 46.
19. Advancing the hypothesis that defense was dominant before and during World War I are Van Evera, "Offense, Defense, and the Causes of War"; Glaser and Kaufmann, "What Is the Offense-Defense Balance?"; Van Evera, "The Cult of the Offensive and the Origins of the First World War"; and Snyder, *Ideology of the Offensive.* For a dissenting argument, see Shimshoni, "Technology, Military Advantage, and World War I." For a more limited critique, see Scott D. Sagan, "1914 Revisited: Allies, Offense, and Instability," *International Security,* Vol. 11, No. 2 (Fall 1986), pp. 151–176.
20. Shimshoni, "Technology, Military Advantage, and World War I." Historians and political scientists alike have noted that had the Schlieffen Plan succeeded—and this is not a ridiculous counterfactual—we would call World War I objectively offense dominant. See Sagan, "1914 Revisited." For other examples of this debate, see L.C.F. Turner, "The Significance of the Schlieffen Plan," in Paul M. Kennedy, ed., *The War Plans of the Great Powers, 1880–1914* (Boston: Allen and Unwin, 1979), pp. 203–204; and Martin Van Crevald, *Supplying War: Logistics from Wallenstein to Patton* (Cambridge: Cambridge University Press, 1977), p. 116. I thank Warner R. Schilling and Daniel H. Nexon for helpful discussions on this topic.

Although this conclusion may seem bizarre, I have trouble seeing how Glaser and Kaufmann would refute it. A possible response might be that while my example focuses on tactics, they claim to measure the offense-defense balance at the strategic level of conflict. This argument is problematic for two reasons, however. First, as Glaser and Kaufmann note, "a change that shifts the balance in a given direction at one level will usually also shift it in the same direction at all higher levels."[21] Therefore "a change that makes tactical offense harder will usually also make operational offense harder, which in turn makes strategic offense more difficult."[22] Indeed, tactical innovations at the Somme in 1918 made offense easier at the operational and strategic levels for the Germans, and later the Allies.[23] Furthermore, even if distinguishing tactical from strategic shifts in the offense-defense balance is theoretically possible, Glaser and Kaufmann provide insufficient guidance on how this could be accomplished. In fact, their own net assessment techniques rely on both tactical and strategic calculations to measure the offense-defense balance.[24]

In sum, Glaser and Kaufmann's assumption of "optimality" neither controls for military skill nor allows for ex ante assessment of the offense-defense balance. Indeed, the optimality assumption means that one would inherently code periods in terms of skill, measuring the offense-defense balance in terms of the most successful strategy employed. In further research, this dilemma could be addressed by assuming symmetrical doctrine. Obviously, whether doctrine is sufficiently symmetrical is ultimately an empirical question. This assumption, however, would allow theorists to compare the efficacy of offensive and defensive strategies while controlling for skill and doctrine.

Conclusion

Although offense-defense theory is riddled with methodological problems, the hypotheses put forth by these theorists are both empirically plausible and policy relevant. The Van Evera and Glaser and Kaufmann articles deserve attention and scrutiny for these reasons. Critical revision of the theory is clearly in order; to this end, I offer two suggestions.

First, a definition of the offense-defense balance that avoids subsuming power and doctrine needs to be constructed. The balance cannot be represented as the probability of taking territory, and should not incorporate the absolute value of an attack. An example of such a definition is one that describes offense dominance as a situation in which it costs less in terms of lives and territory to attack a state than it does to defend against it. Analogously, defense dominance would imply that attacking costs more in terms of lives and territory than defending against an attack. The offense-defense balance is represented as the cost difference between the two. Not only is this particular definition a comparison of the relative efficacy of the strategies, it says nothing about the ultimate outcome of a war, and thus avoids incorporating power into the concept.

21. Glaser and Kaufmann, "What Is the Offense-Defense Balance?" p. 73.
22. Ibid.
23. Ropp, *War in the Modern World*, p. 267.
24. Glaser and Kaufmann, "What Is the Offense-Defense Balance?" p. 75.

Moreover, this conceptualization allows that the offense-defense balance might vary from state to state, given that it is framed in terms of a subjective utility function.

Second, empirical tests of the offense-defense balance must be far more rigorous. Coding periods as offense or defense dominant by looking at the outcomes of war does not tell us about the efficacy of strategies and could easily be representative of balance-of-power factors. To avoid this, one might consider cases where states faced the possibility of conflict with each other over an extended period of time. Finding time periods in which the offense-defense balance varied, while power and doctrine was relatively constant, would be difficult; however, this would allow the variables of power and doctrine to be controlled.

—*Stacie E. Goddard*
New York, N.Y.

Stephen Van Evera Replies:

James Davis, Bernard Finel, and Stacie Goddard raise a number of questions about offense-defense theory. Here I focus on three that seem most important, briefly address five others, and let the rest pass without comment, except to express a general dissent from their arguments.

First, can we characterize specific military technologies or force postures as defensive or offensive? Or is everything dependent on the context of combat?

Bernard Finel takes the latter view, arguing that "the offense-defense balance is . . . an outgrowth of strategic interaction."[1] He much overstates a good point. As he notes, the capacity of a military force can depend on the forces and strategies of its opponent. Some forces are better at attack than defense against some opponents, while being better at defense than attack against others. But many forces are inherently optimized for offense or defense, in a way that applies across opponents.

A secure nuclear deterrent is fundamentally defensive. It makes its owner essentially unconquerable. At the same time, it cannot conquer other states that possess a secure deterrent. These facts apply regardless of the opponent's strategy. The defensive character of the nuclear revolution, which stems from these realities, is the defining feature of modern international relations. It is the single most important aspect of post–World War II and post–Cold War international affairs.

Because nuclear deterrents are defensive, forces designed to counter them are essentially offensive. These forces include offensive strategic nuclear counterforce systems (e.g., accurate intercontinental ballistic missiles and strategic antisubmarine warfare systems) and area ballistic missile defenses.

Modern guerrilla war has defended many countries and conquered none. It is a fundamentally defensive form of warfare. States would pose little threat to one another if all relied on citizen guerrilla defenses.

The accurate repeating rifles, machine guns, barbed wire, railroads, and entrenchments of the western front in World War I were fundamentally defensive, as the

1. Bernard I. Finel, "Taking Offense at Offense-Defense Theory," *International Security*, Vol. 23, No. 3 (Winter 1998/99), p. 183.

repeated failed offensives of 1914–17 demonstrated. They were eventually overcome: first by the German infiltration tactics in 1918, and later by German blitzkrieg concepts of armored war. But for three years they gave dominance to the defense.

Many other developments in military history can be characterized the same way. The weaponry and tactics of the late Middle Ages in Europe advantaged the defense; the gunpowder revolution then overcame these weapons and tactics, restoring the offense; innovations in fortification by Vauban and others restored the defense in the late seventeenth century; and Napoleonic warfare by popular mass army then strengthened the offense. In the twentieth century, German armored blitzkrieg tactics also bolstered the offense. The effects of these modes of warfare varied only modestly with the nature of their opposition. They had an inherent propensity to ease the defense or the offense.

In sum, technology and force posture do, on important occasions, have innate defensive or offensive properties and implications.[2] Nuclear weapons are the most important recent example, but they are only one among many.

Second, does offense-defense theory lack parsimony? Does it commit the sin of explaining by complexifying?

Finel claims that it does. Its independent variable—the ease of conquest—includes factors drawn from the military, diplomatic, geographic, and social spheres. These factors are, says Finel, a "vast, unrelated grab bag of conditions and variables." The use of such an unwieldy variable leads to a "very significant loss in parsimony."[3]

Finel has things backward. Offense-defense theory is elegant. It is parsimonious. It orders and thereby simplifies a previously disordered mélange of phenomena.

To judge Finel's charge, we first must ask: What is parsimony? What provides it, and what detracts from it?

A theory is not shown to lack parsimony simply by demonstrating that its concepts include a diverse range of lesser-included concepts, because this is true of every concept. All concepts are aggregations of lesser concepts. For example, national power is a concept that aggregates national military power, economic power, and the power to marshal allies. But thinking about strategy would be more complicated, not simpler, if we dropped national power from our lexicon and discussed only its component parts. Military power, a main component of national power, is an aggregate of airpower, naval power, and ground power, as well as of material resources, skill, and willpower. Discussion would be far harder if we had to address these components separately each time military questions were at issue. Airpower, a component of military power, aggregates the quality and quantity of aircraft, the quality and quantity of air force personnel, and the quality of air doctrine. The quality of aircraft in turn is an aggrega-

2. A good survey of the history of the interaction of military technology, geography, and the offense-defense balance is Charles L. Glaser and Chaim Kaufmann, "What Is the Offense-Defense Balance and Can We Measure It?" *International Security*, Vol. 22, No. 4 (Spring 1998), pp. 44–82 at 61–66.

3. Finel, "Taking Offense at Offense-Defense Theory," pp. 187, 188. Stacie Goddard likewise complains that "myriad variables" are subsumed in my definition of the offense-defense balance. "One is left wondering what factors could be excluded from this definition." Goddard, "Taking Offense at Offense-Defense Theory," *International Security*, Vol. 23, No. 3 (Winter 1998/99), p. 191. Arguing in the same vein are Glaser and Kaufmann, "What Is the Offense-Defense Balance?" pp. 60, 68–70.

tion of the speed, range, payload, maneuverability, stealthy characteristics, avionics, and durability of the aircraft. At every level, we find that concepts are composed of more concepts from the level below.

My rule of thumb is to judge new concepts by asking if they are theoretically useful. In political science this standard usually requires that concepts somehow correspond to phenomena in the real world. They must fit the way things work, or the way we think. A concept that fails to do this is artificial and only clutters discussion. But a concept that combines other concepts while also capturing reality simplifies discussion.

When considering national security problems, leaders often ask if others can conquer them or if they can conquer others. These questions lie at the core of many past and present foreign and security policy debates. And in asking if conquest is possible, leaders aggregate the same military, diplomatic, geographic, and social factors that I aggregate to capture the ease of conquest. Leaders aggregate because they must; because these factors together decide if they can conquer or be conquered. Aggregation is not easy and cannot be precise, but leaders do it because otherwise they cannot understand their national security situation. As Bismarck planned his wars of German unification, he recurrently asked if Germany was in a position to wage aggressive war successfully. To answer, he had to weigh military, diplomatic, and geographic considerations in some combination. Later, Germans who warned of German insecurity weighed these same factors together. Americans did the same when assessing their national security during the Cold War. Former President Herbert Hoover opposed the U.S. troop deployment to Europe in 1951 because he thought conquest was difficult: "This Hemisphere can be defended from Communist armies come what will. . . . Communist armies can no more get to Washington than any allied armies can get to Moscow."[4] Hoover had to aggregate military, geographic, and diplomatic factors to reach this conclusion. Oppositely, the authors of NSC-68 feared in 1950 that the United States faced a grave threat to its security, in large part because they believed that conquest was easy. They reached this conclusion by aggregating both military and diplomatic factors.[5]

The concept of ease of conquest, then, corresponds to the way policymakers think. It captures the way they organize the world. If it did not, it would be a complicating distraction. But it does. And in so doing, it simplifies our discussion of security problems and policies.

Offense-defense theory achieves other simplicities as well. It argues that a number of important war causes—expansionism, fierce resistance to others' expansion, first-strike advantages, windows of opportunity and vulnerability, faits accomplis, negotiation failures, secrecy, arms races, and "chain gaining" in alliances—that were once

---

4. Quoted in Hugh Ross, ed., *The Cold War: Containment and Its Critics* (Chicago: Rand McNally, 1963), p. 17.

5. See NSC-68, excerpted in Thomas H. Etzold and John Lewis Gaddis, eds., *Containment: Documents on American Policy and Strategy, 1945–1950* (New York: Columbia University Press, 1978), pp. 414, 416 (suggesting that nuclear weapons are offensive), and pp. 427, 430 (suggesting that states tend to bandwagon with threats). Also relevant is p. 396, suggesting that the Soviet empire was vulnerable to Western offensive action, for essentially social reasons.

viewed as independent stem from a single cause. This simplifies the problem of power and war. We see that a number of disparate dangers are fed by a single taproot.

In short, offense-defense theory explains a wide range of phenomena with a parsimonious framework. Far from complexifying, it streamlines our understanding of the war problem.

Third, is offense-defense theory testable? James Davis complains that my formulation of offense-defense theory is too imprecise and incoherent to be tested.[6] Specifically, he notes that I consider two variants of offense-defense theory—an objective variant and a perceptual variant—but I infer and test only one set of predictions from these two variants. Instead, he argues, each variant must have its own distinct set of predictions. Davis makes a good point, but his conclusion is overdrawn. He shows that I failed to explain myself, but does not show that offense-defense theory is untestable or is otherwise flawed.

Let me clarify the confusion that Davis rightly identifies. Offense-defense theory has an objective and a perceptual variant. The objective variant frames the effects of the actual offense-defense balance. The perceptual variant frames the effects of the perceived offense-defense balance. Both the objective and the perceived offense-defense balances are indicators of the other: the objective offense-defense balance influences—and thus indicates—the perceived offense-defense balance; and the perceived offense-defense balance is influenced by—and thus indicates—the objective offense-defense balance. Therefore both the objective and the perceptual variants of offense-defense theory make predictions about the correlates of both the objective and the perceptual offense-defense balances. I conflated these two sets of predictions in my article, offering a single unified forecast about how things would appear if the offense were dominant *or* if it were believed dominant. I did this because it works in this instance: both variants of the theory make parallel predictions about both objective and perceived reality. We need not distinguish these forecasts because they are essentially the same. Two variants of a theory usually produce two divergent sets of predictions, but not in this case.

What do we conclude when the objective and perceived offense-defense balances differ, as in 1914, when the objective balance favored the defense and the perceived balance favored the offense? Both variants of the theory make stronger predictions about the correlates of the perceived balance than the objective balance, so tests that look to the perceived balance are stronger. For example, in the 1914 case both variants predict that perceptions of offense dominance should correlate with war, even if objective realities favor the defense. The perceptual variant predicts this simply because it deals only with the perceptions, and puts the objective balance aside. The objective variant predicts this because the impact of the objective balance is translated into outcomes through its effect on the perceived balance, as follows: objective offense-defense balance ---> perceived offense-defense balance ---> outcomes. The hypothesis on the right (perceived offense-defense balance ---> outcomes) should operate even when the hypothesis on the left (objective offense-defense balance ---> perceived

6. James W. Davis, Jr., "Taking Offense at Offense-Defense Theory," *International Security*, Vol. 23, No. 3 (Winter 1998/99), pp. 179–182 at 180.

offense-defense balance) does not. Hence even the objective variant forecasts a more certain correlation between perceptions of offense dominance and war than between objective offense dominance and war.

I saved my readers this detail because I feared that their eyes might glaze over. But I should have explained it, and Davis is right to complain that my failure to explain is confusing. He is wrong to claim, however, that offense-defense theory is somehow flawed or untestable. I failed to explicate my logic, but I think that logic is sound.

I close with remarks on five other criticisms made by Davis, Finel, and Goddard.

Finel argues that seizing territory does not confer victory in war.[7] He seems to think I believe otherwise, but I agree with him. My view is that the seizing of territory requires victory (not the other way around). Clearly, without a victory of some kind it is impossible to seize and hold another state's territory.

Finel quarrels with my coding of the history of military technology and strategy. He doubts that guerrilla war grew easier with the development of mass-produced modern small arms after 1800, and that accurate repeating rifles, machine guns, and barbed wire made frontal assaults more difficult in the late nineteenth and early twentieth centuries.[8] In so arguing, he takes on many historians in addition to myself. I concur, however, that such questions are not open-and-shut, and we need a detailed study of the history of the offense-defense balance in warfare to help resolve such disputes.

Stacie Goddard argues that I failed to distinguish the offense-defense balance from the balance of power.[9] I certainly meant to distinguish them, and believe I did. In footnote 1, I suggest that the offense-defense balance could be measured by looking at the probability that a determined aggressor could conquer a target state with comparable resources. In other words, this measure asks how often conquest occurs where the balance of power cannot account for the outcome, because the winner starts with no marked resource advantage. It should be clear from this that I am not running together the offense-defense balance and the balance of power.

Goddard and Davis contend that I use the behavior of states to explain the behavior of states.[10] I plead guilty. In fact, the behavior of states often explains the behavior of states. Europe's continental powers reach for hegemony when Europe's offshore balancers (Britain and the United States) are in an isolationist mood, and are more cautious when the balancers are active. If that's how the world works, shouldn't we say so? I don't see a problem.

Davis objects that I define terms in ways that will confine others' analyses.[11] But of course others are free to adopt the definition that best helps them answer their questions, just as I did. I cannot stop them and would not want to.

7. Finel, "Taking Offense at Offense-Defense Theory," p. 184.
8. Ibid., pp. 184–185.
9. Goddard, "Taking Offense at Offense-Defense Theory," pp. 189–190.
10. Ibid., p. 191; and Davis, "Taking Offense at Offense-Defense Theory," p. 180.
11. Davis, "Taking Offense at Offense-Defense Theory," pp. 179–180. Davis's claim that I deviate from Robert Jervis's definition of the offense-defense balance by including diplomacy also seems questionable. Jervis, like me, does mention a diplomatic factor—collective security systems—as an element that affects the security of states. Jervis, "Cooperation under the Security Dilemma," *World Politics*, Vol. 30, No. 2 (January 1978), pp. 167–214 at 176.

Davis, Finel, and Goddard have raised important issues that deserve further attention, but offense-defense theory stands up to their criticisms.

—*Stephen Van Evera*
Cambridge, Massachusetts

Charles L. Glaser and Chaim Kaufmann Reply:

We appreciate the opportunity to respond to the issues raised by Bernard Finel and Stacie Goddard. We believe that there is little real disagreement between us and Goddard—she makes a number of sound arguments, but on some points apparent differences between her position and ours result from her misunderstanding of our views. In contrast, we think that Finel's wholesale condemnation of the entire offense-defense research program is at least premature, based on failures to understand both the purposes of offense-defense theory and the requirements for testing it. Goddard's and Finel's most important points relate to the status of offense-defense theory as a structural theory of international behavior. Therefore we address these issues first, and then turn to several separate issues raised by one or the other correspondent.

As a structural theory, offense-defense theory attempts to predict states' behavior by focusing on the constraints and opportunities presented by their environment. Important constraints include the offense-defense balance and the distribution of resources (power), while among the choices that the theory seeks to predict are decisions about military doctrine and force posture, as well as whether to form alliances and fight wars.

Structural theories of international politics can incorporate various kinds of constraints, some of which are stricter than others. System structure, as defined by Kenneth Waltz, excludes properties of units (typically states). In Waltz's narrow definition, structure consists only of properties that emerge from the relationships of the units to one another and that no individual state can change—international anarchy, for example.[1] Constraints that emerge from purely material facts, which can be properties of states but which states cannot change or evade, we can call material structure; with limited exceptions, geography and weather would qualify. Theories based on these hard constraints are, however, often not satisfying, because they miss too much of interest in most areas of international behavior.

Most structural theories therefore employ a "softer" definition of structure that includes any constraints that states cannot change or evade within the time scales they are likely to consider in planning foreign and defense policy.[2] These may include social facts such as a state's form of government or its level of scientific achievement at a given time. Measurements of power, for example, must include not only a state's purely material resources but also the capacity of the state apparatus to extract resources from

1. Kenneth N. Waltz, *Theory of International Politics* (Reading, Mass.: Addison-Wesley, 1979).
2. Appropriate time scales for different types of foreign policy decisions may vary. For offense-defense theory, which is principally concerned with understanding decisions in peacetime about planning for deterrence or for fighting future wars, we think a relevant time scale is often from several years to ten years, although there could be variations—for example, if a state could be highly confident that it would face no threats for a longer period.

society for military use. Our measure of the offense-defense balance includes whether the state—and its adversaries—are nation-states or multinational empires, as well as many aspects of technology.[3] Including certain unit-level factors does not erase the difference between structural and nonstructural theories. What all structural theories exclude are explanations of the actual decisionmaking process, including the possibility of flaws of perception and judgment, and information about state preferences that is based on their unit-level characteristics.

To understand the impact of constraints, structural theories, including offense-defense theory, must assume that states' policy choices are broadly optimal or rational; subject to the constraints they face, states make effective policy choices for maximizing their interests.[4] Theories that do not assume optimality must include a theory of suboptimal state decisionmaking. In such theories much of the explanation of state behavior is often attributed to domestic political competition and/or to flaws in the decisionmaking process; the impact of environmental constraints on policy choices is weaker and harder to isolate (although those same constraints may still exert a powerful influence on the eventual outcomes of policy choices).

The optimality/rationality assumption in structural theories is useful in three ways: it can provide guidance for making policy; it establishes a baseline against which states' policies can be compared to determine whether they are flawed; and it can help assess the likely impact of flawed policies. By providing a baseline, the optimality assumption in offense-defense theory enables us to separate military skill from the balance. The balance is measured assuming that all countries have high levels of military skill.[5] Choices that diverge from the baseline are suboptimal and indicate low skill.

Goddard objects to our inclusion of an optimality assumption in offense-defense theory, arguing that it leaves us unable to separate military skill from the offense-defense balance.[6] The problem, as she sees it, is that the optimal strategy can be determined only "after observing which strategies succeed during a war," which means that we cannot determine the balance ex ante. We agree that the balance should be measured ex ante—this is required by the theory. Goddard's objection is based on an exaggeration of the standard of optimality required for purposes of the theory. She apparently takes optimality to mean the absolute best choice the state could make given not just the (limited) information available at the time but also the information that would be provided by future actions. With this understanding of optimality, it would

3. Constraints could include facts that decisionmakers theoretically could change, but only at costs they would likely consider unacceptable. For instance, a multinational empire could convert itself into a nation-state by giving up its imperial territories and subjects, but this would likely violate the rulers' sense of the identity of the state, as well as reduce its resources.

4. Except in formal rational choice–oriented work, the assumption is not usually that actual state decisionmaking processes meet normative standards of rationality, but rather that decisionmakers act "sensibly": given the information available to them, decisionmakers usually make choices that do not vary much in substance from those that would be made by purely rational actors.

5. Charles L. Glaser and Chaim Kaufmann, "What Is the Offense-Defense Balance and Can We Measure It?" *International Security,* Vol. 22, No. 4 (Spring 1998), pp. 44–82 at 55–56.

6. Stacie E. Goddard, "Taking Offense at Offense-Defense Theory," *International Security,* Vol. 23, No. 3 (Winter 1998/99), pp. 193–194.

indeed follow that optimal force posture, doctrine, and strategy cannot be known until they are revealed by war outcomes, and that ex ante measurement of the offense-defense balance would be impossible.[7] However, this is not what we mean by optimality; the standard that we employ is that states choose optimally "within reasonable limits of analysis" given the information available to them at the time; to impose a stricter standard would be unreasonable for a theory intended to predict actual behavior.[8] Our standard is the same optimality/rationality assumption that appears in most structural theories.[9]

Based on our understanding of structure, we also reject two of Finel's key points. First, he argues that "the offense-defense balance is not a structural variable. Rather, it can be influenced by immediate decisions about deployments and employment strategies." This is incorrect. The offense-defense balance is a constraint, not a measure of the effectiveness of actual deployed forces for either offense or defense. It answers the question: How secure can states be, assuming that both they and their opponents make optimal choices? The offense-defense balance, in combination with power, determines how well a state *can* do; state decisions in combination with structural constraints determine how well a state will *actually* do. Suboptimal decisions reduce the state's military capability compared to the best that it could be, but do not influence the balance itself.[10] For example, if one state deploys nuclear weapons in vulnerable basing modes, then an attacker's prospects for significantly limiting damage (and therefore for a successful offensive attack) will be much greater than if the state had made better deployment decisions. However, the offense-defense balance for this example remains defined by the *best* retaliatory capability that the state could achieve given both sides' resources and available technology.

Second, Finel argues that the offense-defense balance is not a structural variable because it is an "outgrowth of strategic interaction," so that "the optimal choice depends on the opponent's decisions."[11] In response, we would first like to point out that a great many military policy decisions, especially at the levels of doctrine and force posture, are pure optimization problems involving no interaction. Before World War I, all armies would have been better off deploying more machine guns and less cavalry—regardless of what anyone else did. Between the two world wars, all navies would have been better off investing less in battleships. Finel's observation that ancient armies used widely varying force combinations does not imply, as he suggests, that the best force

7. Actually, optimal strategy in Goddard's sense cannot be reliably determined even after a war. Ex post we know that the victor's choices were successful, but not necessarily whether they were optimal. The victor could have won despite suboptimal strategy because of even worse suboptimal choices made by the loser. Alternatively, even if the loser's choices were optimal and the victory was the result of superior power, it is possible that the victor could have succeeded even more easily with a better strategy that was not tried.
8. Glaser and Kaufmann, "What Is the Offense-Defense Balance?" p. 55.
9. This does mean that when information that would affect states' strategies is unavailable, decisionmakers' estimates of the offense-defense balance may differ from the estimates they would have made had they had the additional information. A possible example is the difficulty in 1939 of estimating whether atomic weapons would be developed in time for use in World War II.
10. Glaser and Kaufmann, "What Is the Offense-Defense Balance?" p. 55.
11. Bernard I. Finel, "Taking Offense at Offense-Defense Theory," *International Security*, Vol. 23, No. 3 (Winter 1998/99), pp. 183, 184.

posture generally depends on the force posture of the opponent. Nearly all of the most successful ancient armies were based on cores of heavy infantry, with lesser investment in cavalry and light troops. When armies were organized on other principles, this was usually because of limitations imposed by social systems or by terrain, not by the nature of the opponent's army. When structural constraints are strong, not only doctrinal but also some wartime strategy choices can become noninteractive. For example, at the start of World War I, all European states would have been better off if they had scaled down their offensive plans, regardless of what other states did.

Strategic interactions do occur. When they do, whether Finel's point is valid depends on exactly what we mean by "interaction." One sense of interaction would mean that each state must optimize its military doctrines and strategies not only subject to those structural constraints that influence it directly, but also subject to the knowledge that their opponents will also be attempting to optimize their own choices subject to the constraints facing them. The offense-defense balance is estimated assuming that both sides do the best they can, each knowing that the other side is also doing the best it can. For example, a state attempting to develop an effective nuclear damage-limitation capability against a particular opponent would have to take into account not only the technical, geographical, and other limits on its counterforce capabilities, but also its opponent's best options for improving the survivability of its strategic forces. If missile accuracy is low, the opponent can deploy survivable forces with little effort, and the balance will strongly favor defense. If accuracy is high and the opponent lacks the technology for survivable launchers such as mobile missiles and nuclear-powered submarines, it will have to spend much more to achieve a robust retaliatory capability, and the balance will be more favorable to offense. In any case, the resulting measure of the balance is fully structurally determined (provided that states do not make suboptimal choices). To the extent that each state responds not only directly to structural factors but also to the other's behavior, it is simply incorporating additional structural effects mediated indirectly through their pressure on the opponent.

A second sense of "interaction" would be a situation in which structural constraints exert only a weak influence on strategy choice, so that each side has two or more options that are equally good from a structural point of view, and that are different enough from each other that the best counterstrategies against each are quite different. Thus success for each side would depend in large part on correctly guessing the other side's choice, but there is no way to guess except by having some insight into the opponent's decisionmaking process, or by luck. In Finel's terms, this would be a true "rock-scissors-paper" situation where there is no one optimal choice, which would indeed mean that the offense-defense balance could not be measured exactly, but would spread out into a band of uncertainty whose width would reflect the impact on war outcomes of different combinations of a state's right or wrong guesses about its opponent's choices.

Such situations do occur in war, although they are more common at the operational and tactical levels, where individual decisions have smaller effects on final war outcomes than do choices at the strategic and grand strategic levels. An example might be German attempts in 1944 to estimate whether the Allied invasion of Europe would come at Normandy or Calais. Various constraints effectively ruled out sites either further west or east, but this still left the Allies (and thus the Germans) with a choice to make between the two remaining options. Had the Germans guessed correctly, the

Allies' 1944–45 campaign would have been noticeably more expensive, but it is quite unlikely that the outcome of the war would have changed or that its length would have been affected very much. By comparison, strategic and grand strategic choices, which can exert larger effects on war outcomes, are usually heavily constrained by factors such as geography; states' political, social, and material resource endowments; and so forth. Therefore strategic and grand strategic choices are often more similar to the mutual optimization model discussed above than they are to guessing games. For example, in World War II the Allies could not defeat Germany without moving huge amounts of men and matériel from the United States to Europe, and their only practical method was by sea. The Germans' most efficient method of opposing this flow was by submarine. Both sides' plans took these facts into account, but this was a mutual optimization problem, not a problem of guessing the other's intentions. Thus, although strategic interaction can create situations where our best measure of the offense-defense balance becomes a band rather than a point, it is not clear that these bands are often wide. This is, however, a worthwhile avenue for empirical research.

Beyond these points that are closely connected to the concept of structure, Goddard and Finel address some additional important points. First, Goddard argues that in adopting a broad definition of the offense-defense balance by including nationalism and force size, we cannot separate the balance from power, because these factors influence power. In fact, nationalism can affect both power and the offense-defense balance, in different ways. As Goddard notes, nationalism can augment a state's power by increasing its ability to extract resources. However, nationalism also affects extraction capabilities differentially, increasing them more when the state is trying to protect territory that is understood to be part of the national homeland than when it seeks to take territory that is not part of this homeland. Nationalism makes it easier to translate aggregate power into the ability to defend the national group's homeland. It makes it harder to translate power into the ability to conquer territory that is perceived to be outside the national homeland. This differential effect cannot be incorporated into our standard notions of power; rather, it is best understood as a shift in the offense-defense balance in favor of defense.

Goddard is correct that force size can reflect power, but it should also be included among the factors that influence the balance. Some states simply lack the resources to deploy forces of the size that would create an offense-defense balance best matched to their goals. States then do the best they can with the resources available, that is, within the constraints imposed by their power, which determines the size of their forces. For example, the impact of conventional force size on the balance results from an interaction between power and geography. High force-to-space ratios tend to favor defense more than do low force-to-space ratios. As a result, two states whose mobilization potential is relatively large in relation to length of front will face a balance more in favor of defense than if both were weaker. For example, even if technology had remained constant, the balance between Germany and France in 1914 would have been more favorable to defense than it was in 1870. Both states were able to mobilize greater resources, which shifted the offense-defense balance as well as the balance of power. If we consider only power, we would predict the wrong impact on the outcome of war. Although from the 1870s onward Germany's power grew faster than France's, its chances of conquering France declined. We could have lumped this effect under the general heading of "geography," but we consider it important enough in its own right to mention separately.[12]

A main theme of Finel's letter is that the offense-defense balance cannot be measured: the factors that influence the balance cannot be reliably coded, and the results of war games are not reproducible. We agree with Finel that whether the balance can be measured is central to the utility of the theory. In our article we explain, first, why the tools of military net assessment are well matched to measuring the balance.[13] Also, although we offer reasons for optimism, we stress that our article does not settle the question of how closely the balance can be measured. In fact, the article concludes by calling for research into whether net assessment has been feasible in the past and whether net assessment techniques can be further improved. Second, our article explains how offense-defense theory can often make useful predictions even when net assessment is difficult and therefore substantial uncertainty about the value of the balance exists. Thus Finel's criticism would be telling only if reliable net assessment can never be done. To accept Finel's pessimism would mean, for example, that during the Cold War we did not and could not have had any real idea of Soviet prospects for a successful offensive on the central front, and that the German military before World War I lacked the analytic resources to detect the flaws and uncertainties in the Schlieffen Plan. If Finel holds these views, he could have contributed to the debate by spelling out why we should accept them.

Finel also objects to our version of offense-defense theory because he says it wrongly treats seizure of territory as the principal method of victory in war. This would indeed be a mistake, and would skew net assessments and thus estimates of the offense-defense balance—but this is not our position. Finel has simply misread us, confusing the *measure* of success with the *means* for achieving success. For purposes of the theory, we treat changes in political control of territory as the measure of success in war outcomes (i.e., we use the standard Clausewitzian definition). A military offensive should be considered a success if, at the end of the war, political control has changed to the advantage of the attacker—either because the attacker has gained full or partial control of territory it did not control before, or because it has undermined an opponent's control of territory.[14] The two most important means of achieving military victory are (1) destruction of enemy forces or mobilization potential, undermining their ability to contest territorial control; and (2) credible threats of such severe punishment that the opponent would rather concede the territory than continue the conflict.[15] It is true, however, that

12. Concern over nuclear proliferation illustrates a similar interaction between power and the offense-defense balance. Opponents of proliferation argue that new nuclear states will be unable to build the large forces required to provide adequate retaliatory capabilities; as a result, although nuclear weapons provided the superpowers with highly effective deterrent capabilities, they will not do the same for small countries with limited resources. In other words, according to this argument, the offense-defense balance for these less powerful countries is more favorable to offense than it was for the superpowers because these weaker countries will have smaller forces.

13. Glaser and Kaufmann, "What Is the Offense-Defense Balance?" pp. 74–78.

14. Ibid., pp. 54–55.

15. There is some evidence that, at least in conventional conflicts, punishment is less likely to be decisive than is destruction of forces. Robert A. Pape, *Bombing to Win: Air Power and Coercion in War* (Ithaca, N.Y.: Cornell University Press, 1996).

who controls certain territories during a war can sometimes affect victory to the extent that the territory can be used to mobilize additional capabilities with which to pursue one or both of these approaches.[16] Thus the offense-defense balance, which is defined as the cost of the means of victory relative to the cost of the means of denying victory, reflects the impact of territory only when it influences war outcomes.

Finally, Finel argues that even if the offense-defense balance could be adequately specified and measured, the cost in terms of complexity would outweigh the insights that offense-defense theory can provide. We agree that measuring the offense-defense balance is complex. Nevertheless, three points suggest that Finel is again too negative. First, as we argue in our article, key variables in other structural theories of international relations, such as power in balance-of-power theory, also become difficult to operationalize if we take the task seriously. This is simply a general problem in international relations theory. Second, as we discuss in our article, in cases where complexity makes complete net assessment intractable, analysts can often simplify their task while preserving reasonable confidence in their estimate of the balance. One possibility is to focus on particular theaters or campaigns that are expected to have a decisive impact on the overall war outcome. For example, in the 1980s, net assessments of a conventional World War III in Europe concentrated on estimating the prospects of success of the Warsaw Pact's initial offensive into Western Europe. Given NATO's much larger mobilization potential and control of the oceans, the Warsaw Pact could not hope to win a war if the initial offensive was not successful.[17] Third, and perhaps most important, the predictions made by offense- defense theory often diverge significantly from standard power-based structural theories, which means that if the offense-defense balance can be measured, the payoff is likely to be worth the effort.

In closing, we continue to believe that our article has placed offense-defense theory on a firmer foundation. Further research is warranted, because offense-defense theorists have established powerful deductive arguments showing that power alone is insufficient to explain state decisions about military forces, strategy, and war, and that offense-defense variables should influence these decisions. It is too early to render a verdict from empirical testing of offense-defense theory, which is still in its infancy. Further testing is the key.

—*Charles L. Glaser*
Chicago, Illinois
—*Chaim Kaufmann*
Bethlehem, Pennsylvania

16. See our discussion of cumulativity of resources: Glaser and Kaufmann, "What Is the Offense-Defense Balance?" pp. 67–68.
17. We discuss other types of simplifications in our article; see ibid., p. 61.

Must War Find a Way?

A Review Essay

Richard K. Betts

Stephen Van Evera, *Causes of War: Power and the Roots of Conflict*
Ithaca, N.Y.: Cornell University Press, 1999

War is like love, it always finds a way.
—Bertolt Brecht, *Mother Courage*

Stephen Van Evera's book revises half of a fifteen-year-old dissertation that must be among the most cited in history. This volume is a major entry in academic security studies, and for some time it will stand beside only a few other modern works on causes of war that aspiring international relations theorists are expected to digest. Given that political science syllabi seldom assign works more than a generation old, it is even possible that for a while this book may edge ahead of the more general modern classics on the subject such as E.H. Carr's masterful polemic, *The Twenty Years' Crisis*, and Kenneth Waltz's *Man, the State, and War.*[1]

Richard K. Betts is Leo A. Shifrin Professor of War and Peace Studies at Columbia University, Director of National Security Studies at the Council on Foreign Relations, and editor of Conflict after the Cold War: Arguments on Causes of War and Peace *(New York: Longman, 1994).*

For comments on a previous draft the author thanks Stephen Biddle, Robert Jervis, and Jack Snyder.

1. E.H. Carr, *The Twenty Years' Crisis,* 2d ed. (New York: Macmillan, 1946); and Kenneth N. Waltz, *Man, the State, and War* (New York: Columbia University Press, 1959). See also Waltz's more general work, *Theory of International Politics* (Reading, Mass.: Addison-Wesley, 1979); and Hans J. Morgenthau, *Politics among Nations: The Struggle for Power and Peace,* 5th ed. (New York: Knopf, 1973). After these standard works, it is difficult to select the most important from recent decades. Among them would be Quincy Wright, *A Study of War,* 2 volumes (Chicago: University of Chicago Press, 1942); and Robert Gilpin, *War and Change in World Politics* (New York: Cambridge University Press, 1981). Most intense examination of the causes of war has come from scholars in the realist tradition. For a provocative attack on the main arguments in this tradition, see John Mueller, *Retreat from Doomsday: The Obsolescence of Major War* (New York: Basic Books, 1989). For a liberal perspective, see Michael W. Doyle, *Ways of War and Peace: Realism, Liberalism, and Socialism* (New York: W.W. Norton, 1997). For the best overall review of contemporary literature, see Jack S. Levy, "The Causes of War: A Review of Theories and Evidence," in Philip E. Tetlock, Jo L. Husbands, Robert Jervis, Paul C. Stern, and Charles Tilly, eds., *Behavior, Society, and Nuclear War,* vol. 1 (New York: Oxford University Press, 1989).

There is one particular book to which Van Evera's especially begs comparison. Apparently unconcerned about having order clerks in college bookstores confuse his own book with a contemporary classic, he chose the same title as historian Geoffrey Blainey. In this generation, Blainey's is the book most similar in scope, although different in approach and style, and it is one that will ultimately last at least as well. Blainey examined and debunked more than a dozen popular notions about why wars happen and, by process of elimination, settled on one main conclusion: "Wars usually begin when two nations disagree on their relative strength." At least one of the nations in conflict must miscalculate who would succeed in a test of arms, or the weaker would yield without a fight. Thus a clear pecking order in international relations may not necessarily produce justice, but it promotes peace. A roughly even balance of power, in contrast, makes miscalculation easier. The key to peace is clarity about the distribution of power.[2] Blainey arrived at these spare and powerful conclusions inductively, and his book is a tapestry of unconventional questions, analytical excursions and examples, and ironic observations that social scientists would consider literate and lively but unsystematic.

Van Evera travels a different route, one that proceeds more methodically and deductively and aims to honor the canons of social science. He arrives at a place apparently different from Blainey's, but actually similar. Van Evera argues for five hypotheses but concentrates on one: "War is more likely when conquest is easy."[3] This seems to contradict Blainey's view that imbalance favors peace because weakness encourages compliance with the stronger power. In the course of the book, however, Van Evera makes clear his conviction that the main problem is quite different from the one stated in the initial hypothesis: it is not that conquest actually is easy, but that most often it is not, yet is mistakenly *perceived* to be easy. This amendment is quite compatible with Blainey's bottom line.

Van Evera also has an ambitious aim. Where Blainey remains strictly empirical, Van Evera seeks to provide prescriptive analysis that can show policymakers how to manipulate causes of war and deploy countermeasures against them. As a good realist, he does not claim that he can make war obsolete, but he does imply that if statesmen understood the analysis in the book and acted on it, searches for security could be relaxed and more wars could be foregone.

---

2. Geoffrey Blainey, *The Causes of War*, 3d ed. (New York: Free Press, 1988), p. 293; see also pp. 109–114.
3. Stephen Van Evera, *Causes of War: Power and the Roots of Conflict* (Ithaca, N.Y.: Cornell University Press, 1999), p. 4. Subsequent references to this work appear parenthetically in the text.

Because it is mainly mistaken beliefs about threats to security that account for war, education—reading this book—should contribute to a solution.

This is a good book. It covers a lot of territory in a clear and organized manner. It focuses on a coherent set of issues and schools of thought and evaluates them systematically. It distills much of the conceptual and theoretical apparatus that evolved in strategic studies during the Cold War and applies the ideas broadly. It makes strong arguments. The book takes off from the essential premise of structural realism—that international anarchy is the permissive cause of war—and investigates more specific ways in which beliefs about the options provided by certain kinds of power may yield sufficient causes. Van Evera does not demonstrate that any single cause of war is sufficient, nor does he claim that any, except perhaps one—the ease of conquest—is likely to be. His analysis does provide reasonable grounds to believe that *combinations* of some of the causes he examines may well be sufficient in many cases. The book's reach turns out to exceed its grasp, but it is better to aim high and fall short than to fulfill a trivial mission.

Remember all this as you read on. A review essay cannot justify article length if it simply lauds its subject. It must do one of two things. It can use the assignment as a pretext for an excursion of the reviewer's own, and ignore the work that is supposed to be examined. This common ploy is irresponsible. The alternative is to dwell on the limitations of the work under review. That is what this essay does. This emphasis would be unfair to Van Evera if readers forget the generally favorable regard within which my criticisms are wrapped.

Social science has more inherent limitations than natural science, and all important works have important limitations. Because it is easier for academics to score points by finding such limitations than by developing unassailable theories of their own, every major book becomes a target. This one goes out of its way to draw fire, however, by claiming to offer a "master theory." The claim overreaches, for four main reasons.

First, the volume aims for great prescriptive utility, but its conceptual apparatus is deeply rooted in two events of one century (our own): World War I, and the maturation of nuclear deterrence between the United States and the Soviet Union. This may not limit the lessons for the twenty-first century if we are concerned only about interstate wars among great powers, but it is less certain that this frame of reference is the best guide to understanding the majority of contemporary wars: intrastate struggles among groups vying to determine the scope of communities and the character of regimes.

Second, the book does too much. The master theory (offense-defense theory) turns out to roll so many things into the definition of one master variable (the offense-defense balance) that the distinction between the author's theory and those he seeks to surpass erodes. The improvement over theories that focus on power in general, such as Blainey's, becomes hard to discern. Van Evera's book does not make the concept of offense-defense balance identical to relative power—his definition does leave out some significant aspects of the latter—but it comes too close for comfort.

Third, the book does too little. Like most realist theories, it avoids considering the substantive stakes of political conflict. Such issues may be secondary to the main causes that lie in the distribution of power, but they are still too important to be excluded from a master theory, especially one that deals with cost-benefit calculations of attackers and defenders. Van Evera piles his most important arguments onto the question of when and why states believe that conquest is "easy." This is a vague adjective on which to rest a master theory. It would be more specific and relevant to say instead: when states believe that conquest of a *desired objective* is achievable at *acceptable cost*. That in turn depends not only on the odds of military success but on the value that the government or group places on the objective. As with structural realists of other stripes, there is an apolitical flavor to some of the assumptions that animate the analysis.

The fourth significant limitation is the book's normative bias in favor of the status quo and the prevention of war. Perhaps all civilized people in the enlightened and satisfied West can agree on these aims. They contaminate analysis, however, when they spill over into an empirical assumption that states normally seek security and cause problems for others because of misguided notions that security requires expansion. This premise is really a hypothesis that remains to be demonstrated, and contradicts the traditional realist view that states' ambitions grow as their power expands.[4] It is all the more significant because Van Evera's fundamental diagnosis is like Pogo's ("We have met the enemy and he is us"). Given that real insecurity is rare, misperception to the contrary becomes "a self-fulfilling prophecy, fostering bellicose policies."

4. See Fareed Zakaria, "Realism and Domestic Politics," *International Security*, Vol. 17, No. 1 (Summer 1992), pp. 191–192, 196; and Zakaria, *From Wealth to Power: The Unusual Origins of America's World Role* (Princeton, N.J.: Princeton University Press, 1998). Van Evera cites the United States as a good test of his theory's prediction that states with high security will be less aggressive. He dismisses Zakaria's alternative explanation (weakness of the state apparatus) by saying that it does not explain the "decline of U.S. activism after 1815 or after 1991" (pp. 182–184, 184 n. 224). U.S. activism, however, did *not* decline in these periods. In the decades after 1815, the United States doubled in size, taking half of Mexico. Washington has been remarkably active since disposing of the Soviet empire, fighting two medium-sized wars in less than a decade since the Cold War ended, engaging in several smaller interventions on three different continents, and sporting the mantle of world orderer with alacrity.

As a result, "the prime threat to the security of modern great powers is . . . themselves" (pp. 191–192; ellipsis in original).

"Security" is a polymorphic value, as hopelessly slippery as "national interest." Van Evera does not define "security" in detail, but he implicitly identifies it with the political status quo, especially in terms of what states control what territory. Dispossessed groups with grievances, however, have no reason to identify security that way, and no more reason to see resort to force for revising an unjust status quo as illegimate than Van Evera has to see use of force for defending the status quo as legitimate. Or greedy groups may want to take what others hold irrespective of whether what they hold themselves is secure.

The Cold War hard-wired the value of stability in the minds of Western strategists, but stability is as normatively charged as any value. Sanctifying stability, as E.H. Carr argued, represents the international morality of the "haves" against the "have-nots."[5] If war is always a greater evil than injustice, and if prevention of war is the only value at issue, the solution is simple: pacifism. In an important sense, as Clausewitz notes, wars are begun not by invaders but by defenders who resist them.[6] (As George Quester once pointed out to me, this is why we say World War II began in September 1939, when Poland resisted the German invasion, rather than in March 1939, when Czechoslovakia did not.) If the aim of avoiding war is subordinate to some other normative value, and if a book on causes of war is to be one for all seasons, it is a leap of faith to accept that the uniform answer is the one Van Evera prefers: securing stability through recognition that defense is easier than attack.

The book proceeds by examining five hypotheses. (A handy appendix, pp. 259–262, lists the numerous corollaries that go with the main hypotheses.) Each hypothesis gets a chapter, which reviews the arguments, relates Van Evera's impressions of what his years of research have suggested to him about the bulk of evidence that bears on the hypothesis, and presents a few case studies to test the hypothesis. Four of the five hypotheses are dealt with in the first half of the book. The core of the book is the second half—the exploration of the fifth hypothesis and the body of offense-defense theory associated with it. This essay comments on the arguments in the book in the same order and proportion.

---

5. Carr, *Twenty Years' Crisis*, pp. 53–55, 79–84.
6. Carl von Clausewitz, *On War*, ed. and trans. Michael Howard and Peter Paret (Princeton, N.J.: Princeton University Press, 1976), p. 377.

The First Four Hypotheses

The early chapters present a good survey of how several aspects of power and perceptions—especially what Van Evera calls the "fine-grained" as opposed to gross structure of power—create incentives for war. The author strives to encase his arguments in social scientific rigor, but he does not shrink from recognizing some of the limitations of evidence that make historians skeptical of sleek generalizations. Indeed, Van Evera's forthrightness in laying out the logic, and the illustrations that he believes test it, eases the critic's task.[7]

The first four hypotheses (p. 4) are that war is more likely when:

- "states fall prey to false optimism about its outcome."
- "the advantage lies with the first side to mobilize or attack." (This hypothesis encompasses "stability theory" and concerns about how incentives for preemptive attack can be inherent in certain configurations of capability.)
- "the relative power of states fluctuates sharply—that is, when windows of opportunity and vulnerability are large." (This encompasses concerns about hegemonic transitions and impulses to preventive war.)
- "resources are cumulative—that is, when the control of resources enables a state to protect or acquire other resources."

The analysis of the fourth hypothesis, on cumulativity of resources, is most convincing. It engages the question of economic costs and benefits of war and the logic of imperialism, and rests on the notion that if the cost of conquering territory to extract resources exceeds the utility of the resources, "cumulativity is negative" (p. 106), and incentives for war are reduced. In agricultural eras cumulativity was high, but in recent times cumulativity has declined because knowledge-based economies are harder to loot than are smokestack economies

---

7. There are questions about Van Evera's method that this essay does not have room to explore in detail. He sets out criteria for case selection and theory testing that are somewhat confusing as stated in the book (e.g., saying that the case studies "proceed by comparing the case to normal conditions," without indicating what a normal condition is supposed to be), notes that orthodox social science methodology would count his tests as weak because they do not adhere to certain rules, and dismisses the criticism by saying that he has "never found these rules useful," but without saying why (p. 12). He also reports that he surveyed thirty wars (mostly modern ones) in background research (p. 13 n. 24), but does not indicate why he chose those particular thirty. (Why, for example, the Falklands/Malvinas War but not the much larger Chaco War? or the Indochina War but not the more consequential Russian, Spanish, or Chinese Civil Wars?) For clarification of Van Evera's views on methodology, see his *Guide to Methods for Students of Political Science* (Ithaca, N.Y.: Cornell University Press, 1997).

(pp. 114–115). Van Evera presents a balanced and reasonable argument that conquest still pays sometimes, but less often than in the past.

The first three hypotheses are more problematic. They are compromised by underwhelming evidence, or prove less convincing as explanations of war than of strategy and timing—explanations of "how" and "when" more than of "whether" to fight.

UNSURPRISING OPTIMISM

Van Evera's first hypothesis says that governments are more likely to fight if they believe they will win handily, and that they are often mistaken in this belief. This is convincing but not surprising, especially for one who has read Blainey. Yes, false optimism "preceded every major war since 1740" (p. 16). It must also have preceded most wars before 1740, but so did valid optimism. When a war has a winner and a loser, Van Evera's proposition and its opposite are both true by definition: the victor's optimism proves correct and the loser's mistaken. If neither side is optimistic about its chances, on the other hand, war is less probable, because neither side sees much to gain from starting it.

The hypothesis is more relevant for understanding pyrrhic victories or stalemates that expose false optimism on *both* sides, or cases in which attackers start wars that they expect to win but wind up losing. The hypothesis is important if such cases are more typical of war than cases in which one side wins at acceptable cost, that is, if most cases are ones in which optimism by *either* of the contenders is invalid. Determining that would require a substantial empirical project. World War I is a case of that sort, and it is omnipresent in Van Evera's theoretical apparatus and empirical analysis, but being an important case does not make it typical.

How much does it help to cite a case like Frederick the Great's invasion of Austria (p. 17), where the defender's optimism proves wrong and the attacker's right? That combination confirms common sense. Or what about cases where indications of false optimism are awash in other evidence of pessimism, such as U.S. decisions on Vietnam (pp. 17, 23)?[8] A problem that afflicts analyses of government decisions in many cases is that one is likely to find ample

8. Van Evera says that "U.S. officials recurrently underestimated their opponents in Vietnam," but the record in general does not support this. See Leslie H. Gelb with Richard K. Betts, *The Irony of Vietnam* (Washington, D.C.: Brookings, 1979); and H.R. McMaster, *Dereliction of Duty* (New York: HarperCollins, 1997). Van Evera cites a Joint Chiefs of Staff estimate that 205,000 U.S. troops would be needed to achieve goals in 1961, but does not mention that U.S. troops were not committed in force until four years later, when communist forces were stronger and army and marine corps estimates of needed U.S. manpower ranged between 500,000 and 700,000.

evidence of *both* optimism and pessimism, as policymakers confronting a clash of incentives and constraints are racked by ambivalence. When historians know that whatever sample of available evidence is incomplete, how firm a lesson can we draw from the many cases in which, say, five statements reflecting optimism can be stacked against three reflecting pessimism?

STABILITY THEORY

Exploring the second hypothesis, Van Evera amends and salvages "stability theory" and its emphasis on "first-move advantages." He argues that incentives to mobilize or strike first do not cause war in the manner of Thomas Schelling's image of "reciprocal fear of surprise attack,"[9] but that first-move advantages cripple peace-seeking diplomacy by leading countries to conceal their grievances, capabilities, and plans (p. 39).

Van Evera asserts that real first-move advantages are rare, while the illusion of such advantage is common (p. 71). This could be true, but the point is not demonstrated by the analysis. In one paragraph, he gives a list of cases he believes support his assertion (pp. 71–72), but does not present the bases for that conclusion. Instead he presents three case studies in detail that together do as much to question as to confirm his arguments that first-move advantages are rare, and that belief in them prompts war: World War I, China's entry into the Korean War, and the 1967 Arab-Israeli War. In these three cases, there *were* real first-move advantages. Van Evera reports that on the eve of World War I French General Joseph Joffre warned "that France would lose fifteen to twenty kilometers of French territory for each day mobilization was delayed" (p. 71 n. 31), as if Joffre was wrong. If a mobilization lag would make no difference in where the lines wound up, Van Evera does not say why. As it was, by September 1914 the Germans penetrated close to 100 kilometers into France—about 20 kilometers from Paris—and they managed to fight the rest of the war on French territory. That seems to be a significant advantage from moving first and fast. Similarly, in the other cases used to explore the hypothesis, the Chinese army rocked the 8th Army back into the longest retreat in U.S. military history in 1950, and Israel clobbered Egypt and Syria in 1967.

Perhaps Van Evera's point should be a different one—that first-move advantages rarely are great enough to win a war. If so, it would be clearer to focus on *decisive* first-move advantages. The Germans did lose World War I four

9. "If surprise carries an advantage it is worthwhile to avert it by striking first. Fear that the other may be about to strike in the mistaken belief that we are about to strike gives us a motive for striking, and so justifies the other's motive." Thomas C. Schelling, *The Strategy of Conflict* (New York: Oxford University Press, 1963), p. 207.

years after they overran northern France. But the Israelis won decisively in 1967, and the Chinese succeeded in settling the war in 1953 on terms that secured their gains on the battlefield and ratified the expulsion of United Nations (UN) forces from North Korea (which they had almost entirely liberated before the Chinese threw them back). To his credit, Van Evera recognizes that first-move advantages were real in at least the 1950 and 1967 cases, but his argument still claims more than he demonstrates. Additional data might support his views, but the evidence selected for presentation in the chapter does not demonstrate that even decisive first-move advantages are rare.

Whether imagined or decisive, first-move advantages are hard to cite as causes of two of the three wars Van Evera examines under this hypothesis. Rather they were causes of *strategy*. The Chinese would not have refrained from intervening in Korea if General Douglas MacArthur had not made it easier for them by splitting his forces in the advance to the Yalu. Van Evera offers a convoluted rationale for why Beijing could have decided to stay out if U.S. signals had been different, but notes nevertheless that Mao Zedong decided to intervene as soon as U.S. forces entered North Korea, before they approached the Yalu (pp. 56–61). Similarly he says, "Egypt and Israel had many reasons for war in June 1967 and probably would have come to blows absent a first-strike advantage" (p. 68).

WINDOWS

The third hypothesis is that fluctuations in power cause war by creating windows of opportunity and vulnerability and thus incentives for preventive or preemptive action. Declining states are tempted to launch preventive wars and take bigger risks; diplomatic alternatives to war are weakened because declining states cannot make credible offers and rising states conceal their grievances to avoid triggering preventive war; and if the balance of power shifts peacefully, the new disequilibrium may produce war if the fallen state refuses to yield privileges. This chapter produces a generally sensible discussion.

Again, however, the three cases Van Evera chose to test his hypothesis provide very mixed evidence; two of the three do not provide good support. The best of the three, and a quite powerful one, is Japan's 1941 attack on the United States (pp. 89–94). Without the windows created by Washington's combination of escalating demands and lagging mobilization, Japan might never have struck first. In regard to Germany in 1939, however, "window theory" proves convincing as an explanation of the timing rather than of the decision for war. The book provides no reason to disbelieve the widely held view that

Hitler wanted war from the beginning. If Hitler was determined to have a war, his strategic decisions were about when, not whether. Van Evera argues (pp. 94–99) that this was true only of German designs in the East, that Hitler's aims did not require war against France and Britain. The Allies' guarantee to Poland, however, meant that Hitler could not take Poland without a war in the West, and could not move against the Soviet Union without first defeating France. *Alliances* account for his decisions to attack in the West; windows may account for when he did so. A cause of strategy is not the same as a cause of war.

The third case is the U.S. military buildup in the early 1950s, which Van Evera discusses as a cause of increased American belligerence. This confuses cause and effect. The United States did act more cautiously when it was weak in 1950–51, and more toughly when a tripled defense budget gave it more power a few years later. It did not, however, become more belligerent because it became more powerful, but the reverse. The Korean War prompted the militarization of containment. The desire for better strategic options to support deterrence produced the buildup, which enabled prior policy to be pursued with more confidence. Actions could catch up with aims. This example is not a case where a window caused war, or even one where it caused belligerence, but where the buildup was caused by the desire to avoid war—and succeeded![10] This example belongs in a different volume, *Causes of Peace.*

Master Theory? Overmilking Nuclear Deterrence and 1914

The heart of Van Evera's book is the promulgation of "offense-defense theory" (henceforth, ODT) as a "master key to the cause of international conflict," a "master theory" that "helps explain other important causes" and provides "*the most powerful and useful Realist theory* on the causes of war" (pp. 190, 117; emphasis added). So much for Thucydides, Machiavelli, and Hobbes, or Carr, Morgenthau, and Waltz! This is no mean claim. The claim is all the more bold because it feeds on a theory conceptually rooted in very recent times. ODT has been overwhelmingly influenced by the frame of reference that developed around nuclear weapons, and the application of deterrence theorems to the conventional case of World War I.

---

10. Van Evera suggests a more dangerous result when he cites evidence that U.S. leaders thought about preventive war in this period (p. 100). On this, see especially Marc Trachtenberg, *History and Strategy* (Princeton, N.J.: Princeton University Press, 1991), pp. 103–107. Indications that officials ruminated about preventive war does not mean that it reached the point of being considered a serious option for deliberation and decision at the highest levels.

NUCLEAR THINKING

Military historians, and budding arms controllers in the interwar period, have written on the import of shifting advantages of attack and defense. The recent evolution of theory on the subject, however, reflects a systematic and sustained focus on the idea of strategic stability, which crystallized only in the nuclear deterrence theory of the Cold War. American deterrence theorists became preoccupied with how specific configurations of weaponry and targeting options might autonomously induce war or peace. The key was how combinations of technology could provoke or constrain decisionmakers by the prospect of how a nuclear engagement would turn out, depending on who fired first.

The nuclear framework provided a simple model of mutual deterrence based on a few variables—types of targets, types of forces, and their relative vulnerability—and clear distinctions between offensive and defensive operations. None of the weapons that figured in standard scenarios and calculations were dual-purpose. For example, intercontinental ballistic missiles (ICBMs) could be fired only to attack the adversary's territory, not to intercept an adversary's attack on one's own territory. Attacking and defending forces could support each other (counterforce first strikes could boost the effectiveness of antiballistic missile (ABM) defenses by drawing down the number of the victim's retaliating forces that had to be intercepted), but the scenarios of targeting and force interaction that dominated nuclear theorizing were few and uncomplicated compared to the historical variety of conventional warfare. The literature on nuclear strategy drew analysts toward a fixation on autonomous military causes of war and peace, and away from attention to the substantive political issues and motives in international disputes that had traditionally preoccupied diplomatic historians and political scientists. If the seedbed of deterrence theory does not account for the production of recent ODT, it certainly accounts for the intellectual receptivity of its consumers in the peak period of attention in the 1980s.

Common to academic analysis of the nuclear era were convictions that strategic stalemate was inherent in the nature of military technologies and force structures of the times; resort to war (at least nuclear war) was irrational because it was tantamount to suicide; recognition of this reality should prevent war; the prime danger to peace was misperception and miscalculation; and stability might be enhanced through measures of arms control that reinforced confidence in mutual deterrence. Bits and pieces of such ideas can be unearthed from earlier writing about military matters, but coherent application of them to conventional strategy became prominent only in the surge of debate about the balance between the North Atlantic Treaty Organization (NATO) and the Warsaw Pact during the Cold War.

Van Evera and other contemporary exemplars of ODT—George Quester, Robert Jervis, Jack Snyder—were all steeped in nuclear deterrence theory before venturing into analysis of conventional military strategy.[11] The companion image of the "security dilemma" also figured heavily in ODT (it is even evident in post–Cold War applications to unconventional civil wars).[12] The Cold War agenda made ODT especially appealing. Most obviously in the form presented by Jervis, the theory offered a way out of the security dilemma.[13] If capabilities could be configured to the advantage of defense, and Washington and Moscow could recognize and codify the situation, they would be less susceptible to inadvertent war or expensive arms competitions arising from a spiral of misperceptions. If only the conventional balance could be made like the nuclear balance, competition could be stabilized.

Given the strength of mutual nuclear deterrence, the main debates about stability in the Cold War devolved onto the conventional military balance between NATO and the Warsaw Pact. Doves wanted to find an excuse for less reliance on the threat of deliberate nuclear first use to deter Soviet conventional attack, and warmed to analyses suggesting that NATO's conventional forces were less hopelessly incapable of holding the line than typically assumed; hawks wanted stronger military capabilities across the board and denied the defensive adequacy of Western forces on the central front. ODT reinforced one

---

11. George Quester, *Offense and Defense in the International System*, 2d ed. (New Brunswick, N.J.: Transaction, 1988); Robert Jervis, "Cooperation under the Security Dilemma, *World Politics*, Vol. 30, No. 2 (January 1978), pp. 167–214; and Jack L. Snyder, *The Ideology of the Offensive: Military Decision Making and the Disasters of 1914* (Ithaca, N.Y.: Cornell University Press, 1984). These versions are more articulated than other forms of offense-defense theory that predated the nuclear era. For examples of how frameworks redolent of deterrence theory were superimposed on World War I, see the section of articles on "The Great War and the Nuclear Age" in *International Security*, Vol. 9, No. 1 (Summer 1984). Scott Sagan also argued within the terms of reference of deterrence theory to criticize interpretations by Van Evera and others who idealized defense dominance, and to argue for benefits of offense dominance. Sagan, "1914 Revisited," *International Security*, Vol. 11, No. 2 (Fall 1986), pp. 151–176. It is also important to note that contributors to ODT do not all agree about many things. When this essay makes general statements about ODT implications, it is (unless otherwise specified) referring to implications consistent with Van Evera's.

12. Barry R. Posen, "The Security Dilemma and Ethnic Conflict," *Survival*, Vol. 35, No. 1 (Spring 1993), pp. 27–57; and Barbara F. Walter and Jack Snyder, eds., *Civil Wars, Insecurity, and Intervention* (New York: Columbia University Press, forthcoming).

13. Jervis, "Cooperation under the Security Dilemma," pp. 186–214. In principle, ODT is independent of normative preferences for stability and defense dominance. Its association with these preferences may have been stronger during the Cold War, given the prospective costs of war between the superpowers, than it will be in the future.

side of the debate by offering ideas for affordable ways to make a Soviet armored offensive as unpromising as a nuclear strike.

Events of 1914–17 seemed to illustrate a configuration of capabilities that should have done this had the defense-dominant nature of that configuration been recognized. Failure of the Schlieffen Plan, years of inconclusive carnage on the western front, and costs far greater than the belligerents would have chosen to bear had they foreseen them showed that understanding of the offense-defense balance should have led to restraint in the crisis of 1914. If the chances of successful conventional offensives on the central front in Europe in the Cold War could be reduced to what they had been in 1914, and this could be demonstrated to political leaders, East and West could compete in peace. Implicit in the promise of ODT was the hope for a technical fix to the Cold War military competition.

The fascination of many realists with ODT was not shared by all, especially Reaganite idealists who saw the Cold War problem as Soviet aggressive motives rather than unfounded mutual fear. To them, seeing the security dilemma as the essence of the problem implied moral equivalence. The strategic task was not to reassure the Russians but to press them. These skeptics rejected the canonical wisdom about the benign nature of mutual assured destruction (MAD) and promoted nuclear counterforce capabilities and across-the-board increases in conventional forces. This view carried the day in the policy arena, but was a minority position in academia, where satisfaction with stable nuclear deterrence prevailed.

As social science theories go, nuclear deterrence was exceptionally elegant, so the ODT about conventional military competition that fed on it evolved into a hot genre in strategic studies. As such, it naturally spawned criticism. One skeptical article by Jack Levy appeared at the same time as Van Evera's initial analysis of World War I.[14] Levy faulted the genre for the variety of terms in which the offense-defense balance had been defined: "the defeat of enemy armed forces, territorial conquest, protection of population, tactical mobility, the characteristics of armaments, attack/defense ratios, the relative resources expended on the offense and defense, and the incentive to strike first." He also faulted ODT for tautological hypotheses, circular definitions, confused concepts, and other flaws, and concluded that "the concept of the offensive/defensive balance is too vague and encompassing to be useful in theoretical

14. Jack S. Levy, "The Offensive/Defensive Balance of Military Technology," *International Studies Quarterly*, Vol. 28, No. 2 (June 1984), pp. 219–238; and Stephen Van Evera, "The Cult of the Offensive and the Origins of the First World War," *International Security*, Vol. 9, No. 1 (Summer 1984), pp. 58–107.

analysis." Critics chipped away at the technological formulations of the offense-defense balance, especially challenging the notion that particular weapons supported either attack or defense as consistently as they did in nuclear theory.[15] Others counterattacked on behalf of ODT.[16] Van Evera's book engages only some of the critiques, and usually in brief asides.

One problem in this body of theory is confusion of dimensions of analysis in the definition of the offense-defense balance. Before the nuclear era, most conceptions of the balance were in terms of *military outcomes*—which side would have the operational advantage when attackers crashed into defenders in a combat engagement. Offense was identified as *attack* by ground, naval, or air forces against an adversary's territory (either to seize ground or to destroy assets behind the lines), and defense as *blocking* the attacking forces (holding ground or intercepting bombers or missiles before they can destroy their targets). A balance was considered "offense dominant" if, other things being equal, attacking forces were more likely to succeed in advancing or destroying their target than defending forces were to keep them from doing so.

The newer variants of ODT confuse the question by sometimes shifting the definition of offense-defense balance to the level of *political outcomes*—the effects of operational advantages on national decisions about whether to initiate war. Confusion is most obvious in the inversion of terms when nuclear weapons are considered. Oxymoronically, most contemporary exponents of ODT cite the nuclear offense-defense balance as defense dominant—even though no effective defenses against nuclear attack yet exist—because mutual vulnerability to the operational dominance of offensive forces prevents political leaders from deciding to start an engagement. (The traditional approach calls the nuclear balance offense dominant because when fired in an engagement, attacking missiles would accomplish their mission and helpless defenders would be destroyed.) Even if the oxymoronic definition were reasonable for the nuclear case considered alone, it muddies the conceptual waters. It is inconsistent to characterize both the 1914 conventional balance and the Cold War nuclear balance as defense dominant—which most offense-defense theorists do—because that is true in terms of combat interactions for the first but

15. Levy, "Offensive/Defensive Balance," pp. 222–223, 226–227, 229.
16. Sean M. Lynn-Jones, "Offense-Defense Theory and Its Critics," *Security Studies*, Vol. 4, No. 4 (Summer 1995), pp. 660–691. Some versions of ODT, especially recent ones that have absorbed earlier critiques, do not assume that particular weapons can be categorized as inherently either offensive or defensive. For the most recent critiques and responses, see James W. Davis, Jr., Bernard I. Finel, Stacie E. Goddard, Stephen Van Evera, and Charles L. Glaser and Chaim Kaufmann, "Correspondence: Taking Offense at Offense-Defense Theory," *International Security*, Vol. 23, No. 3 (Winter 1998/99), pp. 179–206.

not the second, or in terms of political effect for the second but not the first.[17] It would be more accurate to call the nuclear situation "peace dominant" if the criterion is decisions in peacetime rather than results in wartime. Moreover, the inverted conception has other limitations.

None of those who purveyed the inverted definition in the 1980s approved of its purest application to conventional warfare: Samuel Huntington's proposal to shift the logic of NATO's conventional deterrent from denial to punishment, from a defensive blocking strategy to a retaliatory counteroffensive. Huntington transposed the scorpions-in-a-bottle logic of nuclear deterrence directly to conventional strategy, arguing that a threat to retaliate against Soviet assets in Eastern Europe was a more potent deterrent to Soviet adventurism than an uncertain capability to hold the inner-German border against Soviet tanks.[18] Proponents of ODT were uniformly aghast at Huntington's proposal and insisted on a NATO conventional strategy that was literally rather than oxymoronically defensive.

ODT's benign view of the effects of nuclear weapons on the danger of war rests on an old belief that MAD guarantees military restraint, a belief that was never universally shared among analysts or policymakers. Those who lacked confidence in MAD's pacifying effect focused on the logical tension between confidence in mutual deterrence between the superpowers at the nuclear level and confidence in mutual deterrence at the conventional level. This debate about the credibility of threats to escalate bedeviled NATO strategy for more than three decades and was never resolved. If mutual nuclear deterrence was so solid that neither side would ever strike first, then the world was safe for conventional war: the side with conventional superiority could win while both sides' nuclear forces held each other in check. To threaten deliberate escalation, even unto intercontinental strikes, in order to deter conventional attack—NATO's official doctrine for decades—mutual nuclear deterrence must have an exception. Either mutual fear of nuclear annihilation is so strong that neither side would ever strike first, or it is not. Theorists and policy analysts spilled gallons of ink in arcane and contorted attempts to square this circle, and never

17. The balance in 1914 was defense dominant only in operational terms, because decisionmakers were *not* constrained by accurate anticipation of what would happen when the forces collided. The reality became evident only with the test of war. One might then say that the offense-defense balance should be defined in terms of perception—what decisionmakers believe would happen—which is indeed the basis for characterizing the nuclear balance, given that there has been no test of war for this case. If belief rather than demonstrated reality is the criterion, however, the 1914 balance must be cited as offense dominant.

18. Samuel P. Huntington, "Conventional Deterrence and Conventional Retaliation in Europe," *International Security*, Vol. 8, No. 3 (Winter 1983/84), pp. 32–56.

succeeded without resorting to more oxymoronic solutions like "the rationality of irrationality," or simple assertions that the "balance of resolve" would favor the side defending the status quo.[19]

ODT glosses over the fact that the view of MAD as pacifying or defense dominant prevailed only among academics and diplomats, and lost where it counted—within policy circles that determined procurement, force posture, and war plans.[20] (This will be a problem for students reading Van Evera's book. A decade past the Cold War, few of them now know any more about nuclear strategy than they read in the works of theorists, and believe that MAD captures all they need to understand about the subject.) The glossing may reflect Van Evera's barely concealed contempt for the foolishness of those who have sought alternatives to reliance on MAD, a contempt that occasionally leads him to obiter dicta or errors—for example, his charge that Air Force Chief of Staff Curtis LeMay "urged general nuclear war during the missile crisis" (p. 141 n. 90).[21]

The biggest disappointment of this volume is the brevity and pugnacity of Van Evera's chapter on nuclear weapons, which is far less scrupulous in argumentation than the rest of the book. It reads like a polemical bit of arm waving written for some other purpose and tacked on to the book. This chapter ignores all the old unresolved debates about nuclear strategy and conventional stability, says next to nothing about real dilemmas, and simply asserts the ineluctability of MAD. It adds nothing to the literature on nuclear strategy or to the arguments in the rest of the book. For example, Van Evera proclaims without any consideration of counterarguments that "conquest among great powers is almost impossible in a MAD world. . . . States also can better defend third parties against aggressors under MAD. . . . the United States deployed troops to Germany as a tripwire during the Cold War" (p. 246). It is as if history stopped in 1957, before the long and intense debates over extended deterrence,

---

19. For criticism of this assertion, see Richard K. Betts, *Nuclear Blackmail and Nuclear Balance* (Washington, D.C.: Brookings, 1987), pp. 133–144.
20. Aaron Friedberg, "A History of U.S. Strategic 'Doctrine'—1945–1980," *Journal of Strategic Studies*, Vol. 3, No. 3 (December 1980), pp. 37–71; Desmond Ball, "U.S. Strategic Forces: How Would They Be Used?" *International Security*, Vol. 7, No. 3 (Winter 1982/83), pp. 31–60; Fred Kaplan, *The Wizards of Armageddon* (New York: Simon and Schuster, 1983); Desmond Ball and Jeffrey Richelson, eds., *Strategic Nuclear Targeting* (Ithaca, N.Y.: Cornell University Press, 1986); and Scott D. Sagan, *Moving Targets: Nuclear Strategy and National Security* (Princeton, N.J.: Princeton University Press, 1989). The benign view dominant among academics was far from a consensus among analysts in general (especially ones involved in government). Among those dissatisfied with MAD were Fred Iklé, Paul Nitze, Albert Wohlstetter, Colin Gray, Donald Brennan, and William Van Cleave.
21. He cites Raymond Garthoff, *Reflections on the Cuban Missile Crisis*, rev. ed. (Washington, D.C.: Brookings, 1989), p. 94n. This passage actually refers to LeMay's support for the *conventional* air-strike option against bases in *Cuba*.

flexible response, the credibility of NATO doctrine, and prolonged conventional defense—which the author does not even mention let alone analyze.

OFFENSE-DEFENSE THEORY OPPOSES DEFENSE

Ironically, Van Evera is most adamant against the idea of *real* nuclear defense dominance—that is, a world in which defense dominates at the operational level of combat effect. He coins the cute acronym BAD (for "Both Are Defended") to contrast this sort of world to MAD, and opines that "if BAD is ever achieved, it will not last long. One side will soon crack the other's defenses." Perhaps, but the same could as easily be said of any defense dominance constructed at the level of conventional combat. Van Evera then rails that "the lives of states in BAD will be brutal and short. They will conquer and be conquered at a fast pace" (p. 251). As to why, he gives not a clue. The cavalier quality that diminishes his offensive against nuclear defenses is particularly unfortunate, because academic conventional wisdom opposing ballistic missile defense is now on the losing end of policy debate; it dominated policy in the 1970s, and fought a successful delaying action against Ronald Reagan's Strategic Defense Initiative in the 1980s, but has lost its influence in the 1990s. Even Bill Clinton—who populated his foreign policy team with liberals and arms controllers who would fall on their swords to save the ABM treaty—has accepted that the United States should deploy antimissile defenses.

There are still convincing reasons for continuing to oppose investment in active defenses (although there is no excuse for opposing passive civil defense).[22] For example, no one has yet demonstrated how any system designed to date will overcome efforts of a major power to saturate it with attacking forces, or of a minor power to circumvent it by using alternate means of delivery. The case is not nearly as clear, however, as it was during the Cold War. For example, the United States and Russia might hypothetically agree to field limited and comparatively cheap active defenses that would not be sufficiently effective to negate their deterrents against each other (given that each could retain 1,000 or more deliverable weapons), but that could blunt attacks by third countries with small numbers of missiles. North Korea, Iraq, Iran, or Libya will not have the resources to field more than a few ICBMs. Even China, in 1999, has fewer than twenty ICBMs; to deploy enough to saturate a U.S. or Russian defense might require a tenfold increase, a daunting expense

22. Richard K. Betts, "The New Threat of Mass Destruction," *Foreign Affairs*, Vol. 77, No. 1 (January/February 1998), pp. 36–40.

for a country with other priorities. This sort of speculation may not justify missile defense programs, but Van Evera does not consider any of the arguments for them in the chapter that condemns them out of hand.

If old-fashioned mainstream academic deterrence theory is all we need, maybe we can rest easily with the oxymoronic notion that lack of defenses constitutes defense dominance. Such faith smacks of hubris. The strength of nuclear deterrence theory came from its parsimony. That in turn reflected the theory's blindness to big problems revealed by a focus on different variables. Bruce Blair, John Steinbruner, Desmond Ball, and Paul Bracken showed how fundamental assumptions about stability in mainstream deterrence theory were thrown into serious doubt if analytical focus was shifted from the vulnerability of forces to the vulnerability of command systems.[23] These conceptual revisionists took a fundamental assumption of standard deterrence theory (the physical capacity to effect retaliation if the victim's forces survived attack), subjected it to empirical analysis, and showed it to be shaky.

Finally, the combination of political and military criteria exemplified in definitions of the nuclear offense-defense balance raises questions about consistency in operationalization of the concept. Van Evera writes, "I use 'offense-defense balance' to denote the relative ease of aggression and defense against aggression" (p. 118 n. 2). "Aggression" is a value-laden term implying *unprovoked* attack, which does not cover all cases of first-strike or invasion. Which ones it does cover depends on evaluation of motives and attributions of illegitimacy to particular attacks. Unless Van Evera really meant to say just "initial uses of force" rather than aggression, this definition of the offense-defense balance mixes normative political and empirical military criteria in ambiguous ways. Was an offense-defense balance that enabled a defensively motivated Israel to save itself from destruction in 1967 by initiating a devastating conventional offensive one that was offense dominant or defense dominant? All in all, it is better for analytical accuracy, clarity, and simplicity to return to strictly military operational criteria for categorizing the offense-defense balance, and to keep separate the purposes to which the operational advantages may be put or the decisions they may prompt.

---

23. Bruce G. Blair, *Strategic Command and Control* (Washington, D.C.: Brookings, 1985); Blair, *The Logic of Accidental Nuclear War* (Washington, D.C.: Brookings, 1993); John D. Steinbruner, "National Security and the Concept of Strategic Stability," *Journal of Conflict Resolution*, Vol. 22, No. 3 (September 1978), pp. 411–428; Desmond Ball, *Can Nuclear War Be Controlled?* Adelphi Paper 169 (London: International Institute for Strategic Studies, 1981); and Paul Bracken, *The Command and Control of Nuclear Forces* (New Haven, Conn.: Yale University Press, 1983).

HOW MUCH DOES WORLD WAR I PROVE?

The utility of offense-defense theory for stabilizing the Cold War was underwritten by another wave of analysis of the dynamics of the 1914 crisis. Had decisionmakers in European capitals then understood that the military equation of the time favored the defense, rather than assuming the reverse, peace might have endured. The diplomatic events of 1914 and the military events of 1914–17 inspired theoretical excursions that were applied more generally to conventional military strategy. From extensive reading and thinking, Van Evera has concluded that "perceived offense dominance is pervasive and it plays a *major* role in causing *most* wars. Knowing exactly which wars would still have erupted in its absence requires a close analysis of each case—impossible here. But the evidence does indicate that it had a vast role" (p. 185, emphasis added).

It is frustrating to hear the author say that he cannot show the evidence, but then hear him report what the evidence shows. Apart from one case, the book presents scant evidence of instances in which policymakers decided to initiate war because of perceived offense dominance—that is, where it is clear that the perception was a principal cause rather than one among a dozen on a laundry list of reasons. The one exception is World War I, which gets a whole chapter of its own as well as space in other chapters. Although he notes five other analysts of World War I who doubt that a "cult of the offensive" caused it, Van Evera reports that the "test" case of World War I confirms twenty-four of twenty-seven predictions from ODT, and concludes, "We have a verdict: war is markedly more likely when conquest is easy, or is believed easy" (p. 233; see also p. 238 n. 180).

Eyebrows might be raised when we see the case that inspired much of recent ODT being presented as the test of the theory, especially when two pages later Van Evera mentions offhandedly that some other cases show offense dominance as conducive to peace. Similarly, the book has a short section on "Qualifications: When Offensive Doctrines and Capabilities Cause Peace" (pp. 152–160), but gives little indication of how heavily these qualifications should weigh. Implicitly, they are minor exceptions that do not vitiate ODT's presumption in favor of defense. World War II, however, is a rather big exception. As Van Evera recognizes, "The Western allies of the 1930s needed offensive capabilities and might have deterred the war had they adopted more offensive strategies"; belief in defense dominance contributed to war (pp. 154, 235). Why should we be sure that we have more to learn from 1914 than from 1939?

Fine Grain or Gross? A Bloated Concept of Offense-Defense Balance

The most powerful concepts are ones that can be defined simply and operationalized clearly. Most attempts to promulgate simple and powerful concepts get battered if not discredited by the complexity of real life. This did not happen with nuclear deterrence theory because reality never provided any test cases (nuclear wars in which the conditions of deterrence failure and the results of combat engagements could be studied). Because of the volume of evidence on conventional warfare, in contrast, the simplest versions of ODT ran aground. Recent debate highlights the trade-off between accuracy and elegance in the theory.

CONCEPTUAL ENLARGEMENT

Refining conventional offense-defense theory proved to be trickier than the thought experiments and exchange calculations of nuclear strategy, where only a few scenarios were at issue and weapons could be used operationally only for attack or defense rather than for both. Critics raised questions about how far the simplicity of nuclear terms of reference could be transferred to conventional war, where attack and defense are advantages that shift in the course of a war; the same weapons can be used for both purposes; multiple weapons can be reorganized and combined with doctrines in different integrated systems with different implications; the tactical offensive and strategic defensive (or the reverse combination) can be combined in campaigns; or even tactical offensive and defensive actions can reinforce each other within the same engagement. For example, in May 1940 the Germans expanded bridgeheads by using "interior lines and the ability to concentrate against one face of their salients while standing on the defensive on the other two";[24] in 1973 the Egyptians used air defense missiles offensively to screen the advance of ground

24. Archer Jones, *The Art of War in the Western World* (Urbana: University of Illinois Press, 1987), p. 531. ODT promoters also sometimes cite the authority of Clausewitz when he proclaims that "*the defensive form of warfare is intrinsically stronger than the offensive*" (emphasis in original), but neglect to mention that Clausewitz says this to endorse tactical exploitation of defense in the service of a strategic offensive. Given that defense has a negative object, it "must be abandoned as soon as we are strong enough to pursue a positive object. When one has used defensive measures successfully, a more favorable balance of strength is usually created; thus the natural course of war is to begin defensively and end by attacking. *It would therefore contradict the very idea of war to regard defense as its final purpose*" (emphasis added). Clausewitz, *On War*, p. 358; see also pp. 370, 600.

forces, and the Israelis used attack aircraft and tanks defensively to hold their line; and so forth.

ODT supporters have generally responded by introducing qualifications that reduce the emphasis on particular military technological determinants and take the theory further from parsimony. Some buttress the theory by adding and subtracting components to cope with anomalies, and making the theory's claims more limited. In an otherwise impressive article, Charles Glaser and Chaim Kaufmann complicate the specifications of the theory so much that a reader would need to keep a cue card just to be able to think about how to apply it. They also introduce simplifying assumptions that ease theorization but do not comport with important aspects of the empirical record of modern conventional war, and that essentially eliminate strategy from the strategic equation. They insist that "the offense-defense balance should be assessed assuming optimality—that is, countries choose the best possible strategies and force postures," and they eliminate differences in "military skill" from consideration.[25] By this restricted definition, the offense-defense balance must have favored France over Germany in 1940, Egypt over Israel in 1967, Argentina over England in 1982, and Iraq over the United States in 1991! The authors appropriately keep their claims restrained, saying that ODT "does not claim that the offense-defense balance is in general a more important determinant of military capabilities than is power or skill."[26]

Van Evera reacts to critics' challenges about how much ODT can explain in a different way, and in the bargain sidesteps debates about what elements should be included in the concept. Instead of refining the definition of the offense-defense balance, he broadens it;[27] instead of resolving debates about what should be included, he includes almost everything that has ever been suggested; and instead of modifying claims, he expands them. Although he criticizes other realists for misdirected attention to "gross" power rather than the "fine-grained" structure of power, he crams so many things into the definition of the offense-defense balance that it becomes a gross megavariable. His offense-defense balance includes "military technology and doctrine, geography, national social structure, and diplomatic arrangements (specifically,

25. Charles L. Glaser and Chaim Kaufmann, "What Is the Offense-Defense Balance and Can We Measure It?" *International Security*, Vol. 22, No. 4 (Spring 1998), pp. 46, 48, 55.
26. Ibid., pp. 48–49.
27. This point leaps out. It is also made by all three critics responding independently to a preview of Van Evera's book. See Davis, Finel, and Goddard, "Correspondence: Taking Offense at Offense-Defense Theory," pp. 180, 187, 190–191.

defensive alliances and balancing behavior by offshore powers)," as well as manpower policy (the *levée en masse*), strategic choices in campaigns (the maldeployment of UN forces in the advance to the Yalu), popularity of regimes, and the breakdown of collective security organizations (the League of Nations) (pp. 122, 160–166). Phew! If we could get him to throw in population and financial resources, we could almost call the offense-defense balance "relative power" and be back to basic realism. Van Evera does not eliminate the distinction between offense-defense balance and power, but he blurs it enough to make the difference in their implications unclear.[28]

Van Evera would have none of this, because he sees his approach as a basic variant of structural realism differentiated from others' focus on polarity. He distinguishes four schools (p. 10): Type I, classical realism; Type II, structural realism, which focuses on polarity; his own preferred Type III ("fine-grained structural realism"); and Type IV ("misperceptive fine-grained structural realism"). He does not believe that herding a horde of variables into the offense-defense balance washes out the fine grain. Indeed, he insists that his expansive definition is very parsimonious because it "achieves simplicity, binding together a number of war causes into a single rubric. Many causes are reduced to one cause with many effects" (p. 190). In a response to critics he argues, "A theory is not shown to lack parsimony simply by demonstrating that its concepts include a diverse range of lesser-included concepts. . . . All concepts are aggregations of lesser concepts."[29] But why is it more useful to consider all the variables lumped in the definition above as "lesser-included" elements of an offense-defense balance than it is to consider a more coherently specified balance as a lesser-included concept of "power"? It might be if we had a formula that converts all the "fine-grained" elements of Van Evera's balance to a metric that is more illuminating than relative power as a whole. But there is no indication of how all these elements in the concept should be aggregated, and thus no clear way to *measure* the offense-defense balance. Van Evera is reduced to relying on "author's estimates" when he comes to categorizing whether the balance in actual periods of history favored offense or defense.

28. There is also a problem of virtual circularity in the theory. A big advantage for defenders is that other status quo powers often align with them. Keeping alignment out of the equation makes the offense-defense balance a much weaker predictor. Putting it in the equation, however, makes strategic behavior both a cause and an outcome. See Davis, ibid., p. 180; and Goddard, ibid., p. 191. On whether he uses the behavior of states to explain the behavior of states, Van Evera says, "I plead guilty. . . . I don't see a problem." Ibid., p. 199.
29. Ibid., p. 196.

How then can other analysts apply this concept? Without a clear formula, it is hard to see how far we are left from Blainey's conclusion that war hinges on disagreements (miscalculations) about relative power.[30]

DISTINGUISHING THE OFFENSE-DEFENSE BALANCE FROM THE BALANCE OF POWER
There is a difference between the offense-defense balance and power as a whole that is clearer in more limited formulations of ODT. Blainey views parity of power as destabilizing because he does not see decisionmakers differentiating the convertibility of resource inputs for combat outputs—that is, he does not consider that governments might take fewer risks if they understood that equal resources would produce unequal effectiveness in attack and defense. Rough parity lets governments miscalculate easily in Blainey's world because either side can convince itself that a modest edge in the balance of power gives it an opening to gain victory. Power parity promotes peace in a recognized defense-dominant world, in contrast, because both sides know they would need a much bigger edge than they have in aggregate resources to overcome the other's defense.

Van Evera preserves the crucial criterion that the balance should be some sort of pound-for-pound ratio, rather than a measure of aggregate probability of combat success, when he formulates the concept as "the probability that a determined aggressor could conquer and subjugate a target state *with comparable resources*" (p. 118 n. 2, emphasis added). But in the next breath, he subverts it by admitting alliance shifts to the calculus, because realignment can overturn the balance of resources—indeed, that is its purpose. This is why the focus on "gross" power (polarity) of the Type II structural realists from whom Van Evera wants to differentiate himself is likely to remain most important. If not, one should expect that Van Evera's Type III focus on the fine grain should yield different or more useful explanations and predictions about who would win wars. But does it? Would his framework have told statesmen more than "gross" power analysis about what they would have wanted to know before going to war in the Persian Gulf or Kosovo? Perhaps, but it is hard to know without a formula that converts the offense-defense megabalance into a sort of "unit-price" measure of relative combat efficiency.

30. One problem is uncertainty about what aggregate "power" means, or whether there is such a thing. Realists tend to see power as a conglomeration of material capabilities, mainly economic and military, that is fungible. Others who do not see power as fungible insist that one must ask, *power to do what?*—which is what an offense-defense balance is about. See David A. Baldwin, "Power Analysis and World Politics: New Trends vs. Old Tendencies," *World Politics*, Vol. 31, No. 2 (January 1979), pp. 161–194.

One reason Van Evera focuses on the fine-grained structure of power is that he considers it "more malleable than the gross structure; hence hypotheses that point to the fine-grained structure yield more policy prescriptions" (p. 8). Others such as Quester also see the benefit of focusing on the offense-defense balance as its comparative malleability (although he keeps the concept more manageable, by restricting it to the level of military capabilities and techniques).[31] It is indeed sensible to concentrate on causes that can be manipulated, and a central mission of strategic studies is to direct attention to the causal significance of military force structures and strategies in international relations. There are two problems, however, in touting the malleability of the balance.

First, if we keep the concept more coherent than Van Evera does and we focus on the military elements of the offense-defense balance, we can see that they are more manipulable than some factors such as geography, but not necessarily more manipulable than others such as alliances or diplomatic strategies. For example, what was crucial in setting off World War II in Europe was not any notion of a military offense-defense balance or changes that the Germans could make in it. Rather it was the stunning German-Soviet nonaggression pact, which solved Hitler's problem of a potential war with great powers on two fronts. "Fine-grained" aspects of power (such as innovations in doctrine for armored warfare) may explain why the Germans did so well in early phases of the war, but the gross distribution of power (poles and their alignment) tells us more about how the war got started.[32]

Second, if we accept Van Evera's sweepingly inclusive definition (which simply incorporates the Hitler-Stalin pact in the offense-defense balance), it becomes more important to remember that manipulability works two ways. Van Evera is interested in presenting a menu of variables that can be manipulated to reinforce peace, but there is nothing that ordains that status quo powers will shape the balance to their purposes more successfully than revisionist powers will shape it to theirs. Manipulability is necessarily conducive to peace only if we assume that the problem lies entirely in the security dilemma, that countries in conflict will welcome a structured stabilization of capabilities that appears to make conquest hard for either of them.

Packing many elements into the offense-defense balance concept does not alter the fact that the elements are very different things. Whether "lesser-

31. Quester, *Offense and Defense in the International System*, p. 10.
32. See Randall L. Schweller, *Deadly Imbalances: Tripolarity and Hitler's Strategy of World Conquest* (New York: Columbia University Press, 1998), chaps. 4–5.

included" or not, this variety means that there are numerous points of entry for a revisionist strategist to try to overturn a benign balance. This does not mean that it can always be done. In some conventional military equations, structural constraints may prove irremediably favorable to the defense. But the more complicated the equation, the more variables and the more potential combinations a revisionist strategist has to work with. Organizations and doctrines may be developed to change the effects of combined arms operations, and force structures may be adapted to apply existing technologies in novel combinations; or innovation and research may be channeled to develop technologies to counter tactical defensive advantages; or strategic plans may be found to capitalize on tactical advantages of defense for strategically offensive purposes; or allies may be acquired or detached from adversaries; and so forth. Indeed, this is what *strategy* is supposed to be all about. Jonathan Shimshoni illustrates how successful militaries do this, how they strive to make balances dynamic and evolving, rather than static and confining, to give them an edge.[33] An assumption that countries will seek stability if only its foundations can be made clear to them is an assumption about motives, not capabilities.

Beyond Power

Over time observers have cited many motives for war: God, gold, and glory, or fear, honor, and interest.[34] Like most realists, Van Evera has little to say about any of these but fear and interest.[35] Indeed, because his premise is that most wars result from security seeking, he focuses primarily on fear, and how to relieve it by adjusting the structure of power and suppressing misperceptions of it. Van Evera shares with Blainey a lack of interest in the political substance of issues at stake in conflicts among states—territorial claims, religious or ideological disputes, economic interests, and so forth. To Blainey, understanding aims and motives may provide a theory of rivalry but not of war.[36] If rivalry is a prerequisite for war, however, reducing causes of rivalry will tend to reduce the probability of war.

33. "A military entrepreneur must constantly ask two questions. . . . (1) what would war look like if fought today, and (2) how can I 'engineer' the next war away from (1) so as to maximize my relative advantages. . . . " Jonathan Shimshoni, "Technology, Military Advantage, and World War I: A Case for Military Entrepreneurship," *International Security*, Vol. 15, No. 3 (Winter 1990/91), p. 199.
34. The latter three are emphasized by Thucydides. See Donald Kagan, *On the Origins of War and the Preservation of Peace* (New York: Anchor Books, 1995), pp. 7–8 and passim.
35. For systematic attempts to integrate the issues, stakes, or political motives of conflict in theories of war, see John A. Vasquez, *The War Puzzle* (New York: Cambridge University Press, 1993); and Kalevi J. Holsti, *Peace and War* (New York: Cambridge University Press, 1991), chap. 10.
36. Blainey, *The Causes of War*, pp. 149–150.

POLITICAL STAKES AND OBJECTIVES

To Van Evera, the beauty of a benign offense-defense balance is that it can prevent war irrespective of motives, because states can see that attack cannot succeed.[37] The common sense that appeals to Blainey is also implicit here: if a prospective war seems unwinnable, no one will start it. One reason to pay attention to political incentives as well as to military constraints is that this starkly utilitarian common sense does not always govern decisions on war and peace. For honor or other reasons that seem quaint to economists, countries may fight in the face of probable or even certain defeat.[38]

There is also reason to consider motives within the utilitarian calculus. Sometimes the problem is, as Blainey argued, that contenders of roughly equal power do not see that they will not be able to use force effectively, especially if they believe they can exploit novel stratagems. A defense-dominant balance might solve this problem, but it would have to be lopsided; Van Evera's concept of the balance includes so much that it would be hard to find many situations in which some elements of the conglomeration did not seem to cancel out other defense-dominant ones, leaving an opening for successful attack—for example, where geography, or social structure, or alliance choices conducive to offensive action overrode military technology conducive to defense.

Where the answer to which side would prevail in combat is a matter of probability rather than certainty, risk propensity and willingness to sacrifice enter the equation. These can depend in turn on the value of the stakes. "A place in the sun" may strike Van Evera, a humane and modern man, as no decent reason to launch a war that courts catastrophe. The value of that aim was far different to the Kaiser's Germany, however, when it was coming late to the imperialist scramble at a time when the world was made of empires. A difference in the relative value of the stakes to two contestants may also affect their relative will to sacrifice, the amount of their power that they commit to the contest, and their chances of winning. North Vietnam had far less power than the United States, but the United States had far less stake in the outcome of the war, so committed a much smaller fraction of its power. Van Evera examines the costs of war intently, but neglects the benefits, which require examination of stakes and motives.

37. Van Evera, "Cult of the Offensive," p. 105.
38. Robert Jervis, "War and Misperception," in Robert I. Rotberg and Theodore K. Rabb, eds., *The Origin and Prevention of Major Wars* (New York: Cambridge University Press, 1989), pp. 103–104.

Realists usually ignore the stakes in dispute for theoretical purposes because they assume that conflict is more or less a constant. Something always comes up to put countries in conflict, but many conflicts persist without war. The question is what pushes conflict over the brink to combat. For a highly general theory of average outcomes over time in international politics—what interests Kenneth Waltz—this lack of attention to the substantive issues of disputes makes sense. For any particular case, however, the fact remains that there are two necessary conditions for war: a distribution of power that makes war appear an effective option, and a conflict of interest. The first without the second will not produce war. The United States has more than enough power to take Canada, but lacks sufficient political incentive to do so.

War between countries in conflict can be avoided through either of two ways: constraining their power or resolving their conflict. ODT sees more promise in the first because it seems more manipulable. But is this necessarily so? If the offense-defense balance includes geography, social structure, and other such weighty things, as in Van Evera's formulation, why is it more manipulable than attitudes toward stakes? Washington and Moscow dueled inconclusively for two decades over nuclear arms control and the Mutual and Balanced Force Reduction talks, but the Cold War ended more because of a political decision to resolve the conflict than because of any beliefs about an offense-defense balance between East and West. Van Evera argues that Mikhail Gorbachev was willing to make concessions because he adopted ODT (p. 119). If we look at all that happened at the end of the 1980s, however, it is hard to believe that whatever thought Gorbachev gave to ODT was more than a tactical rationalization for much grander political decisions. Gorbachev did not bargain with the West to negotiate arms reductions or diplomatic deals. He simply gave up the global ideological contest that was the subject of the Cold War. Realist ODT was less evident in his thinking than liberal institutionalist nostrums about a "common home." The Soviets did not exchange concessions with the West. They made unilateral concessions, which the West pocketed. The Cold War ended not because the Russians bought some academic idea about how much easier military defense of their interests could be, but because they gave away those interests, surrendered their empire, and left the West with nothing to fight about.

SUBCONVENTIONAL WAR

More attention to political stakes would also help if we want a theory about the causes of war in general, rather than the subset of wars that interests Van Evera and most international relations theorists: big, world-shaking, international wars among great powers. Those sorts of wars are most important, and they could occur again. They have not been representative of the phenomenon of war since 1945, however, and with at least a temporary respite from serious great power conflict after the Cold War, hot wars may be of other types for some time to come.

Most recent wars have been intranational, internal, civil wars. Most international wars are about which countries control which territories; most intranational wars are about which group gets to constitute the state. Van Evera's research surveyed thirty wars, 90 percent of which were international and 10 percent intranational (p. 13 n. 24). Since 1945 the proportions of war that have actually occurred are closer to the reverse. In civil wars the relative salience of concerns about political values, as opposed to material power, is usually greater than in international wars.

Some wars could count in both categories because they involve interventions by outside countries in civil wars, and some categorizations could be disputed (e.g., the Korean War could be considered a civil war between northern and southern Koreans who considered their country to be temporarily divided, or an international war between two Korean states—as American white southerners regarded what northerners called the Civil War). In terms of military character, some civil wars are similar to international ones, if the contending groups are geographically separated and they marshal conventional armies. Many contemporary internal wars differ significantly, however, from the conventional terms of reference by which recent ODT has developed. When contending groups are divided by ideology, religion, or ethnicity they are often intermingled, and combat is not carried out on one front with conventional lines of operation. Rather the front is everywhere, as groups compete to dominate villages, provinces, and loyalties throughout a country.

Irregular or guerrilla warfare does not follow the operational logic of ODT grounded in modern conventional war and high force-to-space ratios that are possible on a linear front. Irregular war is more similar to that of ancient and medieval times, where low force-to-space ratios made raids the usual form of combat. The rule of thumb purveyed by counterinsurgency theorists during the Cold War was that a government defending against guerrilla attacks needed a ten-to-one superiority in forces. This is thirty times the legendary one-to-three ratio in the rule of thumb for the tipping point between attack and defense in conventional warfare. When guerrillas can be engaged in a conventional battle where they fight on the defense, they can be annihilated. Insurgents who hide among the civilian population, in contrast, can concentrate at will to raid specific objectives, destroy pockets of defending forces, and retreat to avoid battle with the government's main forces. The government must disperse its forces to defend everywhere. The ability of the insurgents to hide in the population, in turn, is determined by political and social factors.

Van Evera does not agree, because he thinks about international war. In answer to a critic he writes, "Modern guerrilla war has defended many countries and conquered none. *It is a fundamentally defensive form of warfare.*"[39] But the communist guerrillas in Greece, Malaya, and South Vietnam, or the anti-communist mujaheddin in Afghanistan and Contras in Nicaragua were attacking native governments. Foreign intervention on behalf of such governments internationalized the wars precisely because of the guerrillas' *offensive* success.

A final reason to look beyond power, and especially beyond the offense-defense balance, is the conservative presumption in Van Evera's theory. The book implies that war is justified only to defend what one has, not to conquer what others hold. If we were to leave open the normative question of whether conquest or coercion are ever desirable, we would leave open the question of whether it is desirable to institutionalize defense dominance. In principle, however, there is no automatic reason why it should be any more legitimate to use force on behalf of what is than on behalf of what should be. Once what should be becomes admissible as a warrant for force, there can be no consistent preference for defense over attack, because there is no automatic congruence between what should be and the status quo. Revolution may be justified in an unjust society, and international revisionism may be justified in an unjust international order. If the norm is to be that a just cause can warrant resort to force, whether peace should be preferred to war is an issue, not a premise. On this matter Blainey's analysis is more detached and value free, because he regards peace as no more natural than war.[40]

Perhaps it sounds perverse or insidious to raise this issue. But is it any accident that American or other Western analysts favor peace, when the United States and other Western great powers are status quo states satisfied with the international pecking order as it is, a bunch of prosperous, righteous countries calling the tune for others? When an unjust status quo catches the eye of Western leaders, however, they do not hesitate to see the virtue in attacking it. Witness the assault launched against Serbia to wrest an oppressed Kosovo from its control. Bill Clinton and Tony Blair were certainly glad to enjoy offense

---

39. Van Evera, "Correspondence: Taking Offense at Offense-Defense Theory," p. 195 (emphasis added).
40. Blainey, *The Causes of War,* chaps. 1, 16.

dominance, as NATO air forces could range across Serbia's skies unchallenged, methodically destroying the country with impunity.

War Will Usually Find a Way

Offense-defense theory is in many respects quite hardheaded and realistic about what military power can accomplish, but some aspects of it are non-Clausewitzian. ODT is most concerned with *stability,* how to prevent war by making it ineffective as an instrument of policy; Clausewitz thinks in terms of *strategy,* how to make war serve policy. Most devotees of the theory would like to paralyze strategic options for using conventional forces as thoroughly as MAD seems to paralyze nuclear strategy. The ideal offense-defense balance is implicitly some benign form of military stasis. Clausewitz, in contrast, pays little attention to the causes of war. He focuses on its nature and purpose. The strategic orientation he represents is dynamic: how to scan the environment and history to find a way around obstacles that protect the adversary, how to find a way to make war work for one's purposes.

Some carefully exclude strategy from ODT; in contrast to Van Evera's excessive inclusiveness, this narrows the concept too much. Shimshoni's critique that strategic entrepreneurship can overcome a constraining offense-defense balance is dismissed by Glaser and Kaufmann as flawed because it "relaxes the assumption of optimality. All of his examples . . . hinge on states having significant advantages in military skill over their opponents."[41] This protects their conceptualization of an offense-defense balance, but does not give any reason to look to such a balance as a reliable bar to war. As Levy noted, the eighteenth century has generally been regarded as defense dominant, but the campaigns of Frederick the Great were an exception. Analysts do not call the offense-defense balance offense dominant in this period because Frederick's success came from tactical and strategic innovation rather than technology. "Still, it cannot be denied that Frederick demonstrated what was possible given the technology of the time."[42]

Strategic schemes help war find a way. ODT looks for means to bar the way, to neutralize strategy or freeze it. This proves harder in conventional warfare than in the nuclear deterrence theory in which contemporary ODT germinated.

---

41. Glaser and Kaufmann, "What Is the Offense-Defense Balance?" p. 55 n. 40.
42. Levy, "The Offensive/Defensive Balance," pp. 231–232. See also Finel and Goddard, "Correspondence: Taking Offense at Offense-Defense Theory," pp. 183–184, 193.

When Waltz quoted the line from Brecht that appears at the beginning of this essay, he followed up with the point that "for half a century, *nuclear* war has not found a way."[43] This led Waltz, John Mearsheimer, and a few others to view the spread of nuclear weapons as a welcome force for peace. Whether valid or not, this idea appeals only to a handful of hyper-realists and has no future as a policy prescription.[44]

If ODT is to help bar the way to war, it has to do so at the conventional level of combat. In theory, this might be accomplished by finding defensive technologies as enduringly superior as offensive systems have been in nuclear forces, or by freezing defense-dominant configurations in arms control deals. So far, however, no formulation of defense dominance for conventional forces has proved immune to strategy. The stalemate of 1914–17 came closest to such robust defense dominance, but it too was eventually overcome by innovations. German infiltration tactics on the one hand, and British deployment of tanks on the other, made attack more effective and unfroze the lines. If strategic effort can find ways to overcome an inhibiting offense-defense balance, and if political stakes and motives are strong enough to provide an impetus, war will still find a way.

That is not to say that victory will necessarily find its way. There is no reason to expect that miscalculation or plain foolishness will become less common. Nor do the various problems discussed in these pages mean that the idea of an offense-defense balance is wrong or useless. The question is still open, however, as to how best to conceptualize the balance and how much to claim for it. If it is to be analytically useful, the concept's scope should be kept limited to the dimension of military operations: relative combat effectiveness, on a pound-for-pound basis, when attackers crash into defenders in a tactical engagement or strategic campaign. There is no consensus yet on a precise definition in these terms—any one yet offered seems to run up against some question or scenario that makes it problematic—but we could do worse than to work for particular purposes with some of those that have emerged from methodological wrestling. For example, Levy's: "the offensive/defensive balance is inversely proportional to the ratio of troops needed by an attacker to overcome an enemy defending fixed positions"; or Lynn-Jones's: "the amount of resources that a state must invest in offense to offset an adversary's investment

43. Waltz, "Waltz Responds to Sagan," in Scott D. Sagan and Kenneth N. Waltz, *The Spread of Nuclear Weapons: A Debate* (New York: W.W. Norton, 1995), p. 93 (emphasis in original).
44. As James Kurth says, "There probably has not been a single foreign policy professional in the U.S. government who has found this notion to be helpful." Kurth, "Inside the Cave: The Banality of I.R. Studies," *National Interest*, No. 53 (Fall 1998), p. 33.

in defense. . . . the offense/defense investment ratio required for the offensive state to achieve victory"; or Glaser and Kaufmann's: "the ratio of the cost of the forces that the attacker requires to take territory to the cost of the defender's forces."[45]

It will not do, as Van Evera has done, to swallow up contradictions and confusions brought out in recent debates by rolling all manner of things that make conquest easy into the offense-defense balance. Including factors that affect the total resources available to contestants makes the balance almost congruent with a comprehensive net assessment of which side would win a war, and too close to just another term for how to parse "power." To make good use of the concept of the offense-defense balance, it is necessary to return to more limited and coherent formulations of it. The last word on this subject remains to be written.

Van Evera makes an important point that is buried in a footnote early in his book: "Most of the important contributing causes of war are neither necessary nor sufficient" (p. 41 n. 19). The works of Kenneth Waltz concentrate on one of the main necessary or permissive causes of war: international anarchy. Blainey concentrates on another: disagreements about relative power. But necessary or permissive causes are not enough to understand *specific* wars, as any statesman will insist that a theory aspiring to prescriptive utility must do. For that, looking back toward classical realism (e.g., Carr's emphasis on the interactions among grievance, satisfaction, and power) helps more.

More interest in efficient or sufficient causes will lead us to look more at motives and stakes than Waltz, Blainey, or Van Evera has yet done. The subjects that Van Evera announces he will deal with in a planned second volume get closer to the efficient causes. He plans to explore four explanations, that: (1) militaries cause war "as an unintended side-effect of their efforts to protect their organizational welfare";[46] (2) states "infuse themselves with chauvinist myths" and "underestimate their own role in provoking others' hostility"; (3)

45. Levy, "Offensive/Defensive Balance," p. 234; Lynn-Jones, "Offense-Defense Theory and Its Critics," p. 665; and Glaser and Kaufmann, "What Is the Offense-Defense Balance?" p. 46.

46. It is interesting, however, that in the vast region where militaries were most dominant in politics—South America in much of the twentieth century—there was almost no interstate war. Felix Martín shows that neither main paradigm of international relations theory—realism or liberalism—explains this, a much longer peace than the vaunted Long Peace in Europe during the Cold War. (Causes of war identified by realism existed in South America, but war did not occur, and causes of peace identified by liberalism did not exist, yet peace endured.) If anything, organizational interests and transnational professional identification—in effect, an epistemic community of military officers in the region—discouraged a resort to force across borders. Felix Martín-Gonzalez, "The Longer Peace in South America, 1935–1995," Ph.D. dissertation, Columbia University, 1997.

states "often misperceive because they lack strong evaluative institutions"; and (4) states avoid defining their national strategies clearly, so "official thinking deteriorates," states are less able to assess each others' intentions, and they become "blind to the other's concerns" (pp. 256–258).

Whether we are concerned with necessary, permissive, efficient, or sufficient causes, there is not likely to be a breakthrough on the first-order issues of debate about what causes war. The debate has cycled back and forth for centuries in many variations on a few themes in the realist and liberal traditions of international relations theory. This elemental debate is no more likely to be resolved in the next century than in the last one, if only because each tradition has a purchase on different facets of the problem, and the glass that is half full for one will always be half empty for the other.

At less cosmic levels of debate, where most research takes place, progress is more plausible even if it is likely to be more halting and inconclusive than ardent social scientists expect. There will be room for revisionist research and creative revisitations of old issues as long as historians keep political scientists honest by disciplining their oversimplifications about what past cases prove. There will be plenty of work to do, for a long time, on the question that is among the handful of fundamental issues in international relations: What causes wars? Van Evera's book does some of that work, imperfectly like all good books, but usefully. It need not make good its claim to a master theory to merit a place at the table.

Grasping the Technological Peace

The Offense-Defense Balance and International Security

Keir A. Lieber

Offense-defense theory argues that international conflict and war are more likely when offensive military operations have the advantage over defensive operations, whereas cooperation and peace are more likely when defense has the advantage. According to the theory, the relative ease of attack and defense—the offense-defense balance—is determined primarily by the prevailing state of technology at any given time. When technological change shifts the balance toward offense, attackers are more likely to win quick and decisive victories. This prospect of quick and decisive warfare exacerbates the security dilemma among states, intensifies arms races, and makes wars of expansion, prevention, and preemption more likely. When technological innovation strengthens the defense relative to the offense, states are more likely to feel secure and act benignly.[1]

Offense-defense theory has deep roots, but has become increasingly popular in international relations scholarship and foreign policy analysis in recent years. The idea that the nature of technology affects the prospects for war and

Keir A. Lieber is a Research Fellow in the Foreign Policy Studies Program at the Brookings Institution and a doctoral candidate in the Department of Political Science at the University of Chicago.

For helpful comments and discussions, I thank Stephen Biddle, Meredith Bowers, Jasen Castillo, Alexander Downes, David Edelstein, Benjamin Frankel, Charles Glaser, Robert Lieber, John Mearsheimer, Robert Pape, Jordan Seng, Stephen Walt, Paul Yingling, two anonymous reviewers, and participants in the Program on International Politics, Economics, and Security at the University of Chicago. I would also like to thank the John D. and Catherine T. MacArthur Foundation, the Andrew Mellon Foundation, and the Smith Richardson Foundation for their generous support of this research.

1. The foundational works on offense-defense theory are Robert Jervis, "Cooperation under the Security Dilemma," *World Politics*, Vol. 30, No. 2 (January 1978), pp. 167–214; George H. Quester, *Offense and Defense in the International System* (New York: John Wiley and Sons, 1977); and Stephen Van Evera, *Causes of War: Power and the Roots of Conflict* (Ithaca, N.Y.: Cornell University Press, 1999), especially chap. 6. For crucial theoretical developments, refinements, and extensions, see Stephen Van Evera, "Offense, Defense, and the Causes of War," *International Security*, Vol. 22, No. 4 (Spring 1998), pp. 5–43; Charles L. Glaser and Chaim Kaufmann, "What Is the Offense-Defense Balance and Can We Measure It?" *International Security*, Vol. 22, No. 4 (Spring 1998), pp. 44–82; Sean M. Lynn-Jones, "Offense-Defense Theory and Its Critics," *Security Studies*, Vol. 4, No. 4 (Summer 1995), pp. 660–691; and Charles L. Glaser, "Realists As Optimists: Cooperation As Self-Help," *International Security*, Vol. 19, No. 3 (Winter 1994/95), pp. 50–90.

peace is simple, powerful, and intuitively plausible. Thus the offense-defense balance concept has been used to address a variety of important historical, theoretical, and policy questions even when scholars have not adopted the basic assumptions and logic of the theory.[2] Perhaps the most important reason offense-defense theory continues to appeal to scholars is that it offers a compelling argument for why intense security competition among states is not an inevitable consequence of the structure of the international system. Specifically, for realists who believe that threats are more important than raw material power in explaining state behavior, the offense-defense balance appears to provide a systematic method of predicting when the balance of power is threatening and when it is not.[3]

Offense-defense ideas also continue to shape contemporary foreign policy debates on arms control, conventional and nuclear deterrence and force posture, the prevention of civil and ethnic conflict, and the so-called revolution in military affairs. On the latter issue, for example, the *Economist* recently proclaimed that the world is in the early stages of a new military revolution that will strengthen the offense relative to the defense and thus create "a strong incentive to strike first."[4] The most policy-relevant conclusion offered by offense-defense theory is that arms races, conflict, and war may be prevented through carefully designed arms control agreements that either deliberately shift the balance of technology toward defense or seek to correct misperceptions of the balance.[5]

---

2. The offense-defense balance has been used to explain the origins of interstate war, ethnic and civil conflict, arms control, arms racing, alliance behavior, military doctrine, the consequences of revolutions, grand strategy, and the structure of the international system. See discussion and references in Glaser and Kaufmann, "What Is the Offense-Defense Balance?" pp. 44–45, nn. 2, 3; and Lynn-Jones, "Offense-Defense Theory and Its Critics," pp. 660–662, nn. 3, 4.
3. For defensive realists, the degree to which one state threatens another is a function of the relative distribution of power (i.e., capabilities) filtered through the offense-defense balance. If power is distributed roughly equally and the balance of technology does not heavily favor the offense, states may feel secure and signal their peaceful intentions. If the balance heavily favors offense, however, states will face strong incentives to build offensive forces and fight preemptively. On the relationship between offense-defense theory and structural realism, see Van Evera, *Causes of War,* pp. 7–11, 117, 255–256; Glaser and Kaufmann, "What Is the Offense-Defense Balance?" pp. 48–49; Lynn-Jones, "Offense-Defense Theory and Its Critics," pp. 660, n. 1, 664–665; and Glaser, "Realists As Optimists," pp. 54, 60–64.
4. "Select Enemy. Delete," *Economist,* March 8, 1997, pp. 21–24 at p. 21.
5. See Jervis, "Cooperation under the Security Dilemma," pp. 199–201; Glaser and Kaufmann, "What Is the Offense-Defense Balance?" p. 44; and Van Evera, "Offense, Defense, and the Causes of War," p. 40. Other proponents of offense-defense theory are less sanguine about the capacity of arms control to reduce the likelihood of war, pointing out that arms control may be possible only when it is not necessary.

In this article I argue that the central concept of offense-defense theory—the offense-defense balance of technology—is deductively and empirically flawed. My analysis conclusion follows from two basic questions: First, is there an offense-defense balance of technology that can be used to predict military outcomes? Second, do perceptions of the offense-defense balance affect political decisions to initiate conflict? I conclude that scholars have overstated both the degree to which the balance of technology shapes battlefield outcomes and the influence that beliefs of offense or defense dominance have on political or strategic behavior.

The article is organized as follows. First, I examine the offense-defense balance concept; I present the basic definitions and assumptions required to operationalize the balance for empirical evaluation and discuss why the "core" version of the balance (which focuses solely on technology) is better than the "broad" conception (which incorporates additional variables). Next, I identify and evaluate the most frequently employed criteria for classifying how technology gives a relative advantage to offense or defense. Finally, I assess how offense-defense explanations fare against the record of four watershed technological innovations since 1850. These illustrative cases are the emergence of railroads in the nineteenth century, the artillery and small arms revolution of the late nineteenth and early twentieth centuries, the innovation of the tank in the first half of the twentieth century, and the nuclear revolution of the latter half of the twentieth century. The evidence suggests that although technology can occasionally favor offense or defense, perceptions of a technological balance have little effect on the likelihood of war.

The Offense-Defense Balance

The label "offense-defense theory" refers to the body of work that explores how changes in the offense-defense balance shape state behavior in international politics. Scholars working in the field, however, have conceptualized, operationalized, and employed the balance differently.[6] This diversity notwithstanding, there are several central features of the balance common to almost all approaches.

6. See Sean M. Lynn-Jones, "Realism, Security, and Offense-Defense Theories: The Implications of Alternative Definitions of the Offense-Defense Balance," paper presented at the annual meeting of the American Political Science Association, Boston, Massachusetts, September 3–6, 1998.

SHARED DEFINITIONS AND ASSUMPTIONS

The offense-defense balance denotes some measure of the relative ease of attacking and taking territory versus defending territory. "Relative ease" refers to the relative costs and benefits of attacking versus defending. The terms "offense" and "defense" refer to actual military actions, not the political intentions, goals, or objectives that motivate military action. Specifically, offense means the use of military force to attack, seize, and hold a portion or all of a defender's territory. Defense involves using military force to prevent an attacker from seizing territory.

The causal logic of offense-defense theory is based on the relative ease of offense and defense at the strategic level of war, not the operational or tactical level. The strategic level pertains to the highest levels of war planning and direction and the achievement of ultimate war goals, the operational level deals with the conduct of specific campaigns within a theater of operations, and the tactical level concerns actions taken within a particular battle.[7] The theory ultimately aims to explain decisions to initiate war; what matters is leaders' expectations of final war outcomes based on their perceptions of the strategic balance. Of course, the feasibility of strategic offense and defense depends on the success of operational and tactical offense and defense, and thus the nature of warfare at these levels is highly relevant to understanding the overall offense-defense balance.

Proponents of the theory have struggled to offer a more precise definition of the balance than just the relative ease of attack and defense.[8] Perhaps the most popular definition of the balance is cast in terms of a cost or investment ratio required for offensive success: the ratio of the amount of resources that an attacker must invest in offensive forces to offset the amount of resources a defender has invested in defensive forces.[9] Although measuring such a ratio (especially in historical cases) may be impossible, for present purposes it stipulates that the offense-defense balance is a continuous, not a dichotomous,

---

7. Edward Luttwak and Stuart L. Koehl, *The Dictionary of Modern War* (New York: Gramercy, 1998), pp. 568, 442, 598.
8. Jack S. Levy has effectively highlighted the logical and methodological problems of several previous attempts to define the balance in Levy, "The Offensive/Defensive Balance of Military Technology: A Theoretical and Historical Analysis," *International Studies Quarterly*, Vol. 28, No. 2 (June 1984), pp. 222–230.
9. For variations on this definition, see Jervis, "Cooperation under the Security Dilemma," p. 188; Glaser, "Realists As Optimists," p. 61; Lynn-Jones, "Offense-Defense Theory and Its Critics," p. 665; Glaser and Kaufmann, "What Is the Offense-Defense Balance?" pp. 3, 7–10; and Robert Gilpin, *War and Change in World Politics* (Cambridge: Cambridge University Press, 1981), p. 62.

variable. What matters most for empirical evaluation is not whether the balance in any given period favors offense or defense in absolute terms—in fact, it is almost always easier to defend than to attack—but how and to what degree the balance has shifted in either direction.

Finally, the offense-defense balance should be distinguished from two other key variables in international politics: power and skill. First, the balance must be defined independently of the balance of power among states. Battlefield outcomes clearly depend on a host of factors other than the offense-defense balance, such as the relative distribution of military forces and resources. Thus the success of any given offensive or defensive strategy is not necessarily indicative of the balance. For example, it might be misleading to say that offense is relatively easier than defense when a given state easily conquers another state, because this situation could simply have resulted from an overwhelming numerical disparity in the balance of forces, rather than from an offensive advantage. Similarly, a situation in which only one state in a conflict has acquired a new technology effectively represents a change in the balance of power, not in the offense-defense balance. The effects of a new technology on the balance can best be assessed in a conflict between two roughly equal-sized military forces employing the technology. Although the effects of the offense-defense balance can theoretically overcome disparities in material resources in determining war outcomes, the two variables are analytically distinct.

Second, the offense-defense balance should be defined independently of large disparities in the level of skill between the attacker and the defender. The relative ease of attack and defense is a concept meant to capture the objective effects of military technology on war and politics. To understand these baseline effects, one must assume that states make reasonably optimal or rational decisions about force posture, doctrine, and strategy. The standard of optimality employed by proponents does not require that attackers and defenders make the absolute best strategic choices, which might be impossible to determine in any case. Instead, optimality in this context assumes that states make reasonably intelligent decisions about how to employ existing technologies and forces given prevailing knowledge at the time.

CORE AND BROAD VERSIONS OF THE BALANCE

The "core" version of offense-defense theory looks almost exclusively at changes in military technology as the cause of shifts in the balance between offense and defense.[10] Some scholars, however, incorporate a host of factors in addition to technology when operationalizing the balance. Proponents of the "broad" version include some or all of the following factors: geography; the cumulativity of resources (the ease of exploiting resources from conquered territories); nationalism; regime popularity; alliance behavior; force size; and military doctrine, posture, and deployment.[11] For example, according to proponents of the broad approach, nationalism tends to favor defense relative to offense because people are more likely to fight harder when they believe they are defending their rightful homeland from foreign invaders.[12] Incorporating a host of geographic, social, political, and military factors into the balance clearly makes the theory more complex, but proponents believe the core version is otherwise incomplete. The relative ease of attack or defense is determined by a set of basic causal factors, they argue, and omitting these factors is unlikely to result in accurate explanations or predictions.

Although proponents of the broad version of the offense-defense balance believe their approach strengthens offense-defense theory, there are at least three important theoretical and practical advantages to focusing solely on the core balance of technology. First, technology is the one determinant of the balance common to all versions of the theory, the most significant factor shaping the balance, and often the only factor analyzed in any detail by scholars. None of the factors identified by the broad approach have such wide applicability and importance. Thus offense-defense theory's contribution to the conceptual toolbox of international relations largely turns on the role that the technological balance plays in shaping state behavior.

Second, because the broad offense-defense balance incorporates factors unique to particular states (such as geography, nationalism, and regime popularity), it is not a systemic variable, and the resulting theory is no longer

---

10. Adopting the core approach are George Quester, Robert Jervis, and Sean Lynn-Jones. Although Jervis cites technology and geography as the two main factors that determine whether offense or defense has the advantage, technology is far more important in his work. See Jervis, "Cooperation under the Security Dilemma," pp. 194–196.

11. Adopting the broad version are Stephen Van Evera, Charles Glaser, Chaim Kaufmann, and, to some extent, Ted Hopf and Jack Snyder. For discussions of these factors, see Van Evera, "Offense, Defense, and the Causes of War," pp. 17, n. 23, 19–22; Glaser and Kaufmann, "What Is the Offense-Defense Balance?" pp. 41, 64–70; Lynn-Jones, "Offense-Defense Theory and Its Critics," p. 669; Ted Hopf, "Polarity, the Offense-Defense Balance, and War," *American Political Science Review,* Vol. 85, No. 2 (June 1991), pp. 477–478; Jack Snyder, *The Ideology of the Offensive: Military Decision Making and the Disasters of 1914* (Ithaca, N.Y.: Cornell University Press, 1984), pp. 21–26; and Jervis, "Cooperation under the Security Dilemma," p. 195.

12. Glaser and Kaufmann, "What Is the Offense-Defense Balance?" pp. 66–67. The authors also note exceptions under which nationalism favors offense.

structural. Offense-defense theory claims to share the appeal of other structural theories in international relations because it focuses on the war-causing effects of a variable that is essentially exogenous to states. Technology, in principle, provides similar constraints and opportunities for all states in a given international system.[13] Thus comparable patterns of state behavior should arise under similar technological balances in history. At best, the broadly defined balance might shed light on a specific military conflict, but its effects are not generalizable across space and time.

Finally, and most important, the core version makes for a much more parsimonious theory than the broad version. All things being equal, simpler theories are both easier to measure (and test) and more intuitively appealing. First consider the issue of measurement. Although measuring the balance of military technology is extremely complicated, it is certainly more feasible than measuring a balance that incorporates a host of complex, ambiguous, and sometimes cross-cutting variables. Even if one were able to accurately assess the impact of the broad factors on the relative ease of attack and defense, one would still have to weigh the relative importance of each factor and aggregate them into a single value of the balance. If the balance cannot be measured, the theory has little explanatory power and prescriptive utility.

In addition, a theory that uses few variables to explain a class of phenomena in the real world is more satisfying than a theory built on all possible causes. Adopting the broad balance renders offense-defense theory atheoretical; it becomes a grab bag of variables employed in a purely post hoc descriptive enterprise. Although each of the factors identified in the broad version may shape the relative ease of attack and defense, a laundry-list explanation is not intuitively appealing. The outbreak of violence in any single case results from an inevitably complex set of opportunities and constraints, motives and goals, and decisions and actions. A more interesting theoretical issue, however, is whether there is something about military technology itself that affects the likelihood of war and peace.

How Does Technology Affect the Offense-Defense Balance?

What are the criteria used to identify how technology gives a relative advantage to offense or defense at any given time? Without such coding criteria, we have no theoretical guidance for judging which factors contribute dispro-

13. See Lynn-Jones, "Offense-Defense Theory and Its Critics," p. 668.

portionately to offense or defense on the battlefield and, thus, cannot determine the offense-defense balance. Proponents have struggled to provide objective and consistent criteria for distinguishing between offensive and defensive technologies.[14] The few explicit discussions of differentiation criteria have often been supported by ambiguous arguments or contradictory examples.[15]

Despite the high degree of confusion, there appears to be at least some consensus that mobility innovations favor offense, whereas firepower innovations favor defense. Not all offense-defense proponents make these claims, and the large majority who do would not argue that these are concrete laws of military history. Nevertheless, the mobility and firepower criteria are the most useful, clearly articulated, and frequently employed hypotheses.

HYPOTHESIS 1: MOBILITY-ENHANCING TECHNOLOGIES FAVOR OFFENSE

Almost all proponents of offense-defense theory believe that new or improved technologies that enhance mobility contribute relatively more to offense than defense.[16] In military terms, mobility is the ability of troops and equipment to

---

14. Critics of offense-defense theory often argue that it is impossible to determine how technology affects the balance because it is very difficult to categorize weapons as offensive or defensive. See Levy, "Offensive/Defensive Balance," pp. 219–238; John J. Mearsheimer, *Conventional Deterrence* (Ithaca, N.Y.: Cornell University Press, 1983), pp. 25–27; Samuel P. Huntington, "U.S. Defense Strategy: The Strategic Innovations of the Reagan Years," in Joseph Kruzel, ed., *American Defense Annual, 1987–1988* (Lexington, Mass.: Lexington Books, 1987), pp. 35–37; Jonathan Shimshoni, "Technology, Military Advantage, and World War I: A Case for Military Entrepreneurship," *International Security*, Vol. 15, No. 3 (Winter 1990/91), pp. 190–191; and Colin S. Gray, *Weapons Don't Make War: Policy, Strategy, and Military Technology* (Lawrence: University Press of Kansas, 1993), chap. 2. Proponents claim that the logic of the theory depends not on the ability to classify weapons as entirely offensive or defensive, but on whether given weapons make offense or defense easier. The disagreement is primarily semantic and, in any case, proponents contend that in practice offensive and defensive weapons and force postures are generally distinguishable. Lynn-Jones, "Offense-Defense Theory and Its Critics," pp. 674–677; and Glaser and Kaufmann, "What Is the Offense-Defense Balance?" pp. 79–80.

15. Given the prominence and influence of offense-defense theory, the paucity of coding criteria is remarkable. For example, Jervis was pessimistic about the ability to define in theory or identify in practice the offense-defense variables that shape the severity of the security dilemma, and it is difficult to find any coding criteria in his seminal piece on offense-defense theory. Jervis, "Cooperation under the Security Dilemma." Similarly, Van Evera states that "military technology can favor the aggressor or the defender," but provides no criteria for deciding the issue anywhere in his book. Van Evera, *Causes of War*, p. 160. Charles Glaser and Chaim Kaufmann provide the most explicit discussion in Glaser and Kaufmann, "What Is the Offense-Defense Balance?" For a review and critique of attempts to classify the technological characteristics of offense and defense, see Keir A. Lieber, "Offense-Defense Theory and the Prospects for Peace," Ph.D. dissertation, University of Chicago, forthcoming, chap. 2.

16. Glaser and Kaufmann, "What Is the Offense-Defense Balance?" pp. 61–63; and Quester, *Offense and Defense*, p. 3.

move from one place to another. There are essentially three types of mobility: strategic, operational, and tactical.

Strategic mobility is the ability to transport military forces from the homeland to a theater of operations, or from one theater to another. Offense-defense theorists argue that greater strategic mobility allows the attacker to expeditiously transport and supply its forces far from its own borders, thus negating the defender's geographic advantage. Operational mobility is the ability to move forces within a theater. According to proponents, greater operational mobility allows the attacker to concentrate forces quickly to achieve a numerical advantage on a small portion of the front, rapidly exploit weak points in a defender's line, or outflank a defender's position altogether. Tactical mobility is the ability to move forces on the battlefield, in the face of enemy fire. Offense-defense proponents argue that greater tactical mobility reduces the number of casualties suffered by an attacker because these losses are partly a function of the amount of time that forces are exposed to enemy fire in an assault.

There are several counterarguments to the mobility-favors-offense explanation. First, in terms of strategic mobility, it is not clear why the ability to transport and supply forces far from the homeland gives an attacker an *advantage* over a defender who already has this capability. Once an attack is under way, in fact, the defender depends more than the attacker on the ability to quickly move forces to that theater. Moreover, the impact of strategic mobility appears indeterminate when the defender relies more heavily than the attacker on reinforcement from overseas territories and allies.[17] Second, in terms of operational mobility, the attacker depends more on the element of surprise than on mobility to achieve a successful breakthrough of a defender's front line, whereas the defender places a premium on mobility to reinforce threatened points in the front.[18] Unless a breakthrough, penetration, or envelopment occurs so rapidly that the defender never has a chance to react and counterattack, the defender would also seem to profit more from mobility once an advance penetration is under way. Third, greater tactical mobility may be more advantageous to the defender than the attacker in several ways. Tactical mobility allows defenders to trade space for time through a series of tactical withdrawals to fortified positions where they can continue to fire on attacking forces. In addition, greater offensive tactical mobility may actually *increase*

17. Glaser and Kaufmann, "What Is the Offense-Defense Balance?" p. 63, n. 59.
18. Mearsheimer, *Conventional Deterrence*, pp. 25–26.

attacker casualties, as the greater speed of an assault often comes at the price of reconnaissance, protection, and preparatory artillery fire.[19] Finally, tactical mobility is advantageous for the defender because the defender often must seize the tactical counteroffensive to avoid defeat.

HYPOTHESIS 2: FIREPOWER-ENHANCING TECHNOLOGIES FAVOR DEFENSE

The other plausible criterion for assessing the offense-defense impact of new military technologies is the characteristic of firepower. Firepower is a measure of the destructive power of the weapons or array of weapons available to sides in a conflict. Firepower consists of not only explosive power, but also range, accuracy, and rate of fire.

According to most proponents, technological innovations that enhance firepower capability are disproportionately advantageous to the defense.[20] First, firepower allows the defender to threaten the attacker's concentration of forces before an attack. An attacker typically needs a local advantage of combat power to pierce the defender's forward defenses. Numerical superiority requires density; but the greater density of forces provides more targets for defensive fire, and thus more attacker casualties. Second, firepower favors defense because it reduces the mobility (i.e., offensive power) of the attacker. In the face of greater defensive fire, an attacker must seek more armored protection, cover, concealment, and dispersal—all of which slow the attacker's advance. Finally, defensive firepower forces the attacker to provide its own covering fire in the advance, which slows the attack because of added weight and time required to reposition covering fire.

There are several reasons to believe, however, that firepower is as crucial in the attack as it is in defense. The attacker relies heavily on suppressive or covering fire to neutralize or inhibit defender forces, weapons, or reconnaissance. Suppressive fire by an attacker reduces the amount of fire faced by advancing forces and can pin down defender forces until they can be overrun and destroyed. More important, most successful offensives require preparatory bombardments before the attack. Preparatory barrages can shatter the morale of defenders, destroy defensive positions, and disrupt defender reinforcements and communication. Finally, just as the defender uses firepower to disrupt

19. See Stephen Biddle, "The Determinants of Offensiveness and Defensiveness in Conventional Land Warfare," Ph.D. dissertation, Harvard University, 1992, pp. 68–77.
20. The most clear-cut logic behind the firepower hypothesis is found in Glaser and Kaufmann, "What Is the Offense-Defense Balance?" p. 64.

attacker concentrations of forces before an attack, the attacker depends on firepower to disperse defender forces into greater depth away from the frontline. Because firepower, especially artillery, does the greatest damage to forces that are grouped together, dispersal is the wisest option. When a defender disperses, however, the force-to-force ratio shifts in the attacker's favor, making offensive breakthroughs more likely.

Mobility, Firepower, and International Security

The mobility and firepower criteria, whatever their logical weaknesses, provide a clear blueprint for case selection and empirical evaluation of offense-defense theory. I consider the four biggest technological innovations in mobility and firepower in modern history: railroads, the artillery and small arms revolution, tanks, and nuclear weapons.[21] In each case, I focus on two questions. First, what impact did the innovation have on military outcomes? The relevant issue is whether a mobility innovation shifted the offense-defense balance toward offense, resulting in more quick and decisive victories for the attacker, or whether a firepower innovation shifted the balance toward defense, resulting in longer, indecisive battles of attrition. Second, what impact did the innovation have on political outcomes? Here we need to ask if and how decisionmakers thought about these revolutionary innovations in offense-defense terms and whether these perceptions made war more likely. Specifically, did leaders believe that the attacker or the defender was privileged by the innovation? Were leaders more inclined to initiate war when they believed offense was favored?

EMERGENCE OF RAILROADS

The introduction of steam-powered railroads in the second half of the nineteenth century perhaps marked the greatest revolutionary development in military mobility since the wheel. Armies were suddenly able to move and sustain huge forces across vast distances at up to ten times the speed of marching troops. The first practical locomotive appeared in 1825, railroads spread rapidly across the European continent in the 1830s and 1840s, and by

21. With the exception of the gunpowder revolution in the fifteenth and early sixteenth centuries, technological progress in warfare before the nineteenth century was more gradual and evolutionary than revolutionary. See Martin van Creveld, *Technology and War: From 2000 B.C. to the Present* (New York: Free Press, 1991); and Bernard Brodie and Fawn M. Brodie, *From Crossbow to H-Bomb* (Bloomington: Indiana University Press, 1973).

1850 all the major powers had conducted field exercises in moving and supplying troops by rail.[22]

MILITARY OUTCOMES. The advent of railroads coincided with several relatively short and decisive conflicts between 1850 and 1871, and thus would appear to support the hypothesis linking mobility improvements with offensive advantages. A closer look at the evidence, however, reveals that these battlefield outcomes resulted primarily from large asymmetries in power and skill, rather than from the offensive advantages of railways. Moreover, although no wars occurred among the European great powers between 1871 and 1914, World War I suggests that railroads if anything favored the strategic defender.

In 1850, in one of the earliest strategic uses of railroads, Austria quickly mobilized and transported 75,000 soldiers by rail to Bohemia, forcing Prussia to back down in an escalating crisis. Prussia had a small rail network and poor administrative organization at the time, however, and bungled its own mobilization to the front.[23] The war between Austria and France in northern Italy in 1859 saw the large-scale use of railroads for strategic concentration, operational reinforcement, and even tactical movement of troops. The French were able to deploy 120,000 soldiers in eleven days and, once in the theater of conflict, use railroads to quickly and unexpectedly shift forces to defeat the Austrians. In this case, however, the Austrians were guilty of incompetent preparation, mobilization, and transportation, and probably would have been defeated by the French even in the absence of rail transport.[24]

---

22. Some proponents of offense-defense theory view the railroad case as an exception to the mobility-favors-offense prediction. They argue that railway mobility is more useful for defenders because rail networks can be destroyed by retreating defenders more easily than they can be extended by advancing attackers. Van Evera, "Offense, Defense, and the Causes of War," pp. 16–17; Glaser and Kaufmann, "What Is the Offense-Defense Balance?" p. 63; and Quester, *Offense and Defense*, chap. 8. This remains an important test case, however. First, according to the general logic employed by proponents, railroad mobility should favor offense at the *strategic* level, where the attacker can more quickly concentrate forces at the front to surprise and/or overwhelm the defender. Second, our confidence in the mobility-favors-offense hypothesis would be considerably diminished by an empirical finding that railroads actually favored defenders on the whole, given that railroads marked such a revolutionary improvement in mobility. Finally, even if military outcomes demonstrate the defensive advantages of railroads, we can still evaluate offense-defense theory based on how leaders' perceptions of the impact of railroads affected their behavior.

23. Edwin A. Pratt, *The Rise of Rail-Power in War and Conquest, 1833–1914* (Philadelphia: J.B. Lippincott, 1916), p. 8; and Dennis E. Showalter, *Railroads and Rifles: Soldiers, Technology, and the Unification of Germany* (Hamden, Conn.: Archon, 1975), pp. 37–38.

24. Pratt, *Rise of Rail-Power*, pp. 9–13; John Westwood, *Railways at War* (San Diego, Calif.: Howell-North, 1980), pp. 14–16; Brodie and Brodie, *From Crossbow to H-Bomb*, p. 149; and Michael Howard, *War in European History* (Oxford: Oxford University Press, 1976), pp. 97–98.

Prussia's quick and decisive victories in the Wars of German Unification—against Denmark (1864), Austria (1866), and France (1870–71)—have commonly been attributed to the offensive power of the Prussian railroads. In fact, railroads had much less impact on the conduct of these wars than did Prussia's superior doctrine, organization, and material power.

Against Austria, the Prussian attackers made extensive use of their railways to mobilize and transport an unprecedented 200,000 troops to the theater of operations within three weeks. But after initial deployments, the Prussians ran into great trouble supplying and sustaining their offensive beyond the railheads. Prussian forces quickly outran their supply convoys, leaving food and fodder rotting at hopelessly congested railheads. From the crossing of the Austrian border to the decisive battle of Königgrätz, railways were irrelevant to the outcome of the war.[25]

The Austrians were decisively defeated by Prussia because of crucial asymmetries in power and skill, not because of the inherent offensive power of railroads. The Austrians had only one major railroad line leading into the theater of war, while Prussia had five. Because of this superior railway network, as well as the excellent planning of the Prussian general staff, Prussia was able to mobilize and deploy more battle-ready troops to the field than the Austrians. The rapid Austrian defeat was also facilitated by the technical edge of the Prussian infantry's breech-loading rifle compared to the Austrian muzzle-loader, and Prussia's superior doctrine of maneuvering forces to assume the tactical defensive to take advantage of modern firepower.[26]

The Franco-Prussian War reveals a similar story, though in this case the planned French offensive was defeated quickly and decisively. The French mobilization and concentration of forces was utterly incompetent, despite its strategically superior rail network, and was the single most important cause of France's defeat. Although Prussia's mobilization of roughly 400,000 troops by railroad was organized and efficient, the concentration of their forces on the French frontier was relatively inept. These problems exposed the Prussians to potential defeat by a better organized and capable defender.[27]

---

25. Martin van Creveld, *Supplying War: Logistics from Wallenstein to Patton* (London: Cambridge University Press, 1977), pp. 83–85; and Pratt, *Rise of Rail-Power,* pp. 104–105. Ironically, the unexpected Prussian freedom of movement created by the need to live off the land instead of depending on supplies from railheads helped them win a decisive victory. Westwood, *Railways at War,* p. 57.
26. Showalter, *Railroads and Rifles,* pp. 14–15, 59–68; and Larry H. Addington, *The Patterns of War since the Eighteenth Century* (Bloomington: Indiana University Press, 1994), pp. 53–54, 94–97.
27. Thomas J. Adriance, *The Last Gaiter Button* (New York: Greenwood, 1987), pp. 47–54; and Pratt, *Rise of Rail-Power,* pp. 110–115.

After the initial deployment of forces, railroads played virtually no important role in the Prussian offensive into France. The extended railway lines were jammed with traffic and vulnerable to French attacks, the railheads could not keep up with the advancing forces, and the Prussians faced enormous difficulties in getting supplies from the railheads to the front. The heavy artillery, ammunition, and forces conveyed by railroads did make possible the siege and bombardment of Paris, but this occurred well after the decisive mobile phase of the campaign was over.[28] Other sharp disparities in military skill and capability compounded France's dismal performance, including the French army's flawed command and staff system, lack of reserves, and unsuitable tactical doctrine of massed frontal attacks against the Prussian Krupp steel, rifled, breech-loading artillery.

The American Civil War (1861–65) offers additional evidence that railroad mobility had not shifted the balance toward offense. The control of railroads was crucial for both Union and Confederate forces given the vast territorial scale of military operations. At critical times, both sides used railroads to concentrate forces at strategic points to hold off enemy offensives.[29] Although Union forces also often relied on long rail lines for communications and supplies as they advanced deep into the South, railroads were on balance more useful for the Confederates fighting on the strategic defense. Outnumbered and outgunned, the Confederates depended on rapidly concentrating separated forces against key segments of the Union army. Railroads prolonged the Civil War and made it more difficult to fight quick and decisive campaigns.[30]

The impact of railroads in World War I clearly contradicts the mobility-favors-offense hypothesis. By the time of the war, all sides in the conflict had a good understanding of railroad technology, had adopted appropriate doctrines for its use, and had a generally equal level of skill in its employment. At the outbreak of war, railroads moved soldiers, weapons, and supplies at an

---

28. Van Creveld, *Supplying War,* pp. 96, 104; and Westwood, *Railways at War,* p. 66.
29. See, for example, the Confederate use of railroads at First Bull Run and Chickamauga and the Union use of railroads to reinforce troops on the Chattanooga front after Chickamauga. George Edgar Turner, *Victory Rode the Rails: The Strategic Place of the Railroads in the Civil War* (Lincoln: University of Nebraska Press, 1992), chaps. 7, 21; Robert C. Black, III, *The Railroads of the Confederacy* (Chapel Hill: University of North Carolina Press, 1998), pp. 184–191; and Thomas Weber, *The Northern Railroads in the Civil War, 1861–1865* (Bloomington: Indiana University Press, 1952), pp. 180–181.
30. Westwood, *Railways at War,* pp. 17, 29; and Christopher R. Gabel, *Railroad Generalship: Foundations of Civil War Strategy* (Fort Leavenworth, Kans.: U.S. Army Command and General Staff College, 1997).

unprecedented pace and scale. The enhanced strategic mobility conferred by the railroad did not translate, however, into quick and decisive battlefield outcomes, as was demonstrated by the French use of railroads to shift resources to halt the Schlieffen Plan, the initial German offensive, and the German use of railroads to shift forces from the western to eastern front to defeat the Russian offensive.[31]

POLITICAL OUTCOMES. Railroads did not confer an advantage on the attacker. Did political and military leaders believe and act as if they did?[32] In fact, the historical record flips the standard offense-defense hypothesis on its head. Conventional wisdom between 1850 and 1871 (when wars were more frequent) held that railroads favored the defender, while the dominant view after 1871 (when wars were infrequent) held that railroads favored the attacker.

By the mid-nineteenth century, although some commentators warned that the building of railroads would only facilitate a foreign invasion of the homeland, the prevailing military view was that railroads would favor the defender by greatly improving the defender's ability to shift troops to counter any threatened sector of the frontier. Theoretical writings, most notably by the economist Friedrich List, even surmised that the defensive advantages of railroads would bring perpetual peace to the European continent.[33]

Military leaders in Prussia, which was surrounded by potential enemies, were especially quick to see the defensive benefits of railroads. Helmuth von Moltke, chief of the general staff beginning in 1857, thought that Prussia would eventually be attacked and believed that the defensive mobility provided by an extensive network of railroads could counterbalance Prussia's disadvantage in sheer number of forces.[34] After building such a comprehensive railway network, and despite the perception that railroads favored the defender, Prussia promptly provoked three wars in less than a decade, waging some of the most decisive offensive campaigns in history. In fact, it was largely *because* the railroad made the defense of Prussian territory easier—that is, troops could be deployed rapidly from the center to any front or redeployed from front to front—that Prussia was able to act aggressively toward its neighbors.

---

31. See Westwood, *Railways at War*, p. 143; and van Creveld, *Supplying War*, pp. 111–140.
32. Some offense-defense proponents claim that while actual offense dominance has been rather rare, "perceived offense dominance is pervasive, and it plays a major role in causing most wars." Van Evera, *Causes of War*, p. 185.
33. Showalter, *Railroads and Rifles*, pp. 18–35; and Westwood, *Railways at War*, pp. 8–12, 91, 197.
34. Van Creveld, *Supplying War*, p. 88; and Showalter, *Railroads and Rifles*, pp. 18, 28, 43, 56.

After the Franco-Prussian War, every state in Europe quickly concluded that railroads favored the attacker and strove to copy Prussian institutions for the use of railways.[35] All the European general staffs believed that quick and decisive victory would come to the side that mobilized and concentrated its troops the fastest, and the railroads were thought to provide the key. Despite the new dominant view that railroads favored the attacker, however, no war occurred on the continent for the next forty years.

SMALL ARMS AND ARTILLERY REVOLUTION

A technical revolution occurred in the late nineteenth and early twentieth centuries with the development of rifled, breech-loading small arms and artillery of unprecedented range, accuracy, and rate of fire. The combined effect was an enormous increase in firepower that armies could bring to bear on the battlefield.

MILITARY OUTCOMES. The firepower revolution rendered massed frontal assaults exceedingly difficult, and many conflicts were marked by costly battles of attrition. On balance, therefore, it is fair to say that the new technologies shifted the offense-defense balance toward the defender.

As early as the Crimean War (1854–56), rifles showed the potential to be highly effective defensive weapons against attacking infantry. But the American Civil War, Wars of German Unification, Russo-Turkish War (1877–78), Anglo-Boer War (1899–1902), Russo-Japanese War (1904–05), and World War I provide the best evidence that defenders armed with modern rifles, machine guns, and artillery had gained an enormous advantage against assaulting infantry.

The defensive advantage conferred by firepower has often been exaggerated, however. Two sets of evidence are notable. First, the tactical impasses created by firepower technologies did not necessarily translate into strategic or operational deadlock. The clearest example is the Prussian method of using strategic envelopment and flanking maneuvers to place forces where they could employ tactical defensive firepower against the enemy rear or flank. This fusion of the strategic offensive with the tactical defensive contributed to Prussia's quick and decisive victories against Austria in 1866 and France in 1870–71.[36] In the

35. Westwood, *Railways at War*, p. 197; Quester, *Offense and Defense*, pp. 77–83; Howard, *War in European History*, p. 101; and van Creveld, *Technology and War*, p. 159.
36. See Gunther E. Rothenberg, "Moltke, Schlieffen, and the Doctrine of Strategic Envelopment," in Peter Paret, ed., *Makers of Modern Strategy from Machiavelli to the Nuclear Age* (Princeton, N.J.: Princeton University Press, 1986), pp. 296–325.

American Civil War, Union forces were able to bring the war to an end more quickly when they learned to employ a strategic offensive/tactical defensive doctrine in pursuit of the Confederates.[37] In fact, every major war between 1861 and 1905 was ultimately decided by strategic offensive maneuvers.[38] In World War I, the Schlieffen Plan was modeled on earlier Prussian victories and almost succeeded—recall the French "miracle on the Marne."

A second reason to question the degree to which technology favored defense in this period was the success of the Germans and then the Allies in carrying out a series of offensive breakthroughs in the later stages of World War I using innovative infantry and combined arms tactics. As far back as the American Civil War, infantry learned with some success to break up from waves of attacking troops into small groups that alternated advancing with providing covering fire while others moved forward. But it was the German army's decision in 1917 to introduce new "infiltration tactics" that provided a real tactical solution to the stalemate of trench warfare. These tactics called for a brief surprise artillery bombardment aimed at disrupting narrow weak points in the enemy line, followed by the quick penetration by small independent groups of storm troops who were to bypass points of strong resistance and advance as far as possible. The Germans employed infiltration tactics with great success in late 1917 and, especially, in the spring of 1918 with the famous Ludendorff offensives.[39] Despite no major changes in technology, the Ludendorff offensives achieved significant and unprecedented breakthroughs followed by deep advances behind the Allied lines.[40] These offensives, of course, ultimately failed on the strategic level, as the Germans lacked the transportation and logistical capabilities necessary to follow up on their tactical successes, but the Allies adopted similar tactics in their own offensives until the end of the war.

Massed frontal assaults in the face of modern firepower were clearly not the optimal method of attack, but it is impossible to say whether warfare would

37. Trevor N. Dupuy, *The Evolution of Weapons and Warfare* (Indianapolis: Bobbs-Merrill, 1980), pp. 201–202.
38. Van Creveld, *Technology and War*, p. 177.
39. Timothy T. Lupfer, *The Dynamics of Doctrine: The Changes in German Tactical Doctrine during the First World War* (Fort Leavenworth, Kans.: U.S. Army Command and General Staff College, 1981); Dupuy, *Evolution of Weapons and Warfare*, pp. 225–229; and Hew Strachan, *European Armies and the Conduct of War* (London: Routledge, 1983), pp. 142–149.
40. Stephen Biddle analyzes the first of the Ludendorff offensives as a test of offense-defense theory in Biddle, "Recasting the Foundations of Offense-Defense Theory," paper presented at the annual meeting of the American Political Science Association, Boston, Massachusetts, September 3–6, 1998.

have looked drastically different had infiltration tactics been introduced earlier in World War I. At a minimum, the German offensives suggest that the defensive advantage of firepower was not as intrinsic to the prevailing technology as is often portrayed.

POLITICAL OUTCOMES. In the decades before World War I, according to offense-defense proponents, European statesmen and military leaders erroneously believed that attackers would benefit most from the vast increases in firepower, and wars would thus be short and decisive. This "cult of the offensive," proponents argue, was a principal cause of World War I.[41]

Europeans did embrace offensive strategies before World War I, but this had little to do with beliefs in the offensive advantages of technology. Instead the dominant preference for offensive strategies sprung from a host of organizational, social, political, and psychological causes. Rather than address these causes of the cult of the offensive, which have been well documented,[42] I focus here on whether perceptions of the nature of the firepower revolution dampened or promoted conflict.

Prussian leaders were aware of the defensive impact of firepower before initiating the Wars of German Unification. As early as 1858, Moltke argued that any potential enemy should be forced by maneuver into taking the tactical offensive against Prussian defensive firepower.[43] After the costly attacks by his forces against Denmark in 1864, Moltke concluded that in the age of the breech-loading rifle, no combination of bravery and superior numbers could overcome the problem of attacking frontally over open ground against modern firepower. "The attack of a position," Moltke wrote in 1865, "is becoming notably more difficult than its defense."[44] This view of the

41. Van Evera, *Causes of War,* chap. 7; Quester, *Offense and Defense,* chap. 10; Jervis, "Cooperation under the Security Dilemma" pp. 190–192; Stephen Van Evera, "The Cult of the Offensive and the Origins of the First World War," *International Security,* Vol. 9, No. 1 (Summer 1984), pp. 58–107; and Snyder, *Ideology of the Offensive.* For a critique of the "cult of the offensive" argument, see Marc Trachtenberg, *History and Strategy* (Princeton, N.J.: Princeton University Press, 1991), chap. 2. The firepower revolution and World War I is a crucial case for offense-defense theory. James D. Fearon and Richard K. Betts note that the war has served as the principal source for generating offense-defense hypotheses and at the same time as the principal empirical test of these hypotheses. Fearon, "The Offense-Defense Balance and War since 1648," paper presented at the annual meeting of the International Studies Association, Chicago, Illinois, February 21–25, 1995, p. 2; and Betts, "Must War Find a Way? A Review Essay," *International Security,* Vol. 24, No. 2 (Fall 1999), pp. 166–198, at p. 184. Contradictory findings would thus be especially problematic for the theory.

42. Van Evera, "Cult of the Offensive"; and Snyder, *Ideology of the Offensive.*

43. Strachan, *European Armies,* p. 115.

44. Quoted in Showalter, *Railroads and Rifles,* p. 125. Prussian military instructions and training manuals reflected this belief as well.

increased power of the defense did not dissuade Prussia from initiating war with Austria in 1866 and France in 1870. Instead, Moltke adopted an offensive strategy that sought to capitalize on the firepower of a tactical defensive.

Even after the spectacular offensive successes by Prussia, no country denied the impact of firepower.[45] Technical arguments that improvements in firepower had benefited offense over defense did appear, but were mainly promulgated to justify offensive doctrines already deemed necessary for political and organizational reasons. In short, perceptions of the offensive advantages of firepower did not lead states to adopt offensive strategies; rather, the bias in favor of offensive strategies made military thinkers concentrate on ways to use firepower in the attack.

Germany's role in the outbreak of World War I contradicts offense-defense predictions. German military leaders evaluated the technical realities of firepower more objectively than all other European general staffs at the time, and thus were fully aware of the increased power of the defense. Yet Germany's war plan since 1891 consistently called for a decisive offensive envelopment against France before rapidly shifting forces against Russia. Neither Alfred von Schlieffen, his successor Moltke (the younger), nor Germany's civilian leaders envisioned that conquest would be easy. Germany's geostrategic position, combined with its foreign ambitions, demanded a quick and decisive victory at the outset of an expected two-front war.[46]

Offense-defense predictions about European security between 1890 and 1914 face other major anomalies. Consider Stephen Van Evera's claim: "Belief in the power of the offense increased sharply after 1890 and rose to very high levels as 1914 approached. . . . [This belief] peaked in 1914 in Europe, and Germany had the largest offensive opportunities and defensive vulnerabilities among Europe's powers. Offense-defense theory therefore forecasts that war should erupt in Europe in about 1914, authored largely by Germany."[47] Note first that if war broke out in 1890, 1905, or 1912, Van Evera could still make the same claim of theory validation because beliefs in offense were steadily rising and thus potentially always "peaking." Second, if offense was

45. According to one historian, "nobody was under any illusion, even in 1900, that frontal attack would be anything but very difficult and that success could be purchased with anything short of very heavy casualties." Michael Howard, "Men against Fire: Expectations of War in 1914," *International Security,* Vol. 9, No. 1 (Summer 1984), p. 43.
46. See Rothenberg, "Moltke, Schlieffen, and the Doctrine of Strategic Envelopment."
47. Van Evera, *Causes of War,* pp. 193, 199, n. 25.

perceived to have had an advantage since 1890 (if not since 1871), why did war *not* break out in 1890, 1905, or 1912? Germany's alleged belief in the technical supremacy of the offensive, combined with its sheer military advantage over a weakened France and Russia, ought to have led it to attack France in 1905 or Russia in 1909, when Germany faced real windows of opportunity.[48] Moreover, if the technological revolution in firepower was thought to favor offense, why did no real competitive arms racing on land occur before 1912? In short, offense-defense theory does not appear capable of explaining the outbreak and timing of World War I.

TANKS

The character of land warfare was transformed by the mechanization and motorization of armies from the end of World War I through World War II. The most important military innovation in this period was the tank. The combination of technological advances in the internal combustion engine, armored protection, and radio communication greatly increased operational mobility on the battlefield.

MILITARY OUTCOMES. Proponents of offense-defense theory believe that the incorporation of tanks into the European armed forces resulted in greater offense dominance.[49] In World War I and the interwar period, however, tanks had a negligible affect on operational outcomes. In World War II, the most relevant evidence does not show the offensive superiority of tank forces.

In World War I tanks occasionally contributed to tactical breakthroughs and penetrations, but ultimately could not translate any tactical successes into operational victories. In fact, the most spectacular breakthroughs of the war, such as the 1918 Ludendorff offensives, were made possible by new infantry tactics, not tanks.[50] The wars fought between 1919 and 1939 were primarily

48. David G. Herrmann persuasively argues that the mere existence of a window of opportunity was not enough to prompt preemptive German strikes in the decade before 1914. Herrmann, *The Arming of Europe and the Making of the First World War* (Princeton, N.J.: Princeton University Press, 1996). See also Richard Ned Lebow, "Windows of Opportunity: Do States Jump through Them?" *International Security*, Vol. 9, No. 1 (Summer 1984), pp. 147–186.
49. Jervis, "Cooperation under the Security Dilemma," p. 197; Van Evera, *Causes of War*, pp. 123, 162; Lynn-Jones, "Offense-Defense Theory and Its Critics," p. 676; and Glaser and Kaufmann, "What Is the Offense-Defense Balance?" p. 63.
50. On the role of tanks in World War I, see Williamson Murray, "Armored Warfare: The British, French, and German Experiences," in Murray and Allan R. Millett, eds., *Military Innovation in the Interwar Period* (Cambridge: Cambridge University Press, 1996), pp. 6–49; J.P. Harris, *Men, Ideas, and Tanks: British Military Thought and Armoured Forces, 1903–1939* (Manchester: Manchester Uni-

either civil wars or colonial conflicts, and involved unevenly matched adversaries or forces that were not well equipped with tanks. When tanks were used in battle, as in the continuing French effort to extend control over Morocco (1908–34), the Italian war against Ethiopia (1935–36), the Spanish Civil War (1936–39), and the Russo-Japanese border clashes in Manchuria (1938–39), the results were not illuminating.[51]

The revolutionary potential of the tank was first demonstrated by Germany's rapid envelopment of Polish forces in 1939, quick and decisive defeat of France in 1940, and invasion of the Soviet Union in 1941. These campaigns offer inadequate evidence that tanks conferred a decisive advantage on the offense, however, because the victories resulted more from German material and doctrinal superiority than from the balance of military technology. In September 1939, German forces were better trained, better equipped, and far larger than the Polish army.[52] The root cause of the German victory over France lies with Germany's far superior strategy, tactics, and organization, rather than in the nature of its military hardware.[53] It is highly unlikely that Germany could have achieved its stunning offensive in France if not for glaring Allied weaknesses and if the Allies had been more adept at using armored forces.[54] The German invasion of the Soviet Union was enormously successful, as panzer and motorized divisions achieved spectacular encirclements of entire Soviet armies. Once again, however, the key to German success lay in Josef Stalin's blunders and Red Army failings.[55]

---

versity Press, 1995); and Spencer C. Tucker, *The Great War, 1914–18* (Bloomington: Indiana University Press, 1998).

51. See Archer Jones, *The Art of War in the Western World* (New York: Oxford University Press, 1987), pp. 497–507; and Addington, *Patterns of War,* pp. 191–194.

52. The German armored and motorized divisions played an important role in overwhelming the Polish frontlines and encircling Polish forces, but the German victory was a foregone conclusion, one only hastened by Polish weaknesses and mistakes. See Matthew Cooper, *The German Army, 1933–1945: Its Political and Military Failure* (New York: Stein and Day, 1978), pp. 169–176; B.H. Liddell Hart, *History of the Second World War* (New York: G.P. Putnam's Sons, 1971), pp. 27–32; and Jones, *Art of War,* pp. 508–509.

53. The German armored forces were neither more numerous nor technically superior to Allied forces. In fact, the Allies possessed a slight numerical advantage in tanks (at a ratio of 1.3 to 1) and manpower (1.2 to 1). Barry Watts and Williamson Murray, "Military Innovation in Peacetime," in Murray and Millett, *Military Innovation in the Interwar Period,* pp. 372–373.

54. For the ledger of German strengths and Allied weaknesses, see Jones, *Art of War,* pp. 510–544; Larry H. Addington, *The Blitzkrieg Era and the German General Staff, 1865–1941* (New Brunswick, N.J.: Rutgers University Press, 1971), pp. 101–123; Cooper, *The German Army,* pp. 214–215; Liddell Hart, *History of the Second World War,* chap. 7; and Robert Allan Doughty, *The Breaking Point: Sedan and the Fall of France, 1940* (Hamden, Conn.: Archon, 1990), chap. 1.

55. On the Red Army's deficiencies on the eve of the war, see Alan Clark, *Barbarossa: The Russian-German Conflict, 1941–45* (New York: William Morrow, 1965), chap. 2.

The best evidence with which to evaluate the offense-defense impact of tanks comes from military operations later in World War II, when all sides in the conflict had become relatively adept at armored warfare.[56] The most pertinent period is from operations on the eastern front in the winter of 1943 through the final German offensive on the western front at the end of 1944. By the winter of 1943, after two years of painful lessons, the Russians had developed the necessary organizational and doctrinal expertise to conduct offensive and defensive armored warfare at a reasonably proficient level. After 1944 the impact of tanks on the offense-defense balance was eclipsed by the sheer imbalance of material power between Germany and its adversaries.

The evidence from this period demonstrates that the mobility conferred by tanks did not favor offense. After the Russian encirclement of an entire German army at Stalingrad in late 1942, the Soviets went on the strategic offensive, and the Germans fell back on a fundamentally defensive strategy. Time after time, however, the German army relied on its armored forces to halt and defeat major Soviet offensives. German army doctrine regarded speed, mobility, and counterattack to be the decisive elements of defense, and tanks provided the perfect tool. The Germans discovered early on that the best way to defeat a Soviet armored penetration was by immediate counterattack with tanks against the flanks of the spearhead. Thus, as one panzer general noted, "the armored divisions, originally organized as purely offensive formations, had become [by early 1943] the most effective in defensive operations."[57]

The power of a tank-oriented defense was best displayed by Field Marshal Erich von Manstein's operations against major Soviet offensives in southern Russia and the Ukraine from January to March 1943. Though lacking the forces necessary to fight a true mobile defense, Manstein allowed Soviet penetrations in some sectors, ordered stubborn positional defense in a few other sectors, and rapidly shifted and assembled panzer units for counterattacks against the most threatening breakthroughs. Against a numerical balance of seven to one, Manstein stabilized the southern front and prematurely ended the Soviet winter offensives.[58] Mobile armored units proved uniquely suited for these delicate and demanding defensive operations.

56. Agreeing are Glaser and Kaufmann, "What Is the Offense-Defense Balance?" p. 56.

57. Quoted in Timothy A. Wray, *Standing Fast: German Defensive Doctrine on The Russian Front during World War II, Prewar to March 1943* (Fort Leavenworth, Kans.: U.S. Army Command and General Staff College, 1986), p. 170.

58. These operations are particularly informative because they were conducted largely free of Hitler's rigid no-retreat policy, which elsewhere prevented the German army from conducting a

The Germans were not alone in using the inherent mobility of tank forces to stop enemy armored offensives. The final German strategic offensive of the war on the eastern front aimed at pinching off and destroying the Soviet forces in the Kursk salient in July 1943. This was a far more limited goal than the deep penetrations and multiple encirclements of the earlier German offensives, but Hitler intended to use the same successful formula of massing tanks for a lightning blow against the Red Army.[59] The battle of Kursk, the greatest armored battle in history, was indeed a quick one, but resulted in a clear and decisive victory for the defender. The German armored spearheads were rapidly worn down by antitank defenses and then crushed by the Soviet armored reserves. The defeat marked the first time a German offensive had been halted before it could break through enemy defenses into the strategic depths beyond.[60]

After Kursk, Germany was pushed back along a broad front in an unrelenting series of Soviet offensives. Despite facing a growing numerical imbalance in forces with a deteriorating army, the Germans fought a skillful withdrawal, shuttling dwindling armored reserves back and forth for effective counterattacks on Soviet armored breakthroughs.[61] The war ended on the eastern front with the Russians never having conducted any large strategic encirclements comparable to the German victories of 1939 to 1941.

The war on the western front provides generally less suitable evidence for exploring the offense-defense impact of tanks, primarily because of the Allied preponderance of power.[62] In December 1944 Hitler launched his last major offensive of the war in the Ardennes forest. The German armored spearheads penetrated deep into the allied rear before they were halted by skilled armored

---

flexible, maneuver-oriented defense to which it was best suited and which would have been more effective against Soviet armored offensives. For detailed accounts of these operations, see David M. Glantz, *From the Don to the Dnepr: Soviet Offensive Operations, December 1942–August 1943* (London: Frank Cass, 1991); Erich von Manstein, *Lost Victories* (Novato, Calif.: Presidio Press, 1982), F.W. Von Mellenthin, *Panzer Battles* (New York: Ballantine, 1956); and Wray, *Standing Fast*, pp. 155–164.

59. Albert Seaton, *The Fall of Fortress Europe* (New York: Holmes and Meier, 1981), p. 55.

60. David M. Glantz and Jonathan M. House, *When Titans Clashed: How the Red Army Stopped Hitler* (Lawrence: University Press of Kansas, 1995), p. 167.

61. For an account of these counterattacks, particularly in November and December 1943, see ibid., pp. 174–175.

62. There are, however, some noteworthy instances of failed attempts to emulate the armored offensives of 1939 to 1941. For example, Operation Cobra (the Allied breakout from Normandy in July and August 1944) saw armored breakthroughs and deep penetrations, but the Allies were unable to encircle the bulk of the German forces, mainly because the tank forces had to operate in close conjunction with supporting infantry, artillery, and air forces to avoid destruction by other tanks and antitank forces.

(and tactical air) maneuvers and counterattacks on the flanks of the German bulge. In sum, the evidence suggests that the spectacularly successful armored offensives from 1939 to 1941 were an aberration not to be repeated against opponents skilled at armored warfare.

POLITICAL OUTCOMES. Did a belief in the offensive superiority of tanks contribute to the outbreak of World War II? Adolf Hitler may have been undeterrable, of course, but we may still be able to assess whether his eagerness to take the offensive was influenced by perceptions of the mobility-enhancing potential of tanks. Offense-defense proponents claim that Hitler was more willing to attack his neighbors in 1939–41 because he believed offense was dominant. In particular, proponents argue that Hitler's decision to attack the Low Countries and France in May 1940 (instead of in the fall of 1939 and winter of 1940) can be explained by his recognition that armored forces, combined with the *blitzkrieg* doctrine, had greatly strengthened the offense.[63]

The interwar period witnessed a tremendous debate about how tanks should be integrated into the armed forces, but few experts concluded that tanks would have a revolutionary impact on warfare. In the 1920s a small group of British military thinkers led by Major-General J.F.C. Fuller and Sir Basil Liddell Hart argued that the inherent mobility of tanks could restore offensive superiority to the battlefield. British experiments with armor never lived up to expectations, however, and by the mid-1930s Fuller and Liddell Hart had lost their enthusiasm for tanks as revolutionary offensive weapons.[64]

From the end of World War I through the attack on Poland in September 1939, few German leaders perceived the operational or strategic significance of tanks. In the 1920s the German army had returned to its traditional military doctrine of seeking quick and decisive victories through highly mobile offensive warfare.[65] Most Germans, nevertheless, discounted the combat potential of tanks based on their experience in World War I, where tanks were seen to have a "moral effect" on unprepared troops but could be easily defeated by countermeasures.[66]

---

63. Van Evera, *Causes of War*, pp. 123, 175, 177.
64. On Fuller's views, see J.F.C. Fuller, *The Reformation of War* (New York: E.P. Dutton, 1923), pp. 152–169; Brian Holden Reid, *J.F.C. Fuller: Military Thinker* (New York: St. Martin's, 1987), pp. 57, 183, and chap. 7; and Harris, *Men, Ideas, and Tanks*, chaps. 6–8. On Liddell Hart, see John J. Mearsheimer, *Liddell Hart and the Weight of History* (Ithaca, N.Y.: Cornell University Press, 1988), pp. 36–45, 105–123.
65. See Cooper, *The German Army*.
66. Wray, *Standing Fast*, p. 6.

A myth persists that during the 1930s the German army developed a new doctrine of warfare—the *blitzkrieg*—based on the revolutionary potential of armored forces to achieve a quick and decisive victory for the attacker, which they then employed with great success in 1939–41.[67] It is true that Heinz Guderian, a captain and then general in the German army, was the driving force behind the development of tank forces and leading advocate of the idea of using large, independent armored formations to break through the enemy's front and conduct deep strategic penetrations. Guderian's ideas, however, met with much skepticism, resistance, and outright subversion by the senior leaders of the army throughout the 1930s.[68] After all, the German attack on Poland was based not on a *blitzkrieg* strategy, but on the traditional German strategy of a combined-arms attack on the flanks in search of a decisive envelopment of enemy forces. Hitler eagerly attacked Poland absent any revelations of the offensive power of tanks.

Hitler was not deterred from attacking France in late 1939 and early 1940, and planned to do so without any new model for employing tanks. At the conclusion of the Polish campaign, Hitler met with his military commanders and announced that he had decided to attack in the west as soon as possible. The operational plan drawn up by the German army and endorsed by Hitler in October 1939 called for an attack through the Netherlands, Belgium, and Luxembourg to defeat as much of the French and Allied forces as possible and to capture a large portion of the English Channel coast for subsequent operations against Britain and the remainder of French territory. German military leaders, as well as Hitler himself, realized that this plan could achieve only a limited territorial objective and would probably lead to a war of attrition. Most important, Hitler did not believe that tank forces offered the potential for a decisive victory.[69]

Despite Hitler's own reservations, and the determined opposition of the German military leadership, he gave his full approval to the plan to attack in the west and fixed November 12 as the date for the beginning of the offensive. This attack was postponed because of poor weather conditions, as were a series

67. For an excellent summary and critique of this view, see J.P. Harris, "The Myth of Blitzkrieg," *War in History*, Vol. 2, No. 3 (1995), pp. 335–352.
68. Cooper, *The German Army*, pp. 148–158. See also Heinz Guderian, *Panzer Leader* (New York: Da Capo, 1996) and Barry R. Posen, *The Sources of Military Doctrine: France, Britain, and Germany between the World Wars* (Ithaca, N.Y.: Cornell University Press, 1984), pp. 205–219.
69. Cooper, *The German Army*, pp. 179–215; Doughty, *The Breaking Point*, pp. 19–25; and Mearsheimer, *Conventional Deterrence*, chap. 4.

of rescheduled offensives through December. A final date for the attack was set for January 17, 1940, but on January 10 a German plane carrying secret documents relating to the offensive was forced to land in Belgium, thus compromising German intentions. The unlikely sequence of weather delays, the plane crash, and the onset of winter forced Hitler to postpone his attack until the spring of 1940 and made drastic operational changes more attractive. Hitler eventually accepted an alternative plan of concentrating panzer forces for a surprise attack through the Ardennes.[70] In sum, the evidence of German planning and decisionmaking before World War II indicates that the political decision to initiate military conflict preceded any perceptions of the great offensive potential of tank technology.

NUCLEAR WEAPONS

Proponents of offense-defense theory argue that the nuclear revolution strongly shifted the offense-defense balance toward defense.[71] The case of nuclear weapons is unique, of course, and the traditional concepts of offense and defense do not translate easily from the conventional to the nuclear level. Moreover, the coding of nuclear weapons as defense dominant flows from a more complicated logic than the enhanced-firepower criterion. These differences notwithstanding, offense-defense theory offers a valid explanation for nuclear defense dominance and yields concrete and testable predictions about the political effects of nuclear weapons.

MILITARY OUTCOMES. The impact of nuclear weapons on battlefield outcomes must be based on logical deduction, not empirical evidence. According to offense-defense proponents, when all sides in a conflict possess a secure second-strike nuclear capability (i.e., when no side can launch an attack that is successful enough to prevent retaliation from the other), the defender has an enormous advantage over the attacker. This conclusion is counterintuitive and requires clarification because under conditions of mutual assured destruction (MAD), no side can defend against a nuclear attack, strictly speaking.

Offense-defense theory codes nuclear weapons as defense dominant because it is relatively easier and less costly for states to maintain a retaliatory capabil-

70. Some argue that even the attack launched in May 1940 was based on essentially traditional operational principles and methods, and was not motivated by the belief that armored forces had transformed warfare. See Harris, "The Myth of Blitzkrieg"; Cooper, *The German Army;* J.P. Harris and F.H. Toase, *Armoured Warfare* (New York: St. Martin's, 1990), pp. 64–69; and Doughty, *The Breaking Point*, p. 323.

71. Robert Jervis, *The Meaning of the Nuclear Revolution: Statecraft and the Prospect of Armageddon* (Ithaca, N.Y.: Cornell University Press, 1989), chap. 1; Charles L. Glaser, *Analyzing Strategic Nuclear Policy* (Princeton, N.J.: Princeton University Press, 1990); and Van Evera, *Causes of War,* chap. 8.

ity than to build a force capable of taking away another's retaliatory capability. States are thus deterred from attacking one another in a nuclear world; and deterrence is the functional equivalent of defense. The theory essentially aims to explain when states feel secure and when they do not or, alternatively, when they can deter attacks and when they cannot. When states rely on deterrence for their security, forces that enhance deterrence are essentially defensive. In a world of conventional arms, deterrence becomes easier as the defender is increasingly capable of denying territorial gains to the attacker. In the nuclear world, deterrence rests on the defender's ability to punish the attacker with unacceptable costs for attempted aggression. The only way to take territory at an acceptable cost in the nuclear world is by eliminating the defender's second-strike capability. This is very difficult to do, however, because it is much easier to enhance one's own deterrent forces than to strengthen forces that threaten an adversary's deterrent forces.[72] Thus nuclear weapons favor the defender by greatly improving the ability to deter by punishment.

POLITICAL OUTCOMES. The consequences of nuclear warfare in a MAD world are easy to comprehend and extremely difficult to change. Large and drastic shifts in the offense-defense balance, such as has occurred with the nuclear revolution, should have significant effects on international politics. The most fundamental prediction of offense-defense theory is that war among nuclear powers should not occur. The theory makes two additional predictions that can be evaluated against the historical record: arms racing beyond robust MAD levels should not occur, and security competition over distant territory should not be intense.

First, consider the no-war prediction. According to offense-defense theory, the prospect of devastation in a nuclear conflict is enough to deter even the most highly expansionist country, and the robust security provided by nuclear weapons virtually eliminates fears that might lead status quo states to launch preventive or preemptive wars. In short, the implausibility of obtaining military victory—not to mention a quick and decisive one—makes war among the nuclear powers virtually obsolete.

War among the major powers has not occurred since the introduction of nuclear weapons. Indeed, it is hard to imagine any great power armed with thermonuclear weapons and advanced delivery systems being attacked and conquered in the traditional sense. Alternative explanations for the "long

72. Jervis, "Cooperation under the Security Dilemma," p. 198; Glaser, *Analyzing Strategic Nuclear Policy*, p. 96; and Van Evera, *Causes of War*, pp. 177–178. On deterrence by denial and deterrence by punishment, see Glenn H. Snyder, *Deterrence and Defense: Toward a Theory of National Security* (Princeton, N.J.: Princeton University Press, 1961).

peace" abound, however; specifically, that either bipolarity, economic integration, or political/normative changes were responsible for preventing the Cold War from becoming hot.[73] More important, one could point to important cases that cut against the logic of the no-war prediction. For example, a state armed with nuclear weapons has been attacked (Israel in 1973), a state has intervened in a war against a nuclear power (China in Korea against the United States in 1950), and two states possessing nuclear weapons have fought each other (the major armed clashes between China and the Soviet Union in 1969).[74] Furthermore, the military conflict between Pakistan and India, two nuclear powers, in the Kargil mountains of Kashmir in the spring of 1999 was a short war by traditional casualty measures.[75] Nuclear weapons, nevertheless, have been a major force for preventing war.

Second, according to offense-defense theory, arms racing should not occur once states believe they have acquired the capability for assured nuclear retaliation.[76] This prediction has both a quantitative and qualitative element. In quantitative terms, adversaries will not be too concerned with comparing the relative size of their nuclear arsenals because even large shifts in relative force levels pose little threat to the "weaker" side's ability to retaliate and inflict unacceptable damage. The actual size of the superpower arsenals in the Cold War, however, far exceeded any reasonable estimate of the capabilities required for assured destruction or deterrence. By the early 1960s, both U.S. and Soviet leaders perceived that the United States could not effectively disarm the Soviets with a first strike.[77] In terms of U.S. requirements for deterrence, the minimum force levels thought necessary to inflict unacceptable damage on the Soviet Union in a retaliatory strike were well in hand by 1964.[78] By this time

73. See John Lewis Gaddis, *The Long Peace: Inquiries into the History of the Cold War* (New York: Oxford University Press, 1989).
74. The severity of the 1969 Sino-Soviet conflict has been underappreciated. See Raymond L. Garthoff, *Detente and Confrontation: American-Soviet Relations from Nixon to Reagan*, rev. ed. (Washington, D.C.: Brookings, 1994), pp. 228–242; and Henry Kissinger, *White House Years* (Boston: Little, Brown, 1979), pp. 183–194.
75. The best estimates are roughly 1,200 killed in the ten-week conflict. The Indian army and air force suffered 474 killed. Government of India, *Report of the Kargil Review Committee* (New Delhi: Government of India, March 2000), Executive Summary, p. 10. An estimate of regular Pakistani army casualties is 700 killed. *Report of the Kargil Review Committee*, p. 75.
76. See Jervis, "Cooperation under the Security Dilemma," pp. 188, 198; and Van Evera, *Causes of War*, pp. 244–245.
77. David Alan Rosenberg, "The Origins of Overkill: Nuclear Weapons and American Strategy, 1945–1960," *International Security*, Vol. 7, No. 4 (Spring 1983), pp. 38–44; and David Holloway, *The Soviet Union and the Arms Race* (New Haven, Conn.: Yale University Press, 1983), chap. 3.
78. Most analysts cite Secretary of Defense Robert McNamara's famous criteria for an assured destruction capability developed in 1963, which assumed the United States would need about

the United States had deployed 4,718 strategic nuclear warheads on a triad of bombers, intercontinental ballistic missiles (ICBMs), and submarine-launched ballistic missiles (SLBMs), while the Soviet Union had deployed almost 800 warheads. Yet the Soviet Union had begun a massive military buildup of both nuclear and conventional forces, and the United States quickly followed with a vast increase of its own warheads. The United States deployed 6,135 deliverable warheads in 1970; 10,768 in 1980; 12,304 in 1990; and almost 7,000 in 1999—a decade after the end of the Cold War. The Soviet Union deployed 2,327 warheads in 1970; 7,488 in 1980; 11,252 in 1990; and Russia still possessed almost 5,500 warheads in 1999.[79]

The qualitative aspect of the no-arms-race prediction is that once states find themselves in a MAD world, they should not attempt to gain an advantage at the nuclear level by building offensive counterforce weapons (i.e., forces aimed at destroying an adversary's strategic nuclear weapons). Possession of an assured destruction capability already provides states with a high degree of security; a first-strike advantage is virtually unattainable, impossible to maintain, and thus irrational to pursue. The evolution of U.S. and Soviet nuclear strategies from the early 1960s was characterized by a persistent interest in escaping from MAD through the deployment of sophisticated counterforce weapons systems. Both countries threatened each other's retaliatory capabilities, aimed at limiting damage to their own forces and society, and generally sought to prevail in the event of nuclear war. In the U.S. case, policymakers declared a nuclear doctrine consistent with MAD, but actually embraced a counterforce posture and strategy.[80] U.S. counterforce planning and targeting began in the late 1950s even though it was apparent that the Soviet Union would soon acquire a secure second-strike

---

1,500 warheads. For discussions of McNamara's criteria, see Alain C. Enthoven and K. Wayne Smith, *How Much Is Enough? Shaping the Defense Program, 1961–1969* (New York: Harper and Row, 1971), pp. 175, 207–210; Fred Kaplan, *The Wizards of Armageddon* (Stanford, Calif.: Stanford University Press, 1983), chap. 22; and Scott D. Sagan, *Moving Targets: Nuclear Strategy and National Security* (Princeton, N.J.: Princeton University Press, 1989), pp. 32–34.

79. All figures are from Robert S. Norris and Thomas B. Cochran, *US and USSR/Russian Strategic Offensive Nuclear Forces, 1945–1966* (Washington, D.C.: National Resources Defense Council, 1997); and John Pike, "Nuclear Forces Guide," Federation of American Scientists, http://www.fas.org/nuke/guide/summary.htm (data as of January 1999).

80. See Aaron L. Friedberg, "The Evolution of U.S. Strategic Doctrine, 1945–1980," in Samuel P. Huntington, ed., *The Strategic Imperative: New Policies for American Security* (Cambridge, Mass.: Ballinger, 1982), pp. 53–99; Desmond Ball, "The Development of the SIOP, 1960–1983," in Ball and Jeffrey Richelson, eds., *Strategic Nuclear Targeting* (Ithaca, N.Y.: Cornell University Press, 1986), pp. 57–83; Sagan, *Moving Targets;* and Eric Mlyn, *The State, Society, and Limited Nuclear War* (Albany: State University of New York Press, 1995).

capability.[81] This strategy was accelerated in the 1960s, when more than 90 percent of Soviet bloc targets in the U.S. nuclear war plan were counterforce targets.[82] In the 1970s, despite having concluded agreements with the Soviet Union to limit strategic defense, the United States continued to enhance its counterforce arsenal by building highly accurate weapons capable of destroying hardened Soviet targets.[83] In the 1980s the Reagan administration took counterforce to an extreme by pursuing effective strategic defenses and offensive counterforce programs.[84] The Soviet Union also did not regard the possession of an assured destruction capability as sufficient to maintain its security. The Soviets began to build an antiballistic missile (ABM) system around Moscow in the mid-1960s and, after signing the ABM treaty in 1972, continued to develop its strategic defense capabilities through air defense programs against bombers and civil defense efforts aimed at protecting Soviet leadership. More important, the Soviets embarked on a massive nuclear counterforce buildup that stressed heavy, accurate missiles specifically designed to destroy the U.S. ICBM force.[85] Both the United States and the Soviet Union recognized their mutual vulnerability to nuclear destruction, but were driven by the goal of winning a nuclear war and accordingly based their nuclear strategies on robust counterforce arsenals.

A final prediction offense-defense theory makes about behavior under nuclear defense dominance is that states should not compete or fight too intensely over territory beyond the homeland or the homeland of close allies. Nuclear

81. Rosenberg, "The Origins of Overkill"; Friedberg, "Evolution of U.S. Strategic Doctrine"; and Robert Jervis, *The Illogic of American Nuclear Strategy* (Ithaca, N.Y.: Cornell University Press, 1984), p. 44.
82. Ball, "Development of the SIOP," pp. 66–67.
83. Among other steps, the United States built Minuteman ICBMs and Polaris SLBMs, added threatening multiple independently targetable reentry vehicles (MIRVs) to its ICBMs and SLBMs, and decided to upgrade its Minuteman III ICBMs and deploy highly lethal and accurate Peacekeeper MX ICBMs, Trident D-5 SLBMs, and Pershing II medium-range ballistic missiles—all of which threatened the Soviet ability to retaliate in a nuclear exchange.
84. Strategic defense efforts fell under the Strategic Defense Initiative (SDI) and a host of air-defense, early-warning, and civil defense programs. Offensive programs included accelerating the Trident D-5 program and building new bombers, cruise missiles, warheads, and sensors. Barry R. Posen and Stephen Van Evera, "Defense Policy and the Reagan Administration: Departure from Containment," *International Security*, Vol. 8, No. 1 (Summer 1983), pp. 3–45; and Desmond Ball and Robert C. Toth, "Revising the SIOP: Taking War-Fighting to Dangerous Extremes," *International Security*, Vol. 14, No. 4 (Spring 1990), pp. 65–92.
85. See Robert P. Berman and John C. Baker, *Soviet Strategic Forces: Requirements and Responses* (Washington, D.C.: Brookings, 1982), chap. 3; Holloway, *Soviet Union and the Arms Race*, chap. 3; William T. Lee, "Soviet Nuclear Targeting Strategy," in Ball and Richelson, *Strategic Nuclear Targeting*; and David Miller, *The Cold War: A Military History* (New York: St. Martin's, 1998), pp. 98–102.

weapons devalue traditional concerns over geographic depth; in other words, buffer zones and distant bases are less important in a nuclear world because nuclear retaliation can be assured in their absence.[86] In the Cold War, offense-defense theory predicts minimal intervention and competition between the superpowers in the third world: "Nuclear weapons make conquest much harder, and vastly enhance the self-defense capabilities of the superpowers. This should allow the superpowers to take a more relaxed attitude toward events in third areas, including the Third World, since it now requires much more cataclysmic events to shake their defensive capabilities. Whatever had been the strategic importance of the Third World in a nonnuclear world, nuclear weapons have vastly reduced it."[87]

Although a full exploration of the nature of U.S. and Soviet intervention and competition in the third world is beyond the scope of this article, it is fair to say that the superpowers had anything but a "relaxed attitude." In fact, most crises between the United States and the Soviet Union occurred in the third world, as each superpower resorted to the whole range of economic, political, and military means to advance its own interests or block the influence of its rival.[88] The most relevant and striking evidence is the relationship between the Soviet achievement of nuclear parity and its increased level of intervention in the third world. As several scholars have noted, the Soviets were constrained from too overtly challenging the United States in the third world early in the Cold War because of U.S. nuclear hegemony. The arrival of strategic nuclear parity, however, coincided with a much more assertive role for the Soviets, as demonstrated by their actions in the Middle East (1970–73), Angola (1975–76), Ethiopia (1977–78), Yemen (1978–79), and Afghanistan (1979).[89] The emergence of nuclear parity did not mitigate, and may have aggravated, the superpower competition for influence in the third world.

The United States intervened with its own military forces in the third world throughout the Cold War as well, including in Korea (1950–53), Egypt (1956),

86. Van Evera, *Causes of War*, p. 245.
87. Posen and Van Evera, "Defense Policy and the Reagan Administration," p. 33.
88. See Garthoff, *Detente and Confrontation*, pp. 732–745.
89. See Bruce D. Porter, "Washington, Moscow, and Third World Conflict in the 1980s," in Huntington, *Strategic Imperative*, pp. 258–259; and Coit Blacker, "The Kremlin and Detente: Soviet Conceptions, Hopes, and Expectations," in Alexander George, ed., *Managing U.S.-Soviet Rivalry* (Boulder, Colo.: Westview, 1983), pp. 127–128.

Lebanon (1958 and 1982), Thailand (1962), Laos (1962–75), Vietnam (1964–73), Congo (1964 and 1967), the Dominican Republic (1965), Cambodia (1970), Libya (1981 and 1986), Grenada (1983), and Panama (1989).[90] Because great powers have interests beyond defense of the homeland that may require intervention and competition abroad, a superpower military presence and some competition for power in the third world do not necessarily undermine offense-defense theory. The fact that the superpowers were consistently drawn into very real and frequently costly conflicts in areas of little economic or strategic value, however, is disconfirming evidence.

Conclusions

Offense-defense theory contends that the relative ease of attack and defense often plays a major role in causing instability and war in international politics. The theory holds that states will tend to seek security through aggression when offensive advantages render the capabilities of others more threatening, whereas peace is more likely when defensive advantages make changes in the balance of military power less worrisome. The theory also claims prescriptive utility because misperceptions and miscalculations of the balance often lead states to initiate conflict when they otherwise might feel secure with the status quo; these misperceptions can sometimes be ameliorated through arms control and confidence-building measures.

Given the pervasive influence of the offense-defense balance in international security scholarship, the logical consistency and empirical validity of the theory deserve rigorous evaluation. The lion's share of past criticism has been directed at the conceptual and operational problems endemic to "broad" versions of offense-defense theory—those versions that define the balance to include a laundry list of factors in addition to technology. Although proponents of the broad theory believe that they are buttressing the explanatory power of the theory with a more sophisticated approach, in practice the balance becomes an ad hoc collection of variables employed pell-mell to account for empirical anomalies and logical qualifications.

The "core" version of offense-defense theory, which focuses almost exclusively on how technology shapes the relative ease of attack and defense, is the potentially more fruitful approach. The core theory offers two basic criteria for

90. Garthoff, *Detente and Confrontation*, p. 744; and Ellen C. Collier, *Instances of Use of United States Forces Abroad, 1798–1993* (Washington, D.C.: Congressional Research Service, 1993).

judging how a given technology affects the offense-defense balance and thus military outcomes: mobility-improving innovations generally favor offense and result in more quick and decisive victories for the attacker, whereas firepower-enhancing innovations typically strengthen defense and lead to more indecisive warfare. In terms of political outcomes, the theory predicts that states are more likely to initiate conflict when they perceive that the offense-defense balance favors offense.

This article "tested" offense-defense propositions using four illustrative case studies, chosen as the most important mobility and firepower innovations in modern military history. The impact of railroads on military outcomes was mixed, with attackers benefiting from greater strategic mobility and defenders profiting from better operational mobility. The quick and decisive character of some wars at the time primarily arose because of great asymmetries in military strength and doctrine. More important, more wars were initiated when railroads were perceived to favor the defender than when railroads were thought to favor the attacker. The small arms and artillery revolution shifted the offense-defense balance toward defenders. Perceptions of the military consequences of the revolution in firepower technology, however, did little to dampen conflict. In fact, leaders in Prussia and Germany were more cognizant of the technical realities of firepower than those in any other European state and yet provoked all of the major power wars on the continent at the time. Tanks had an indeterminate effect on the offense-defense balance. The conventional view that tanks favored offense is based largely on Germany's stunning operations from 1939 to 1941 and is undermined by the evidence when all sides were adept at armored warfare. Moreover, Hitler attacked his neighbors absent any belief in the great offensive potential of tanks. Finally, the nuclear revolution offers only mixed evidence for offense-defense theory. The logic for coding nuclear weapons as defense-dominant is sound, and nuclear war has not occurred. Although leaders correctly perceived the military consequences of a nuclear conflict, however, the United States and the Soviet Union engaged in an intense and costly arms race and competed hard in regions of secondary importance.

The evidence suggests that scholars have overstated both the degree to which the nature of technology shapes military outcomes and the influence that beliefs of offense or defense dominance have on political and strategic decisions. Understanding the limitations of offense-defense theory might help scholars develop more nuanced causal explanations, as well as construct more precise empirical tests of these hypotheses. Ultimately, however, the theory may not provide enough analytical leverage for understanding international politics, especially given the complexity of operationalizing and measuring the offense-defense balance. In any event, the relationship between technological change and international security is too important and fascinating a subject to abandon based on the flawed concept of an offense-defense balance of military technology.

Attack and Conquer?

International Anarchy and the Offense-Defense-Deterrence Balance

Karen Ruth Adams

Scholars and strategists have long argued that offense is easier than defense in some periods but harder than defense in others.[1] In the late 1970s, George Quester and Robert Jervis took the first steps toward systematizing this claim.[2] Since then, the debate about the causes and consequences of the offense-defense balance has been one of the most active in security studies.[3]

If the relative efficacy of offense and defense changes over time, states should be more vulnerable to conquest and more likely to attack one another at some times than at others. Specifically, when offense is easier than defense, defenders' military forces should be more likely to collapse or surrender when attacked, and defenders' political leaders should be more likely to surrender sovereignty in response to military threats. Thus states in offense-dominant eras should be conquered—involuntarily lose the monopoly of force over all of their territory to external rivals—more often than states in defense-dominant eras.[4] Given their heightened vulnerability to conquest, states in offense-

Karen Ruth Adams is Assistant Professor of Political Science at Louisiana State University.

For detailed comments, I thank Richard Betts, Stephen Biddle, Avery Goldstein, Miles Kahler, Susan Martin, Christopher Muste, Mark Peceny, Kenneth Waltz, James Wirtz, and Eugene Wittkopf. For research assistance, I thank Nicole Detraz, Katia Ivanova, and Aneta Leska.

1. For an annotated bibliography, see Michael E. Brown, Owen R. Coté Jr., Sean M. Lynn-Jones, and Steven E. Miller, eds., *Offense, Defense, and War* (Cambridge, Mass.: MIT Press, forthcoming).
2. George H. Quester, *Offense and Defense in the International System* (New York: John Wiley and Sons, 1977); and Robert Jervis, "Cooperation under the Security Dilemma," *World Politics*, Vol. 30, No. 2 (January 1978), pp. 167–214.
3. For summaries of the debate, see Jack S. Levy, "The Offensive/Defensive Balance of Military Technology: A Theoretical and Historical Analysis," *International Studies Quarterly*, Vol. 28, No. 2 (June 1984), pp. 219–238; and Sean M. Lynn-Jones, "Offense-Defense Theory and Its Critics," *Security Studies*, Vol. 4, No. 4 (Summer 1995), pp. 660–691.
4. My definition of conquest derives from Max Weber's definition of a state as "a human community that (successfully) claims the *monopoly of the legitimate use of physical force* within a given territory." Weber, "Politics as a Vocation," in H.H. Gerth and C. Wright Mills, eds. and trans. *From Max Weber: Essays on Sociology* (New York: Oxford University Press, 1958), p. 78 (emphasis in original). A state exists when a particular community, defined by its political structure, monopolizes force over some territory. It dies when it is no longer able to do so, either because its political structure collapses or because external or internal rivals render it powerless through conquest, union, revolution, or disintegration. Conquest may occur directly, through the collapse or surrender of a state's military forces in response to attacks by an external rival, or indirectly, through the surrender of leaders in response to an external rival's military threats or operations. When a state loses the monopoly of force over part of its territory, it is dismembered, not conquered. Karen Ruth

dominant eras should also be more likely to act on the doctrine that the best defense is a good offense. That is, they should attack one another—conduct offensive military operations against states that have not previously attacked them—more often than states in defense-dominant eras.

Although these points have been widely discussed, neither proponents of offense-defense arguments[5] nor their critics have tested the effects of the offense-defense balance on the historical incidence of attack and conquest.[6] Instead, they have examined its effects on the offensive or defensive character of military doctrine;[7] the extent of arms racing, prevalence of cooperation, and nature of alliances;[8] and especially the incidence, severity, and duration of war.[9]

Adams, "State Survival and State Death: International and Technological Contexts," Ph.D. dissertation, University of California, Berkeley, 2000.

5. Although offense-defense arguments are generally described as theories, they are actually collections of hypotheses (if-then statements about lawlike regularities—e.g., "if offense is dominant, offensive operations are likely"), not theories (statements that explain laws). The theory behind offense-defense arguments is structural realist theory, which explains that offensive operations are likely when offense is dominant because there is no international sovereign to rule them out. Ernest Nagel, *The Structure of Science* (New York: Harcourt, Brace, and World, 1961), pp. 80–81; Alan C. Isaak, *Scope and Methods of Political Science* (Homewood, Ill.: Dorsey, 1969), pp. 138–139; Kenneth N. Waltz, *Theory of International Politics* (New York: McGraw Hill, 1979), pp. 5–6; Kenneth N. Waltz, *Man, the State, and War: A Theoretical Analysis* (New York: Columbia University Press, 1959), pp. 230–238; Lynn-Jones, "Offense-Defense Theory and Its Critics," pp. 664–665; and Richard K. Betts, "Must War Find a Way? A Review Essay," *International Security*, Vol. 24, No. 2 (Fall 1999), p. 168.

6. Peter Liberman, whose dependent variable is expansionism, is the exception. Liberman, "The Offense-Defense Balance, Interdependence, and War," *Security Studies*, Vol. 9, Nos. 1–2 (Summer and Fall 1999), pp. 59–91. Some scholars have suggested that the number (and, by implication, the size) of states in the international system is a function of the offense-defense balance. But because the number of states could fall through conquest, union, or collapse, studies that simply note the different numbers or sizes of states in different periods are not directly testing this hypothesis. Stanislav Andreski, *Military Organization and Society*, 2d ed. (Berkeley: University of California Press, 1968), chap. 3; Richard Bean, "War and the Birth of the Nation State," *Journal of Economic History*, Vol. 33, No. 1 (March 1973), pp. 203–221; Quester, *Offense and Defense in the International System*, pp. 8, 17; and Robert Gilpin, *War and Change in World Politics* (Cambridge: Cambridge University Press, 1981), pp. 61–62, 66.

7. Stephen Van Evera, "The Cult of the Offensive and the Origins of the First World War," and Jack Snyder, "Civil-Military Relations and the Cult of the Offensive, 1914 and 1984," *International Security*, Vol. 9, No. 1 (Summer 1984), pp. 58–107 and pp. 108–146; Scott D. Sagan, "1914 Revisited: Allies, Offense, and Instability," *International Security*, Vol. 11, No. 2 (Fall 1986), pp. 151–175; Jack Snyder and Scott D. Sagan, "Correspondence: The Origins of Offense and the Consequences of Counterforce," *International Security*, Vol. 11, No. 3 (Winter 1986/87), pp. 187–198; and Jonathan Shimshoni, "Technology, Military Advantage, and World War I: A Case for Military Entrepreneurship," *International Security*, Vol. 15, No. 3 (Winter 1990/91), pp. 187–215.

8. Jervis, "Cooperation under the Security Dilemma"; George W. Downs, David M. Rocke, and Randolph Siverson, "Cooperation and Arms Races," *World Politics*, Vol. 38 (1985), pp. 118–146; Stephen M. Walt, *The Origins of Alliances* (Ithaca, N.Y.: Cornell University Press, 1987); Thomas J. Christensen and Jack Snyder, "Chain Gangs and Passed Bucks: Predicting Alliance Patterns in Multipolarity," *International Organization*, Vol. 44, No. 2 (Spring 1990), pp. 137–168; Robert Powell, "Guns, Butter, and Anarchy," *American Political Science Review*, Vol. 87, No. 1 (March 1993), pp. 115–132; James Morrow, "Arms versus Allies: Trade-Offs in the Search for Security," *International Orga-*

Attention to war is easy to understand, for it is one of the most destructive human activities. But testing offense-defense arguments on the incidence of war is problematic because these arguments suggest that wars in offense-dominant eras should have many attacks while those in defense-dominant eras should have few. Moreover, they suggest that attacks in offense-dominant eras should frequently culminate in conquest while those in defense-dominant eras should rarely do so. In defense-dominant eras, states with state-of-the-art capabilities should be able to declare war then wait to counterattack without imperiling their survival. Thus, although there may be many declarations of war, there should be few attacks. Moreover, attacks should not result in conquest unless states have outdated capabilities or strategies. In offense-dominant eras, by contrast, security should come from attacking first. Instead of declaring war, states should engage in surprise attacks, wars should involve a number of attacks by a variety of states, and attacks should frequently result in conquest.

Because offense-defense arguments suggest there should be less variation in war than in attack and conquest, the most direct way to test the core claims of these arguments is to examine the effects of the balance on the incidence of attack and conquest. That is my purpose in this article. But constructing an effective test requires more than applying existing arguments to new dependent variables. It also requires redefining the concept of the balance to distinguish between defense and deterrence dominance, as well as operationalizing the historical balance deductively, at the operational level of analysis, and in purely technological terms.

I begin by explaining the need for these refinements to the logic and application of offense-defense arguments. Then I elaborate and operationalize an ar-

---

nization, Vol. 47, No. 2 (Spring 1993), pp. 207–233; Charles L. Glaser, "Realists as Optimists: Cooperation as Self-Help," *International Security,* Vol. 19, No. 3 (Winter 1994/95), pp. 50–90; and Thomas J. Christensen, "Perceptions and Alliances in Europe, 1865–1940," *International Organization,* Vol. 51, No. 1 (Winter 1997), pp. 65–97.

9. On interstate war, see Ted Hopf, "Polarity, the Offense-Defense Balance, and War," *American Political Science Review,* Vol. 85, No. 2 (June 1991), pp. 475–493; James D. Fearon, "The Offense-Defense Balance and War since 1648," paper presented at the annual convention of the International Studies Association, Chicago, Illinois, February 21–25, 1995; Stephen M. Walt, *Revolution and War* (Ithaca, N.Y.: Cornell University Press, 1996); Stephen Van Evera, "Offense, Defense, and the Causes of War," *International Security,* Vol. 22, No. 4 (Spring 1998), pp. 5–43; Stephen Van Evera, *Causes of War: Power and the Roots of Conflict* (Ithaca, N.Y.: Cornell University Press, 1999), chaps. 6, 7; Betts, "Must War Find a Way?"; Keir Lieber, "Grasping the Technological Peace: The Offense-Defense Balance and International Security," *International Security,* Vol. 25, No. 1 (Summer 2000), pp. 71–104; and Stephen Biddle, "Rebuilding the Foundations of Offense-Defense Theory," *Journal of Politics,* Vol. 63, No. 3 (August 2001), pp. 741–774. On ethnic conflict and civil war, see the annotated bibliography in Brown et al., *Offense, Defense, and War.*

gument about the technological sources of the offense-defense-deterrence balance and the balance's effects on state vulnerability to conquest and propensity to attack other states. Next, I derive hypotheses about the historical incidence of attack and conquest from this argument, as well as from alternative arguments about technology, relative capabilities, duration of great power status, and audience costs.

Then, using the quantitative methodology of event history analysis and new data on attacks by and conquests of great powers and nuclear states, I test the explanatory power of these hypotheses from 1800 to the present. I find that hypotheses derived from my technological argument are strongly supported by historical trends in both attack and conquest. In offense-dominant eras, great powers are significantly more likely to be conquered, to attack other great powers, and to attack non-great powers than they are in defense- and especially deterrence-dominant periods. But the offense-defense-deterrence balance is not the only cause of attack and conquest. Relative capabilities also matter. Specifically, the more capable a great power is, the less likely it is to be conquered and the more likely it is to attack other great powers. Moreover, the imbalance of power between great powers and non-great powers tempts great powers of all capabilities to exercise and expand their influence even in defense- and deterrence-dominant eras.

Finally, I explore the theoretical and political implications of these findings. To summarize, because states are less vulnerable to conquest and less likely to attack one another in defense- and especially deterrence-dominant eras, it is not the case, as offensive realists claim, that states must always act aggressively to survive. Neither is it the case, as defensive realists suggest, that states act aggressively only when their security is threatened. Instead, as structural realists argue, states are sensitive to environmental constraints and opportunities but, given international anarchy, can do as they like, especially when they are strong. This suggests that although deterrence dominance makes nuclear states more secure than great powers have historically been, contemporary nuclear states may attack and conquer nonnuclear states. Given its unrivaled power, the United States is especially likely to do so. Thus if other nuclear states, feeling secure from attack and conquest, move slowly to protect vulnerable states or balance U.S. power, the spread of nuclear weapons is likely to accelerate.

What about Deterrence Dominance?

Although most offense-defense scholars acknowledge the deterrent effects of nuclear weapons, they treat deterrence as a special case of defense. Jervis, for example, argues that "concerning nuclear weapons, it is generally agreed that defense is impossible—a triumph not of the offense, but of deterrence." Yet each of Jervis's "four worlds" of offense-defense advantage and offense-defense differentiation is possible in the nuclear era.[10] Similarly, Stephen Van Evera argues that nuclear weapons make "conquest among great powers . . . virtually impossible," whereas previous military revolutions simply "strengthened the defense." Yet Van Evera's characterization of "military realities" is the same before 1792, from 1816 to 1856, and from 1871 to 1918 as it is from 1945 to 1990.[11]

There are two reasons to distinguish between defense and deterrence dominance. First, defensive and deterrent operations are distinct. As Glenn Snyder puts it, "Deterrence means discouraging the enemy from taking military action by posing for him a prospect of cost and risk outweighing his prospective gain. Defense means reducing our own prospective costs and risks in the event that deterrence fails." Deterrent operations entail punishment; defensive ones, damage limitation.[12] Second, attack and conquest should occur more often in defense-dominant eras than in deterrence-dominant ones. Attack is more likely in defense-dominant eras because, in the absence of the survivable and deliverable "absolute weapons" (such as nuclear weapons) that make deterrence dominant by dramatically increasing the costs of miscalculation, states face fewer costs for playing the odds.[13] Moreover, in defense-dominant eras, states face incentives to prepare for future revolutions in military affairs by extending their perimeters to make it hard for other states to conquer them when offense regains the advantage; when deterrence is dominant, states have less need to expand because the difficulty of defending against absolute weapons makes it unlikely that deterrence dominance will be overturned. Conquest is also more likely in defense-dominant eras, both because states are more

10. The four worlds describe strategic situations ranging from "doubly dangerous," in which "there is no way to get security without menacing others," to "doubly stable," in which defensive advantage encourages states to adopt policies that pose no threat to others, and offensive-defensive differentiation enables them to correctly assess their competitors' intentions. Jervis, "Cooperation under the Security Dilemma," pp. 198, 211–214.

11. Van Evera, "Offense, Defense, and the Causes of War," pp. 17, 24, and 33.

12. Glenn H. Snyder, *Deterrence and Defense: Toward a Theory of National Security* (Princeton, N.J.: Princeton University Press, 1961), p. 3.

13. An absolute weapon has "enormous destructive potency" that "concentrate[s] . . . violence in terms of time." Because it has immediate and devastating effects on whole geographic regions, not just other weapons, its damage cannot be limited by offense or defense. Bernard Brodie, "War in the Atomic Age" and "Implications for Military Policy," in Brodie, ed., *The Absolute Weapon: Atomic Power and World Order* (New York: Harcourt, Brace, 1946), pp. 28–29, 71. On the deterrent effects of biological weapons, see Susan Martin, "The Role of Biological Weapons in International Politics: The Real Military Revolution," *Journal of Strategic Studies*, Vol. 25, No. 1 (March 2002), pp. 63–98.

likely to attack one another and because offense dominance is not the only cause of conquest; attrition can also lead to total military collapse or surrender. Given these differences between the requirements and expected outcomes of defense and deterrence dominance, I redefine the offense-defense balance as the offense-defense-deterrence balance.[14]

Operationalizing the Balance

Defining the balance is easy: The offense-defense-deterrence balance is the relative efficacy of offense, defense, and deterrence given prevailing conditions. Operationalizing the balance—explaining how to measure it so that its effects can be examined—is harder because it entails decisions about which level of analysis to use (i.e., strategic, operational, or tactical),[15] whether to assess the historical balance inductively or deductively, and which variables to consider.

LEVEL OF ANALYSIS

I operationalize the balance at the operational level, for three reasons.[16] First, specifying the balance at the operational level acknowledges that defenders may allow tactical conquest to achieve operational gains (i.e., trade space for time) and that, even if offense is dominant at the tactical level, operational gains may be elusive. It also acknowledges that offense dominance at the operational level can overwhelm variables that favor defense at the strategic level, such as alliances or economic disincentives for conquest. Second, specifying the balance at the operational level makes it possible to avoid the tautology of coding offense as dominant when many states are conquered, then arguing that offense dominance affects state vulnerability to conquest. Instead the claim is that the strategic outcome of conquest (the involuntary loss to an external rival of a state's entire territory) is more likely when offensive operations dominate.[17] Third, the aims of offensive and defensive strategies (winning

14. Henceforth, I use "offense-defense arguments" to refer to work that does not distinguish between deterrence and defense, and by "the balance," I mean the offense-defense-deterrence balance.
15. The strategic level of analysis refers to national war plans and outcomes. By contrast, the operational level involves decisions and outcomes in specific campaigns of a war, and the tactical level refers to objectives and techniques adopted in engagements within a campaign.
16. For a different argument about the relevance of the operational level, see Biddle, "Rebuilding the Foundations of Offense-Defense Theory," pp. 747–748. For the argument that all three levels should be considered, see Charles L. Glaser and Chaim Kaufmann, "What Is the Offense-Defense Balance, and Can We Measure It?" *International Security*, Vol. 22, No. 4 (Spring 1998), pp. 73–74.
17. Scholars have avoided tautology by assuming that conquest is easier in some periods than in

wars) differ from those of deterrent strategies (avoiding wars). Thus strategic assessments of the balance can occur only at a high level of abstraction that blurs the distinction between offense and defense dominance. But deterrence dominance requires absolute weapons (force lethality) that can survive first strikes (force protection) and be delivered to their targets (force mobility). Thus deterrence is easily compared to offense and defense at the operational level of analysis using the same, tangible criteria that analysts readily apply to them.

INDUCTION VERSUS DEDUCTION

To determine whether offense, defense, or deterrence is dominant in particular eras, I deduce the potential lethality, protection, and mobility of state-of-the-art offensive, defensive, and deterrent operations from prevailing conditions. This approach is contrary to current, inductive efforts to count attacker casualties per defender casualties or kilometer conquered, compare the budgets required for state-of-the-art attackers to prevail against state-of-the-art defenders, and compare the costs of conquest to the value of territory seized.[18] When assessing the aggregate effects of the historical balance across the conventional and nuclear eras, induction falls short. Robust evaluation of the historical balance would entail either extensive reenactments or complicated computer models of past campaigns, both of which would require more extensive military, economic, political, and social data than are available. Moreover, inductive measurement of the systemic, post–World War II balance would entail direct assessment of the battlefield effects of nuclear weapons. Thus scholars who pursue the inductive approach must limit their inquiries to contemporary, nonnuclear state dyads and assume that the deterrent effects of other states' nuclear weapons do not extend to conflicts among such states.[19] Because I am interested in aggregate historical outcomes, I take a deductive approach. In doing so, I try to be as explicit and systematic as possible.[20]

---

others, then examining the effects of this supposed fact on the incidence of war and other international political outcomes. When conquest is the dependent variable, one must be more deliberate. Defining the balance at one level and the outcome at another is the best way I have found to do so.

18. On measuring the balance in terms of casualties, see Quester, *Offense and Defense in the International System*, p. 2; and Biddle, "Rebuilding the Foundations of Offense-Defense Theory," pp. 748–749. On value of territory, see Gilpin, *War and Change in World Politics*, pp. 62–63. On economic costs, see Jervis, "Cooperation under the Security Dilemma," p. 188; Glaser, "Realists as Optimists," pp. 61–62; Van Evera, "Offense, Defense, and the Causes of War," p. 5, n. 1; and Glaser and Kaufmann, "What Is the Offense-Defense Balance?" pp. 50–51.

19. Glaser and Kaufmann, "What Is the Offense-Defense Balance?" pp. 57–58.

20. Scholars interested in aggregate outcomes often advocate induction, then abandon it in favor of ad hoc description. For example, although Van Evera operationalizes the balance in terms of

VARIABLES

Although most scholars argue that prevailing technology is among the most important sources of the balance, they assume that technology alone has little explanatory power.[21] By contrast, I adopt a purely technological notion of the balance. Instead of lumping technology with other variables such as military doctrine, force posture and deployments, geography, regime popularity, collective security systems, defensive alliances, and balancing behavior by neutral states to arrive at a "master cause" called the offense-defense balance,[22] I treat technology as the sole determinant of the balance. This does not preclude consideration of the effects of other variables. In fact, it makes it possible to disentangle the effects of many variables while testing the most fundamental argument about the balance.

The Offense-Defense-Deterrence Balance, 1800 to the Present

In this section, I elaborate and operationalize a technological argument about the offense-defense-deterrence balance. My argument is that prevailing technology affects the relative efficacy of offensive, defensive, and deterrent military operations and thus the incidence of attack and conquest. Specifically, attack and conquest should occur more often when technology favors offensive operations than when technology favors defensive and, especially, deterrent operations.

---

economic investment, in his periodization of the historical balance he simply refers to variables that "bolster the defense" or restore "the strength of the offense." Van Evera, "Offense, Defense, and the Causes of War," pp. 16–18. Jervis does refer to relative costs in discussing antitank, antiaircraft, and nuclear weapons, but his discussion of pre–World War II technologies is, like Van Evera's, simply in terms of advantage or superiority. Jervis, "Cooperation under the Security Dilemma," pp. 196–199.

21. The best indication of this is that scholars treat technology as just one of many sources of the balance. For more direct criticisms, see Fearon, "The Offense-Defense Balance and War since 1648," pp. 31–32; and Biddle, "Rebuilding the Foundations of Offense-Defense Theory," p. 753.

22. Van Evera, *Causes of War,* pp. 5, 160–166. On technology, geography, force size, nationalism, and resource cumulativity, see Glaser and Kaufmann, "What Is the Offense-Defense Balance?" pp. 61–68. On force employment, force size, and technology, see Biddle, "Rebuilding the Foundations of Offense-Defense-Theory," pp. 749–756. Biddle's flow chart of "orthodox offense-defense theory" illustrates the tendency to treat the offense-defense balance as an intervening variable between technology, geography, force size, and patterns of diplomatic behavior, on the one hand, and political and military outcomes on the other. Biddle, "Rebuilding the Foundations of Offense-Defense Theory," p. 745. For criticisms of the large number of variables that scholars argue affect the balance, see James W. Davis Jr., Bernard I. Finel, Stacie E. Goddard, Stephen Van Evera, and Charles L. Glaser and Chaim Kaufmann, "Correspondence: Taking Offense at Offense-Defense Theory," *International Security,* Vol. 23, No. 3 (Winter 1998/99), pp. 180, 185–187, 191, 197–198; and Betts, "Must War Find a Way?".

To operationalize this argument, I define offensive, defensive, and deterrent operations as follows:

- *Offensive operations* are actions in which a state uses force to attack another state's military or nonmilitary assets to conquer its territory or compel compliance with policy directives (impose its will on the other state).[23]
- *Defensive operations* are actions in which a state uses force against another state's military assets to repel and limit damage from that state's attacks to retain control of its territory and avoid having the other state impose its will upon it.
- *Deterrent operations* are actions in which a state prepares to use force or demonstrates its ability to use force to attack another state's nonmilitary assets to deter that state from attacking it or to deter it from further attacks once a war has begun.[24]

The technologies relevant to the balance are the "methods, skills, and tools" that affect states' abilities to conduct offensive, defensive, and deterrent operations.[25] These technologies may be strictly military, or they may have nonmilitary origins or applications. Either way, they define the most effective military operations in a particular historical era.[26]

To determine whether and when offensive, defensive, or deterrent operations dominated, I deduce the relative lethality, protection, and mobility of state-of-the-art offensive, defensive, and deterrent operations from what is known about the potential and interactions of technologies that prevailed from

23. Military assets include a state's military forces and the agencies, industries, and infrastructure that command and support those forces. Nonmilitary assets include citizens, cities, cultural artifacts, natural resources, and nonmilitary industries, agencies, and infrastructure.
24. Deterrent *operations* (preparations to retaliate and demonstrations of the ability to do so) always involve punishment; that is what distinguishes them from offensive operations to conquer or compel and defensive operations to deny victory and limit damage. Snyder, *Deterrence and Defense,* p. 14. By contrast, deterrent *outcomes* (decisions not to attack) can occur either because one state fears another's capability to punish or because it fears the other's capability to deny victory. Demonstrations of a state's ability to punish attackers can take various forms, including weapons testing and demonstration strikes. In a demonstration strike, a state uses force against the nonmilitary targets of a state that has already attacked it to remind the attacker of the state's ability to retaliate more extensively should attacks continue. Demonstration strikes in deterrent operations differ from attacks in offensive and defensive operations in that their purpose is to demonstrate the ability to retaliate—not to limit damage, deny victory, conquer territory, or compel compliance with policy directives other than stopping attacks on a state or its allies.
25. Jacques Perrin, "The Inseparability of Technology and Work Organization," *History and Technology,* Vol. 7 (1990), p. 2.
26. For the argument that such technologies are neither domestic nor international variables but contextual or environmental ones, see Adams, "State Survival and State Death," chap. 4.

1800 to the present.[27] When I identify a technological change that transformed the balance, I date the change with the development of that technology[28]—neither its first use in war nor the continued procurement or deployment of men and matériel in ways consistent with that technology. In other words, I specify the balance in terms of what was technologically possible at the operational level, not in terms of the strategic, operational, or tactical moves states made.[29]

FORCE LETHALITY

Lethality is the "effectiveness capability" of operations, "based upon such considerations as range, rate of fire, accuracy, reliability, radius of damage, etc."[30] On this dimension of the offense-defense-deterrence balance, no operational mode was dominant from 1800 to 1945, and deterrence was dominant after 1946.

In assessing the relative lethality of offensive, defensive, and deterrent operations in particular historical eras, one must determine whether there was an absolute weapon—a weapon capable of destroying not just other weapons but also entire geographical areas, such as cities. When such a weapon exists, states

27. Considering the lethality, protection, and mobility of each type of operation is vital because it acknowledges the role of mobility in the success of defensive and deterrent operations (and of force protection in the success of offensive operations), as well as the effects of the operational context in which particular weapons are used. Yet many scholars simply argue that mobility favors offense while firepower favors defense or that certain weapons are inherently offensive or defensive. For a sophisticated example of the argument about mobility and firepower, see Glaser and Kaufmann, "What Is the Offense-Defense Balance?" pp. 61–64. For criticisms of this argument, see Lieber, "Grasping the Technological Peace." For an argument about weapons, see Marion William Boggs, *Attempts to Define and Limit "Aggressive" Armament in Diplomacy and Strategy,* The University of Missouri Studies, Vol. 16, No. 1 (Columbia: University of Missouri, 1941). Critics of this approach include Quincy Wright, *A Study of War,* 2d ed. (Chicago: University of Chicago Press, 1965), pp. 805–810; and John J. Mearsheimer, *Conventional Deterrence* (Ithaca, N.Y.: Cornell University Press, 1983), pp. 25–27. For the argument that offense-defense arguments do not hinge on whether weapons can be classified as offensive or defensive, see Lynn-Jones, "Offense-Defense Theory and Its Critics," pp. 672–677; and Glaser and Kaufmann, "What Is the Offense-Defense Balance?" pp. 79–80.

28. Specifically, I consider one offense-defense-deterrence era to end the year that a balance-transforming technology is developed and a new one to begin the following year.

29. This facilitates a more robust periodization than that articulated by Quester, who asserts that offense dominated from 1789 to 1849, when railroads and breech-loading rifles appeared on the scene, and that defense dominated from 1850 to 1939, when tanks and dive bombers emerged, but does not explain why he chose these years. Quester, *Offense and Defense in the International System,* chaps. 8, 12. It also makes for a less tautological periodization than that offered by Van Evera, who dates changes based on behavioral variables such as demobilization and "defense-enhancing diplomacy." Van Evera, "Offense, Defense, and the Causes of War," pp. 17, 27.

30. Trevor N. Dupuy, *The Evolution of Weapons and Warfare* (New York: Da Capo, 1984), p. 92.

can punish their opponents without first achieving victory through offensive or defensive operations. Thus deterrence dominates. From 1800 to 1944, force lethality grew markedly,[31] but no absolute weapon existed. Thus deterrence was dominated by offense or defense. But once the atomic bomb was developed in 1945, deterrent operations became far easier and more robust than ever before. Because nuclear weapons greatly increase the costs states might have to pay for attacking others' territory and vital interests, they make deterrent operations dominant.[32]

Before the nuclear revolution, advances in explosives technology were neutral (although they are generally thought to have favored the defense) because they could be exploited by either defenders or attackers, depending on their ability to protect and deliver their weapons. Frederick the Great's offensive breakthroughs of the late eighteenth century occurred with the same muskets and rifles used by his opponents; the difference lay in Prussia's tactics of marching in step, firing in unison, and reloading quickly, as well as its use of horse-drawn artillery.[33] Similarly, although artillery inflicted more than half of the battle casualties suffered by Napoleon's opponents, it was the technology that made artillery mobile, not the inherent lethality of French guns, that facilitated offensive lethality.[34] Finally, although the increased lethality of small arms made possible by the 1849 invention of the conoidal bullet (minié ball) for rifled muskets contributed greatly to the subsequent power of the defense,[35] defense dominance would not have been possible without the concurrent vulnerability of offensive forces. When covering fire tactics, motorized armor, and aircraft were developed in the twentieth century, state-of-the-art small arms were effectively employed by the offense as well.

---

31. Ibid.
32. Although deterrent operations did not dominate in the prenuclear era, such operations did occur. For example, to deter other states from attacking them at all or to deter them from further attacks once a war had begun, states planned and carried out "terror bombing" attacks on opponents' population centers. But in the absence of an absolute weapon, such operations were less likely to deter attacks than state-of-the-art offensive or defensive operations were to conquer territory or repel invaders. Ibid., pp. 12–18. George H. Quester, *Deterrence before Hiroshima: The Airpower Background of Modern Strategy* (New Brunswick, N.J.: Transaction, 1986); Mearsheimer, *Conventional Deterrence;* and Barry R. Posen, *The Sources of Military Doctrine: France, Britain, and Germany between the World Wars* (Ithaca, N.Y.: Cornell University Press, 1984), pp. 231–232.
33. Dupuy, *The Evolution of Weapons and Warfare,* pp. 144–152.
34. Ibid., p. 158; and Robert L. O'Connell, *Of Arms and Men: A History of War, Weapons, and Aggression* (New York: Oxford University Press, 1989), p. 178.
35. Dupuy, *The Evolution of Weapons and Warfare,* pp. 170–171, 190–191, 292–293; O'Connell, *Of Arms and Men,* p. 191; and Bernard Brodie, *From Crossbow to H-Bomb* (Bloomington: Indiana University Press, 1973), pp. 131–132.

FORCE PROTECTION

An arsenal of lethal weapons has little utility if it cannot be sheltered from a disabling first strike. Thus whether offense, defense, or deterrence is dominant depends on the extent to which offensive, defensive, and deterrent operations can be protected from one another. On this dimension of the balance, technology favored offensive operations from 1800 to 1849, defensive operations from 1850 to 1933, offensive operations from 1934 to 1945, and deterrent operations from 1946 to the present.

Here again, the first step in assessing the balance is to determine whether an absolute weapon exists. If so, protection of military forces alone is insufficient to shield a state from devastating reprisals for attacks on other states' territory and vital interests because just one nuclear weapon can do profound damage. Only perfect early-warning, interdiction, and interception technologies could protect states from their effects. Such technologies have not been and are unlikely to be developed, for three reasons. First, nuclear weapons are relatively small, thus easy to hide. Second, tactics and technologies for making missiles, bombers, submarines, and other delivery vehicles stealthy are robust. Finally, marginal production costs for the third, fiftieth, and one-hundredth weapon make it possible to build redundant forces without bearing exorbitant costs.[36] Because nuclear weapons are much less vulnerable than state-of-the-art defensive and offensive forces, deterrence has been dominant since 1946 on this dimension of the balance, as well.

To assess force protection during the conventional era, one must consider the ease with which offensive forces could bypass or destroy defensive ones given force mobility (discussed below), as well as the depth of force that attackers needed to evade defensive fire and the depth of force that production technologies enabled them to amass. From 1800 until the development of the conoidal bullet, which enabled defenders to harden positions through constant and increasingly lethal defensive fire, the requisite depth was not very great, so the relative strength of offense and defense hinged on the size of force (especially infantry) that attacking states could amass. Because state-of-the-art pro-

36. This is not to defend procurement of the one-thousandth (or twenty-thousandth) weapon or to deny that nuclear weapons development and deployment entail considerable economic, social, and environmental costs. Neither is it to suggest that all states will find these costs worth bearing. Arjun Makhijani, Howard Hu, and Katherine Yih, eds., *Nuclear Wastelands: A Global Guide to Nuclear Weapons Production and Its Health and Environmental Effects* (Cambridge, Mass.: MIT Press, 1995); and Stephen I. Schwartz, ed., *Atomic Audit: The Costs and Consequences of U.S. Nuclear Weapons since 1940* (Washington, D.C.: Brookings, 1998).

duction technologies made it possible to take hundreds of thousands of men from the fields, put them in uniforms, and give them weapons, offensive operations had the protection advantage.[37] After 1850, however, even sizable offensive forces were vulnerable to defensive fire. This remained the case until 1933, when the development of swift, well-armed and armored tanks and dive bombers enabled attackers to be stealthy until the last minute and shielded from defensive fire once discovered.[38]

FORCE MOBILITY

Mobility refers to the ease with which offensive, defensive, and deterrent forces can deliver their weapons. On this dimension of the balance, prevailing technology favored offensive operations from 1800 to 1835 and 1930 to 1945, defensive operations from 1836 to 1929, and deterrent operations after 1946.

Until the mid-nineteenth century, the speed at which men and horses could run, artillery could be towed, and ships could be rowed defined the limits of mobility. If attacking forces could move faster than defensive forces could be deployed to eliminate them, offensive operations had the advantage. Late eighteenth-century improvements in mobile artillery combined with cumbersome communication and transportation technologies to make offense dominant from 1800 to 1835.[39] Beginning in 1836, the invention of the telegraph and

---

37. Brodie, *From Crossbow to H-Bomb,* pp. 106–107; Quester, *Offense and Defense in the International System,* p. 66; and O'Connell, *Of Arms and Men,* pp. 174–175.

38. Tanks were first built in 1915 and were first used in World War I. Early tanks were lightly armed and armored, with 12–15 millimeter plates that could be penetrated by field artillery, heavy machine guns, or even rifles firing heavy ammunition. During the interwar years, tanks carried increasingly lethal weapons, but models made of thicker (40–60 mm) cast armor that was less vulnerable to defensive fire were not developed until the mid-1930s. The first such tank was the French R-35, which was initiated in 1933. Thus I use 1933 as the end of defense dominance in protection. "War, Technology of," *New Encyclopædia Britannica,* Vol. 29 (Chicago: Encyclopædia Britannica, 2002), pp. 583–585, 667; Dupuy, *The Evolution of Weapons and Warfare,* pp. 221, 231; Kenneth Macksey, *Technology in War: The Impact of Science on Weapon Development and Modern Battle* (New York: Prentice Hall, 1986), pp. 91–95, 103–104, 112–113; and Richard M. Ogorkiewicz, *Armor: A History of Mechanized Forces* (New York: Praeger, 1960), pp. 177–178. The first (naval) dive bomber was developed by the United States in 1929. A prototype of the first dive bomber designed for direct support of ground operations (the German Ju-87 Stuka) was tested in 1935. "War, Technology of," *New Encyclopædia Britannica,* Vol. 29, p. 612; Richard R. Muller, "Close Air Support: The German, British, and American Experiences, 1918–1941," in Williamson Murray and Allan R. Millett, eds., *Military Innovation in the Interwar Period* (New York: Cambridge University Press, 1996), pp. 178–180; and Dupuy, *The Evolution of Weapons and Warfare,* p. 233.

39. The Gribeauval system, developed in 1774, enhanced the mobility of artillery by reducing the length and weight of gun barrels; outfitting carriages with iron axles and large, rugged wheels; improving cannonball production; and developing lightweight, prefabricated cartridges. Dupuy, *The*

steam-powered railroads and ships enabled defenders to obtain early warning of attacks, mount swift and coordinated responses, and sustain large armies in wars of attrition.[40] From 1930 to 1945, improvements in tanks, planes, and especially radio tipped the scales toward offense dominance once again.[41] Since 1946, the small size of nuclear weapons and advances in transportation and

---

Evolution of Weapons and Warfare, p. 158; Brodie, *From Crossbow to H-Bomb*, pp. 102–103; and O'Connell, *Of Arms and Men*, pp. 178–179. Those who claim that the Napoleonic revolution was the result of nationalism and strategic genius overlook these technological developments. Even Napoleon recognized the importance of technology. According to him, "A good infantry is without doubt the soul of an army; but, it cannot long maintain a fight against a superior artillery, it will become demoralized and then destroyed. . . . The fate of a battle, of a state, often follows the [route] taken by the artillery." Quoted in James R. Arnold, *Napoleon Conquers Austria: The 1809 Campaign for Vienna* (Westport, Conn.: Praeger, 1995), p. 198.

40. Although steam locomotives were invented by the British in 1803 and could outpace a horse team by 1814, a reliable locomotive was not invented until 1829, and modern rail, spikes, and roadbed did not appear until 1830. Steamships emerged a few years later. Because the telegraph repeater (the first telegraph capable of sending signals over long distances) actualized the potential for defensive mobility afforded by steam transportation, I date the end of offense dominance with its invention in 1835. Bruce Wetterau, *The New York Public Library Book of Chronologies* (New York: Prentice Hall, 1990), pp. 182, 204; and "Transportation," *New Encyclopædia Britannica*, Vol. 28 (Chicago: Encyclopædia Britannica, 2002), pp. 793–794. Railroads gave the advantage to the defense because "invaders could not use the defenders' railroads (given that railroad gauges differed across states, and defenders destroyed rail lines as they retreated) while the defenders had full use of their own lines." Van Evera, "Offense, Defense, and the Causes of War," p. 17. Steamships favored defense because of their coaling needs. On railroads and steamships, see Quester, *Offense and Defense and the International System*, chaps. 8, 9. On railroads and the telegraph, see Martin Van Creveld, *Technology and War from 2000 B.C. to the Present* (New York: Free Press, 1991), pp. 153–170, 213–214.

41. The first tanks were slow, subject to mechanical breakdown, vulnerable to defensive fire, and unable to communicate with one another. Moreover, the inability of artillery and infantry to keep pace with tanks made it difficult to convert breakthroughs (as at Cambrai in 1917) into operational or even tactical success. After World War I, tanks (as well as motorized artillery and personnel carriers) were faster and more reliable, but attackers' ability to carry out adaptive maneuvers remained limited by communication problems. Similarly, although by 1918 planes reached top speeds of 145 miles per hour and carried more than 3,000 pounds of cargo, communication difficulties limited their ability to give close support to ground operations and deliver men and matériel where it was needed. Improvements in radio technology in the 1920s overcame these barriers, replacing the bulky radios and unreliable and insecure transmissions of World War I with small, high-frequency and frequency modulated (FM) radios and cyphering machines. I date the end of defense dominance with the development of FM radio in 1929, which reduced the effect of ignition and other vehicle noises on the strength of radio transmissions. FM radio was developed six years after the development of the first cyphering machine (the German Enigma, in 1923) and nine years before the design and manufacture of the earliest line of portable high-frequency radios suitable for both land and air forces (by Germany, in 1938). Williamson Murray, "Armored Warfare: The British, French, and German Experiences," in Murray and Millett, *Military Innovation in the Interwar Period*, pp. 6–7, 26, 35; "War, Technology of," *New Encyclopædia Britannica*, Vol. 29, pp. 583–585, 624; Dupuy, *The Evolution of Weapons and Warfare*, pp. 221–223, 231; Wetterau, *The New York Public Library Book of Chronologies*, pp. 185, 222; Macksey, *Technology in War*, pp. 91–95, 103–104, 107–108, 112; and Van Creveld, *Technology and War from 2000 B.C. to the Present*, p. 190. In the 1930s and early 1940s, offense dominance was enhanced by the development of amphibious landing

missile technologies have meant that nuclear weapons can be easily delivered to their targets, whether by bomber, land- or sea-based missile, ship, truck, or even airline passenger. The plethora of options makes deterrence dominant.

SUMMARY OF THE HISTORICAL BALANCE

My coding of the overall offense-defense-deterrence balance from 1800 to the present is shown in Table 1. To summarize, offensive operations dominated from 1800 to 1849 due to lack of defensive lethality, industrial technologies facilitating offensive depth, and improvements in mobile artillery. After 1850, when the telegraph, railroad, and conoidal bullet enabled defenders to respond both quickly and lethally to attacks, defense dominated. With improvements in tanks, planes, and radio, offensive operations regained the mobility advantage in 1930, but the overall balance continued to favor defense until 1933, when the development of well-armed and armored tanks and dive bombers gave state-of-the-art offensive operations dominance in protection as well. Offensive operations then dominated from 1934 to 1945. After 1946, the lethality, survivability, and deliverability of nuclear weapons made deterrence dominant.

Hypotheses about the Effects of the Balance on Attack and Conquest

Conquest is most likely in offense-dominant eras because prevailing technology favors offensive operations. Attack is most likely in such eras as well because states are likely to respond preemptively to the threat of conquest. Attack and conquest are less likely in defense-dominant eras, yet more likely than in deterrence-dominant eras, for two reasons. First, although defense dominance makes it difficult to conquer states through offensive military operations, conquest through attrition is still possible. Second, states are unlikely to be completely dissuaded from attacking one another because the costs of miscalculation are not significantly higher than in offense-dominant eras, and the chance of an offensive revolution in military affairs provides incentives for states to expand. Attack and conquest are least likely in deterrence-dominant eras because states may punish attackers with absolute weapons.

---

craft capable of carrying large numbers of troops and motorized vehicles and landing in shallow waters. Quester, *Offense and Defense in the International System*, pp. 150–151; Van Creveld, *Technology and War from 2000 B.C. to the Present*, p. 214; and Allan R. Millett, "Assault from the Sea: The Development of Amphibious Warfare between the Wars," in Murray and Millett, *Military Innovation in the Interwar Period*, pp. 78–85.

Table 1. Dimensions of the Offense-Defense-Deterrence Balance and Assessment of the Overall Balance, 1800–Present.

Lethality	Protection	Mobility	Overall Offense-Defense-Deterrence Balance[a]
1800–1945 no mode dominant due to the absence of an absolute weapon and the ability of both attackers and defenders to use state-of-the-art weapons, depending on their ability to protect and deliver them	1800–49 offense dominant due to the lack of defensive lethality and agricultural and industrial technologies facilitating offensive depth	1800–35 offense dominant due to the lack of defensive lethality and improvements in mobile artillery	1800–49 offense dominant
	1850–1933 defense dominant due to the lethality afforded by the conoidal bullet and later advances in firearms and artillery, as well as improvements in fortification technology	1836–1929 defense dominant due to the invention of telegraph, railroad, and steamship	1850–1933 defense dominant
	1934–45 offense dominant due to the development of well-armed and armored tanks and dive bombers	1930–45 offense dominant due to improvements in tanks, planes, and especially radio	1934–45 offense dominant
1946–present deterrence dominant due to the absolute nature of nuclear weapons	1946–present deterrence dominant due to the lethality of nuclear weapons and the ease with which they can be hidden, delivered, and produced	1946–present deterrence dominant due to the small size of nuclear weapons and the variety of means for delivering them	1946–present deterrence dominant

[a]In assessing the overall balance during the two "mixed" periods (1836–49 and 1930–33), I gave protection greater weight than mobility because mobility has little utility if soldiers, weapons, and societies cannot be sheltered from defensive, offensive, or retaliatory strikes.

Thus my assessment of the historical offense-defense-deterrence balance yields the following hypotheses about the incidence of attack and conquest:

H1. *States should attack one another most often in the offense-dominant eras (1800–49 and 1934–45), less often in the defense-dominant era (1850–1933), and rarely in the deterrence-dominant era (1946 to the present).*

H2. *States should be conquered most often in the offense-dominant eras (1800–49 and 1934–45), less often in the defense-dominant era (1850–1933), and rarely in the deterrence-dominant era (1946 to the present).*

Alternative Predictions about Trends in Attack and Conquest

To test an argument, one must examine its explanatory power relative to that of other arguments and theories. Alternative explanations of attack and conquest abound. Here I focus on those most often discussed in the offense-defense literature. These arguments suggest that variation in the incidence of attack and conquest is affected by technological variables other than the offense-defense-deterrence balance, states' relative capabilities, the length of time states have been great powers, and audience costs.[42]

OTHER TECHNOLOGICAL VARIABLES

There are two technologically based alternatives to my argument about the offense-defense-deterrence balance. The first is that deterrence is "the equivalent of the primacy of defense."[43] This argument, which underlies current conceptions of the offense-defense balance, suggests that:

H3. *States should attack one another more often in the offense-dominant eras (1800–49 and 1934–45) than in the defense- and deterrence-dominant eras (1850–1933 and 1946 to the present).*

H4. *States should be conquered more often in the offense-dominant eras (1800–49 and 1934–45) than in the defense- and deterrence-dominant eras (1850–1933 and 1946 to the present).*

42. For a comparison of the effects of the offense-defense-deterrence balance, polarity, great power decline, prevailing economic technologies, and international norms, see Karen Ruth Adams, "Conquering Myths: Testing Realist, Liberal, and Constructivist Arguments about State Vulnerability to Conquest," paper presented at the annual meeting of the American Political Science Association, Boston, Massachusetts, August 29–September 1, 2002.
43. Jervis, "Cooperation under the Security Dilemma," p. 198.

The second technological alternative posits a distinction between the conventional and nuclear eras. This argument, exemplified by James Fearon's statement that "the technological/organizational offense-defense balance is simply not a very important cause of war . . . except in the case of nuclear weapons,"[44] suggests that:

H5. *States should attack one another more often in the conventional era (1800–1945) than in the nuclear era (1946 to the present).*

H6. *States should be conquered more often in the conventional era (1800–1945) than in the nuclear era (1946 to the present).*

Because H5 and H6 simply distinguish between the pre- and post-1945 periods, testing them will also illuminate the explanatory power of arguments about the war-inhibiting effects of other post–World War II variables, such as bipolarity and territorial sovereignty norms.[45]

RELATIVE CAPABILITIES

Carl von Clausewitz's argument that "defense is always the stronger form of war" might be interpreted to predict no systematic variation in attack and conquest over time.[46] The same might be said of Kenneth Waltz's structural realist theory and John Mearsheimer's offensive realist theory, which suggest that attack and conquest are always possible because there is no international sovereign to rule them out.[47] Clausewitz's argument, however, is that "defense is easier than attack, assuming both sides have equal means."[48] Similarly, Waltz and Mearsheimer argue that although anarchy makes all states vulnerable to conquest and other undesirable outcomes, more capable states are less vulnerable than others.[49] Thus these arguments suggest that:

H7. *Less capable states should be conquered more often than more capable states.*

---

44. Fearon, "The Offense-Defense Balance and War since 1648," pp. 31–32.
45. On bipolarity, see John J. Mearsheimer, *The Tragedy of Great Power Politics* (New York: W.W. Norton, 2001), pp. 337–344. On territorial sovereignty norms, see Robert H. Jackson and Mark W. Zacher, "The Territorial Covenant: International Society and the Legitimization of Boundaries," paper presented at the annual meeting of the American Political Science Association, San Francisco, California, August 29–September 1, 1996.
46. Carl von Clausewitz, *On War,* ed. and trans. Michael Howard and Peter Paret (Princeton, N.J.: Princeton University Press, 1976), p. 358.
47. Waltz, *Man, the State, and War,* p. 232; and Mearsheimer, *The Tragedy of Great Power Politics,* p. 3.
48. Clausewitz, *On War,* p. 357.
49. Waltz, *Theory of International Politics,* p. 194; and Mearsheimer, *The Tragedy of Great Power Politics,* p. xi.

Clausewitz, Waltz, and Mearsheimer also expect relative capabilities to affect the incidence of attack, yet here their expectations diverge. For offensive realists, anarchy means that great powers "have to seek more power if they want to maximize their odds of survival." They "are always searching for opportunities to gain power over their rivals, with hegemony as their final goal." Thus, great powers are "primed for offense"—likely to adopt strategies such as buck passing, balancing, blackmail, and especially war (i.e., attack).[50] Because the states of greatest capability have the most resources to pursue such strategies, offensive realism predicts that:

H8. *More capable great powers should attack other great powers more often than less capable great powers.*

Structural realism, by contrast, is indeterminate about the effects of relative capability on great power attacks on other great powers because great powers, even more than other states, may do as they like. More capable great powers could act as offensive realism predicts, for three reasons. First, their power "gives [them] a big stake in their system and the ability to act for its sake."[51] Second, balancing the power of powerful states is the primary system-maintaining action. Finally, anarchy means there is nothing to keep states from attacking one another. Yet more capable great powers may choose not to attack their less powerful counterparts because it is rarely necessary to use force to maintain the status quo.[52]

Despite their different expectations about great power attacks on other great powers, structural and offensive realists agree that great powers are more likely than other states to attack non-great powers. Yet they base their conclusions on different considerations. Here again, structural realists focus on the balance of power, arguing that states are more likely to succumb to the temptation of imperialism "where gross imbalances of power exist" because "weakness invites control [and] strength tempts one to exercise it, even if only for the 'good' of other people."[53] Offensive realists, by contrast, explain imperialism with reference to great power competition fueled by hegemonic ambitions.[54] Either way,

---

50. Mearsheimer, *The Tragedy of Great Power Politics*, pp. 3, 21, 29, and chap. 5. See also Fareed Zakaria, "Realism and Domestic Politics: A Review Essay," in Michael E. Brown, Sean M. Lynn-Jones, and Steven E. Miller, *The Perils of Anarchy: Contemporary Realism and International Security* (Cambridge, Mass.: MIT Press, 1995), p. 479, n. 43.
51. Waltz, *Theory of International Politics*, p. 195.
52. Ibid., p. 185.
53. Ibid., pp. 26–27.
54. Mearsheimer, *The Tragedy of Great Power Politics*, p. 355.

H9. *More capable great powers should attack non-great powers more often than less capable great powers.*

Like offensive realists, Clausewitz would expect H8 to find support. Unlike them, however, he would disagree with H9. For Clausewitz, as well as for Geoffrey Blainey and power transition theorists such as A.F.K. Organski, when the policies of two states conflict, war erupts to resolve the difference only if it is unclear who will win, for "war between states of markedly unequal strength" is "absurd" and, in theory, "impossible."[55] From this point of view, attacks are more likely among states of relatively equal power, such as great powers, than between great powers and less powerful states.

Defensive realists, by contrast, would expect H8 and H9 to find support only in offense-dominant eras, when mutual security is elusive. These scholars take the structural realist expectation that the states that survive will tend to seek security and turn it into an assumption that most states seek security.[56] Then, because states are less vulnerable to conquest in defense- and deterrence-dominant eras, defensive realists expect states to attack one another rarely, if at all, in such periods.[57] Structural realists, by contrast, argue that anarchy's effects on international affairs persist even when prevailing technology yields high levels of mutual security, for it is always possible for states to misperceive the conditions in which they operate. Moreover, states can do as they like, especially when they are strong.[58]

DURATION OF GREAT POWER STATUS

Defensive realists suggest that states that approve of the status quo are more concerned with security and thus less likely to attack other states. Assuming that new great powers are less likely than older powers to find the status quo to their liking:

---

55. Clausewitz, *On War,* p. 91; Geoffrey Blainey, *The Causes of War,* 3d ed. (New York: Free Press, 1988), p. 114; and A.F.K. Organski and Jacek Kugler, *The War Ledger* (Chicago: University of Chicago Press, 1980), p. 19.
56. According to Glaser, "Structural realism assumes that, in an anarchic system, security is the end to which states will give priority." Glaser, "Realists as Optimists," p. 71. On defensive realism, see Jack Snyder, *Myths of Empire: Domestic Politics and International Ambition* (Ithaca, N.Y.: Cornell University Press, 1991), pp. 11–13; and Jeffrey W. Taliaferro, "Security Seeking under Anarchy: Defensive Realism Revisited," *International Security,* Vol. 25, No. 3 (Winter 2000/01), pp. 128–161.
57. According to Jervis, "If the defense has enough of an advantage and if . . . states are of roughly equal size, not only will the security dilemma cease to inhibit status-quo states from cooperating, but aggression will be next to impossible, thus rendering international anarchy relatively unimportant." Jervis, "Cooperation under the Security Dilemma," p. 187.
58. Waltz, *Theory of International Politics,* pp. 92, 172, 186–187.

H10. *New great powers should attack other great powers more often than old great powers.*

H11. *New great powers should attack non-great powers more often than old great powers.*

Furthermore, assuming it is easier to uphold the status quo than it is to change it,

H12. *New great powers should be conquered more often than old great powers.*

Although Waltz's argument that powerful states rarely need to use force to uphold the status quo suggests these three hypotheses, structural realist theory is indeterminate about whether the length of time a state has been a great power affects its vulnerability to conquest and propensity to attack other states. Here again, the important point for structural realists is that anarchy enables states to do as they like. States that have recently attained great power status may be content with the status quo. Moreover, older great powers may cease to maintain state-of-the-art capabilities, use force to uphold the status quo when it is not necessary to do so, or grab power rather than settle for security.

AUDIENCE COSTS

Fearon argues that war has become less likely since the French Revolution because of rising "audience costs," that is, the effects of direct costs such as battle deaths on the domestic legitimacy of rulers. According to Fearon, after the French Revolution "swept away . . . the divine right of kings" and the Congress of Vienna enshrined sovereigns as guardians of their states, "monarchs . . . had to worry that costly and/or losing efforts at war would spell the end of their rule, or at least pressure for reform from domestic liberalizers." Since then, audience costs have become increasingly influential in decisionmakers' war calculi as industrialization and communications technologies have created ever larger and better informed "political classes [who pay] attention to and [care] about" their state's participation in war. Assuming that audience costs began to rise with the French Revolution and rose steadily each year,[59]

H13. *States should attack one another less each year from 1800 to the present.*

From Fearon's expectations about war, one might infer that conquest would also occur less often because audience costs would discourage leaders from

59. Fearon, "The Offense-Defense Balance and War since 1648," pp. 33–34.

trying to conquer other states. Yet Fearon identifies a second trend that suggests otherwise. Specifically, he argues that the Napoleonic Wars marked the beginning of "increased offensive advantages . . . [that] increased the variance in military outcomes."[60] Stephen Biddle concurs and elaborates Fearon's point. According to Biddle, since the dawn of modern warfare (which he puts in the 1870s), military technology has been a "a systemic, time-correlated variable: the more recent the operation, the more sophisticated the technology," and "the more extreme the outcome." In other words, "the scale of the victory or defeat that ensues" from war has grown with each passing year. Depending on force employment and force size, attackers either conquer more and more territory or defenders lose less and less.[61] From this, one must conclude that Fearon and Biddle would either expect no systematic variation in the incidence of conquest or concur with the hypothesis that the least capable states are the most likely to be conquered (H7). Nevertheless, because conquest requires attack, it is worthwhile to test the hypothesis that:

H14. *States should be conquered less each year from 1800 to the present.*

Because H13 and H14 predict a steady decline in attack and conquest after 1800, they can be used to test any argument about secular trends over the period, whether the trends are thought to arise from audience costs, learning, or other variables.[62]

Data

To test these hypotheses, I developed data on conquests of and attacks by great powers and nuclear states from 1800 to 1997.[63] Table 2 lists the states included in the data and the years of their inclusion.[64] I restricted the analysis to great powers and nuclear states because, in the prenuclear era, great powers were the states most likely to have state-of-the-art capabilities. Thus they should have been the states most likely to attack other states in offense-dominant eras

60. Ibid., pp. 34–36.
61. Biddle, "Rebuilding the Foundations of Offense-Defense Theory," pp. 748–750.
62. On learning, see Kevin Wang and James Lee Ray, "Beginners and Winners: The Fate of Initiators of Interstate Wars Involving Great Powers since 1495," *International Studies Quarterly*, Vol. 38, No. 1 (March 1994), pp. 139–154; and George Modelski, "Is World Politics Evolutionary Learning?" *International Organization*, Vol. 44, No. 1 (Winter 1990), pp. 1–24.
63. The data set, A&C.gpnp (Version 1), is available on *International Security*'s website at http://www.bcsia.ksg.harvard.edu/IS.
64. For an explanation of my coding of great powers and nuclear states, see Karen Ruth Adams, "Codebook for Data Used in 'Attack and Conquer,'" November/December 2003, available on *International Security*'s website, http://www.ksg.harvard.edu/IS, p. 2.

Table 2. Great Powers and Nuclear States, 1800–1997.

	Great Power	Nuclear State
England/Great Britain/United Kingdom	1495–1945	1953–97
France	1495–1815, 1818–71, 1873–1940	1964–97
Spain	1556–1808	—
Austria/Austria-Hungary	1556–1918	—
Russia/Soviet Union	1721–1989	1949–97
Prussia/Germany	1740–1806, 1813–1919, 1930–45	—
Italy	1861–1943	—
United States	1898–1997	1945–97
Japan	1905–45	—
China	—	1964–97
Israel	—	1967–97
India	—	1974–97
Pakistan	—	1990–97

and least likely to be conquered in defense-dominant ones. In the nuclear era, deterrence hinges less on relative capabilities than on the possession of second-strike nuclear forces. Thus including nuclear states after 1946 provides a more rigorous test of deterrence dominance than simply examining attack and conquest among great powers.

The unit of analysis in the data is the state year. In other words, each year from 1800 to 1997 that a state was a great power or nuclear state is a single case.[65] There are 1,140 cases (state years) in the data. For each case, I coded three dependent variables: whether all of the state's territory was conquered, the number of other great powers the state attacked, and the number of non-great powers it attacked.

I coded a state as conquered if it involuntarily lost its monopoly of force over all of its territory to an external rival and it was a great power or nuclear state when this occurred.[66] To identify great powers that experienced this fate, I relied on my State Survival and Death (SSAD) data on conquest in Europe and the Middle East from 1816 to 1994, as well as Quincy Wright's list of wars from 1800 to 1825 and Jack Levy's list of great power wars from 1495 to 1995.[67] From 1800 to 1997, there were 13 great power conquests (see Table 3).[68]

65. Henceforth, unless otherwise noted, I use "great power" to refer to both great powers and nuclear states.
66. Conquest may occur either directly, through the complete collapse or surrender of a state's military forces, or indirectly, through the surrender of political leaders in response to military threats or operations.
67. The SSAD data are described in Adams, "Conquering Myths," and are available from the author. Wright, *A Study of War,* Table 37; and Jack S. Levy, *War in the Modern Great Power System, 1495–1975* (Lexington: University Press of Kentucky, 1983), pp. 72–73.
68. For references and more detailed information on each conquest, see Adams, "Codebook for Data Used in 'Attack and Conquer,'" found at http://www.ksg.harvard.edu/IS.

Table 3. Conquests of Great Powers and Nuclear States, 1800–1997.

Conquered State and Date of Conquest	Conquered By	Military Occupation of Capital prior to Armistice	Collapse or Surrender of All Troops prior to Armistice
Austria 12/4/1805	France War of the Third Coalition: Austerlitz campaign	yes	yes
Prussia 11/24/1806	France War of the Third Coalition: Jena campaign	yes	yes
Spain 5/10/1808	France Peninsular War: French invasion, forced abdication of Charles and Ferdinand in favor of Napoleon's brother	no	no
Austria 7/10/1809	France Franco-Austrian War: Battle of Wagram	yes	yes
France 4/11/1814	Prussia, Austria, Sweden, Britain War of Liberation: Leipzig campaign	yes	yes, through Napoleon's abdication
France 6/22/1815	Britain, Prussia Hundred Days' War: Waterloo campaign	no	yes, through Napoleon's abdication
France 1/28/1871	Prussia Franco-Prussian War: capitulation of Paris	no, but Paris had been under siege for months	no
Austria 11/3/1918	Italy, France, Britain, United States World War I: Battle of Vittorio Veneto	no	yes
Germany 6/28/1919	United States, Britain, France World War I: Treaty of Versailles accepted by Germany	no	no
France 6/22/1940	Germany World War II: Battle of France	yes	yes
Italy 9/8/1943	United States, Britain, Germany World War II: surrender to Allies, forced disarmament by Germany	no	no
Germany 5/7/1945	Soviet Union, United States, Britain World War II: Soviet winter offensive, Allied Ruhr campaign	yes	yes
Japan 8/14/1945	United States World War II: naval blockade, bombing of mainland, conquest of Okinawa, Soviet invasion of Manchuria	no, but Tokyo was heavily bombed	no

I coded a state as attacking a great power if it was a great power or nuclear state and it conducted offensive operations against military or nonmilitary assets in the territory of a great power or nuclear state that had not previously attacked it.[69] To identify great power attacks on other great powers from 1800 to 1997, I developed a list of great power wars based on Wright's and Levy's lists, as well as the Correlates of War (COW) Project's Interstate War data for 1816 to 1997.[70] Then I used Ernest Dupuy and Trevor Dupuy's *Encyclopedia of Military History* to disaggregate the great power attacks in each war.[71] From 1800 to 1997, I found thirty-one such attacks in thirteen wars (see Table 4).

Counting great power attacks on non-great powers was more difficult because the sources I used to code great power attacks on other great powers lack data on a number of great power/non-great power wars.[72] Moreover, even if accurate data existed, the large number of such wars would make it difficult to disaggregate them into discrete attacks. Thus I estimated the number of such attacks based on data from Wright's and Levy's lists for the period from 1800 to 1815, COW's Interstate and Extra-State War data for 1816 to 1997,[73] and my SSAD data on conquest in Europe and the Middle East from 1816 to 1994.[74] The result is a conservative estimate of 105 great power attacks on non-great powers from 1800 to 1994.

Methodology

To test the hypotheses elaborated above, I assigned numerical codes to the independent variables identified in the hypotheses. Then I used the quantitative technique of event history analysis to examine the relationship between these variables and the historical incidence of attack and conquest among great powers.

I coded the independent variables as follows. To test my hypotheses about the effects of the technological offense-defense-deterrence balance on attack

69. For discussion of this coding rule, see Adams, "Codebook for Data Used in 'Attack and Conquer,'" p. 3.
70. Wright, *Study of War*, Table 37; Levy, *War in the Modern Great Power System, 1945–1975;* and COW Interstate War Data, 1816–1997 (Version 3), available at http://pss.la.psu.edu/DATARES.HTM.
71. R. Ernest Dupuy and Trevor N. Dupuy, *Encyclopedia of Military History from 3500 B.C. to the Present* (New York: HarperCollins, 1993).
72. Adams, "Codebook for Data Used in 'Attack and Conquer,'" pp. 3–4.
73. COW Interstate War Data, 1816–1997 (Version 3), and COW Extra-State War Data, 1816–1997 (Version 3), http://pss.la.psu.edu/DATARES.HTM. COW Interstate War Data, 1816–1997.
74. For an explanation of how I made this estimate and discussion of its reliability, see Adams, "Codebook for Data Used in 'Attack and Conquer,'" p. 4.

Table 4. Great Power Attacks on Great Powers, 1800–1997.

War	Year	Description	Did Attack Culminate in Conquest within One Year?
Franco-British (1803–14)	1804	British attack on French flotilla at the mouth of the Rhine	no
	1805	British attack on Spanish flotilla at Cape Trafalgar	no
Third Coalition (1805–07)	1805	French invasion of Austria	yes
	1806	French invasion of Prussia	yes
Peninsular (1807–14)	1808	French invasion of Spain	yes
Franco-Austrian (1809)	1809	Austrian invasion of Bavaria and Italy (held by France)	no
Franco-Russian/War of Liberation (1812–14)	1812	French invasion of Russia	no
Hundred Days' (1815)	1815	British invasion of France	yes
	1815	Prussian invasion of France	yes
Crimean (1853–56)	1854	British attack on Russia	no
	1854	French attack on Russia	no
Italian Unification (1859)	1859	French invasion of Lombardy (held by Austria)	no
Seven Weeks (1866)	1866	Prussian invasion of Austria	no
	1866	Italian invasion of Austria	no
Franco-Prussian (1870–71)	1870	Prussian invasion of France	yes
World War I (1914–18)	1914	French invasion of Alsace-Lorraine (held by Germany)	no
	1914	Russian invasion of Germany	no
	1914	Austrian invasion of Russian Poland	no
	1914	British attack on Germany navy in German waters	no
	1915	Italian attack on Austria	no
Russo-Japanese War (1938–40)	1938	Japanese attack on the Soviet Union (at Changkufeng Hill)	no
World War II (1939–45)	1939	German attack on British fleet in British waters	no
	1940	German invasion of France	yes
	1941	British invasion of Ethiopia (held by Italy)	no
	1941	German invasion of the Soviet Union	no
	1941	Japanese attack on the United States	no
	1941	Japanese attack on Britain (in Kowloon)	no
	1942	German attacks on the U.S. coast	no
	1943	U.S. invasion of Italy	yes
	1943	German disarmament and imprisonment of Italian forces	yes
	1945	Soviet invasion of Manchuria (held by Japan)	yes

SOURCE: R. Ernest Dupuy and Trevor N. Dupuy, *Encyclopedia of Military History from 3500 B.C. to the Present* (New York: HarperCollins, 1993), pp. 812–1309.

and conquest (H1 and H2), I developed an ordinal variable coded 0 in the deterrence-dominant era (1946–97), 1 in the defense-dominant era (1850–1933), and 2 in the offense-dominant eras (1800–49 and 1934–45). To test the hypotheses that defense and deterrence are functionally equivalent (H3 and H4), I developed a dummy variable coded 0 in the defense- and deterrence-dominant eras and 1 in the offense-dominant eras. To test the hypotheses that attack and conquest are best explained by relative capabilities (H7, H8, and H9), I imported COW's indexed capabilities data for each state.[75] To test the hypotheses that newer great powers should be more likely to attack other states and, thus, to be conquered (H10, H11, and H12), I created a variable for the number of years a state had been a great power or nuclear state.[76] Finally, to test the hypotheses that the historical incidence of attack and conquest have declined steadily over time as audience costs have risen (H13 and H14), I used the case year as an independent variable.

To test the effects of these independent variables, I used multivariate event history models, which are appropriate because I am interested in how these variables influence "the patterns and correlates of . . . events,"[77] namely the incidence of attacks by and conquest of great powers. Specifically, I used the Cox proportional hazards model to determine whether it was appropriate to assume that the baseline hazard rates for attack and conquest were constant from 1800 to 1997.[78] Finding that these rates were quite constant, I then used the discrete-time logit model to test the various hypotheses.[79] Finally, I calcu-

---

75. This index measures each state's overall capabilities relative to those of all other states in the system based on military personnel, military expenditure, energy consumption, iron and steel production, urban population, and total population. J. David Singer, "Reconstructing the Correlates of War Data Set on Material Capabilities of States, 1816–1985," *International Interactions,* Vol. 14 (1988), pp. 115–132. The data are available at http://www.umich.edu/~cowproj/dataset.html#Capabilities. To download the data, I used D. Scott Bennett and Allan Stam's, *EUGene* software, version 2.25, available at http://eugenesoftware.org/. On EUGene, see D. Scott Bennett and Allan Stam, "EUGene: A Conceptual Manual," *International Interactions,* Vol. 26 (2000), pp. 179–204.

76. In coding this variable, I relied on Table 2. Each time a great power regained its independence after conquest, I restarted the clock at zero.

77. Kazuo Yamaguchi, *Event History Analysis,* Applied Social Research Methods Series, Vol. 28 (Newbury Park, Calif.: Sage, 1991), p. 1; Paul D. Allison, *Event History Analysis: Regression for Longitudinal Data,* Quantitative Applications in the Social Sciences Series, Vol. 46 (Newbury Park, Calif.: Sage, 1984); Janet M. Box-Steffensmeier and Bradford S. Jones, "Time Is of the Essence: Event History Models in Political Science," *American Journal of Political Science,* Vol. 41, No. 4 (October 1997), pp. 1414–1461; and D. Scott Bennett, "Parametric Models, Duration Dependence, and Time-Varying Data Revisited," *American Journal of Political Science,* Vol. 43, No. 3 (January 1999), pp. 256–270.

78. From 1800 to 1997, the baseline hazard rates for great power attacks on other great powers were 1.0. Those for great power attacks on non-great powers were 0.96, 0.99 (99 percent of cases), or 1.0. Baseline hazard rates for great power conquest were either 0.99 (99 percent of cases) or 1.0.

79. Logit is appropriate because the baseline hazard rates are steady and the clustering of conquests and attacks in time (the tied nature of the data) reduces the accuracy of the Cox propor-

lated probabilities of attack and conquest based on the independent variables significant in logit analysis.[80]

In analyzing the data, I examined six time spans: 1800 to 1997, for which I have data on conquests and attacks; 1816 to 1993, for which there are COW capabilities data; 1850 to 1993, which contains one full period each of offense, defense, and deterrence dominance; and 1800 to 1945, 1816 to 1945, and 1850 to 1945, which allow for the possibility that the effects of the balance are driven by nuclear weapons or other variables unique to the post–World War II era. The period from 1816 to 1993 begins in the middle of the offense-dominant era, after the conquests and attacks of the Napoleonic Wars, making it a very hard test of the balance. By contrast, the period from 1850 to 1993, which begins with the change from offense dominance to defense dominance, poses a more reasonable test.[81] Thus in discussing my quantitative findings, I focus on the period from 1850 to 1993.

In testing the various hypotheses about conquest, I held constant the number of great power war dyads in which each state was involved each year.[82] Given that states cannot be conquered unless they are at war and that they are most likely to be conquered when they are fighting many great powers, doing so isolates the effects of the balance from those of war participation.

Finally, in testing all of the hypotheses, I adjusted the standard errors for clustering on country codes. This tests the possibility that states' geographic, cultural, and other internal attributes affect their vulnerability to conquest and propensity to attack other states more than contextual variables such as the offense-defense-deterrence balance and relative capabilities.[83]

---

tional hazards model. Yamaguchi, *Event History Analysis*, pp. 16–17. Because logit can underestimate the probability of rare events (those that occur in less than 5 percent of cases), I estimated the model for conquest (which occurred in just 1 percent of cases) and attacks on great powers (3 percent of cases) using Gary King and Langche Zeng's *ReLogit* software, which corrects the bias. Because the coefficients estimated by *ReLogit* were almost always smaller than those estimated by logit and increased the statistical significance the variable in just one case, I report the logit results. King and Zeng, "Logistic Regression in Rare Events Data," *Political Analysis*, Vol. 9, No. 2 (Spring 2001), pp. 137–163. *ReLogit* software is available at http://gking.harvard.edu.

80. To calculate these probabilities, I used Michael Tomz et al.'s *Clarify* software. Gary King, Michael Tomz, and Jason Wittenberg, "Making the Most of Statistical Analyses: Improving Interpretation and Presentation," *American Journal of Political Science*, Vol. 44, No. 2 (March, 2000), pp. 341–355. The software is available at http://gking.harvard.edu.

81. Nevertheless, the balance is a significant predictor of conquest and both types of attack from 1800 to 1997, 1816 to 1993, and 1850 to 1993, as well as in the pre-1946 periods enumerated in note 86.

82. I considered a state to participate in a particular war dyad from the time it declared war on or attacked another state (whichever came first) until it concluded a peace treaty with that state or one side's military forces collapsed, surrendered, or were withdrawn (again, whichever came first).

83. Adjusting for clustering on country codes does not provide a measure of the statistical

Findings

Historical trends in attack and conquest provide strong support for my hypotheses about the effects of the offense-defense-deterrence balance (H1 and H2). As shown in Table 5, from 1800 to 1997, the average annual rates of great power conquest and attack were consistently higher in offense-dominant eras than in defense- and especially deterrence-dominant eras. Moreover, as shown in Table 6, event history analysis indicates that the offense-defense-deterrence balance is a statistically significant predictor of attack and conquest. In fact, the offense-defense-deterrence balance is the only variable to predict great power conquest, great power attacks on other great powers, and great power attacks on non-great powers. Relative capabilities are statistically significant predictors of conquests of and attacks by great powers (confirming H7 and H8), but (contrary to H9) they do not predict great power attacks on non-great powers. Similarly, duration of great power status predicts conquest (confirming H12), but (contrary to H10 and H11) predicts neither type of attack. Finally, contrary to H13 and H14, year predicts neither attack nor conquest.[84]

Separate analyses (not reported in Table 6) reveal that the offense-defense-deterrence balance is not only a more consistent predictor of attack and conquest than relative capabilities, duration of great power status, and year; its predictive power also trumps that of the binary offense-defense balance. Although the binary balance is a significant predictor of conquest and both types of attack, models testing this formulation of the balance explain less of the historical variance in these outcomes than models containing the offense-defense-deterrence balance.[85] Thus, contrary to H3 and H4, defense and deterrence dominance are not equivalent in their effects.

Moreover, when tested on the pre-1946 periods (results not reported in

---

significance of any clustering that does occur; it simply assesses the significance of other variables given that clustering.

84. The number of great power war dyads in which states are involved is also a significant predictor of conquest, but this is to be expected given than conquest requires war and states are most likely to be conquered when they are at war with many great powers.

85. When tested on conquest from 1850 to 1993, the coefficient for the binary offense-defense balance is 2.673, the standard error is 0.819, and the pseudo R^2, which indicates the amount of variance explained by the model, is 0.366. When tested on great power attacks on great powers, the coefficient is 3.136, the standard error is 0.581, and the pseudo R^2 is 0.166. When tested on great power attacks on non-great powers, the coefficient and the standard error for the offense-defense balance are 1.438 and 0.332, and the pseudo R^2 is 0.047. Although the differences between the R^2s for the offense-defense balance and the offense-defense-deterrence balance are presently slight, they will increase with each year that passes without the conquest of or attacks by nuclear states.

Table 5. Incidence of Attack and Conquest among Great Powers and Nuclear States in Offense-, Defense-, and Deterrence-Dominant Eras, 1800–1997.

Technological Environment	Number of Years	Number of Great Powers Conquered	Average Annual Rate of Great Power Conquest	Number of Great Power Attacks on Great Powers	Average Annual Rate of Great Power/Great Power Attack	Estimated Number of Great Power Attacks on Non-great Powers	Average Annual Estimated Rate of Great Power/ Non-great Power Attack
offense dominant							
1800–49	49	6	0.12	9	0.18	20[a]	0.41
1934–45	11	4	0.36	11	1.00	24	2.18
defense dominant							
1850–1933	83	3	0.04	11	0.13	55	0.66
deterrence dominant							
1946–97	51	0	0.00	0	0.00	6	0.12
totals		13		31		105	

[a]This estimate is low because I did not attempt to code great power attacks on non-great powers during the Napoleonic Wars.

Table 6. Logit Results for Conquest of and Attacks by Great Powers and Nuclear States, 1850–1993.

Dependent Variable	Independent Variable	Coefficient[a]	Standard Error
Conquest of great powers	offense-defense-deterrence balance	2.273*	1.070
	indexed capabilities	−15.617**	6.382
	years great power or nuclear state	−0.005**	0.002
	year	−0.013	0.031
	number of war dyads	1.068**	0.298
	constant	18.347	57.827
	log likelihood = −25.582 chi^2 (5df) = 287.00 prob > 5 chi^2 = 0.000 pseudo R^2 = 0.370		
Attacks on great powers	offense-defense-deterrence balance	2.623**	0.605
	indexed capabilities	4.808*	2.850
	years great power or nuclear state	−0.002	0.002
	year	−0.015#	0.009
	constant	21.076	17.538
	log likelihood = −81.167 chi^2 (4df) = 36.32 prob > 5 chi^2 = 0.000 pseudo R^2 = 0.177		
Attacks on non-great powers	offense-defense-deterrence balance	1.032**	0.272
	indexed capabilities	2.553	3.457
	years great power or nuclear state	−0.001	0.001
	year	−0.003	0.007
	constant	2.573	12.844
	log likelihood = −217.733 chi^2 (4df) = 29.75 prob > chi^2 = 0.000 pseudo R^2 = 0.054		

[a]An asterisk after the coefficient indicates that this variable is a statistically significant predictor of the dependent variable when the other variables in the model are held constant. Because most of the hypotheses tested here are directional, I used one-tailed tests of significance. Coefficients marked ** are highly significant ($p < 0.01$), those marked * are significant ($p < 0.05$), and those marked # approach statistical significance ($p < 0.10$). In each of these models, there were 857 cases.

Table 6), the offense-defense-deterrence balance is a significant predictor of conquest and both types of attack.[86] This confirms that offense and defense dominance have different effects. Thus, contrary to H5 and H6, the significance of the balance is not simply the result of nuclear deterrence or any other attack-inhibiting variable unique to the postwar era.

Interestingly, in these tests of the prenuclear era, year is a significant predictor of attacks on great powers from 1816 to 1945. Yet, contrary to H13, more recent years are associated with higher, not lower, probabilities of attack. Because there is no variance in great power attacks on great powers after 1945 (because there have been no such attacks), the near-statistical significance of year from 1850 to 1993 does not reflect the effects of rising audience costs or other time-dependent, attack-inhibiting variables. Instead it suggests a secular increase in attack-promoting variables from 1816 to 1945.

Based on the findings shown in Table 6, I calculated the probabilities that an individual great power would be conquered or attack other states in a given year of the offense-, defense-, and deterrence-dominant eras. As shown in Table 7, great powers were seven times as likely to be conquered in each year of the offense-dominant era from 1934 to 1945 (probability 0.014) as they were in each year of the defense-dominant era from 1850 to 1993 (probability 0.002), and they were twice as likely to be conquered in the defense-dominant era as they were in the deterrence-dominant era from 1946 to 1997 (probability 0.001).[87] Offense dominance affected the probability of great power attacks on other great powers even more strongly. Such attacks were twelve times more likely when offense was dominant (probability 0.156) than when defense was dominant (probability 0.013), and they were more than thirteen times more likely when defense was dominant than when deterrence was dominant (probability 0.001). Great power attacks on non-great powers have been less affected by the balance, decreasing by less than a factor of three from both offense dominance (probability 0.194) to defense dominance (probability 0.077) and from defense to deterrence dominance (probability 0.030). Yet among all of the independent variables tested, only the offense-defense-deterrence balance is a significant predictor of great power attacks on non-great powers. Thus even

86. Specifically, it is a very significant predictor of conquest and great power attack from 1800 to 1945 and of both types of attack from 1850 to 1945. When King and Zeng's *ReLogit* software is used to adjust for rare events data, the balance is also a significant predictor of conquest from 1816 to 1993.

87. To make these comparisons, I divide the probability of conquest in one era by the probability of conquest in another.

Table 7. Annual Probabilities of Conquest of and Attacks by Great Powers and Nuclear States in Deterrence-, Defense-, and Offense-Dominant Eras, 1850–1993.[a]

Probability of	Deterrence-Dominant Era (1946–97)	Defense-Dominant Era (1850–1933)	Offense-Dominant Era (1934–45)
Conquest of great powers	0.001	0.002	0.014
Attacks on great powers	0.001	0.013	0.156
Attacks on non-great powers	0.030	0.077	0.194

[a]This table summarizes the probabilities that an individual great power would be conquered or attack other states in a given year of the offense-, defense-, and deterrence-dominant eras. These probabilities are based on the logit results shown in Table 6, with indexed capabilities, duration of great power status, year, country code, and number of great power war dyads held constant at their means.

states that lack nuclear weapons seem to have benefited from deterrence dominance.

I also used the findings summarized in Table 6 to calculate the annual probabilities of conquest and attack for great powers and nuclear states of different capabilities. As shown in Table 8, irrespective of the offense-defense-deterrence balance, the least capable great powers (those with indexed capabilities in the 10th percentile) from 1850 to 1993 were forty times more likely to be conquered (probability 0.008) than their most capable counterparts (those in the 90th percentile, probability 0.0002) and were two and a half times less likely to attack other great powers (probabilities 0.006 and 0.015, respectively).

If these findings are robust, qualitative analysis of attacks and conquests in different technological eras should confirm the effects of the offense-defense-deterrence balance and relative capabilities. Although detailed analysis of this nature is beyond the scope of this article, it is notable that (as shown in Table 4) great power attacks on other great powers culminated in the conquest of the target state within one year twice as often in the offense-dominant eras as they did in the defense-dominant era.[88] Furthermore, as shown in Table 3, attackers

88. As shown in Table 4, five of nine (56 percent of) attacks on great powers from 1800 to 1849 culminated in the conquest of the target state within one year, and four of eleven (36 percent) did so from 1934 to 1945, compared to one of eleven (9 percent) from 1849 to 1933. But because two attacks in each of the offense-dominant eras were joint attacks and because the Japanese surrender in 1945 reflected both the U.S. atomic bombings and the Soviet invasion of Manchuria, it is more accurate to say that four of nine and two of eleven (44 percent and 18 percent of) great power attacks in the offense-dominant eras culminated in conquest within one year, compared with 9 percent in the defense-dominant era. Thus the ratio of conquest to attack was twice as high in the safest offense-dominant era as it was in the defense-dominant one.

Table 8. Annual Probabilities of Conquest of and Attacks by Great Powers and Nuclear States for Great Powers and Nuclear States of Different Capabilities, 1850–1993.[a]

Probability of	10th Percentile (low capability)	Mean	90th Percentile (high capability)
Conquest of great powers	0.008	0.002	0.000
Attacks on great powers	0.006	0.008	0.015

[a]This table summarizes the probabilities that an individual great power of particular capabilities would be conquered or would attack other great powers in a given year. These probabilities are based on the logit results shown in Table 6, with the offense-defense-deterrence balance, duration of great power status, year, country code, and number of great power war dyads held constant at their means.

occupied the capitals of conquered great powers before an armistice was reached at least half of the time in the two offense-dominant eras but never in the defense-dominant era. Similarly, the militaries of conquered great powers completely collapsed or surrendered at least half of the time in the offense-dominant eras but just one-third of the time in the defense-dominant one. Thus offense dominance does seem to make it easier to take territory than to hold it and, as a result, more likely that states will be conquered.

Furthermore, Prussia's conquest of France in 1871—the one great power conquest during the defense-dominant era that occurred in less than one year and involved the complete collapse or surrender of a great power's military forces—was clearly the result of defense dominance. France failed to develop either a defensive strategy or state-of-the-art defensive capabilities. Moreover, Napoleon III surrendered at Sedan because French troops were unable to retake French territory under Prussian artillery fire. Finally, Paris capitulated only after a siege.[89]

Conclusions and Implications

Historical trends in attack and conquest among great powers and nuclear states from 1800 to 1997 provide strong support for my argument about the effects of the technological offense-defense-deterrence balance. The balance, which I operationalized by deducing the relative lethality, protection, and mobility of offensive, defensive, and deterrent military operations from prevailing

89. Posen, *The Sources of Military Doctrine*, pp. 25–27; and Dupuy and Dupuy, *Encyclopedia of Military History from 3500 B.C. to the Present*, pp. 910–916.

technologies, is a significant predictor of great power vulnerability to conquest and propensity to attack both other great powers and non-great powers. These findings have important implications for future research on the balance, as well as for international political theory and the future of international politics.

IMPLICATIONS FOR FUTURE RESEARCH ON THE BALANCE

My findings about the effects of the technological offense-defense-deterrence balance on attack and conquest belie three prevalent assumptions in the offense-defense literature. First, the incidence of war is not the best dependent variable to use in testing the balance. Although great power wars occurred in both offense- and defense-dominant eras, great powers were less likely to attack one another and to be conquered when defense was dominant. Thus, although the balance may affect the incidence of war, it most directly affects the incidence of attack and conquest.

Second, scholars have too quickly dismissed arguments about the effects of the technological offense-defense-deterrence balance. Military doctrine, geography, force employment, perceptions, and many of the other variables previously invoked in defining the balance no doubt affect the vulnerability of states to conquest and the likelihood that they will attack other states. But differences in the potentials and interactions of offensive, defensive, and deterrent military operations in different technological eras have significant effects of their own.

Third, it is not the case that defense and deterrence are functionally equivalent. Defensive and deterrent operations have different logics and requirements. They also have different international political effects. As shown in Table 7, during the defense-dominant era from 1850 to 1933, great powers were almost twice as vulnerable to conquest, thirteen times more likely to attack other great powers, and more than twice as likely to attack non-great powers as nuclear states from 1946 to 1997. Because the offense-defense-deterrence balance explains more of the variance in conquest than the offense-defense balance (and will explain more with each year that passes without attacks by or conquests of nuclear states)—and because its significance is not driven by nuclear deterrence or other post-1945 attack-inhibiting variables—scholars seeking to explain other outcomes or to explain more about attack and conquest should use this version of the balance.

In explaining other outcomes, it is important to recognize two points. First, because the offense-defense-deterrence balance affects the incidence of attack, states seem to be aware that they are more vulnerable to conquest in offense-dominant eras than they are in defense- and deterrence-dominant ones. Thus the balance may also affect the character of military doctrine, the extent of arms racing, and the prevalence and nature of international cooperation. But, second, although the balance is a significant predictor of attack and conquest, models containing the balance, relative capabilities, and the other independent variables tested here do not explain all of the variance in these outcomes.[90] Because the balance's effects on doctrine, procurement, and cooperation are likely to be more muted than its effects on attack and conquest, the challenge will be to avoid lumping other variables into the balance to increase its explanatory power. Yet only when this is avoided will it be possible to assess the relative importance of each variable. Thus a firm distinction should be drawn between adding variables to the technological offense-defense-deterrence balance, which raises questions about the meaning of the balance and complicates testing, and adding variables to models, which should improve explanatory power while maintaining clarity and specificity.

Efforts to test arguments about other causes of attack and conquest while controlling for the offense-defense-deterrence balance and relative capabilities are also needed, both to explain more of the variance in these outcomes and to explain the gap between attack and conquest in offense-dominant eras. As shown in Table 7, from 1934 to 1945, the probability that one great power would attack another in a given year (0.156) was eleven times greater than the probability that an individual great power would be conquered (0.014). This gap is striking, especially given that (as shown in Table 8) the most capable great powers (those with indexed capabilities in the 90th percentile) have been more than twice as likely to attack other great powers (probability 0.015) as their less capable counterparts (probability 0.006).

Perhaps, as Clausewitz would argue, the gap between attack and conquest in offense-dominant eras can be attributed to a ubiquitous defensive advantage, which is only marginally eroded by offense dominance. Or perhaps, as Waltz and other structural realists would suggest, the failure of most attacks to culminate in conquest, even in offense-dominant eras, is a result of the tendency for states to balance the power of the states most capable of conquering others. Then again, maybe great powers fail to perceive the dominance of offensive operations, adopting defensive doctrines and strategies that reduce the vulnerability of other states. Or perhaps attacking states fail to employ their

90. The amount of variance explained by each model is indicated in Table 6 by the pseudo R^2s.

forces to optimal effect. Finally, perhaps great power wars in offense-dominant eras so quickly exhaust the participants that few are able to mount effective offensives for long. To test these possibilities, efforts to elaborate and test hypotheses about the effects of homeland advantage, power balancing, military doctrine, force employment, and attrition are needed.

IMPLICATIONS FOR INTERNATIONAL POLITICAL THEORY

My findings not only confirm the importance of the offense-defense-deterrence balance. They also provide strong support for structural realist theory while raising questions about the explanatory power of offensive realism, defensive realism, and arguments about the pacific effects of unbalanced power and rising audience costs.

Both structural and offensive realism are supported by the finding that less capable great powers are more likely to be conquered and less likely to attack other great powers than their more capable counterparts. But because great powers are significantly less likely to attack and conquer one another when defense and, especially, deterrence are dominant than when offensive operations dominate, the offensive realist argument that states can be secure only when they are "primed for offense" fails to convince.[91] Defense and deterrence dominance afford high levels of security to states with state-of-the-art defensive and deterrent capabilities.

Yet security is not the only end great powers seek. Although great power attacks on non-great powers are significantly less likely in defense- and deterrence-dominant eras than they are in offense-dominant ones, as shown in Table 7, in all technological eras the probability that a great power will attack a non-great power is higher than the probability that it will attack one of its peers. In fact, in the deterrence-dominant era from 1946 to 1997, the probability that a great power or nuclear state would attack a nonnuclear state (0.030) was thirty times larger than the probability that it would attack another great power (0.001). Moreover, older great powers are no less likely than their younger counterparts to attack other states. Thus the defensive realist claim that states simply seek security also fails to find support. All one can say with any certainty is that states, "at a minimum, seek their own preservation and, at a maximum, drive for universal domination."[92] Because historical trends in attack and conquest indicate both that national security is affected by structural

91. Mearsheimer, *The Tragedy of Great Power Politics,* p. 3.
92. Waltz, *Theory of International Politics,* p. 118.

and technological conditions and that security is not the only end states seek, structural realism explains more than either offensive or defensive realism.

These findings also pose problems for those who expect unbalanced power or rising audience costs to have pacific effects on international affairs. It may be "absurd" for states to attack states with substantially less capability than their own,[93] but strong states frequently do so. Such attacks could reflect lower audience costs for attacks on less powerful states than for attacks on one's peers (because the former are more likely to succeed). But contrary to Fearon's argument that audience costs have risen steadily since the early nineteenth century and that they dissuade states from starting wars they might lose, great powers were increasingly likely to attack other great powers each year from 1816 to 1945. Thus the age-old logic of international relations still applies. Military, economic, political, and other costs of war—as well as self-restraint—may induce the strong to forego attacking the weak, but in the absence of an international sovereign there is nothing to compel them to do so.

IMPLICATIONS FOR INTERNATIONAL POLITICS

The finding that states armed with second-strike nuclear forces are less vulnerable to conquest and less likely to attack other states than erstwhile great powers have been, indicates that there is something to be said for the proposition that more nuclear states "may be better."[94] But whether less powerful states decide that going nuclear is best for them will depend on what today's nuclear states do with the deterrence dividend. If they eschew the temptation to spend it on efforts to expand their power and are more mindful of less powerful states' concern with survival and prosperity than historical great powers have been, less powerful states will continue to enjoy the benefits of deterrence dominance, and nuclear weapons will spread slowly, if at all. But if even one nuclear state, like its great power predecessors, embarks on conquest and other efforts that imperil the survival of less powerful states while its nuclear peers rest comfortably in the knowledge that they are more secure than the most powerful states in the defense- and offense-dominant eras ever were, less powerful states will perceive that deterrence dominance is doubly dangerous—making them, at once, more attractive targets than nuclear-armed great powers and less important allies than their counterparts in offense- and defense-dominant eras. In this case, the spread of nuclear weapons is likely to accelerate.

---

93 Clausewitz, *On War*, p. 91.

94. Kenneth N. Waltz, "More May Be Better," in Scott D. Sagan and Waltz, *The Spread of Nuclear Weapons: A Debate* (New York: W.W. Norton, 1995), chap. 1, especially p. 42.

If the state that spends the deterrence dividend on expansion is the most powerful state in the system and those waiting idly by are the states most capable of balancing its power, deterrence dominance will not only increase the insecurity of nonnuclear states. It will also delay the emergence of a new balance of power capable of mitigating their insecurity. At present, this seems to be what is happening. China, Russia, France, and other nuclear states, knowing that their homelands are secure from attack and conquest by the most powerful state in the system, are sitting on the sidelines, criticizing the U.S. conquest of Iraq. Meanwhile, the United States prepares to pursue other preemptive whims, and less powerful states redouble their efforts to go nuclear so they can avoid becoming targets.[95]

The more the United States expands, the more it will threaten the political and economic interests of other nuclear states. Thus, a new balance of power will surely form somewhere down the road. When it does, nonnuclear states should once again enjoy the benefits of deterrence dominance. But when this balance will form, how many states will decide that the best way to provide for their security in the meantime is to obtain nuclear weapons, and how far the United States will go in trying to maintain and extend its power remain to be seen.

95. Howard W. French, "North Korea Says Its Arms Will Deter U.S. Attack," *New York Times*, April 7, 2003, p. 13; Steven R. Weisman, "New U.S. Concerns on Iran's Pursuit of Nuclear Arms," *New York Times*, May 8, 2003, p. 1; Stephen Fidler, Andrew Jack, and Roula Khalaf, "Russia Considers Building Syrian Nuclear Reactor," *Financial Times*, January 16, 2003, p. 10; and Ewen MacAskill and Ian Traynor, "Saudis Consider Nuclear Bomb," *Guardian*, September 18, 2003, http://www.guardian.co.uk/international/story/0,3604,1044380,00.html.

Suggestions for Further Reading

There is a significant literature on offense-defense theory and the related concept of the security dilemma in international politics. This bibliography includes many of the most important contributions to this literature that are not contained in this volume. To avoid repetition, all the works listed appear only once, even if they make contributions in several categories.

Early Discussions of Offense-Defense Theory and Related Issues

The following works are precursors of contemporary theories of offense and defense in international politics. They distinguish between offensive and defensive capabilities, discuss the implications of variations in the strength of the offense, and consider whether offensive weapons can be identified or limited.

Bloch, Ivan S. *The Future of War.* Trans. R.C. Long. Pref. W.T. Stead. New York: Doubleday and McClure, 1899, pp. xxx–xxxi, lxxix. This book offers one of the earliest arguments that war becomes more likely when conquest is easy.

Boggs, Marion William. *Attempts to Define and Limit "Aggressive" Armament in Diplomacy and Strategy.* University of Missouri Studies, Vol. 16, No. 1. Columbia: University of Missouri, 1941. This monograph is a landmark analysis of the attempts to limit offensive capabilities at the 1932–33 Geneva Disarmament Conference. For a more recent account of the conference, see Marlies ter Borg, "Reducing Offensive Capabilities—the Attempt of 1932," *Journal of Peace Research,* Vol. 29, No. 2 (1992), pp. 145–160.

Hoag, Malcolm W. "On Stability in Deterrent Races," *World Politics,* Vol. 13, No. 4 (July 1961), pp. 505–527. This article offers a brief formulation of the core ideas of offense-defense theory.

Liddell Hart, B.H. "Aggression and the Problem of Weapons," *English Review,* No. 55 (July 1932), pp. 71–78. This article is an early discussion of the distinction between offensive and defensive forces and its implications.

Liddell Hart, B.H. *Memoirs.* London: Cassell, 1965, chap. 8. This book presents additional discussion of the problem of offense and defense during the interwar period.

Wright, Quincy. *A Study of War,* Vol. 1. Chicago: University of Chicago Press, 1942. This massive work contains many scattered references to offensive and defensive capabilities and their relationship to war, but does not fully develop an offense-defense theory. See also Quincy Wright, *A Study of War,* 2d rev. ed., Chicago: University of Chicago Press, 1965, pp. 792–808.

Explications of Offense-Defense Theory

In addition to the selections by Robert Jervis, George Quester, Stephen Van Evera, Charles Glaser and Chaim Kaufmann in this volume, the following works offer detailed explications of offense-defense theory.

Glaser, Charles L. "Realists as Optimists: Cooperation as Self-Help," *International Security,* Vol. 19, No. 3 (Winter 1994/95), pp. 50–90. This article uses offense-defense theory to contribute to defensive realist theory. Glaser integrates offense-defense variables into realist theory to demonstrate that realism does not preclude cooperation.

Glaser, Charles L. "The Security Dilemma Revisited," *World Politics,* Vol. 50, No. 1 (October 1997), pp. 171–201. Glaser offers an explication and analysis of the implications of the ideas developed by Jervis in "Cooperation under the Security Dilemma." He also defends offense-defense theory against some criticisms.

Snyder, Jack. "Perceptions of the Security Dilemma in 1914," in Robert Jervis, Richard Ned Lebow, and Janice Stein, eds., *Psychology and Deterrence.* Baltimore, Md.: Johns Hopkins University Press, 1985, pp. 153–179. Snyder explicates the concept of the security dilemma and explains how the great powers perceived the offense-defense balance on the eve of the First World War.

Van Evera, Stephen. *Causes of War: Power and the Roots of Conflict.* Ithaca, N.Y.: Cornell University Press, 1999. This is a very important book on offense-defense theory and the most detailed presentation of Van Evera's version of the theory. Chapters 6 and 7 offer a comprehensive explication of offense-defense theory and several tests. Chapter 8 discusses the implications of the nuclear revolution. Other chapters present additional explanations for war.

Applications and Tests of Offense-Defense Theory

The following works are a representative selection of how insights from offense-defense theory have been applied to phenomena as diverse as war; the number and size of states in the international system; alliances; internal conflict; nuclear deterrence; and arms races and arms control.

WAR

Hopf, Ted. "Polarity, the Offense-Defense Balance, and War," *American Political Science Review,* Vol. 85, No. 2 (June 1991), pp. 475–494. Hopf argues for a broad conception of the offense-defense balance, including perceptions, technology, and the propensity of states to balance or bandwagon, and finds that variations in this balance explain patterns of peace and war in sixteenth-century Europe.

Posen, Barry R. *Inadvertent Escalation: Conventional War and Nuclear Risks.* Ithaca, N.Y.: Cornell University Press, 1991. Posen applies insights from offense-defense theory to explore potential risks of escalation in U.S.-Soviet Cold War interactions.

Snyder, Jack. *The Ideology of the Offensive: Military Decision Making and the Disasters of 1914.* Ithaca, N.Y.: Cornell University Press, 1984. World War I has been an important case for offense-defense theory. Along with Van Evera's work, Snyder's book is a seminal account of the origins of offensive doctrines and their consequences in 1914.

Walt, Stephen M. *Revolution and War.* Ithaca, N.Y.: Cornell University Press, 1996. This book explains why revolutions often lead to wars by considering how revolutions increase the severity of the security dilemma between revolutionary states and others.

STATE SIZE AND NUMBER

The following three works offer brief arguments that the number and size of states in the international system may depend on the offense-defense balance. With the partial exception of the article by Karen Ruth Adams in this volume, these hypotheses have not been refined and tested as often as offense-defense theory's hypotheses on war, arms races, and alliances.

Andreski, Stanislav. *Military Organization and Society,* 2d ed. London: Routledge and Kegan Paul, 1968, pp. 75–76.

Bean, Richard. "War and the Birth of the Nation State," *Journal of Economic History,* Vol. 33, No. 1 (March 1973), pp. 203–221 at 204–207.

Gilpin, Robert. *War and Change in World Politics.* Cambridge, U.K.: Cambridge University Press, 1981, pp. 61–62.

ALLIANCE FORMATION

Christensen, Thomas J., and Jack Snyder. "Chain Gangs and Passed Bucks: Predicting Alliance Patterns in Multipolarity," *International Organization,* Vol. 44, No. 2 (Spring 1990), pp. 137–168. The authors argue that perceptions of the offense-defense balance influence alliance formation. They suggest that alliances form faster and are tighter when states perceive an offensive advantage and that behavior of the European great powers before the world wars supports this argument.

Walt, Stephen M. *The Origins of Alliances.* Ithaca, N.Y.: Cornell University Press, 1987. In a general sense, Walt's balance-of-threat theory is closely related to offense-defense theory because offensive power is a key factor in explaining alliance formation. This book also contends that variations in the offense-defense balance determine whether balancing occurs rapidly or slowly and whether alliances are tight or loose. See, in particular, pp. 24–25, n. 31, and pp. 165–167.

INTERNAL CONFLICT

Kaufmann, Chaim. "Possible and Impossible Solutions to Ethnic Civil Wars" *International Security,* Vol. 20, No. 4 (Spring 1996), pp. 136–175. Kaufmann uses security dilemma logic to show why internal ethnic conflicts cannot be resolved easily. Contends that ethnic separation may be necessary to achieve peace.

Posen, Barry R. "The Security Dilemma and Ethnic Conflict," *Survival,* Vol. 35, No. 1 (Spring 1993), pp. 27–57. Posen applies core elements of offense-defense theory to analyze the outbreak of conflict between Serbs and Croats in the former Yugoslavia and the absence of war between Russia and Ukraine after the disintegration of the Soviet Union.

Rose, William. "The Security Dilemma and Ethnic Conflict: Some New Hypotheses," *Security Studies,* Vol. 9, No. 4 (Summer 2000), pp. 1–54. Extending and testing hypotheses offered by Posen and Van Evera, this article considers how variations in the intensity of the security dilemma influenced relations between Serbs and Croats in Croatia and between Russians and Ukrainians in the Crimea. Rose finds support for most of Van Evera's hypotheses.

Snyder, Jack, and Robert Jervis. "Civil War and the Security Dilemma," in Barbara F. Walter and Jack Snyder, eds., *Civil Wars, Insecurity, and Intervention.* New York: Columbia University Press, 1999, pp. 15–37. Snyder and Jervis explore how the concept of the security dilemma offers insight into civil wars and consider how outside intervention can overcome the security dilemma.

NUCLEAR DETERRENCE

Feldman, Shai. *Israeli Nuclear Deterrence: A Strategy for the 1980s.* New York: Columbia University Press, 1982. Feldman develops elements of offense-defense theory, including the argument that nuclear weapons and nuclear deterrence make conquest extremely difficult. He then applies this analysis and argues that Israel should adopt a deterrent strategy for its nuclear arsenal.

Glaser, Charles L. *Analyzing Strategic Nuclear Policy.* Princeton, N.J.: Princeton University Press, 1990. Glaser applies many principles of offense-defense theory to his analysis of U.S. policy for its strategic nuclear forces.

ARMS RACES, ARMS CONTROL, AND COOPERATION

Downs, George W., David M. Rocke, and Randolph Siverson. "Arms Races and Cooperation," in Kenneth A. Oye, ed., *Cooperation under Anarchy.* Princeton, N.J.: Princeton University Press, 1986, pp. 118–146. This article considers many strategies that states may use to end arms races. It examines many historical cases and finds that states rarely, if ever, deploy defensive forces in hopes that this policy will defuse an arms race.

Glaser, Charles L. "Political Consequences of Military Strategy: Expanding and Refining the Spiral and Deterrence Models," *World Politics,* Vol. 44, No. 4 (July 1992), pp. 497–538. Glaser applies concepts from offense-defense theory to assess how states can reduce the likelihood of arms races and war by taking into account how military policies influence the political calculations of adversaries.

Powell, Robert. "Guns, Butter, and Anarchy," *American Political Science Review,* Vol. 87, No. 1 (March 1993), pp. 115–132. Powell develops a formal model that incorporates the offense-defense balance and the intensity of the security dilemma. He uses this model to predict how much a state will allocate to its military and whether it will attack other states. In the model, states increase their military allocations when the offense-defense balance shifts toward offense, but such shifts do not necessarily increase the probability of war.

Critiques of Offense-Defense Theory

In addition to the contributions of Richard Betts, Keir Lieber, Scott Sagan, Jonathan Shimshoni, James Davis Jr., Bernard Finel, and Stacie Goddard in this volume, the following works lay out many of the major criticisms of offense-defense theory.

Biddle, Stephen D. "Testing Offense-Defense Theory: The Second Battle of the Somme, March 21 to April 9, 1918." Paper presented at the annual meeting of the American

Political Science Association, September 2–5, 1999. Biddle argues that changes in tactics, not technology, explain why offenses sometimes succeeded during World War I.

Butfoy, Andrew. "Offence-Defence Theory and the Security Dilemma: The Problem with Marginalizing the Context," *Contemporary Security Policy,* Vol. 18, No. 3 (December 1997), pp. 38–58. Butfoy argues that offense-defense theory is too abstract and mechanistic. He claims that the dynamics of the security dilemma and offensive advantages often do not explain the outbreak or absence of war. In his view, perceptions of political intent and other nonmilitary factors often are more important.

Fearon, James D. "The Offense-Defense Balance and War since 1648." Paper presented at the annual convention of the International Studies Association, Chicago, Illinois, February 21–25, 1995. This unpublished but oft-cited paper finds that wars are less frequent in periods when offense has the advantage. To explain this pattern, Fearon offers the deductive argument that states are more cautious when an offensive advantage makes total defeat more likely.

Gray, Colin S. *House of Cards: Why Arms Control Must Fail.* Ithaca, N.Y.: Cornell University Press, 1992, pp. 28, 66–68. This comprehensive critique of arms control includes the argument that offensive and defensive weapons cannot be distinguished.

Gray, Colin S. *Weapons Don't Make War: Policy, Strategy, and Military Technology.* Lawrence: University of Kansas Press, 1993, chap. 2. Gray argues that offensive and defensive weapons cannot be distinguished and that the intentions of states are more important than the characteristics of weapons.

Huntington, Samuel P. "U.S. Defense Strategy: The Strategic Innovations of the Reagan Years," in Joseph Kruzel, ed., *American Defense Annual, 1987–1988.* Lexington, Mass.: Lexington Books, 1987, pp. 23–43 at 35–37. Huntington offers a brief but forceful set of arguments against offense-defense theory, including the claim that offensive and defensive weapons cannot be distinguished.

Levy, Jack S. "The Offensive/Defensive Balance of Military Technology: A Theoretical and Historical Analysis," *International Studies Quarterly,* Vol. 28, No. 4 (June 1984), pp. 219–238. One of the earliest and most important critiques of offense-defense theory, this article elaborates many of the problems of defining and measuring the offense-defense balance.

Liberman, Peter. "The Offense-Defense Balance, Interdependence, and War," *Security Studies,* Vol. 9, Nos. 1 and 2 (Summer and Fall 1999), pp. 59–91. This important article argues that offense-defense theory needs to be qualified because defensive advantages make long wars more likely and the resulting expectations of protracted wars of attrition lead trade-dependent states to conquer economically valuable territories. German and Japanese policies before the world wars support this argument.

Mearsheimer, John J. *Conventional Deterrence.* Ithaca, N.Y.: Cornell University Press, 1983, pp. 24–27. In the context of developing his broader theory of conventional deterrence, Mearsheimer argues that offense-defense theory is flawed because it assumes that offensive and defensive weapons can be distinguished.

Morrow, James. "Arms versus Allies: Trade-Offs in the Search for Security," *International Organization,* Vol. 47, No. 2 (Spring 1993), pp. 207–233. Morrow presents a critique of the arguments of Christensen and Snyder (listed under "alliance formation") on offense-defense theory and alliance formation, contending that domestic politics determines whether states form alliances or build up their own arms.

Russell, Richard L. "Persian Gulf Proving Grounds: Testing Offence-Defence Theory," *Contemporary Security Policy*, Vol. 23, No. 3 (December 2002), pp. 192–213. Russell argues that offense-defense theory fails to explain the Iran-Iraq war and the 1990–91 Gulf War. Criticizing the emphasis on military capabilities and technology in offense-defense theory, he suggests that the political causes of war deserve more attention.

Responses to Critiques of Offense-Defense Theory

Many explications and applications of offense-defense theory contain brief responses to actual and potential objections. The following articles are among the few that offer more detailed rebuttals to the prominent critiques of offense-defense theory. The exchanges of correspondence included in this volume present brief replies to some specific criticisms.

Christensen, Thomas J. "Perceptions and Alliances in Europe, 1865–1940," *International Organization*, Vol. 51, No. 1 (Winter 1997), pp. 65–97. Rebuts James Morrow's critique of Christensen and Snyder with the argument that decisions to form alliances are driven by perceptions of the offense-defense balance and of relative power.

Lynn-Jones, Sean M. "Offense-Defense Theory and Its Critics," *Security Studies*, Vol. 4, No. 4 (Summer 1995), pp. 660–691. This article offers a comprehensive reply to many of the principal criticisms of offense-defense theory.

Extensions, Refinements, and Directions for Future Research

Anderton, Charles H. "Toward a Mathematical Theory of the Offensive/Defensive Balance," *International Studies Quarterly*, Vol. 36, No. 1 (March 1992), pp. 75–99. After reviewing some criticisms of offense-defense theory, Anderton draws on the Lanchester model of conventional war to offer a more rigorous mathematical theory. He suggests future directions for quantitative empirical and theoretical work.

Betts, Richard K. "The Soft Underbelly of American Primacy: Tactical Advantages of Terror," *Political Science Quarterly*, Vol. 117, No. 1 (Spring 2002), pp. 19–36. In the context of an overview of the relationship between U.S. power and terrorist attacks, Betts applies offense-defense theory to terrorism and concludes that offense has the advantage for both sides in the U.S. "war on terror."

Biddle, Stephen D. "Rebuilding the Foundations of Offense-Defense Theory," *Journal of Politics*, Vol. 63, No. 3 (August 2001), pp. 741–774. Biddle argues that the concept of a technologically determined offense-defense balance is problematic. He proposes a new variant of offense-defense theory that argues that national strategic and tactical choices are key determinants of the relative ease of offense and defense.

Lynn-Jones, Sean M. "Does Offense-Defense Theory Have a Future?" Working Paper No. 12, Research Group in International Security, Université de Montréal/McGill University, 2001. This paper offers an overview of the strengths and weaknesses of offense-defense theory. It discusses the implications of broad and narrow definitions of the offense-defense balance. The conclusions suggest potential directions for further research.

International Security

The Robert and Renée Belfer Center for
Science and International Affairs
John F. Kennedy School of Government
Harvard University

Articles in this reader were previously published in **International Security**, a quarterly journal sponsored and edited by the Robert and Renée Belfer Center for Science and International Affairs at the John F. Kennedy School of Government at Harvard University, and published by MIT Press Journals. To receive subscription information about the journal or find out more about other readers in our series, please contact MIT Press Journals at Five Cambridge Center, Fourth Floor, Cambridge, MA, 02142-1493 or at www.mitpress.com.